W0050103

THERMAL CONDUCTIVITY 14

A Continuation Order Plan is available for this series. A continuation order will bring delivery of each new volume immediately upon publication. Volumes are billed only upon actual shipment. For further information please contact the publisher.

THERMAL CONDUCTIVITY 14

Edited by
P. G. Klemens and T. K. Chu

University of Connecticut, Storrs

SPRINGER SCIENCE+BUSINESS MEDIA, LLC

Library of Congress Cataloging in Publication Data

International Conference on Thermal Conductivity, 14th, Storrs, Conn., 1975.
  Thermal conductivity 14.

  Includes bibliographical references and index.
  1. Heat—Conduction—Congresses. I. Klemens, P. G. II. Chu, Tsu-kai, 1933-
    III. Title.
QC320.8.I57 1975   536'.2012   76-10951
ISBN 978-1-4899-3753-7   ISBN 978-1-4899-3751-3 (eBook)
DOI 10.1007/978-1-4899-3751-3

Proceedings of the Fourteenth International Conference on
Thermal Conductivity held in Storrs, Connecticut, June 2-4, 1975

©1976 Springer Science+Business Media New York
Originally published by Plenum Press, New York in 1976
Softcover reprint of the hardcover 1st edition 1976

All rights reserved

No part of this book may be reproduced, stored in a retrieval system, or transmitted,
in any form or by any means, electronic, mechanical, photocopying, microfilming,
recording, or otherwise, without written permission from the Publisher

INTERNATIONAL THERMAL CONDUCTIVITY CONFERENCES

GOVERNING BOARD

R. U. Acton (Chairman)	Sandia Corporation
P. Wagner (Secretary)	Los Alamos Scientific Laboratory
H. R. Shanks (Alternate Chairman)	Iowa State University
C. Y. Ho (Steward)	CINDAS, Purdue University
S. G. Bapat	Southern Research Institute
D. H. Damon	University of Connecticut
G. L. Denman	Air Force Materials Laboratory
P. G. Klemens	University of Connecticut
M. J. Laubitz	National Research Council of Canada
C. F. Lucks	Columbus, Ohio
D. L. McElroy	Oak Ridge National Laboratory
M. L. Minges	Air Force Materials Laboratory
J. D. Plunkett	University of Denver
R. L. Reisbig	University of Houston Victoria Campus
H. J. Sauer, Jr.	University of Missouri - Rolla
R. E. Taylor	CINDAS & School Mech. Eng., Purdue University
R. P. Tye	Dynatech R/D Company
A. E. Wechsler	A. D. Little, Inc.

FOREWORD

It was seven years ago this month when I had the pleasure of writing the Foreword to the Proceedings of the Eighth Conference on Thermal Conductivity hosted by TPRC/ Purdue University in 1968. Since then this Conference has developed to the point where one can say it has just entered a new phase. At its meeting in June 1975, the Board of Governors of the International Thermal Conductivity Conferences passed a resolution which formalizes two main policies that were felt to be desirable for a number of years. A key item of the resolution was for CINDAS/Purdue University to become the permanent Sponsor of the Conferences and in this capacity assist the Conferences in all matters which will result in the effective implementation of its goals and mission. In short, CINDAS will serve as a home base for the Conferences thus providing continuity and a permanent point of contact. CINDAS/Purdue University is pleased to accept this responsibility as it is well within its mission to promote the advancement and dissemination of knowledge on thermophysical properties of matter.

A second important aspect of the Conference resolution was the establishment of a policy to publish the Proceedings of future conferences on a continuing and uniform basis effective with this, the Fourteenth Conference. Of the previous thirteen conferences only four conferences did publish a formal Proceedings; these were:

Conference and Year	Title of Volume	Publisher and Year
7th (1967)	THERMAL CONDUCTIVITY Proceedings of the Seventh Conference	Superintendent of Documents/GPO (1968)
8th (1968)	THERMAL CONDUCTIVITY Proceedings of the Eighth Conference	Plenum Press (1969)
9th (1969)	NINTH CONFERENCE ON THERMAL CONDUCTIVITY	USAEC (1970)
13th (1973)	ADVANCES IN THERMAL CONDUCTIVITY Papers Presented at XIII International Conference on Thermal Conductivity	University of Missouri, Rolla (1974)

Professor P. G. Klemens, Chairman of the Fourteenth Conference and Technical Editor of this Proceedings volume should be congratulated for his leadership and generous contribution in time and effort which has made this volume possible. CINDAS looks forward to working together with future host institutions, chairmen, and technical editors on this most worthwhile and important activity which has become a National Institution.

The International Thermal Conductivity Conferences are an example of how a technical community with a common purpose can transcend the invisible artificial barriers between disciplines and gather together in increasing numbers without the need of national publicity and continuing funding sponsors, when they see something worthwhile going on. I am convinced that this Series will not only grow stronger each succeeding year but will set an example for other groups on how to attack their own problem areas.

Y. S. Touloukian
Director, Center for Information and
Numerical Data Analysis and Synthesis (CINDAS)
and
Distinguished Atkins Professor of Engineering
Purdue University

January 1976
West Lafayette

PREFACE

The Fourteenth International Thermal Conductivity Conferences were held in Storrs, Connecticut, June 2-4, 1975. The Conference was hosted by the University of Connecticut. The Local Organizing Committee were

P. G. Klemens (Chairman)	Physics
D. H. Damon	Physics
H. Hilding	Mechanical Engineering
F. P. Lipschultz	Physics

We are indebted to CINDAS/Purdue University for vital assistance with mailing and publicity, and particularly for making the arrangements for the publication of these Proceedings.

We acknowledge the financial support given by the University of Connecticut Research Foundation, and the sponsorship and timely help given by the Department of Physics and by the Institute of Materials Science of the University. The University's Center for Conferences and Institutes relieved us of many of the cares associated with organizing a meeting.

The present organizers are interested in the thermal conductivity of solids at low temperatures as a tool in the study of lattice defects, and the program is somewhat biased towards this area. We have also attempted to attract papers on the fluid state. Papers on measurement techniques and on technical applications were not as numerous as one would have wished. We hope that this will be corrected at future meetings. The International Thermal Conductivity Conferences are a unique forum for bringing together a variety of interests which are diverse and yet interrelated. Technical needs, measurements, sensitivity to microstructure, and theoretical interpretations all impinge on each other and are of mutual interest.

The reader will note that the final organization and grouping of the papers in this Proceedings Volume do not necessarily parallel, on a one-to-one basis, the sessions of the Conference. Once the final manuscripts were at hand the restructuring was found desirable in order to bring about what was felt to be a more compatible grouping.

In order to maintain the record of these conferences, pertinent data on past Conferences are listed below. One will note that effective with this Conference, the conferences are no longer held each year, and the Thermal Conductivity Award, which is presented at each Conference, is likewise no longer an annual award. The present Conference has selected as recipient of the Award, Dr. M. J. Laubitz of the Division of Physics, National Research Council, Ottawa, for his outstanding and reliable measurements of thermal conductivity of metals at elevated temperatures, and for his work in interpreting their results in terms of the modern theory of electrons in solids.

Conference and Year	Host Organization and Site	Chairman
1st 1961	Battelle Memorial Institute Columbus, OH	C. F. Lucks
2nd 1962	National Research Council (Canada) Ottawa, Canada	M. J. Laubitz
3rd 1963	Oak Ridge National Laboratory Gatlinburg, TN	D. L. McElroy

Conference and Year	Host Organization and Site	Chairman
4th 1964	U.S. Naval Radiological Defense Lab. San Francisco, CA	R. L. Rudkin
5th 1965	University of Denver Denver, CO	J. D. Plunkett
6th 1966	Air Force Materials Laboratory Dayton, OH	M. L. Minges & G. L. Denman
7th 1967	National Bureau of Standards Gaithersburg, MD	D. R. Flynn & B. A. Peavy
8th 1968	Thermophysical Properties Research Center/Purdue University W. Lafayette, IN	C. Y. Ho & R. E. Taylor
9th 1969	Ames Laboratory & Office of Naval Research Ames, IA	H. R. Shanks
10th 1970	Arthur D. Little, Inc. & Dynatech R/D Co. Boston, MA	A. E. Wechsler & R. P. Tye
11th 1971	Sandia Laboratories, Los Alamos Scientific Laboratories and University of New Mexico Albuquerque, NM	R. U. Acton, P. Wagner & A. V. Houghton, III
12th 1972	Southern Research Institute & University of Alabama Birmingham, AL	W. T. Engelke, S. G. Bapat & M. Crawford
13th 1973	University of Missouri - Rolla Lake of the Ozarks, MO	R. L. Reisbig & H. J. Sauer, Jr.
14th 1974	University of Connecticut Storrs, CT	P. G. Klemens

The Conferences have been empowered to name distinguished workers as Fellows of the Thermal Conductivity Conferences. The present Conference is the first occasion for doing so, and the Conference has thus named R. Berman (Oxford University) and R. P. Tye (Dynatech R/D Co.) as Fellows. Dr. Berman has long been active in the study of the thermal conductivity of dielectric solids at low temperatures, having been one of the first to investigate lattice defects through their effect on the thermal conductivity. Dr. Tye has a long and distinguished record of thermal conductivity studies of many materials, first at the (U.K.) National Physical Laboratory, and then at Dynatech Corporation, where he developed methods of measuring heterogeneous and technical materials.

The next Conference will be held in Canada during the summer of 1977. Location and time have yet to be decided. The organizers will be V. V. Mirkovich, Department of Energy, Mines and Resources, Ottawa, and I. D. Peggs, Atomic Energy of Canada, Ltd., Pinawa, Manitoba.

Storrs, Connecticut
January 1976

P. G. Klemens and
T. K. Chu, Editors
University of Connecticut

CONTENTS

CHAPTER I. SOLIDS AT LOW TEMPERATURES

CHAPTER II. SOLIDS AT HIGH TEMPERATURES

Chapter I

Solids at Low Temperatures

THE THERMAL CONDUCTIVITY OF DIAMONDS

R. Berman and M. Martinez*

Clarendon Laboratory

Oxford OX1 3PU, England

The thermal conductivities of a number of gem-quality diamonds have been measured between 2 and 300 K. All the diamonds contained nitrogen in various forms and concentrations (and thus belonged to type I). The results have been analysed in terms of a number of contributions to the scattering of phonons. At the upper end of the temperature range the thermal resistivity due to defects is dominated by a scattering rate proportional to the fourth power of the phonon frequency, which is ascribed to groups of between 4 and 10 nitrogen atoms. At intermediate temperatures a scattering rate proportional to the square of the frequency is important in some specimens and is ascribed to the platelets which have been observed by transmission electron microscopy. Using a scattering power appropriate to stacking faults leads to a platelet area density in reasonable agreement with that deduced from infra-red absorption. Two specimens showed dips in the conductivity curves at 10 K and we very tentatively ascribe this to defects with diameter $\sim$ 5 nm. From infra-red measurements, these two diamonds were found to have an appreciable fraction of their fairly small nitrogen content in a form different from that in which the majority occurs.

---

*   Supported jointly by the Technical Assistance Training Organisation (administered by the British Council) and Consejo Nacional de Ciencia y Tecnologia, Mexico.

The work reported here concerns the presence of nitrogen in most natural gem-quality diamonds, the broadening of the X-ray diffraction spots for many natural diamonds into 'spikes' and the relation between these two observations. The relationship was first thought to be close but has recently been shown not to be a quantitative one. Even the qualitative relationship is at present unclear.

INTRODUCTION

The X-ray spikes were first observed in 1940 and their intensities were measured by Hoerni and Wooster[1]. On the basis of these measurements, Frank[2] proposed that the spikes were caused by the precipitation of impurity in the $\{100\}$ planes, and in 1958 Caticha-Ellis and Cochran[3] deduced that an impurity concentration of $\sim 0.1\%$ would be required to produce the required effects. In the following year Kaiser and Bond[4] found by chemical analysis that nitrogen occurred in gem-quality diamonds at concentrations up to 0.23% and that the nitrogen content correlated with the infra-red absorption at $7.8\ \mu$m. Elliott[5] proposed that the X-ray spikes were caused by platelets of precipitated nitrogen atoms in the cube planes and another model for the platelets was put forward by Lang[6]. In 1962 Evans and Phaal[7] obtained transmission electron microscope pictures from thinned sections of diamonds and observed platelets in the $\{100\}$ planes of type I diamonds (diamonds can be divided into types I and II on the basis of various optical absorption features). About 0.1% nitrogen concentration would be required to account for the size and number of the platelets observed if they were made wholly of nitrogen, so that a large fraction of the nitrogen impurity would be in platelet form. However, in 1968 Sobolev, Livoisan and Lenskaya[8] found that there was no correlation between the total X-ray spike intensity (which was thought to be related to the platelets) and the $7.8\ \mu$m absorption (known to correlate with the total nitrogen present), but that the spike intensity correlated with absorption at $7.3\ \mu$m. In 1971 Evans and Wright[9] found a direct correlation between the integrated spike intensity and the total platelet area per unit volume, and this together with the findings of Sobolev et al. indicated that the total platelet area does not correlate with the total nitrogen content. The platelets could either contain variable numbers of nitrogen atoms per unit area or no nitrogen at all. However, their constitution appeared to be invariant, since the relative shift of the lattice on either side of them was constant. In 1970 Davies[10] had already concluded that the $7.3\ \mu$m absorption was caused by not more than 10% of the nitrogen atoms in a diamond and might, in fact, owing to the experimental inaccuracy in this figure, not be caused by any nitrogen atoms at all.

THERMAL CONDUCTIVITY MEASUREMENTS

The thermal conductivities of several gem-quality diamonds were measured by Hudson[11]. Since then, new measurements have been made and we here report on the interpretation of all the results which have a bearing on the nature of the nitrogen defects.

Columns (c) and (d) of Table 1 show the nitrogen concentration in the seven type I diamonds measured, the total amounts being determined from the 7.8 μm absorption (the division into A and B features will be discussed later). The concentrations marked with an asterisk were too small for accurate determination, while those marked with a dagger were not measured with as refined an optical technique as was used for the other specimens.

Column (d) of Table 2 shows the absorption coefficient at 7.3 μm; four specimens had small coefficients, two were intermediate and one was very large. The conversion of absorption coefficient to platelet area density has at present to be made via the separate correlations between each of these and the spike intensity; unfortunately it is not clear to what extent the units in which the spike intensity was measured were the same in both sets of experiments[8,9], so that the platelet area density cannot be deduced to within better than a factor of two.

Fig.1 shows the conductivity of specimens C5, C6 and C7, together with that of a type IIa (nitrogen-free) diamond for comparison. Of the type I specimens, C7 contains least nitrogen and has, indeed, the highest conductivity except at the lowest temperatures where the state of the surface is all-important in determining the heat flow. The nitrogen contents of C5 and C6 are similar to one another and their conductivities at room temperature are also similar. However, at intermediate temperatures the specimen with more platelets has a lower conductivity, as would be expected from the fact that relatively large defects can scatter effectively even the long wavelength phonons which are then dominant.

Point Defects

The conductivity curves were analysed in the standard way, assuming phonon relaxation rates with various frequency dependences and a Debye frequency spectrum. The fit to the experimental results is shown by the lines in Fig.1. The defect scattering rate dominant at high temperatures is proportional to ω^4 and is ascribed to point defects. If the defect is nitrogen, the concentration required can be deduced from the relative mass difference (2/12) and the relative change in lattice constant (Kaiser

Table 1. Nitrogen in various forms in units of 10^{19} atoms per cm^3

Specimen	From ϰ Small groups		From I.R		From ϰ Medium clusters D~6 nm	
	n = 2 (a)	n = 10 (b)	A form (c)	B form (d)	short λ (e)	long λ (f)
IA - 1	50	9	9	-	-	-
IA - 2	12	2	~1*	~0.6*	0.3	4
IA - 3	60	11	18	-	-	-
IA - 4	10	2	~1*	~0.6*	0.2	2
C5	100	20	18†	?		
C6	110	22	>18†	?		
C7	22	4	~10†	?		

Table 2. Platelet area density in units of 10^{16} nm^2 per cm^3

Specimen	From ϰ Nitrogen platelets	Stacking faults	From 7.3 μm absorption and spikes	Abs coeff. cm^{-1}
IA - 1	0.4	1	~10*	3
IA - 2	-	-	<10*	2.2
IA - 3	0.2	1	~10*	2.9
IA - 4	-	-	~10*	2.8
C5	10	500	~500	38†
C6	1	9	~70	7†
C7	1	10	~80	7†

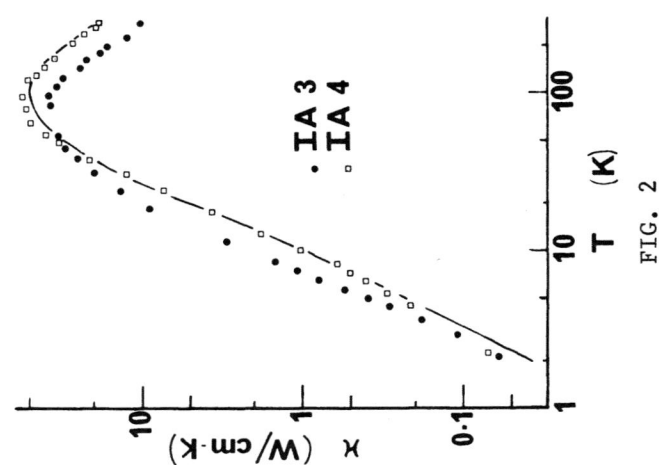

FIG. 2

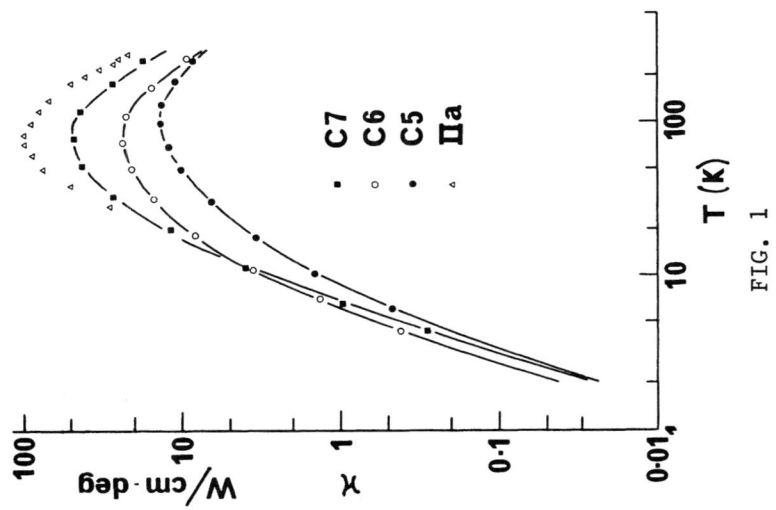

FIG. 1

and Bond[4]).   For such Rayleigh scattering by defects of volume V,
the relaxation rate is proportional to V^2.   For a given concentration of nitrogen in a diamond, the number of defects is proportional to $1/V$, so that the scattering for a fixed nitrogen
concentration is proportional to V and thus to the number of atoms
in a cluster.   From the scattering rate which fits the measured
conductivity, we may then deduce the nitrogen concentration as a
function of the number of atoms, n, assumed to be in a cluster.
Columns (a) and (b) of Table 1 show the concentrations derived by
taking n = 2 and n = 10.   We do not show the values corresponding
to n = 1 since the nitrogen would then be paramagnetic, and none
of the diamonds showed this.   It can be seen that for all specimens the measured nitrogen concentration can account for all the
point defect scattering if n lies between 4 and 10.

       Specimen C5 can be used to throw some light on the constitu-
tion of the platelets.   From Table 2 it can be seen that this
diamond had a very high platelet concentration, and if the platelets were made of nitrogen according to the Lang model[6], then
nearly 90% of the total nitrogen would reside in platelets.   This
would leave only about 2×10^{19} atoms per cm^3 which could scatter
as point defects.   For this small amount of nitrogen to give the
large point defect scattering deduced from the conductivity, the
non-platelet nitrogen would have to be in groups of ~ 100 atoms.
Such large aggregates would not act as point defects except for
long wavelength phonons (such as those dominant below 50 K).
This result, by itself, suggests that little, if any, nitrogen is
in the platelets.   In the other specimens the number of platelets
is very small and could, at the most, only account for $\sim 10\%$ of the
total nitrogen.   The precision in deriving the point defect
scattering is not sufficient to reveal whether the entire nitrogen
content is, to this accuracy, required to account for it.

Platelets

       At the time when it was still thought that the platelets were
composed of nitrogen, Turk and Klemens[12] derived an expression for
the scattering of phonons by them.   The scattering by N_p platelets per unit volume of radius R is almost proportional to $N_p \pi R^2 \omega^2$.
There is a very weak dependence of the scattering on the ratio of
phonon wavelength, λ, to R, but this is not enough to make it
possible to derive really worthwhile information about the average
platelet radius, in isolation from the product $N_p R^2$.   Column (a)
of Table 2 shows values of $N_p \pi R^2$ derived by using the Turk and
Klemens expression.   It can be seen that the actual scattering by
a platelet must be considerably less than they calculated.

Since the suggestion that the platelets are not, in fact, made of nitrogen, we have calculated their total area density by using an expression given by Klemens[13] for scattering by stacking faults (without introducing the R/λ correction term). It can be seen from Table 2 that the agreement with the platelet area density is closer, and indeed very close for the specimen in which platelet scattering is dominant at low temperatures. For the other specimens, boundary scattering is so dominant at low temperatures that the computed conductivity is rather insensitive to changes in the strength of platelet scattering.

We hope that further work on diamonds with large concentrations of platelets will help us to elucidate their scattering power.

A and B Features

In 1954 Sutherland, Blackwell and Simeral[14] found that certain features in the infra-red absorption spectrum of diamond varied together, and they referred to the two groups as A and B features. Davies and Summersgill[15] have suggested that both groups are caused by nitrogen and have given a recipe for determining the relative amounts of nitrogen which give rise to the different features (they showed that the 7.3 μm absorption feature, which is correlated with platelets, does not belong to either group). The division into A and B forms of nitrogen was not looked for in Hudson's specimens C5, 6 and 7, but column (c) and (d) of table 1 show that an appreciable fraction of the small total nitrogen content was in the B form in diamonds IA - 2 and 4. Fig.2 shows the conductivity of specimen IA-4 (IA-2 had very similar conductivity) compared with specimen IA-3 which contains a small concentration of platelets, but no detectable B-type nitrogen. The interesting feature for specimen IA-4 is the extended depression centred at 10 K. We have fitted this by an additional scattering rate which might be expected to be a crude approximation to the scattering by a middle-sized object which is small for some wavelengths and large for others. It is assumed that for frequencies from 0 to ω_o the wavelengths are large enough for Rayleigh scattering to occur with a relaxation rate $\tau^{-1} = E\omega^4$. For frequencies above ω_o the scattering is assumed to be purely geometrical with $\tau^{-1} = E\omega_o^4$. The conductivity is fitted by choosing E and ω_o suitably, with the result shown by the line in Fig.2. To obtain an order of magnitude for the dimensions of the scatterers, they are assumed to be spherical of diameter $D \sim \lambda_o/4$, where λ_o is the wavelength corresponding to ω_o. From the analysis this gives $D \sim 6$ nm.

We have deduced the concentration of atoms which go into the formation of such clusters by assuming that they are made of nitrogen and that each cluster, containing $\sim 2 \times 10^4$ atoms, scatters 2×10^4 more strongly than the same number of isolated nitrogen atoms scattering according to the Rayleigh law. The concentration can also be deduced from the high frequency scattering by assuming that the relaxation rate is $N_B \pi R^2 v$ by analogy with simple kinetic theory, where v is the phonon velocity and N_B the number of scatterers per unit volume. The results of these two ways of calculating the concentration of nitrogen which may be in middle-sized clusters are shown in columns (e) and (f) of Table 1. It can be seen that the values for the concentration of nitrogen in B features, derived from optical absorption, lie somewhere between the two extremes. In spite of this rough agreement, we must emphasise that our suggestion is very tentative; we hope to find other diamonds in which the B features are relatively prominent and the conductivity is not dominated by platelet scattering at intermediate and low temperatures.

SUMMARY

We have shown that most of the nitrogen present in natural diamonds occurs in groups of between 4 and 10 atoms. Evidence from one diamond suggests that very few of the atoms which form a platelet are nitrogen. We make a very tentative suggestion that the B features in the absorption spectrum are due to clusters of nitrogen atoms about 6 nm in diameter.

References

1. J.A. Hoerni and W.A. Wooster, Experientia 8, 297 (1952); Acta Cryst. 8, 197 (1955).
2. F.C. Frank, Proc. Roy. Soc. A237, 168 (1956).
3. S. Caticha-Ellis and W. Cochran, Acta Cryst. 11, 245 (1958).
4. W. Kaiser and W.L. Bond, Phys. Rev. 115, 857 (1959).
5. R.J. Elliott, Proc. Phys. Soc. 76, 787 (1960).
6. A.R. Lang, Proc. Phys. Soc. 84, 871 (1964).
7. T. Evans and C. Phaal, Proc. Roy. Soc. A270, 538 (1962).
8. E.V. Sobolev, V.I. Livoisan and S.V. Lenskaya, Soviet Physics Crystallography 12, 665 (1968).
9. T. Evans and C.E. Wright, Diamond Conference, Cambridge, 1971, Paper 2 (see T.Evans, Diamond Research 1973, p2, Industrial Diamond Information Bureau).
10. G. Davies, Nature 228, 758 (1970).
11. P.R.W. Hudson, Thesis, Oxford University (1972).
12. L.A. Turk and P.G. Klemens, Phys. Rev. B9, 4422 (1974).
13. P.G. Klemens, Can.J.Phys. 35, 441 (1957).
14. G.B.B.M. Sutherland, D.E. Blackwell and W.G. Simeral, Nature 174, 901 (1954).
15. G. Davies and I. Summersgill, Diamond Research 1973, p6.

THERMAL CONDUCTIVITY OF HIGHLY ORIENTED PYROLYTIC BORON NITRIDE

E. K. Sichel and R. E. Miller

RCA Laboratories

Princeton, N. J. 08540

ABSTRACT

We have measured the thermal conductivity of boron nitride in the hexagonal plane over the temperature range 1.5 - 350K. The influence of crystal perfection and the influence of the two dimensional character of the material on the thermal conductivity is discussed.

I. INTRODUCTION

High thermal conductivity materials are increasing in technical importance as heat sinks for solid state devices. By high thermal conductivity, we mean above 1 watt/cm-K at room temperature. Two layer structures fall in this class: graphite and boron nitride. Above 100K, graphite is second only to diamond in its high thermal conductivity parallel to the layers.[1] Values of 20 watt/cm-K are reported[2] at room temperature. Boron nitride which is, unlike graphite, an electrical insulator, forms a hexagonal layer structure similar to graphite. The lattice parameters of boron nitride are a_o = 2.504A, c_o = 6.661A. For graphite, a_o = 2.456A, c_o = 6.696A. Figure 1 is a diagram of the structure of BN. The highest thermal conductivity reported for pyrolytic[3] BN is about 2.5 watt/cm-K at room temperature, measured parallel to the layers.[4] The thermal conductivity measured perpendicular to the layers is a factor of 100 lower.

In view of the structural similarity of graphite and boron nitride, and similarity of atomic mass, we might expect the thermal conductivity of boron nitride to be much higher than reported to

date. In this paper we examine some of the factors influencing
and limiting the thermal conductivity of BN. The experiments
covered a range of low temperatures not previously reported in
the literature.

<h2 align="center">II. EXPERIMENTAL</h2>

The specimens were pyrolytic compression annealed boron
nitride.[3] In appearance they were translucent, almost white, and
looked much like mica. The dimensions of specimen B were X=1.85
cm, Y=1.1cm, Z=0.043 cm and specimen C, X=5.1 cm, Y=0.69 cm,
Z=0.11 cm. The c axis was in the Z direction and the XY plane
was the basal plane. Some relevant sample parameters are given in
Table I. X-ray studies revealed that the samples consist of small
crystallites. The crystallites are platelets in the hexagonal
plane. Platelet thickness along the c-axis was determined from
the line broadening. The long dimension of the crystallites in the
hexagonal plane has not been examined by x-rays. The mosaic spread
in c-axis alignment was determined from the full width at half maximum (FWHM) of the (00.2) Cu $K\alpha$ reflection. For comparison, top
quality pyrolytic graphite[5] may have a mosaic spread of only 0.4°.

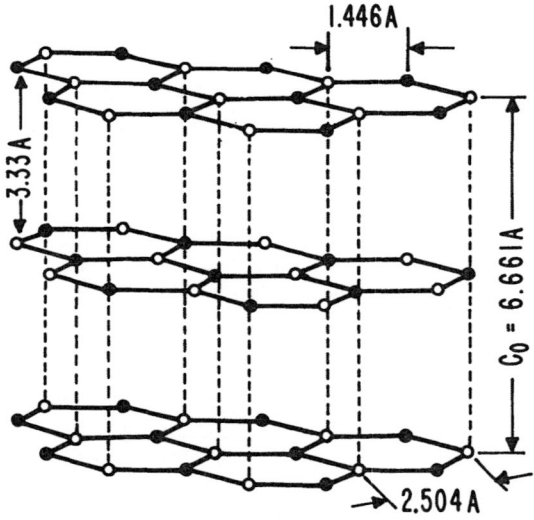

Figure 1 - Structure of hexagonal boron nitride

Table I

HEXAGONAL BORON NITRIDE

specimen	density g/cm^3	FWHM spread in c-axis	crystallite size along c-axis
B	2.00	$12^0 - 14^0$	200 A
C	2.18	2^0	500 - 1000 A

We used a conventional steady state technique to measure the thermal conductivity. The apparatus appears in Fig. 2. A heater, H, was fastened with GE 7031 varnish to one end of the specimen. The other end of the specimen was clamped in a copper holder, part of a copper heat sink. Indium was placed between the specimen and the copper, and the clamp was tightened until the indium flowed. The copper vacuum can acted as a radiation shield. For measurements in the range 80-350K, two 3-mil calibrated chromel-alumel

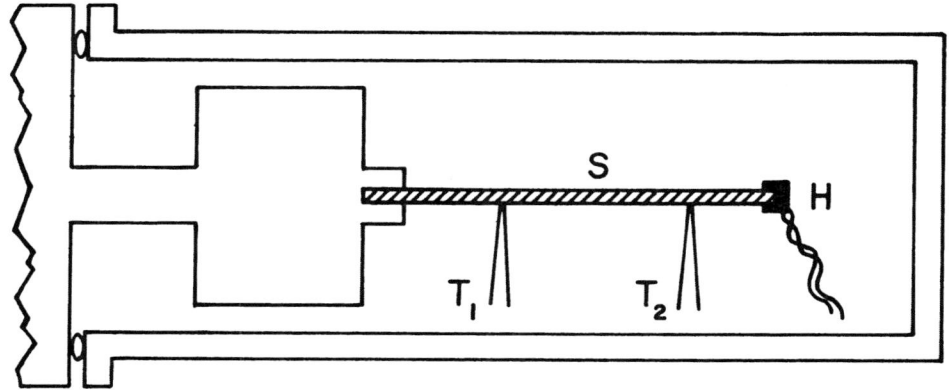

Figure 2 - Schematic diagram of thermal conductivity apparatus. T_1 and T_2: 3-mil chromel-alumel thermocouples. H: sample heater, S: specimen.

thermocouples were fastened to the specimen with GE 7031 varnish.
For measurements below 80K, 1/8 watt Allen-Bradley carbon resistor
thermometers were sanded flat and glued to the specimen with the
varnish. The thermometers were placed 0.8 to 2.3 cm apart. The
temperature gradients under experimental conditions were 0.6-2
K/cm above 80K and 0.02 - 0.3 K/cm below 80 K.

III.   RESULTS AND DISCUSSIONS

Our experimental results are shown in Figs. 3 and 4. The symbol K_a denotes the thermal conductivity in the basal plane. We

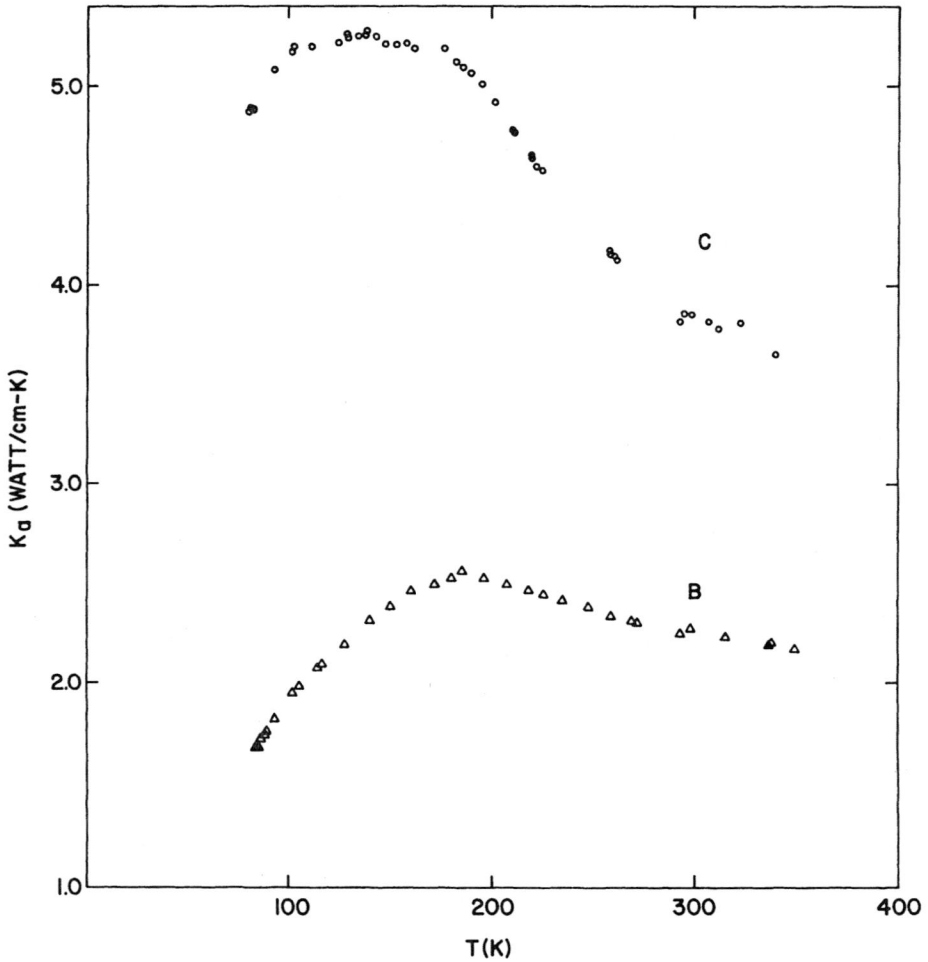

Figure 3 - Thermal conductivity of boron nitride as a function of
temperature in the range 80K-350K for specimens B and C.

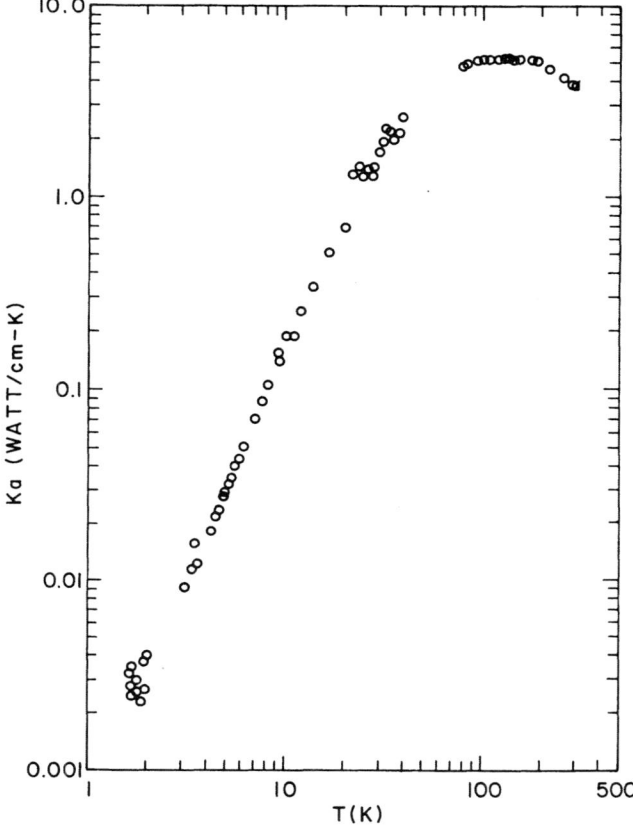

Figure 4 - Thermal conductivity of specimen C as a function of
temperature.

note that specimen C has an appreciably higher thermal conductivity
than specimen B. This is consistent with the finding that C has
better crystallinity than B. In fact, the thermal conductivity
of specimen C at room temperature, 3.9 watt/cm-K, is comparable
to the value for copper. This value is substantially higher than
any previously reported.[4]

At high temperatures, phonon Umklapp scattering limits the
conductivity of electrically insulating solids, whereas at low
temperature, phonon scattering off crystallite boundaries determines the conductivity. This gives rise to the characteristic maximum in the thermal conductivity of solids as a function of

temperature. In the boundary scattering region, the thermal conductivity is given by

$$K = \frac{1}{3} C \, v\ell \, , \tag{1}$$

where K is the thermal conductivity, C is the specific heat, v the average sound velocity, and ℓ the average crystallite size. Thus we would expect the thermal conductivity and heat capacity to have the same temperature dependence. From Fig. 4, we find that the thermal conductivity below 20K varies at $T^{2.4 \pm 0.1}$ and above that temperature, the slope decreases as the curve begins to round off, approaching the maximum. Unfortunately, the only heat capacity data available do not extend to temperatures below 20K. Dworkin et al[6] have shown that the heat capacity of hexagonal BN varies as $T^{2.0}$ in the range 20-65K. Thus we are uncertain whether the heat capacity and thermal conductivity have the same power dependence of the temperature in the boundary scattering regime.

Nevertheless, we have made an order of magnitude calculation to determine ℓ in Eq. (1). Following the suggestion of Klemens[7] in a paper concerned with graphite, we let v be the highest longitudinal sound wave velocity. This number is not available for BN, but we have the value for graphite $v \simeq 2.5 \times 10^6$ cm/sec reported by Nicklow et al.[8] The Debye temperature of hexagonal BN[6] is about 600K compared to about 770K for graphite.[9] Since the lattice spacing and atomic mass of graphite and BN is similar, we assume

$$\frac{\Theta_D^{graphite}}{\Theta_D^{BN}} \simeq \frac{v^{graphite}}{v^{BN}} \, , \tag{2}$$

which gives $v^{BN} = 1.95 \times 10^6$ cm/sec. At 20K we take $K \simeq 1$ watt/cm-K, $C \simeq 0.065$ cal/mole-K from Dworkin et al,[6] and $\rho \simeq 2$ g/cm³. Then in the hexagonal plane, the crystallite dimension is $\ell \simeq 7000$A for specimen C.

IV.   CONCLUSION

Graphite and boron nitride have the same crystal structure and similar atomic masses. In many simple compounds it has been found[10] that for $T > \Theta_D$, $K \propto \bar{M}\delta\Theta_D^3$ where $\bar{M}$ is the average mass of an atom of the crystal, δ is one half the X-ray lattice constant, Θ_D is the Debye temperature. Thus, for $T > \Theta_D$, we would expect the thermal conductivity of boron nitride to be lower than that of a crystal of graphite of comparable quality by a factor of $(\Theta_D^{BN}/\Theta_D^{graphite})^3$, or about 0.5. In view of this, room temperature thermal conductivity of high quality BN could be as high as 10 watt/cm-K. The crystallite size and mosaic spread of our BN specimen indicate a less perfect specimen than the high quality graphites. If the crystal quality of BN could be improved, we expect that the room temperature thermal conductivity would be significantly higher.

V. ACKNOWLEDGEMENTS

A. W. Moore kindly supplied the specimens used in this study. We are grateful to R. Smith for x-ray analysis and to H. Whitaker for density measurements and to B. Abeles and R. W. Cohen for critical readings of the manuscript.

VI. REFERENCES

1. "Thermal Conductivity of Selected Materials, Part 2", C.Y. Ho, R.W. Powell, P.E. Liley, eds., National Standard Reference Data Series, National Bureau of Standards - 16 (1968), p. 82.

2. Ibid,, p. 125.

3. A.W. Moore, Nature 221, 1133 (1969).

4. See, for example, A. Simpson and A.D. Stukes, J. Phys. C4, 1210 (1971).

5. Union Carbide Corp., graphite for x-ray and neutron monochromator gratings.

6. A.S. Dworkin, D.J. Sasmor, and E.R. Van Artsdalen, J. Chem. Phys. 22, 837 (1954).

7. P.G. Klemens, Australian J. Phys. 6, 405 (1953).

8. R. Nicklow, N. Wakabayashi, H.G. Smith, Phys. Rev. 5B, 4951 (1972).

9. W. DeSorbo and W.W. Tyler, Phys. Rev. 83, 878 (1951).

10. G.A. Slack, J. Phys. Chem. Solids 34, 321 (1973).

FREQUENCY DEPENDENCE OF THE SCATTERING PROBABILITY OF NEARLY MONOCHROMATIC PHONONS

W. E. Bron

Indiana University

Bloomington, Indiana

The probability of nearly monochromatic phonons scattering from atomic scattering sites in SrF_2 is examined experimentally and theoretically. Phonon distributions are obtained at three frequencies using successively Sn, $Pb_{0.5}Tl_{0.5}$ and Pb superconducting flourescer films driven by constantan heater films. These generators yield, respectively, phonon distributions with narrow band maxima at 287, 407 and G70 GHz. The phonon distributions are propagated through single crystals of SrF_2 containing 0.1 mole % of Eu^{2+}. Time of flight measurements of nanosecond duration phonon pulses propagated across the crystal are taken with carefully monitored aluminum superconducting bolometers. The ratio of the integrated diffusive component to the integrated ballistic component of the observed signal is taken as a measure of the scattering probability. The scattering probabilities, normalized to the 287 GHz result, vary with increasing phonon frequency as 1:16:66 as compared to that expected from simple Rayleigh scattering, i.e., a ω^4 dependence, of 1:4:30. A comparison will be made between the experimental results and a calculation of the frequency dependence of the scattering probability based on the T-matric formalism for a mass defect with small, local force constant changes leading to a broad scattering resonance in the 2 to 3 THz region.

THE THERMAL CONDUCTIVITY OF 1 MEV ELECTRON-IRRADIATED Al_2O_3 BETWEEN 1 AND 70 K*

D. Strom, G. Wilham, P. Saunders, and F. P. Lipschultz

Dept. of Physics and Institute of Materials Science

Univ. of Conn., Storrs, Connecticut 06268

Preliminary measurements of the thermal conductivity of 1 MeV electron-irradiated Al_2O_3 with about 0.01% impurities have shown that below 50 K the thermal conductivity decreased about 20 - 25% upon irradiation and recovered almost completely upon annealing.

The sample used in this work was Verneuil-grown Al_2O_3 obtained from Linde.[1] The impurity concentration was estimated to be on the order of 0.01%. The sample diameter was 0.32 cm and the spacing between thermometers was typically about 7 cm. The thermal conductivity of the sample was measured in the "as received", "irradiated", and "annealed" states. The sample was irradiated with 1 MeV electrons to a dosage of about $10^{17}/cm^2$; the sample was rotated about the cylindrical axis during radiation to achieve as uniform an irradiation as possible. Following irradiation the sample had a pale tan tinge. The anneal was accomplished by heating the sample to 600 C for 24 hours in flowing helium gas.

The results of thermal conductivity measurements in the three states are shown in Figures 1 and 2. In the "as received" state the sample had a peak thermal conductivity of 68 wts cm^{-1} K^{-1} at 34 K; in the boundary scattering limit the thermal conductivity was approximately 2.9×10^{-2} W cm^{-1} K^{-1} at 1.0 K and proportional to $T^{2.9}$. In Figure 1, where both the "as received" and "irradiated" thermal conductivities are shown, the thermal conductivity of the irradiated sample is reduced by 20 - 25% below 50 K. The peak conductivity was lowered to about 52 W cm^{-1} K^{-1} at 39 K; in the boundary scattering limit the thermal conductivity was approximately 2.3×10^{-2} W cm^{-1} K^{-1} at 1.0 K and again proportional to $T^{2.9}$. The results are comparable to measurements reported by Berman

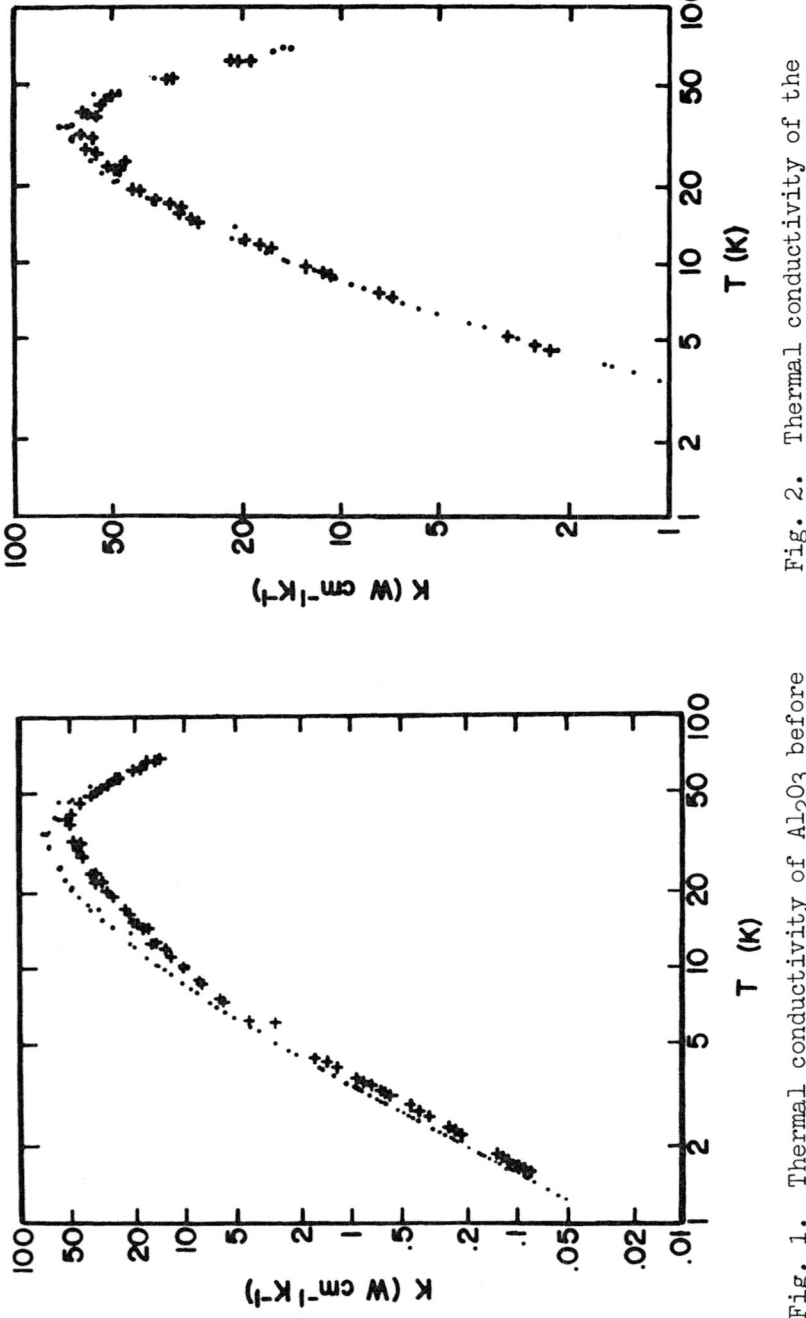

Fig. 1. Thermal conductivity of Al_2O_3 before and after 1 MeV electron irradiation. • unirradiated ("as received"); + irradiated.

Fig. 2. Thermal conductivity of the same sample after annealing at 600°C for 24 hours, compared with the un-irradiated sample. • unirradiated; + annealed.

et al[2,3] and de Goer and Dreyfus[4] (their "pure" sample), but quite different from the results of de Goer and Dreyfus (their "doped" samples) and Brown[5] for γ-irradiated Al$_2$O$_3$.

A comparison of the "annealed" and "as received" stages is made in Figure 2. There appears to be a full recovery of the thermal conductivity in the low temperature region below about 10 K, and nearly full recovery near the maximum in the conductivity. If the remaining thermal resistance is not an artifact of experimental error, it may indicate that some defects of small dimension were produced which are not completely removed by the annealing temperatures used here.

These results are to be contrasted with other results of de Goer and Dreyfus, who also measured γ-irradiated Al$_2$O$_3$ which was doped with chromium and Al$_2$O$_3$ which was doped with manganese. They found the measured increase in thermal resistance of the Cr-doped Al$_2$O$_3$ to be larger than that observed by Berman et al, and that the size of the increase depended on the time between the irradiation and the measurement. The thermal conductivity of de Goer and Dreyfus' Mn-doped sample increased on irradiation. Brown, using a "pure" Al$_2$O$_3$ specimen noted no change in the thermal conductivity when he γ-irradiated this sample. However an Al$_2$O$_3$ specimen doped with magnesium showed an increase in thermal resistance when γ-irradiated at 77 K, though the thermal conductivity recovered to its original value when left at room temperature for a few hours.

Berman et al[3] reported that the additional thermal resistance produced by γ-irradiation did not vary for dosages in the range of 3×10^5 to 10^8 r. Analogous saturation phenomena with γ-irradiation dosage has been reported in optical absorption,[6,7] EPR,[8] and gamma photoconductivity[9] measurements. The dependence of the increase in thermal resistance with electron dosage for the sample used in the present work has not been determined as yet.

Further thermal conductivity measurements will be needed to determine if the additional thermal resistance saturates with electron dosage, as well as identify the scattering mechanisms found by thermal conductivity with defects and impurities determined by other kinds of measurements. Acoustic attenuation studies, which also measure phonon relaxation times, should contribute to an understanding of the mechanism which decreases the low temperature thermal conductivity.

REFERENCES

*   Work supported in part by the United States Army Research
    Office – Durham.
1. Union Carbide, 8888 Balboa Ave., San Diego, California 92123.
2. R. Berman, E. L. Foster, and H. M. Rosenberg, "Defects in
    crystalline solids," Report of Bristol Conference (The Physical
    Society, London, 1954).
3. R. Berman, E. L. Foster, B. Schneidmesser, and S. M. A. Tirmizi,
    J. Appl. Phys. $\underline{31}$, 2156 (1960).
4. A. M. de Goer and B. Dreyfus, Phys. Stat. Solidi $\underline{22}$, 77 (1967).
5. M. A. Brown, J. Phys. C: Solid State Phys. $\underline{6}$, 642 (1973).
6. R. A. Hunt and R. H. Schuler, Phys. Rev. $\underline{89}$, 664 (1953).
7. P. W. Levy and G. J. Dienes, "Defects in crystalline solids,"
    Report of Bristol Conference (The Physical Society, London,
    1954).
8. F. T. Gamble, R. H. Bartram, C. G. Young, and O. R. Gilliam,
    Phys. Rev. $\underline{134}$, A589 (1964).
9. D. J. Huntley and J. R. Andrews, Can. J. Phys. $\underline{46}$, 147 (1968).

THE THERMAL CONDUCTIVITY OF SOLID NEON†

Jim E. Clemans‡

Department of Physics and Materials Research Laboratory

University of Illinois, Urbana, Illinois 61801

ABSTRACT – The thermal conductivity of solid neon samples at constant volume was measured from 0.5 K to 10 K. An isotopically purified ^{20}Ne sample and a sample with the natural isotopic mixture were studied. Measurements were made at molar volumes of 13.4 cc/mole and 12.9 cc/mole. The conductivity of samples grown under pressure shows a very strong increase with decreasing molar volume. For the natural neon samples examined between 5 K and 10 K, $\kappa \propto V^{-11\pm3}$. In the same temperature range, the value of κ is proportional to $T^{-2.2}$. At low temperatures, from 1 K to 0.5 K, the conductivity data of all samples showed a dependence on temperature of approximately $T^{2.7}$. Minimum phonon mean free paths of 0.013 cm, estimated from the conductivity at the lowest temperatures, are two to four times longer than those of earlier experimenters. However, they still indicate a substantial concentration of crystal defects which could not be reduced further by any of the variety of growth procedures used. The high temperature thermal conductivity of the isotopically purified solid neon is observed to have a much stronger temperature dependence than the natural neon samples. For purified samples at the highest temperatures, the thermal conductivity falls below that of natural neon of similar molar volume, in contradiction to the assumption that scattering processes are additive.

I. INTRODUCTION

The thermal conductivity of crystalline neon and other rare gas solids has appeared in recent years as an ideal starting point in the description of thermal conductivity from the fundamental point of view of Leibfried and Schlömann[1]. They use the Boltzmann transport equation and the interatomic potential. Until very recently experi-

23

mental thermal conductivity data for neon has been dominated by
extrinsic phonon scattering processes not characteristic of good
quality samples. The data presented here, together with other
recent data[2,3] mark a large improvement in the experimental knowledge of neon. These independently verified new data give further
theoretical effort a much stronger motivation.

II. EXPERIMENTAL METHOD AND THERMAL CONDUCTIVITY RESULTS

The object of careful sample preparation is to reduce as far
as possible the phonon scattering by crystal defects. Neon samples
were grown in situ from gas with fewer than 10 ppm of any chemical
impurity. Samples were grown from both natural neon (approximately
9.27% neon-22 in neon-20), and an isotopically purified sample of
neon-20, which contained 390 ppm neon-21 and 120 ppm neon-22. In
addition to chemical impurities, mechanical imperfections of the
crystal lattice are an important extrinsic source of phonon scattering. In earlier experiments these defects severely decreased
the magnitude of the observed thermal conductivity. There is evidence from many experiments with neon and other rare gas solids
that the thermal contraction of a sample causes mechanical damage[2,4,5]. In hopes of minimizing this damage, the samples grown
here were crystallized at pressures over 0.7 kbars to give them
molar volumes less than the molar volume at zero pressure and temperature. After constant pressure growth the sample chamber was
sealed by freezing the inlet line. Subsequent handling of the sample was thus accomplished at constant volume. The most delicate
part of this handling was the slow (3 K/hour) cooling from the sol-
idification temperature of 45 K to 10 K where measurements were begun.

The sample chamber was a 1/8" stainless steel tube with a 1/16"
bore. The correction to the thermal conductivity for the heat conduction by the chamber walls was obtained directly by measuring the
conductivity of the empty chamber. At 10 K and higher temperatures
the correction for parallel heat conduction exceeded 50% of the
total conduction and decreased the accuracy of the measurements.

Figure 1 shows samples with molar volumes approximately the
same as that for vapor pressure growth, 13.4 cc/mole, except for
one sample grown at 12.89 cc/mole which will be discussed later.
The low temperature thermal conductivities of samples from the
present work are two to three times higher than the best values
of Kimber and Rogers[2]. However, the mean free paths are still one
tenth of those due to single crystal boundary scattering in the low
temperature limit. In the 5 K to 10 K range, the "Umklapp" range,
the agreement with Kimber and Rogers data is reasonably good, as it
should be in this range where intrinsic processes dominate. This
agreement only emphasizes the extremely large amount of crystal
defects present in the White and Woods[4] sample.

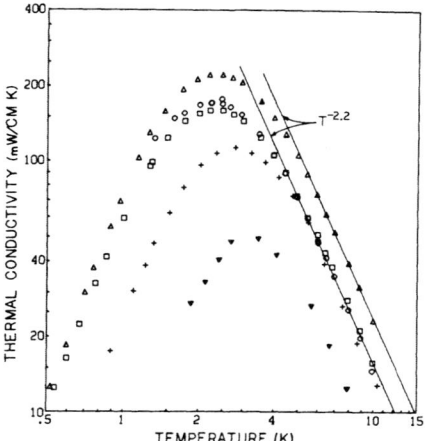

Figure 1 - Thermal conductivity of natural neon samples. O sample 1,
and □ sample 3 from present work, + Kimber and Rogers (1973),   ▼
White and Woods (1958) all with molar volume 13.4 cc/mole; Δ sample
10 from present work at 12.89 cc/mole.

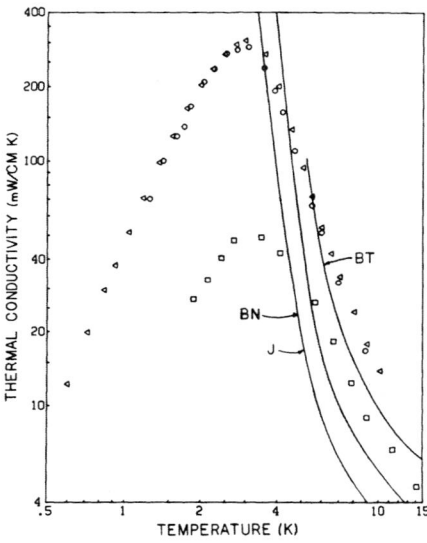

Figure 2 - Thermal conductivity of isotopically purified neon-20 and
"Ziman limit" theoretical results. ◁ sample 18 from present work,
O Kimber and Rogers (1973), □ natural neon White and Woods (1958).
Theoretical curves (J) Julian (1965), (BN) Benin (1968), (BT) Bennett
(1970).

Figure 2 shows the data from the isotopically "pure" neon-20 samples of the present work compared with those of Kimber and Rogers[2]. The agreement of these data in the 4 K to 10 K range is within experimental error. By coincidence, similar mechanical defect concentrations result in similar low temperature conductivities.

III. DISCUSSION

Figure 2 also shows the theoretical curves of Julian[6], Benin[7], and Bennett[8]. These theories all employ the technique of Leibfried and Schlömann[1] in what is known as the "Ziman variational limit" of infinitely fast normal process phonon scattering. Thus all the thermal resistance is the result of Umklapp scattering[9]. According to Benin[10], the position of these curves may be determined more by the parameters of the interatomic potential, than differences in the approximations of the three theories. All of these calculations were made when only the White and Woods data was available. It would be interesting to re-examine them in the light of the present data, using recent modifications of the interatomic potential[11],[12]. Any effect of a finite normal process phonon scattering rate, as well as that of extrinsic scattering due to crystal defects is left out of these theories. There is of course, the phenomenological relaxation time theory of Callaway[13]. It is capable of taking into account all scattering mechanisms by combining them into a total relaxation time. The Callaway theory uses an integral modification of the equation for the thermal conductivity of gases. Callaway expresses the thermal conductivity $\kappa = \kappa_1 + \kappa_2$, where κ_1 would represent the conductivity if Normal process scattering was small, while κ_2 is included when Normal processes are significant.

$$\kappa = \kappa_1 + \kappa_2$$

$$\kappa_1 = GT^3 \int_o^{\theta/T} dx \; J_4'(x)\tau_c$$

$$\kappa_2 = GT^3 \frac{[\int_o^{\theta/T} dx \; J_4'(x)(\tau_c/\tau_N)]^2}{\int_o^{\theta/T} dx \; J_4'(x)(\tau_c/\tau_N\tau_R)}$$

where $G = k^4/\hbar^3 2\pi^2 v$, $x = \hbar\omega/k{\cdot}T$, $J_n'(x) = x^n e^x/(e^x-1)^2$, and v is the volume dependent Debye sound velocity. The combined relaxation rate $\tau_c^{-1} = \sum_i \tau_i^{-1}$, where i ranges over the operative scattering processes. In this case, contributions to the total rate are made by boundary, dislocation, isotope, Normal and Umklapp scattering processes. $\tau_R^{-1} = \sum_{i \neq N} \tau_i^{-1}$, where Normal process rates τ_N^{-1} are omitted from the sum. Although the form of the various relaxation times are not known exactly, the following relaxation rate was used in computer fitting (in cgs units):

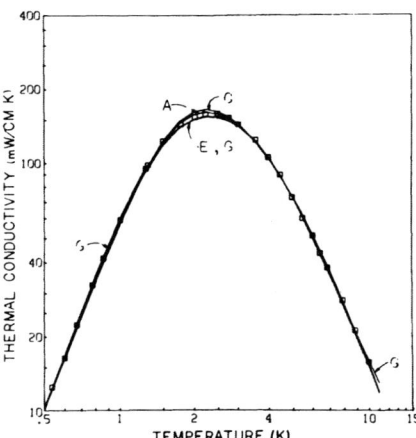

Figure 3 - Callaway theory fits for forms A, C, E, and G to data for natural neon sample 3. Parameters R, N_d, β_N, and β_U varied in each fit.

$$\tau_c^{-1} = \tau_B^{-1} + \tau_D^{-1} + \tau_I^{-1} + \tau_N^{-1} + \tau_U^{-1}$$

$$= 2.014 \times 10^7 \, V^{-2.18}/R + 1.645 \times 10^{-3} N_d xT$$

$$+ 4.75 \times 10^{-3} \, S \cdot V^{7.53} \Gamma x^4 T^4 + \beta_N x^2 T^5 + \beta_U x^2 T^5 e^{-\theta/4T},$$

where R is the average grain size, V is the molar volume, N_d is the dislocation density, Γ is the mass defect $\sum_i f_i (\frac{\Delta M_i}{M})^2$, S = 1.5 represents the enhancement of isotope scattering due to zero point energy differences[14], θ is the volume dependent Debye temperature, and β_N and β_U are adjustable parameters.

Figure 3 shows the results of computer fits to one sample. Almost all samples showed a $T^{2.7}$ dependence between 0.5 K and 1.0 K. A good fit to the data is found by combining the effects of the T^3 boundary scattering and the T^2 dislocation scattering. The values of R and N_d used in the fit are obtained by a least squares fit. Here, R = 0.013 cm, which is one tenth of the sample diameter. The dislocation density N_d is $4 \times 10^8/cm^2$. However, uncertainties in the relaxation time expression suggest that $2 \times 10^8 < N_d < 10^{10}$, which at the highest, may be too high even for a system that is known to be very sensitive to thermal stress[15]. R and N_d could not be measured directly.

Normal and Umklapp scattering processes are dominant in the temperature range above the peak conductivity. Modifications of

Herring's formula $\tau_N^{-1} \propto x^s T^5$ were used for the functional form of the relaxation rates[16]. The exponential $e^{-\theta/bT}$ was added to τ_U^{-1} after Peierls[17]. Several different functional forms of $\tau_N^{-1} + \tau_U^{-1}$ were fit to the same data. The forms tried are listed in Table 1. The parameters obtained in least squares fits using each form are also listed. Forms with x^2 stress the longitudinal phonons whereas forms with x emphasize transverse phonons. To make allowances for the fact that the temperature and frequency regions of interest exceed those for which Herring's calculation is strictly valid, one may reduce the temperature dependence of τ_N^{-1}. This compensates for the reduced density of states at higher frequencies. This may be done by reducing the power of T, as in forms E and G, or, in τ_U^{-1}, by increasing the value of the constant 'b', as in form E. Figure 3 shows that several forms may be used to provide a good fit. Of these forms, only E and G come anywhere near fitting data on isotopically pure neon. Forms A, C and E fit data on the smaller molar volume sample 10. Attempts to derive the explicit volume dependence of β_N and β_U and thus fix these parameters to compare fits at different molar volumes were not successful, but might be worth pursuing.

In figure 1, data are compared for two samples with different molar volumes. In the 5 K to 10 K range the thermal conductivity obeys the empirical relation $K = 2500 \, (T/K)^{-2.2} (V/13.4 \text{ cc/mole})^{-11 \pm 3}$ mW/cm K. This relation is the result of a multitude of physical processes and does not have any particular significance. It is interesting, however, that one can show from simple arguments using $K = \frac{1}{3} \frac{V}{8\pi} \int C_q v_q^2 \tau_q q^2 dq$, the Lennard-Jones potential, and the assumption

TABLE 1

Functional Form of τ_U^{-1} and τ_N^{-1} Used in Computer Calculations		Values of Parameters in Least Squares Fits To Thermal Conductivity Data of Sample 3				
Functional Form $\tau_N^{-1} \quad + \quad \tau_U^{-1}$	Letter Designation	N_d $(10^8/cm^2)$	R (cm)	β_N $(sec^{-1}K^{-5})$	β_U $(sec^{-1}K^{-5})$
$\beta_N x^2 T^5 + \beta_U x^2 T^5 e^{-\theta/4T}$	A	5.20	0.0135	75,600	131,000
$\beta_N x T^5 + \beta_U x^2 T^5 e^{-\theta/4T}$	C	5.82	0.0139	231,000[a]	103,000
$\beta_N x^2 T^4 + \beta_U x^2 T^5 e^{-\theta/6.5T}$	E	3.82	0.0120	203,000[b]	109,000
$\beta_N x^2 T^4 + \beta_U x^2 T^4 e^{-\theta/4T}$	G	3.38	0.0123	229,000[b]	2,000,000[b]

[a] In units of K^{-5}

[b] In units of $sec^{-1}K^{-4}$

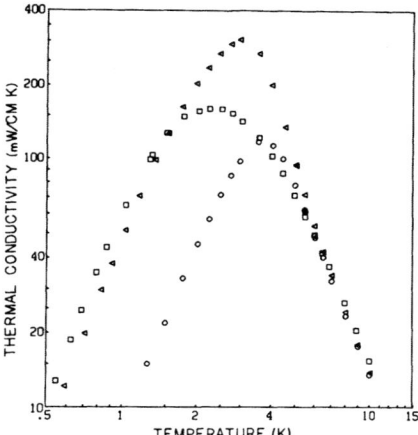

Figure 4 – Isotope effect on thermal conductivity. All molar volumes
are approximately 13.4 cc/mole. □ sample 3, natural neon; O sample
17 and ◁ sample 18, isotopically 'pure'.

that small $\vec{q}$ processes carry most of the heat, that $\kappa(V)$ is approximately proportional to V^{-10} in the "Umklapp" region. In the extremely low temperature range, below the lowest temperatures reported
here, where $C_V \propto (T/\theta)^3$, $\kappa(V) \propto V^5$. In figure 1, the fact that the low
temperature slope of the sample with the smaller molar volume is
steeper indicates qualitatively that the thermal conductivity is
tending toward a reduced V dependence if not toward reversing the
sign of this depencence.

     Although the Callaway theory was not adaptable to samples with
different molar volumes, fitting data on samples which differ only
in the concentration of a well known point defect is one way to use
the theory to good advantage. A sample with the natural isotopic
mixture and isotopically purified samples are compared in figure 4.
Their molar volumes are the same to within 0.4%. This translates
into an uncertainty of 5% in the thermal conductivity. At the
highest temperatures the conductivity of the "pure" sample falls 10%
below that of the natural neon. This is an anomaly because the usual
defect scattering gives the opposite effect. The molar volume error
is not large enough to account for the discrepancy. An effect due
to the 1% larger mass density for natural neon at the same molar
volume may be involved. The possibility of additional scattering
at high temperatures in the "pure" samples due to unknown defects
cannot be ruled out. At the present time, however, this anomaly
prevents the simple application of the Callaway theory to this data.

IV. CONCLUSION

In conclusion, the results reported here agree reasonably well with other recent data on the thermal conductivity of solid neon. They also add new data at a smaller molar volume. It is hoped that these results may encourage a re-examination of the thermal conductivity calculations made on the basis of the interatomic potential of neon.

---

† Work supported by Advanced Research Projects Agency under contract DAHC-15-73-G-10 and the National Science Foundation under Grants GH-33634 and GH-37757.
‡ Present address: Department of Materials Science and Engineering, Cornell University, Ithaca, New York   14853.
1. G. Leibfried and E. Schlömann, Nachr. Adad. Wiss. Göttingen, Math.-physik. Kl. 4, 71 (1954).
2. R.M. Kimber and S.J. Rogers, J. Phys. C 6, 2279 (1973).
3. H.T. Weston and W.B. Daniels, Bull. Amer. Phys. Soc. II 19, 337 (1974).
4. G.T. White and S.B. Woods, Phil. Mag. 3, 785 (1958).
5. F. Clayton and D.N. Batchelder, J. Phys. C. 6, 1213 (1973).
6. C.L. Julian, Phys. Rev. 137, 128 (1965).
7. D.B. Benin, Phys, Rev. Letters 20, 1352 (1968).
8. B.J. Bennett, Solid State Commun. 8, 65 (1970).
9. J.M. Ziman, Electrons and Phonons (Oxford Univeristy Press, Oxford, 1960).
10. D.B. Benin, Phys. Rev. B 5, 2345 (1972).
11. V.V. Goldman and M.L. Klein, J. Low Temp. Phys. 12, 101 (1973).
12. J.M. Farrar, Y.T. Lee, V.V. Goldman and M.L. Klein, Chem. Phys. Letters 19, 359 (1973).
13. J. Callaway, Phys. Rev. 113, 1046 (1959).
14. H.D. Jones, Phys. Rev. A 1, 71 (1970).
15. M.W. Ackerman, Phys. Rev. B 5, 2751 (1972), also K. Ohashi, J. Phys. Soc. Japan 24, 437 (1968).
16. C. Herring, Phys. Rev. 95, 954 (1954).
17. R.E. Peierls, Quantum Theory of Solids (Oxford University Press, Oxford, 1955).

THERMAL CONDUCTIVITY OF A GLASSY METAL AT LOW TEMPERATURES*

J. R. Matey and A. C. Anderson

Department of Physics and Materials Research Laboratory

University of Illinois, Urbana, Illinois 61801

The phonon thermal conductivity of an amorphous PdSi metallic alloy has been measured and found to have the magnitude and temperature dependence characteristic of non-crystalline dielectric materials.

I. INTRODUCTION

Below room temperature all non-crystalline dielectric solids exhibit a small, characteristic thermal conductivity (see Fig. 1) and a large characteristic specific heat which appear to be rather insensitive to the details of the chemical composition [1,2]. Since it is known that thermal transport is provided by phonons [3], these phonons must be scattered by localized sites which presumably are intrinsic to the amorphous structure. Anderson et al. [4] and Phillips [5] have proposed a model in which the scattering sites are localized atoms or groups of atoms which undergo a quantum mechanical transition from one orientation to another with the absorption or emission of a phonon. This tunneling-states model also accounts for the specific heat which is in excess of that associated with the phonons. Experimentally, however, no correlation has definitely been established between the anomalous specific heat and the mechanism which scatters phonons. In the present paper, we explore this problem through measurements of phonon thermal transport in a glassy metal, namely $Pd_{0.775}Cu_{0.06}Si_{0.165}$.

There are three reasons for selecting the PdSi alloy [6,7]. (i) The densities of the glassy and crystalline phases are nearly

the same [8,9] (within 2%), contrary to other amorphous systems
(SiO_2 changes by $\approx 20\%$), and the atomic arrangement in the glassy
state is rather well-defined as a random packing of Pd spheres [10,
11] with the Si occupying the naturally occurring holes [12].
Thus the density of tunneling-states, if they exist, might be
considerably reduced relative to other non-crystalline materials,
and might be amenable to a first-principles calculation. Indeed
Golding et al.[8], in measurements of specific heat, conclude that
there is no excess amount beyond that contributed by the phonons.
(ii) The phonon (acoustic) velocities [8] and specific heats [8,
13] of glassy PdSi alloy have previously been measured. (iii) The
phonon conductivity had not been measured for any glassy metal.
Of the several possible alloys, PdSi has the advantage of being
readily formed into samples of reasonable size. Also the interpretation of the data is not complicated by magnetic or superconducting phenomena (at least above ≈ 0.1 K) as in many other
glassy metals.

    In Section II, we present a brief description of the experi-
mental technique, since it is somewhat unique. The data, and a
discussion of the results, appear in Section III. In brief, we
find the phonon conductivity of glassy PdSi to have the same
characteristic behavior as found in other amorphous materials.

II.   EXPERIMENTAL TECHNIQUE

    The PdSi alloy [14] was in the form of a wire of 0.03 cm
diameter. At temperatures above ≈ 3 K, the thermal conductivity
was measured using one calibrated Ge resistance thermometer and
two electrical heaters. This technique has been described elsewhere [3].

    At lower temperatures a different arrangement was required.
Since the thermal conductance of the sample was so small, electrical leads would thermally shunt the sample and any vibration
would introduce an intolerable heat leak. Therefore the sample
was supported at both ends and heat was produced in the sample
itself by passing through it an electrical current I of known
magnitude [15]. Temperature differences were measured with carbon
resistance thermometers placed on the sample as shown by the inset in Fig. 1. These thermometers were specially prepared with
low thermal conductance leads made by evaporating superconducting
PbSn on a thin Mylar strip [16]. They were calibrated in each run
against a CMN magnetic thermometer and the vapor pressure of
liquid ^{3}He.

    The power dissipation per unit volume, $\dot{Q}$, is given by the
relation $\dot{Q} = I^2 \rho A^{-2}$ where ρ is the measured electrical resistivity
and A is the cross sectional area of the sample. The temperature
distribution must satisfy the differential equation $D(\lambda DT) = -\dot{Q}$,

where D = (d/dx) and x represents a coordinate parallel to the
length of the sample. With ρ being nearly independent of temperature, and treating λ also as a constant for the small temperature
gradients used in the measurements, solution of the differential
equation gives

$$\lambda_t = (\dot{Q}/2) [x_2 x_3 - x_2^2]\{\Delta T_2 - (x_2/x_3)\Delta T_3 - [1 - (x_2/x_3)]\Delta T_1\}^{-1}, \quad (1)$$

where the subscripts refer to the positions indicated in Fig. 1 and
$\Delta T = [T(\text{power on}) - T(\text{power off})]$ for the indicated thermometer.

Equation (1) gives the total thermal conductivity of the
sample, λ_t. From this must be subtracted the electronic contribution, $\lambda_e^t = 3.48 \times 10^{-4} T(\text{W/cm K})$, obtained from the measured
electrical resistivity through the use of the Wiedemann-Franz law.
Since the residual resistivity ratio is ≈ 1, this relation may be
used throughout the temperature range of the measurements. The
result, $\lambda = \lambda_t - \lambda_e$, gives the thermal conductivity contributed by
the phonons.

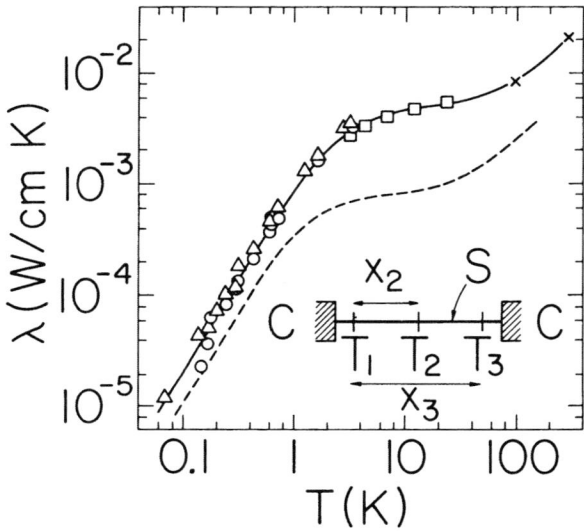

Fig. 1 Phonon thermal conductivity λ vs temperature T. The dashed
curve is representative of phonon transport in a variety of amorphous dielectrics (see Ref. 19). The closed symbols are for three
different samples of glassy PdSi alloy. The points represented by
X are from Refs. 23 and 24. The inset shows the experimental
arrangement for $T \le 3$ K: C-cold finger; S-wire sample; T-carbon
resistance thermometers; $x_2 \approx 1$ cm.

III.   RESULTS AND DISCUSSION

The measured phonon thermal conductivity for the glassy PdSi alloy is shown in Fig. 1. It indeed has the temperature dependence and magnitude which is characteristic of non-crystalline dielectrics, namely λ decreasing with temperature below room temperature, a "plateau" near 10 K, and λ varying roughly as T^2 below ≈ 1 K. However, before we can conclude that the behavior of λ is dictated by the amorphous structure of the alloy, the possible influence of other phonon scattering mechanisms must be investigated. These are (i) boundary scattering from the surface of the wire and (ii) scattering by the electrons.

Boundary scattering would become important for the present sample only at $T \lesssim 0.2$ K and would appear as a deviation from the $\approx T^2$ behavior toward a T^3 dependence. However, it has been shown [17] that, at least for amorphous dielectrics, non-specular boundary scattering of phonons is normally absent at $T \lesssim 1$ K. We would expect the same to be true for this alloy, and indeed there is no convincing evidence for additional scattering near 0.1 K in Fig. 1.

The effect of the scattering of phonons by the electrons can be estimated from information provided in Ref. 18. The maximum influence should occur near 2 K, but would require no significant correction to the present data. This conclusion, however, is open to some criticism since the question of the low temperature electron-phonon interaction is not fully resolved.

Thus it appears that the thermal conductivity of glassy PdSi, as shown in Fig. 1, is determined primarily by an intrinsic scattering process just as in all non-metallic glassy materials that have been investigated. As mentioned in Sec. I, the most successful theoretical model that has been proposed to explain the thermal conductivity (and other anomalous properties) is that based on tunneling states. This model suggests that there should be a correlation between the observed excess specific heat (i.e., in excess of that contributed by the thermal phonons) and the thermal conductivity. That is, a greater number of tunneling sites leads to a larger specific heat and a greater amount of phonon scattering or smaller conductivity.

Usually the total specific heat C of non-crystalline materials is approximated by [1,2]

$$C = C_1 + C_3 = aT + bT^3. \tag{2}$$

A portion of C_3 is due to thermal phonons. The portion in excess of the phonons is given by $C_3 - C_D$ where, in the Debye approximation,

$$C_D = 12.3 \times 10^{17} \ T^3 (\bar{v})^{-3} (\text{ergs/cm}^3 \text{K}) \ . \tag{3}$$

Here $\bar{v}$ is the appropriate average of the acoustic phonon velocities. We use explicitly the specific heat per unit volume so as to obtain information about the <u>number</u> density of the scattering sites. This density would presumably be related to the phonon mean free path ℓ which we define by

$$\lambda = 4.1 \times 10^{10} \ \ell T^3 (\bar{v})^{-2} \ , \tag{4}$$

which is again in the Debye approximation.

The tunneling model was developed to explain the T^2 portion of the curve in Fig. 1, hence we calculate ℓ from the measured λ at 0.2 K where this T^2 dependence is well developed. Table I lists values of $\ell_{0.2\ K}$ for several non-crystalline materials. Also listed are measured values of C_1/T and $(C_3-C_D)/T^3$. (The magnitude of C_3-C_D is found to be roughly the same as C_D.)

A comparison of $\ell_{0.2\ K}$ with C_1/T and with $(C_3-C_D)/T^3$ is provided in Figs. 2 and 3, respectively. There does seem to be a roughly inverse dependence, i.e. $\ell_{0.2\ K} \propto C_1^{-1}$ and $\ell_{0.2\ K} \propto (C_3-C_D)^{-1}$, as indicated by the dashed lines. The main exception is As_2S_3 (point J) which may have an additional contribution to the specific heat due to the presence of water as an impurity [19]. Keeping in mind that the use of Eq. (2) is only a convenient and approximate representation of the measured specific heats, Figs. 2 and 3 suggest that an appropriate linear relation does exist between the number of phonon scattering sites and the excess specific heat [20].

We now wish to ask if our data for glassy PdSi fit within the general description obtained for non-metallic materials as represented by Figs. 2 and 3. Two problems arise. First, the term C_1 is not available since it is masked by the large electronic heat capacity. Second, although two independent measurements of the specific heat produced nearly identical data [8,13], the interpretations applied by the experimentalists were radically different. Golding <u>et al</u>. [8] deduced that there was, within an experimental accuracy of $\approx 2\%$, <u>no</u> excess term (C_3-C_D). Chen and Hammerle [13] on the other hand deduced a finite value which is indicated in Table I and is compared to our value of $\ell_{0.2\ K}$ in Fig. 3 (point M). The difficulty is that the specific heat of glassy PdSi is not a simple function of temperature and is further complicated by the presence of the electrons and, perhaps, magnetic impurities [13]. Nevertheless there is qualitative agreement with the behavior depicted in Fig. 3 in that any excess, $(C_3-C_D)/T^3$, should be quite small based on our measured value of $\ell_{0.2\ K}$.

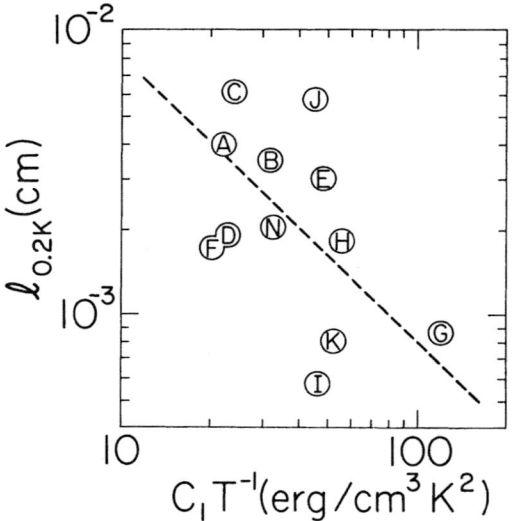

Fig. 2 Phonon mean free path at 0.2 K vs the component in the
excess specific heat which is linear in T. The several amorphous
materials are identified in Table I.

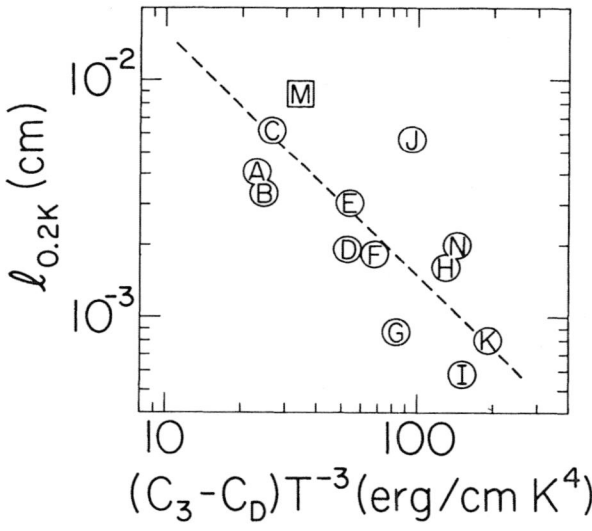

Fig. 3 Phonon mean free path at 0.2 K vs the component of the
excess specific heat which varies as T^3. The several amorphous
materials are identified in Table I. The present result for the
PdSi alloy is indicated by M.

Table I

Excess specific heat and phonon mean free path for several non-crystalline materials.

Code	Material	C_1/T (erg/cm^{3}K^2)	$(C_3-C_D)/T^3$ (erg/cm^{3}K^4)	Ref.	$\ell_{0.2\,K}$ (10^{-4} cm)	Ref.
A	(Pyrex)	22	23	19	39	19
B	GeO_2	32	24	19	35	19
C	SiO_2	24	26	19	61	19
D	B_2O_3	23	53	19	19	19
E	$3SiO_2 \cdot Na_2O$	48	54	19	30	19
F	B_2O_3	21	68	21	18	21
G	$CaK(NO_3)_3$	120	84	19	8.6	19
H	(PMMA)	55	130	19	18	19
I	(PC)	46	150	19	5.7	22
J	As_2S_3	45	96	19	60	19
K	(PS)	51	190	19	8	19
L	(PdSi)	--	$\lesssim 3$	8	88	--
M	(PdSi)	--	34	13	88	--
N	Se	32	140	19	20	19

In conclusion, a dogmatic statement cannot be made concerning the behavior of phonons in, and the excess specific heat of, this glassy metal. However the data that now exist are, in our opinion, consistent with the behavior that has been observed for non-crystalline dielectrics. Even though the PdSi is atomically very dense, with most major holes apparently filled with Si atoms, the matrix continues to behave like other less dense amorphous materials. Thus the physical behavior of these materials is determined primarily by the fact that the structure is amorphous, not by how or of what the matrix is constructed. In addition, it has been shown that an appropriate, approximate relationship may exist between the excess specific heat and the thermal conductivity of non-crystalline materials. This is consistent with the tunneling-states model, or indeed with any model which postulates a relationship between localized states and phonon scattering.

REFERENCES

[*] This research was supported in part by the National Science Foundation Grants DMR 72-03026 and GH 39135.

[1] R. C. Zeller and R. O. Pohl, Phys. Rev. B4, 2029 (1971).

[2] For a general review of the problem, see H. Böttger, Phys. Status Solidi (B) 62, 9 (1974), and papers cited therein.

[3] M. P. Zaitlin and A. C. Anderson, Phys. Rev. Lett. 33, 1158 (1974), and papers cited therein.

[4] P. W. Anderson, B. I. Halperin, and C. M. Varma, Philos. Mag. 25, 1 (1972).

[5] W. A. Phillips, J. Low Temp. Phys. 7, 351 (1972).

[6] P. Duwez, R. H. Willens, and R. C. Crewdson, J. Appl. Phys. 36, 2267 (1965).

[7] H. S. Chen and D. Turnbull, Acta Metall. 17, 1021 (1969).

[8] B. Golding, B. G. Bagley, and F. S. L. Hsu, Phys. Rev. Lett. 29, 68 (1972).

[9] The crystalline "phase" may be heterogeneous. See, for example, C.-P. Chou and D. Turnbull, J. Non-Cryst. Solids 17, 169 (1975), and papers cited therein.

[10] J. Dixmier, J. Phys. (Paris) 35, C4-11 (1974).

[11] S. R. Herd and P. Chaudhari, Phys. Status Solidi A 26, 627 (1974).

[12] D. E. Polk, Scripta Metall. 4, 117 (1970).

[13] H. S. Chen and W. H. Haemmerle, J. Non-Cryst. Solids 11, 161 (1972).

[14] Alloy #4614, Allied Chemical Corporation, Morristown, New Jersey.

[15] We are grateful to M. P. Zaitlin for suggesting this approach.

[16] G. J. Sellers and A. C. Anderson, Rev. Sci. Instrum. 45, 1256 (1974).

[17] M. P. Zaitlin, L. M. Scherr, and A. C. Anderson (to be published).

[18] A. C. Anderson and S. G. O'Hara, J. Low Temp. Phys. 15, 323 (1974).

[19] R. B. Stephens, Phys. Rev. B8, 2896 (1973). For Se we use the acoustic velocities found in J. C. Lasjaunias, R. Maynard, and D. Thoulouze, Solid State Commun. 10, 215 (1972). For As_2S_3 we use the acoustic velocities reported by D. Ng and R. J. Sladek, Phys. Rev. B 11, 4017 (1975).

[20] Figures 2 and 3 also suggest that a rough linear relationship should exist between C_1/T and $(C_3-C_D)/T^3$, which indeed is true.

[21] L. C. Lasjaunias and D. Thoulouze, Solid State Commun. 14, 957 (1974).

[22] M. P. Zaitlin and A. C. Anderson (to be published).

[23] B. S. Berry and W. C. Pritchet, J. Appl. Phys. 44, 3122 (1973).

[24] M. Barmatz and H. S. Chen, Phys. Rev. B9, 4073 (1974).

MEASUREMENT OF THE ELECTRICAL AND THERMAL

CONDUCTIVITY COEFFICIENTS OF As_2Se_3-As_2Te_3 GLASSY ALLOYS

Kotkata M.F., Atalla S.R. and El-Mously M.K.

Physics Department, Faculty of Science

Ein Shams University, Cairo, EGYPT

Eight alloys of general formula $AsSe_{1.5-x}Te (0.0 \leqslant x \leqslant 1.15)$ were prepared in the glassy state. The coefficients of thermal and electrical conductivity of the eight alloys were measured in a range of temperature up to the glass transformation temperature T_g for these glasses. The obtained results revealed the contribution of electrons is much less than that of phonons to the thermal conductivity. The increase of Te-content leads to monotonic increase of σ . The thermal conductivity decreases to a certain minimum (at 30 atm.% Te & Se) and then increases. This behavior can be attributed to a change of the number of scattering centers due to the dissolving of Te atoms or Se atoms in the amorphous As-Se-Te alloys.

INTRODUCTION

The electrical properties of the As_2Se_3 in the solid and liquid states have been studied by Edmond[1]. Kolomietz and coworkers[2] studied the electrical properties of $As_2Se_3.2\ As_2Te_3$ and $2\ As_2Se_3.As_2Te_3$. Other systems containing As_2Se_3 with additions (5-19 atm.%) Cu,Be,Mg,Ca, and Zn have been studied and reported[3]. However, the thermal conductivity of these semiconductors is still much less studied due to the experimental difficulties.

From the previous studies, it was noticed that the addition of Te to the As_2Se_3 alloys leads to an increase of the electrical conductivity and decrease of the activation energy. Its effect upon the thermal conductivity have not been reported before. The As_2Se_3 with Te content from 0-46 atm% have an amorphous structure.

EXPERIMENTAL TECHNIQUE

Elements of the following purities Se (99.999%), Te (99.93%) and As (99.999%) have been used for the preparation of the investigated eight samples of a general formula As $Se_{1.5-x}Te_x$.

The necessary quantities required for certain composition were fused in evacuated clean silica tube under vacuum of 10^{-4} mm Hg. The synthesis of the samples was carried out at 950 ± 20°C for 4-8 hrs[4]. After synthesis, the silica tubes containing the melt were quenched in air at room temperature to get the samples in the glassy state. The glassy structure was checked by means of X-ray investigation, method of fraction and the value of the electrical conductivity. The samples were cut in the form of circular discs of 4-6 mm and 15 mm diameter, suitable for measurement of the electrical and thermal conductivities.

For the electrical conductivity measurements, the samples were introduced between two brass electrodes, with graphite layers to improve the electrical contact quality. The system was put in a temperature regulated oven with ± 0.2°C fluctuation as a maximum value throughout the measuring temperature range, 20-120°C. The electrical conductivity was measured by means of a Vibrating Reed Electrometer with error less than 2.5%.

The construction of the apparatus used for the thermal conductivity measurements is shown in Fig. (1). A steady-state axial heat flow method was used. The sample was sandwitched between two brass rods (5 cm long, 1.5 cm diameter), surrounded by two guard tubes of inner diameters 2.7 and 5.5 cm. The gap between the guard tubes was filled with glass-wool serving as a thermal insulation. Bakelite shielding rings were introduced for centering the rods together with the guard tubes. The heat flow was generated by means of a heater at the far end of the upper rod. The second end of the lower rod was water-cooled. The whole system was put in an automatic-temperature regulated oven. The temperature gradient was measured by means of three copper-constantan integrating thermocouples, fixed in three holes drilled in each rod at 120° apart.

The apparatus was calibrated by means of a fused quartz disc with a well known thermal conductivity coefficient K=0.0146 Cal/cm. sec. deg. at t = 28°C.

The reproducibility of the results for two different specimens of the same composition, and three different runs each, was quite satisfactory. Measurements of the thermal conductivity was carried out through the process of heating and cooling in the measured temperature range 20-120°C.

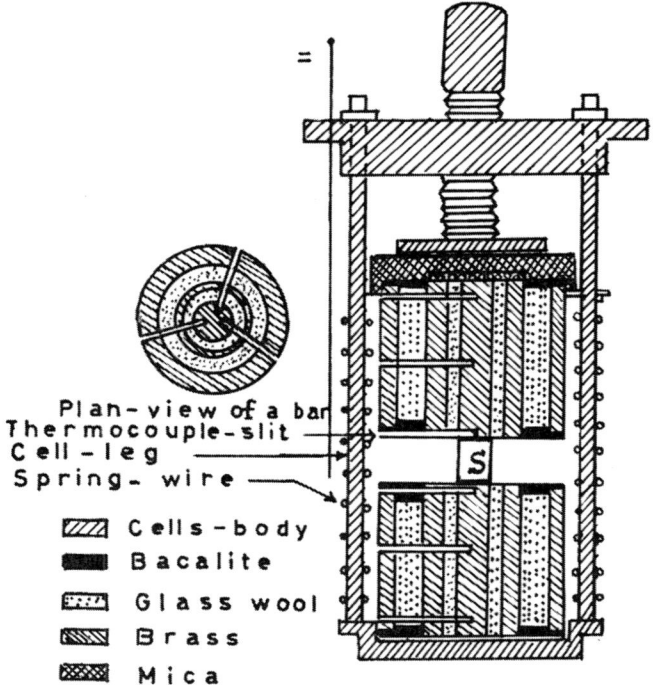

Plan-view of a bar
Thermocouple-slit
Cell-leg
Spring-wire

▨ Cells-body
▰ Bacalite
▱ Glass wool
▨ Brass
▨ Mica

Fig. (1) Schematic diagram of the assembled cell used
for thermal conductivity measurements.

RESULTS AND DISCUSSION

All the measured samples showed a semi-conducting behavior as
the plot log σ —1/T gave a straight line relationship with negative slope. The activation energy E_g and the electrical conductivity at 20°C, log $_{20}$, conductivity at 20°C, log $_{20}$, both show a
linear dependence on the Te concentration in the sample. That is,
as the Te-content increased from 0.0 to 46 atm % - log $_{20}$ decreased from 13.7 to 5.3 ohm^{-1} cm^{-1}, and the activation energy E_g
decreased from 1.87 to 1.22 ev. This can be explained as a result
of the increase in the number of charge carriers.

The coefficient of the thermal conductivity K for the investigated samples was found to increase with temperature as seen in
Fig. (2). Similar temperature dependence of the thermal conductivity have been reported for As_2Se_3,As_2S_3 and As_2S_3.As_2S_3 in the
amorphous phase[5].

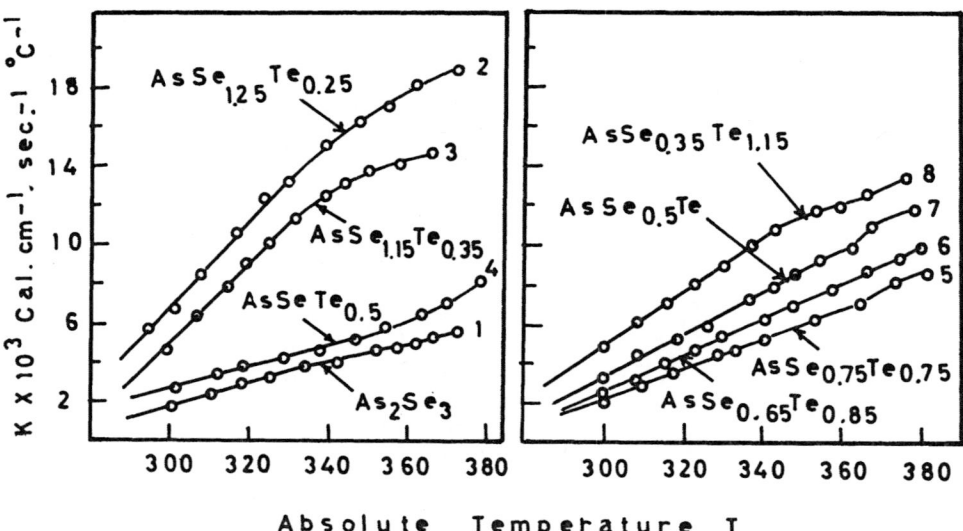

Fig. (2) Dependence of themal conductivity on temperature
for the As_2S_3-As_2Se_3-As_2Te_3 glasses.

In addition to the lattice conductivity in the semiconducting
materials, there can be the electronic contribution to the thermal
conductivity. Here, the photon contribution is excluded in the
measured temperature range[6]. However, the electronic thermal
conductivity is due to thermal motion of free-electrons K_e and to
thermal electron-hole motion (of bipolar type) K_{bp}.

The electronic thermal conductivity K_e is evaluated in terms
of the total electrical conductivity σ from Wiedmann-Franz law

$$K_e = L\ \sigma\ T,$$

where T is the absolute temperature and L is the Lorentz number,
$L = \pi^2/3\ (K/e)^2$. The calculated values of K_e for As_2Se_3 changed
from 1.27×10^{19} at $20°C$ to 5.06×10^{-16} at $100°C$.

The bipolar thermal conductivity K_{bp} can be calculated
according to the formula[7]

$$K_{bp} = \frac{3}{4\ \pi^2}\ L\ \left(\frac{E_g}{KT} + 4\right)^2 \sigma\ T.$$

The calculated values of K_{bp} for the parent glass As_2Se_3 increases
from 3.7×10^{-17} at $20°C$ to 2.03×10^{-13} (CGS) at $100°C$.

The lattice (phonon) conductivity K_L for As_2S_3 have been calculated according to the formula[7]

$$K_L = 2.88 \times 10^{-3} \left[\frac{T_m}{M/n \, (V/n)^{4/3}} \right]^{1/2}$$

where, M= atomic weight, T_m = melting point, T_m = 372°C for As_2Se_3,[8]
V = molar volume at the melting point (taken from the density
measurements), n = number of separate ions per one molecule.

The calculated phonon conductivity was found to be $3.64 \, 10^{-3}$
Cal/cm.sec.deg., which is in agreement with the experimentally
obtained value. So, we can conclude that the main mechanism of
heat transport in the examined glasses is due to phonons.

As seen from Fig. (3), the thermal conductivity at 20°C, $K_{20°C}$
changes with the Te content having minimum value at Se and Te
concentration of equal proportions (30 amt%). Similar results have
been reported for copper telluroide, Cu_2Te[9]. The decrease of the
phonon conductivity K_L was explained by the increase of the number
of scattering centers due to the dissolving of the Te atoms in
amorphous As_2Se_3. As the Te content increase more than 30 atm%,
the Se atoms will play the role of scattering centers in the amorphous As_2Se_3-As_2Te_3 alloy. Additional increase in Te-content leads
to a decrease of the concentration of the Se scattering centers,
leading to an increase of the thermal conductivity.

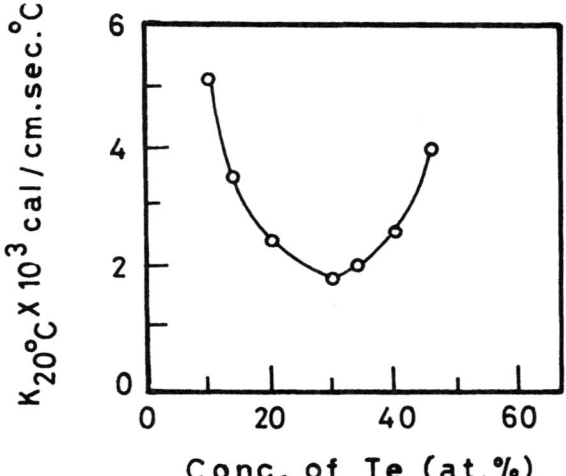

Fig. (3) Dependence of $K_{20°C}$ on Te-content

REFERENCES

1- Edmond J.T., Br. J. of Appl. Phys. $\underline{17}$, 979 (1966)

2- Kolomietz B.T., Phys. Status Solidi, $\underline{7}$, 359 (1964),
$\underline{7}$, 713, (1964).

3- Mott N.F.; Electrons in disordered strucutres.
Advances in Physics (Phil. Mag. Suppl.) $\underline{16}$,
No. 61, 49, (1967).

4- Obraztsov A.A., Borisova Z.V., Isv. Acad. Nauk USSR,
Bullatene of the USSR Acad. of Science, Non-
organic materials, $\underline{VI}$, No.8, (1970).

5- Kolomietz B.T. et al., Soviet Physics-Solid St., 7, $\underline{5}$,
1285 (1965).

6- Glassbrenner, C.J. and Slack G.A., Phys. Rev.,134,
4A, A1058 (1964).

7- Regel A.R., Smirnov I.A., and Shadrichev E.V., J. Non-
cryst. Solids, 8-10, 266 (1972).

8- Thoruburg D.D., J. of Non-cryst. Solids, 11, 2, (1972).

9- Okhotin A.S. et al., Phys. Stat. Solidi, 31, 2, (1969).

THE LOW-TEMPERATURE THERMAL CONDUCTIVITY OF HIGHLY EXTRUDED POLYETHYLENE

S. Burgess and D. Greig

Department of Physics, University of Leeds

Leeds LS2 9JT, England

The temperature dependence of the thermal conductivity, κ, has been obtained from 2K to 100K on specimens of extruded high density (Rigidex) polyethylene, with extrusion ratios, R, as great as 20. The results are compared with our earlier measurements on isotropic polyethylene, and on samples of low density (Hostalen) polyethylene of appreciably smaller extrusion ratios. In all cases the values of κ below 20K are roughly comparable. Above 20K, however, the value of κ parallel to the draw direction, $\kappa_{\shortparallel}$, increases markedly with increasing R. In fact, in the present measurements, for R $\simeq$ 20, κ at 100K is greater than the value in the isotropic Rigidex material by a factor of 8. The thermal conductivity is much higher than normal for polymers or glasses and is more comparable to that of a metallic alloy such as stainless steel. These observations cannot be explained in terms of the models that we have previously used to account for the values of κ in low density polyethylene. In the present instance it is necessary to consider the role of tie molecules that are invoked to explain the exceptionally high mechanical stiffness of these same materials.

In an earlier paper[1] we have reported measurements between 2K and 100K of the temperature dependence of the thermal conductivity, κ, of a number of samples of extruded polyethylene. The particular form of polyethylene used (Hostalen GUR) was of relatively low density and very high molecular weight, so that the extrusion ratio, R, was always limited to rather low values, the greatest of which was 4.4. As a result of our experiments we found that at temperatures $\sim$100K there was a large anisotropy in κ which was

independent of the value of R.   The magnitude of κ in the
extrusion direction was roughly 5 times greater than in the perpendicular direction, and this was explained in terms of the
orientation of the crystallites on extrusion.

        In the present paper we report the results of a further
study of κ in the same temperature range for higher density polyethylene of lower molecular weight in which much higher values of
R can be obtained.   The aim of the work was to test the model
successfully employed in our earlier experiments on specimens in
which, as the result of the high degree of extrusion, the elastic
modulus was known to be an order of magnitude greater[2].

SPECIMENS AND EXPERIMENTAL DETAILS

        For the present measurements the basic material was a linear
polyethylene, Rigidex 50, of average density 0.97g cm^{-3}.   One
specimen, SRO, was machined from this in the form of a cylinder,
1 cm diameter by ∿5 cm long.   The other two specimens SR1 and SR2,
were produced by extrusion on a Fielding-Platt press, and had
values of R of 9.3 and 20 respectively.   The extrudate was in the
form of thin rods of 0.25 cm diameter, and was used for these
experiments without further machining.

        The equilibrium technique employed in the measurement of κ
has been described in detail in earlier papers[1,3], and will only be
mentioned briefly here.   Unfortunately the shape of the extrudate
meant that the specimen geometry was not ideal but reasonable
measurements were possible as the conductance of the samples was
high for polymers.   In the case of specimen SRO the specimen
attachments were identical to those described in these references,
but as the thermal resistances of the relatively thin samples, SR1
and SR2, were rather high, particular care had to be taken with
them.   For these specimens the 1.2kΩ heater consisted of ∿8m of
46swg manganin wire, while a further 80cm of this was used in the
leads to minimise heat losses by conduction.   The rods SR1 and
SR2 were drilled perpendicular to their central axes and tapped at
both ends to receive 10BA screws which were in contact with the
helium can and heater respectively.   To receive the thermocouple
thermometers, two # 80 holes were drilled, 1.5cm apart, in the
central section of the rods.

        The total error involved in the case of specimen SRO was ∿3%.
Three factors contributed roughly equally to this error and these
were, (a) the accuracy of the geometrical factors, (b) the non-
uniformity of heat flow, and (c) heat losses by conduction.   In
the case of the thin rods SR1 and SR2 the error from (c) is much
larger below ∿30K where the data are erratic and unreliable.

RESULTS AND DISCUSSION

The measurements of κ parallel to the extrusion direction are shown in figure 1. Important points to note are: (1) In both extruded specimens the conductivity is significantly higher than in SR0. (2) For the specimen of higher extrusion ratio there is an additional increase in the values of κ. By ∿100K the conductivity of SR2 is greater than that of SR0 by a factor of 8, and is the same as that of stainless steel.

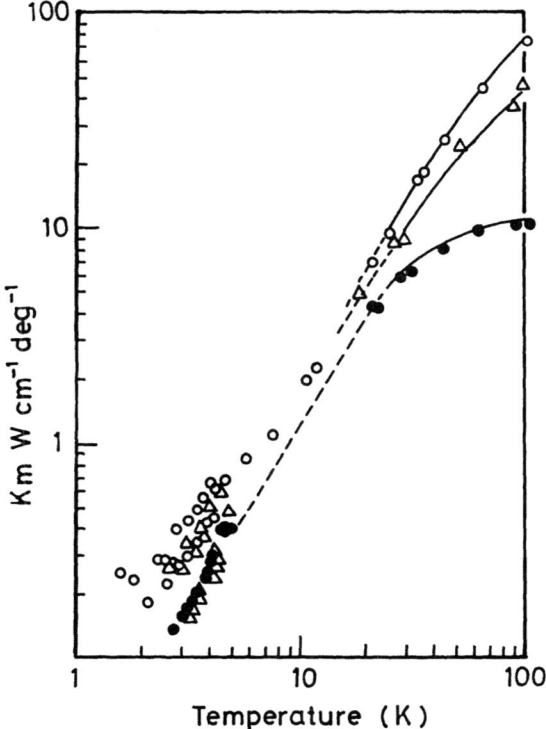

Figure 1:

Experimental temperature variation of κ.● , SR0 isotropic Rigidex 50; Δ , SR1, extrusion ratio 9.3; O , SR2, extrusion ratio 20 (both measured parallel to extrusion direction).

Polyethylene is a semicrystalline polymer with a texture consisting of lamellae ∿150Å thick.   The chains fold within the lamellae to form crystalline regions, and also, to some extent, thread throughout the structure as a whole forming tie molecules and disordered regions between the lamellae[4,5].   In isotropic polyethylene the lamellae are randomly orientated, but when the polymer is extruded this randomness is reduced.   As shown by X-ray diffraction the extruded material has a high degree of orientation[2], and the simplest description is a series arrangement in the extrusion direction of crystalline and amorphous regions, with the chains in the crystallites also sligned in that direction.

Calculations have shown that the values of κ at 100K along the chain axis in a crystallite is ∿50 times greater than that in the amorphous material[1].   Hence in extruded polyethylene κ is determined solely by the interlayer amorphous polymer with the crystalline part shunted out.   In our study of low-density polyethylene the temperature dependence of κ was accounted for in this way, with the calculated values obtained from a recent model by Morgan and Smith[6].   One feature of these experiments was that the conductivity at a given T was independent of the value of R.   A comparison between those values of κ and the measurements on SR1 shows that the new results are almost exactly double those obtained previously.   Since the crystallinity of Rigidex is ∿80% as compared to 59% for Hostalen, we see that the fraction of amorphous material in the present specimens is half that in the lower density specimens, and the doubling of κ is as expected.

The new feature of the present results is the _further_ increase in κ between SR1 and SR2.   This change can be correlated with the continuing increase of elastic modulus referred to earlier[2], and cannot be explained on the basis of further orientation of the crystallites.   The suggestion is that for R > 10, the increase in stiffness may be due to the production of an increasing number of intercrystalline ties, and these would certainly also lead to an enhancement of κ.

We wish to thank Mr. A.G. Gibson for supplying the specimens, and Dr. G.R. Davies and Professor I.M. Ward for discussing recent developments on the morphology of polyethylene.

1. S. Burgess and D. Greig, J.Phys.C: Solid State Phys., $\underline{8}$, 1637 (1975).
2. A.G. Gibson, I.M. Ward, B.N. Cole, and B. Parsons, J.Materials Science, $\underline{9}$, 1193 (1974).
3. S. Burgess and D. Greig, J.Phys.D: Appl.Phys., $\underline{7}$, 2051 (1974).
4. A. Keller, Reports on Progress in Physics, $\underline{31}$, 623 (1968).
5. A. Peterlin, J.Polymer Science, $\underline{C9}$, 61 (1965).
6. G.J. Morgan and D. Smith, J.Phys.C: Solid State Physics, $\underline{7}$, 649 (1974).

THE THERMAL AND ELECTRICAL CONDUCTIVITIES OF SOME CHROMIUM ALLOYS

M. A. Mitchell and J. F. Goff

U. S. Naval Surface Weapons Center, White Oak

Silver Spring, Maryland 20910

and

M. W. Cole

Department of Physics, Pennsylvania State University

University Park, Pennsylvania 16802

The thermal and electrical conductivities of some
Cr-Fe alloys will be presented in context with several
non-ferrous alloys whose transport properties are now
rather well understood. It will be seen that the magnetic moment of the Fe atom greatly complicates the
behavior of the ferrous alloys.

I. INTRODUCTION

The primary purpose of this paper is to make an initial presentation of the thermal and electrical properties of some Cr-Fe
alloys. This presentation will be made in context with the anomalous
nature of Cr and some simpler Cr alloys, which are now in large
part understood. It will be seen that the Cr-Fe alloys are greatly
complicated by the magnetic moment associated with the Fe atom.

The thermal and electrical conductivities of Cr show at least
three anomalies which are of interest here. The first is the well-
known Neél point anomaly at 312 K where Cr changes from a low
temperature antiferromagnetic state to a higher temperature paramag-

netic one.[1] The result, a relative minimum in the electrical resistivity[2] and the Lorenz No. L,[3-5] can be explained by a change in
the band structure which arises from a BCS-like gap's opening over
a portion of the Fermi surface.[6-10] The second anomaly appears as
an enhanced value of L at all temperatures greater than about 100 K.[7]
It has been argued that this anomaly is due either to an enhanced
value of the electronic thermal conductivity λ_e[7,8,10] or to the
lattice thermal conductivity λ_g.[11,12] Thirdly recent measurements
of λ_g for several Cr alloys[13] indicate that it itself is anomalously
small; electronic anomalies can be seen in the thermal conductivity
even at the lower temperatures.

The electronic transport properties of the BCC transition metals
are dominated by large, overlapping hole and electron surfaces.[14]
For metals like Cr, which have an even number of electrons, the Fermi
level lies near a minimum in the density-of-states. The effect of
alloying with elements of different relative valences is to enhance
one or the other of these surfaces, to change the density-of-states
at the Fermi level, and to change the distribution of states about
the Fermi level.[15] The hole portion of these surfaces is slightly
larger than the electron one[16,17] so that the effect of alloying is
not symmetrical with valence. Thus, the antiferromagnetism of Cr,
which depends upon an interaction of these two surfaces is weakened
by solutes that increase the hole surface and strengthened by those
that decrease it.

For these reasons we shall present the results of our Cr-Fe
measurements as compared with those that we have obtained for Cr-Mo
because Cr and Mo are isoelectronic; and therefore, these band structure and magnetic effects are minimized. There will also be a comparison with a Cr-V alloy because Cr-V becomes paramagnetic at all
temperatures for V concentrations of several percent. There are
three important points of comparison to keep in mind:

(1) For concentrations less than about 10 atomic % Fe and Mo
depress the Neél temperature T_N at the same rate.[9] While Fe should
have eight valence electrons as compared to the six of Mo, it would
appear that its localized moment[18-20] makes it behave isoelectroni-
cally in this respect.

(2) Electronic conduction takes place in at least two electronic groups, one of which develops an energy gap below T_N. The
scattering caused by the localized moment of the Fe solute seems to
affect the paramagnetic electron group which has no gap more than
it does the antiferromagnetic group.[9]

(3) λ_g for the Cr-Mo alloys increases with increasing Mo concentrations.[13] Since the antiferromagnetism of Cr is weakened by Mo,
there may possibly be a magnetic-phonon interaction that limits λ_g.

II. EXPERIMENTAL

The preparation and heat treatment of four of the samples (in atomic %), 4.57 V, 3.89 Mo, 9.35 Mo, and 9.35 Fe, have been described previously.[9,13] The 17.31 Fe was prepared with the same starting materials as the 9.35 Fe by arc melting several times to insure homogeneity. The sample was ground from the crude ingot to a square bar of approximate dimensions 62 x 4 x 4. The length-to-area ratio was about 25 cm^{-1} (as it was for most of the samples). The distance between the thermometer clamps was measured with a traveling microscope with a precision of 0.01 mm, and the thickness was measured with a micrometer with a precision of 0.003 mm. The error in the geometry factor was less than 1%. The sample was annealed for two hours at 895 $\pm$ 5°C. The preparation of the 50.81 Fe sample was quite similar but with different starting materials. The 15.3 Fe sample was obtained from Cox and Lucke who prepared it along with those reported in their work on the thermoelectric power.[21]

The thermal conductivity measurements were made in a cryostat described by Goff.[7] The measurements have been discussed elsewhere.[13] Typically, in a measurement of this type the largest errors arise not from geometry measurements, but from thermal instabilities in the apparatus. For example, with no heat inputs from the sample heater, if the heat sink is drifting at a rate of 0.1°K/min and if about half the heat that is withdrawn from one of the samples considered here goes toward establishing a temperature gradient, then the temperature drop between the thermometer clamps can be as high as 0.1°K. While generally the thermal drift in the apparatus was held much lower than this, sometimes sudden changes in ambient temperature outside of the cryostat and fluctuations in the power supplied to the inside cryostat heaters could cause a large thermal drift and lead to scatter ($\sim$ 10%) in thermal conductivity data. As a general rule, though, the thermal drift was not a serious factor and the overall accuracy of the data was on the order of 2-3%. Thermal drift errors are the most difficult to estimate and control.

III. DISCUSSION

A. Electrical Resistivity

The electrical resistivity of the seven alloys discussed here is shown in Figure 1. Five of the alloys are antiferromagnetic and have minima in the resistivity at T_N. The rise in resistivity below T_N is a result of the opening of an energy gap in one of two bands as discussed previously.[9] The 4.57 V and 50.81 Fe are paramagnetic, the latter strongly so. The 4.57 V has a room temperature resistivity which is actually less than that of pure Cr. The reason for

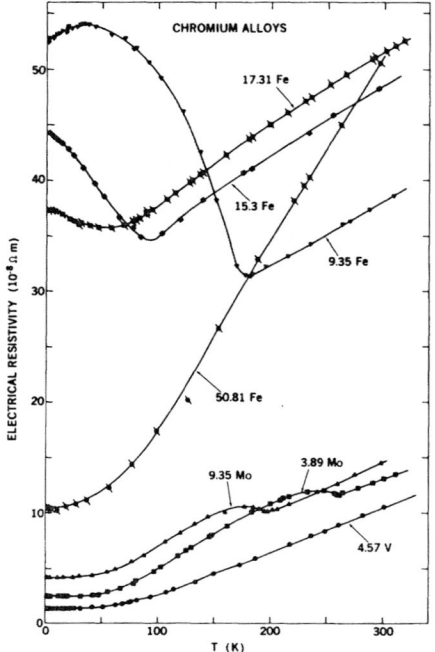

Figure 1. Electrical Resistivity of Chromium Alloys. The 4.57 V and 50.81 Fe are paramagnetic; the rest are antiferromagnetic.

this may be involved with the particular electronic structure of Cr. The increase in the size of the hole octahedron of the Fermi surface upon alloying with V is accompanied by an increase in the density-of-states[15] that may lead to the better electrical conduction.

The Cr-Mo resistivity curves resemble pure Cr, except that there is a larger residual resistivity. On the other hand, the three antiferromagnetic Cr-Fe alloys have resistivities which are large in magnitude and which have much more pronounced increases in magnitude below T_N. This is a combination of two factors, one being the existence of spin-disorder scattering from Fe impurities with localized moments and the other being a difference (from Cr-Mo) in proportion between the amount of conduction in the respective paramagnetic and magnetic bands.[9] The temperature-dependent part of the resistivity has a much higher temperature derivative, partially because the Fe impurities lower the Debye temperature so much more in Cr-Fe ($\theta_D = 235°K$ for 9.35 Fe) than in Cr-Mo ($\theta_D = 550°K$ for 9.35 Mo).[15] The 50.81 Fe has a much lower resistivity at T = 0°K, but the temperature coefficient above 100°K is much higher than in any of the other Cr-Fe alloys.

B. Thermal Conductivity

The total thermal conductivity of five of the alloys is shown in Figure 2 and the remaining Cr-Fe alloys are shown in Figure 3. All of the Cr-Fe samples fall between the values for samples 9.35 Fe and 50.81 Fe. The solid lines below the data points of each sample are derived from the Wiedemann-Franz law ($L_o T/\rho$ where L_o is the Sommerfeld value of L) and drawn in for comparison. While some of the features of the electrical resistivity reflected in the Wiedemann-Franz law can also be seen in the thermal conductivity, there is very little structure in λ in the vicinity of T_N. This is consis-

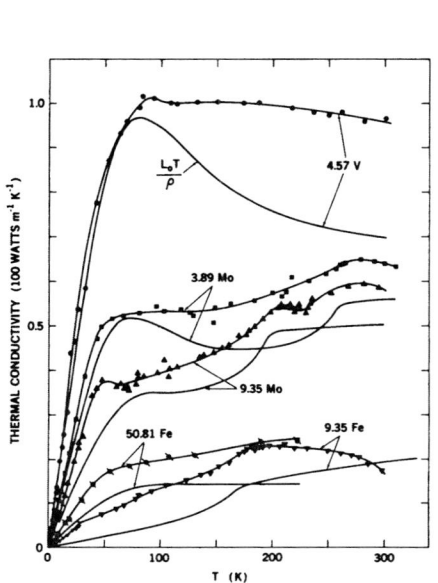

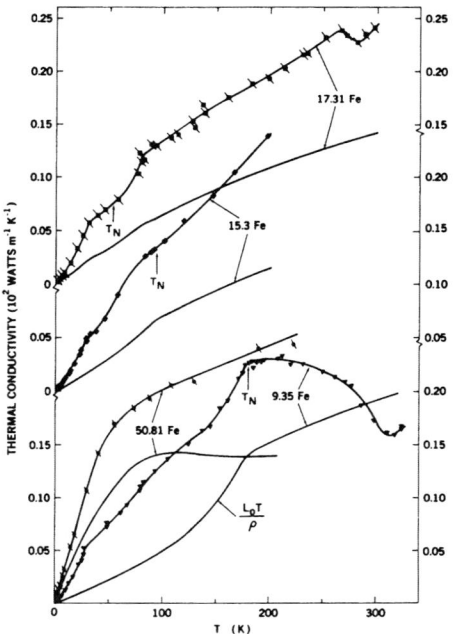

Figure 2. Total Thermal Conductivity. The solid lines below the data points of each sample are the Wiedemann-Franz law.

Figure 3. Total Thermal Conductivity of the Four Cr-Fe Specimens. The solid lines below the data points of each sample are the Wiedemann-Franz law. The vertical arrows indicate T_N.

tent with our view that λ is dominated by λ_e at these temperatures[7,8,10] because the integral which determines the magnitude of λ_e should be insensitive to the band gap that opens at the Fermi surface. However, there is a minimum at about 230 K, or some 33 K above T_N, for sample 9.35 Mo and a knee at T_N for sample 9.35 Fe.

C. Lorenz Number

The Lorenz number of the Cr-Mo and Cr-V samples have been published.[13] The value of L for the four Cr-Fe samples is plotted in

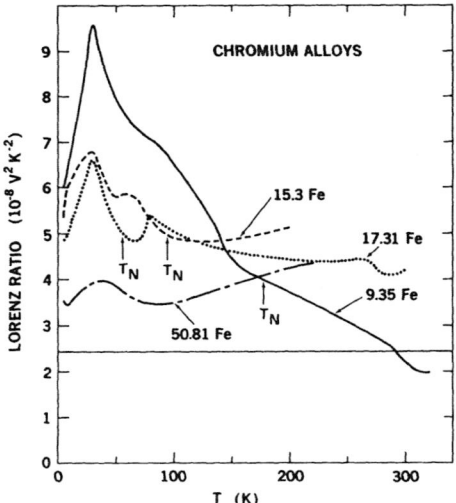

Figure 4. Lorenz Number of the four Cr-Fe samples. All have a peak at about 30 K, as the Cr-Mo also do (see reference 7). The vertical arrows indicate T_N.

Fig. 4. All seven samples have one feature in common: There is a maximum in L at about 30 K. We believe that this peak is caused by λ_g, which peaks at a slightly higher temperature.[13] Except for the 9.35 Mo sample, the values of L for the Cr-Fe samples are higher than in the case of the other two non-ferrous samples. The large values of L at temperatures above 30 K may well indicate electronic effects of the sort that were found in Cr[7,8,10] and the non-ferrous alloys.[13] Sample 9.35 Fe appears anomalous in this respect since L decreases and even becomes smaller than L_O. However, it is known that there is a sharp change in the transport of the Cr-Fe alloys at 10% Fe;[23] and it may be that there is also a difference in the behavior of L.

D. Lattice Thermal Conductivity

The value of λ_g for the non-ferrous alloys was obtained up to about 70 K by a rather complicated analysis that involved four corrections to the thermal resistivity equation: Departures from the additivity assumption for impurity and phonon scattering, changes in the density-of-states at the Fermi level, warping of the Fermi surface, and two band effects.[13] It turns out that the difference $\lambda - L_O T/\rho$ represents this calculation crudely. The principal difference being that the values of λ_g above the maximum are progressively underestimated.

The values of such a calculation for the non-ferrous samples are shown in Figure 5. The lower maxima represent λ_g, and the upper ones represent λ_e. In the case of 9.35 Mo the magnitude of the maximum results for either method of separation are identical. In the case of the 3.89 Mo the complex analysis results are about 25% higher, and in the case of 4.57 V, about 90% higher. In these purer specimens the corrections for ideal thermal resistivity and deviations from Mathiessen's rule are increasingly important. The upper dashed line in Figure 5 is the lattice thermal conductivity that was estimated[12] by using the additive resistance approximation with phonon-electron scattering at low temperatures and Umklapp scattering at high tempera-

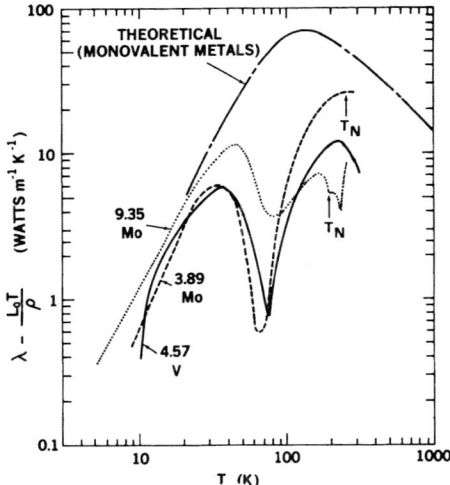

Figure 5. The difference between the total thermal conductivity and the Wiedemann-Franz law. Below about 70 K this provides an estimate of lattice thermal conductivity. The upper dashed line is the theoretical magnitude using the additive resistance approximation with phonon-electron scattering at low temperatures and Umklapp scattering at high temperatures (Leibfried-Schloemann). The vertical arrows indicate T_N.

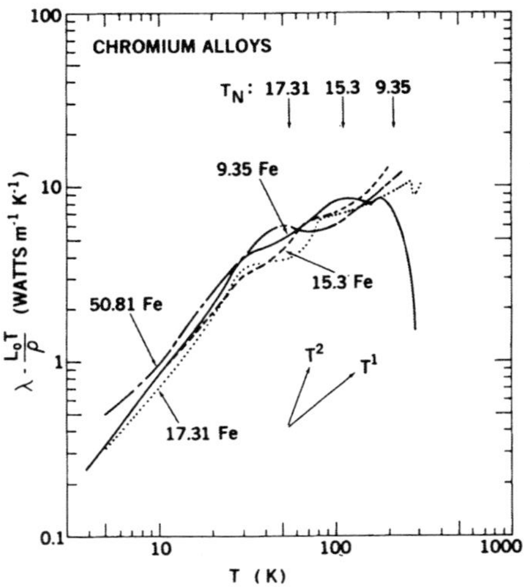

Figure 6. The difference between the total thermal conductivity and the Wiedemann-Franz law. Below 50 K this provides a rough estimate of the lattice thermal conductivity. The vertical arrows indicate T_N.

tures (Leibfried-Schloemann). All experimental determinations of λ_g lead to a much smaller magnitude than one finds from the theoretical estimate.

Similar calculations for the ferrous samples are shown in Figure 6. The interesting feature of these curves is that they all fall within about 25% of one another below 40°K. However, there is the hint of a maximum like that observed in Cr-Mo and 4.57 V only in the 50.81 Fe specimen. The other samples have only an inflection point at about 50°K. In magnitude, $\lambda - L_0T/\rho$ is only about 60-70% of the lattice thermal conductivity of 9.35 Mo. From the results on the Cr-F and Cr-Mo one might speculate that $\lambda - L_0T/\rho$ is mostly lattice thermal conductivity below 50°K, mostly electronic thermal conductivity above 50°K and a mixture in between.

E. Conclusion

The thermal and electrical conductivity data of four Cr-Fe
samples have been presented and compared with three non-ferrous
alloys of Cr with Mo and V whose data is rather well understood.[13]
The differences between the effects of Fe and Mo and V as solutes
in Cr are due mainly to a different balance of effect between the
antiferromagnetic and paramagnetic portion of the Fermi surface in
the two alloy systems. In the Cr-Fe alloys magnetic scattering
seems to be very strong for Fe concentration less than about 20
atomic % and leads to an enormous rise in ρ below T_N. The thermal
conductivity of the ferrous alloys is much lower than that of the
non-ferrous alloys with similar solute concentrations. The lattice
thermal conductivity appears to be considerably lower in both the
ferrous and the non-ferrous alloys than has been predicted theoretically. While there may in fact be significant scattering of phonons
by magnons, the magnon scattering alone does not seem to be large
enough to explain the discrepancy. This follows λ because the difference between the experimental λ_g and the theoretical estimate is
so much greater than the difference between the λ_g in the paramagnetic sample and the magnetic samples.

References

1. C. G. Shull & M. K. Wilkinson, Rev. Mod. Phys. 25 100 (1953).
2. P. W. Bridgman, Proc. Am. Acad. 68 S. 27 (1933).
3. G. T. Meaden, K. V. Rao & H. Y. Loo, Phys. Rev. Letts.23 475 (1969).
4. J. P. Moore, R. K. Williams & D. L. McElroy, Phys. Rev. Letts.
    24 587 (1970).
5. M. J. Laubitz & T. Matsumura, Phys. Rev. Letts. 25, 727 (1970).
6. D. B. McWhan & T. M. Rice, Phys. Rev. Letts. 19, 846 (1967).
7. J. F. Goff, Phys. Rev. B 1, 1351 (1970).
8. J. F. Goff, Phys. Rev. B 2, 3606 (1970).
9. M. A. Mitchell & J. F. Goff, Phys. Rev. B 5, 1163 (1972).
10. J. F. Goff, Phys. Rev. B. 5, 3793 (1972).
11. J. P. Moore, R. K. Williams & D. L. McElroy, 7th Conf. NBS Spec.
    Pub. # 302, p. 297.
12. G. K. White, 10th Conf. Thermal Conductivity (unpublished).
13. M. A. Mitchell & J. F. Goff, forthcoming Phys. Rev. (Sept. 1975).
14. E. C. Snow & J. T. Waber, Acta Met. 17, 623 (1969).
15. F. Heiniger, Phys. Kondens. Materia 5, 285 (1966).
16. W. M. Lomer, Proc. Phys. Soc. 80, 489 (1962).
17. E. B. Amitin & Uy. A. Kovalevskaya, Soc. Phys.-S.S. 10, 1483 (1968).
18. T. Suzuki, J. Phys. Soc. Japan 21, 442 (1966).
19. Y. Ishikawa, S. Hoshino & Y. Endoh, J.Phys.Soc.Jap. 24, 263 (1968).
20. Y. Endoh, Y. Ishikawa & H. Ohno, J.Phys.Soc.Jap. 24, 263 (1968).
21. J. E. Cox & W. H. Lucke, J. Appl. Phys. 38, 3851 (1967).
22. P. G. Klemens, "Thermal Conductivity" (Academic, New York 1969) 1, ρ
23. J. F. Goff, J. Appl. Phys. 39, 2208 (1968).

WIEDEMANN-FRANZ RATIO OF MAGNETIC MATERIALS

K.V. Rao, Sigurds Arajs, and D. Abukay

Department of Physics
Clarkson College of Technology
Potsdam, New York 13676

We demonstrate that in magnetic materials there is
enough evidence to show that some of the physical processes occuring at the onset of magnetic ordering strongly
influence the electrical but not the thermal resistance.
The observed anomalous behaviour of the Wiedemann-Franz
ratio vs temperature is then simply a mathematical artifact manifesting the individual complicated temperature
dependencies in the two resistivities. A new explanation for the observed anomalous Wiedemann-Franz ratio vs
temperature for Cr, and analyses of the thermal and electrical resistivity data of Nd and Tb are given.

It is well known that for magnetic materials, the Wiedemann-
Franz ratio ($W-F \equiv L = \rho/WT$) calculated from the experimental thermal (W) and electrical resistivity (ρ) data, is considerably larger
than the Sommerfeld value $L_\theta = 2.45 \times 10^{-8}$ watt units. Typically,
this can vary from 3 to 8×10^{-8} watt units in rare-earths, transition
metals, and their alloys. As a function of temperature, this ratio
exhibits a very complicated temperature dependence. Since the thermal resistivity is more sensitive to small angle scattering in comparison with the electrical resistivity, it is tempting to attribute
the excess value ($L-L_0$) of this ratio to lattice conductivity and
possibly also to magnon conduction. However, this large excess
value by almost a factor of three would imply that the lattice and/
or magnon contribution to the thermal conductivity in these materials
is considerably larger than the electronic one. This is very unlikely because most of these materials are metallic. Moreover, it
would be a formidable task, indeed, to obtain theoretically the complicated temperature dependencies for these additional conduction
mechanisms which would persist to temperatures higher than the

Debye θ_D, for these materials.   In this paper, we demonstrate that
in magnetic materials there is enough evidence to show that some of
the physical processes occuring at the onset of magnetic transition
strongly influence the electrical resistivity, but not the thermal
resistivity.   The observed W-F ratio then exhibits the mathematical
artifice of reflecting the individual complicated temperature dependencies in the two resistivities.   With a proper understanding of
the mechanisms involved in the ordered state of the magnetic system
or a class of such systems, one can at least qualitatively, account
for the main features manifested in the observed W-F ratio as a function of temperature.   Our analysis on the thermal and electrical
resistivity data for neodymium, terbium, and chromium strongly supports the above observations.

The magnetic properties of light rare earths are of particular
interest because the crystal field splitting of the magnetic energy
levels in their (dhcp) structure are comparable with the exchange
energies.   A consequence of this in neodymium, for example, is the
onset of antiferromagnetic ordering at the hexagonal sites around
19K followed by a similar magnetic structure when the cubic sites
order around 7.5K.   The effect of these transformations on the measured electrical and thermal resistivities and the calculated W-F
ratio from these data in the temperature range 1-40K can be seen in
figures 1, and 2.   Some part of this work has been published before.[1,2]
Figure 1 shows the temperature dependence of the W-F ratio for Nd,

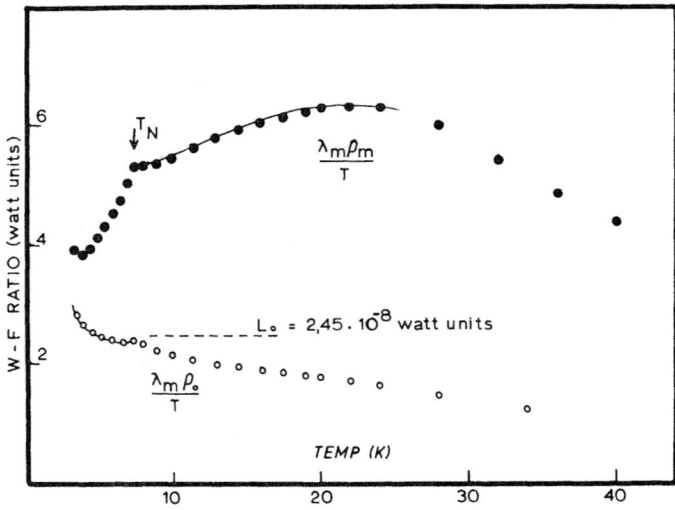

Fig.1:   Temperature dependence of the W-F ratio for Nd, where λ_m
and ρ_m are the measured thermal conductivity and electrical resistivity at temperature T. Open circles: A plot of $\lambda_m \rho_0 T^{-1}$ vs T.

where λ_m and ρ_m are the respective thermal conductivity and the
electrical resistivity measured at temperature T. The value of
this ratio varies from 3.8 to 6.5x10^{-8} watt units in the temperature
range 1-40K. Obviously, the large excess value of this ratio over
the Sommerfeld value, with its complicated temperature dependence,
cannot be totally attributed to lattice conductivity. From fig.2
it is clear that below 7.5K the thermal conductivity data fits only
into an equation of the type

$$\lambda T = a + bT^2. \tag{1}$$

Here, the second term corresponds to the normal electronic conductivity when electron-impurity scattering term is dominant. The coefficient b correlates well with L_0/ρ_0, where ρ_0 is the extrapolated
residual resistivity of this sample. The additional conductivity
term with a T^{-1} temperature dependence cannot be phononic, since at
these temperatures, the phonon contribution to thermal conductivity
of a solid would vary from T^3 to T^2 as the phonon-electron scattering term becomes more effective than the boundary scattering with
increasing temperatures. Hence, we believe that this additional
conduction mechanism is of magnetic origin. An analysis of the electrical resistivity data also shown in fig.2, indicates below 7.5K,
in addition to the residual resistivity ρ_0, a T^2 dependent spin-
disorder scattering term in the antiferromagnetic state. Thus know-
ing the various temperature dependencies in the measured thermal

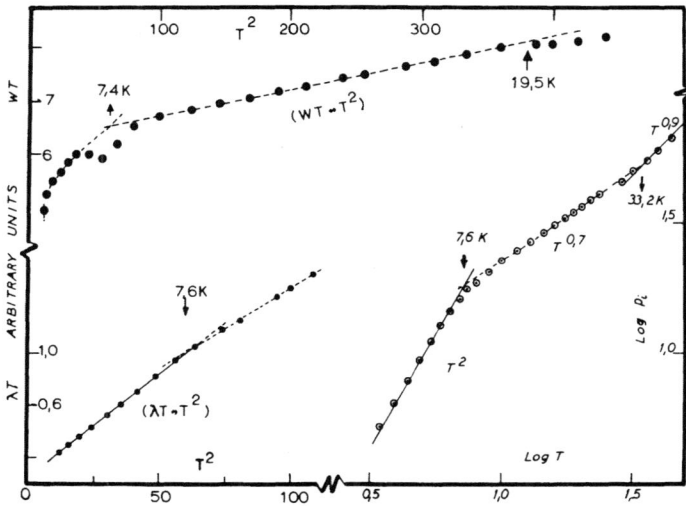

Fig.2: WT vs T^2 for Nd, where W is the thermal resistivity at
temperature T. Bottom left: λT vs T^2 plotted from the thermal conductivity for Nd at a temperature T. Bottom right: log ρ_i vs log T
where $\rho_i = (\rho_m - \rho_0)$.

conductivity λ_m and electrical resistivity ρ_m, it can be easily shown that the calculated W-F ratio below 7.5K satisfies the equation of the form

$$L = \lambda_m \rho_m / T = (aT^{-1} + bT)(\rho_0 + \alpha T^2) \, T^{-1} \qquad (2)$$

The observed excess value for the calculated W-F ratio from the Sommerfeld value therefore need not be solely due to the additional contributions to thermal conductivity alone. In this respect, a plot of $\lambda_m \rho_0 T^{-1}$ vs T, shown by open circles in fig.1, is most revealing. In having used only the residual resistivity term, i.e., having subtracted all known temperature dependences from the total resistivity, this plot now realistically reveals the various contributions to the thermal conductivity alone in this material. The order of magnitude of the excess value of the W-F ratio over the Sommerfeld value below 7.5K is now more reasonable and can be attributed to the additional conduction mechanism discussed above. It is therefore quite reasonable to assume that what one observes in the $\lambda_m \rho_m T^{-1}$ plot is really a mathematical artifact of all the temperature dependencies involved in λ_m and ρ_m respectively.

We now consider terbium, a typical rare-earth ferromagnet. Figure 3 is a plot of the W-F ratio for monocrystal terbium along

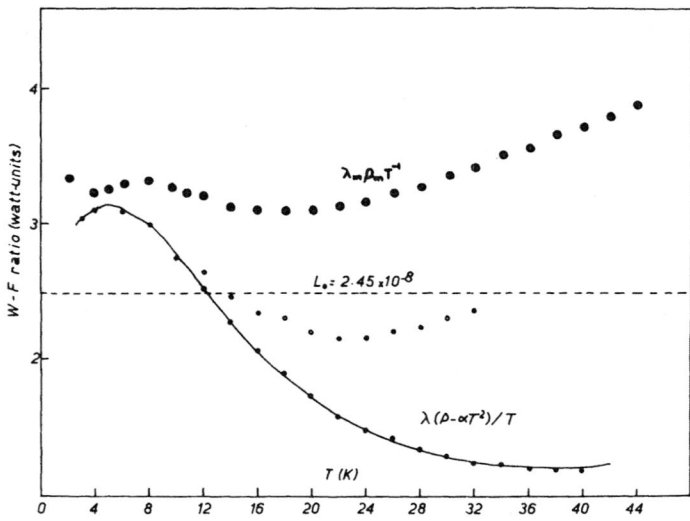

Fig.3: The W-F ratio, $\lambda_m \rho_m T^{-1}$, for Tb monocrystal calculated from the measured thermal conductivity and the electrical resistivity at a temperature. Bottom curve: The W-F ratio using λ_m and the electrical resistivity from which the spin-wave-scattering term, αT^2, has been subtracted. Dotted open circles represent the result when the effect of spin-wave-energy-gap on the αT^2 term is taken into account.

the c-axis calculated from the measured thermal conductivity and electrical resistivity below 44K[3].   Again this ratio is seen to be in excess of the Sommerfeld value with a non-linear temperature dependence.   It is well known that at low temperatures the electrical resistivity is dominated by the residual, ρ_0, and the spin-wave scattering terms (αT^2).   A plot of $\lambda_m(\rho - \alpha T^2)T^{-1}$, shown in fig. 3 reflects more realistically the various contributions to thermal conductivity.   The enhancement in the observed W-F ratio over the Sommerfeld value L_0 with a peak around 8K would indicate a lattice conductivity term in addition to the electronic contribution.   Above 10K the conductivity is more like that for a metal.   This observation is further supported by an independent analyses of the thermal conductivity data.   Figure 4 is a plot of the total thermal conductivity as a function of temperature for a monocrystal terbium along the c-axis.   Assuming the Sommerfeld-Lorenz-number value L_0 for electron-impurity scattering and using the residual resistivity, the non-electronic contribution to the thermal conductivity thus obtained is shown by the broken-line-curve in fig.4.   The low temperature part of this curve has been known to have an initial T^2 dependence which indicates a phonon-electron scattering contribution[4].   Moreover, the peak in this curve lies in a region $\theta_D/12$ expected for a Debye temperature $\simeq 140K$ for terbium.   Clearly, this non-electronic contribution is mainly due to lattice conductivity.   The $\lambda_m \rho_m T^{-1}$ plot for terbium is thus a manifestation of the individual contributions to the thermal and electrical resistivities.

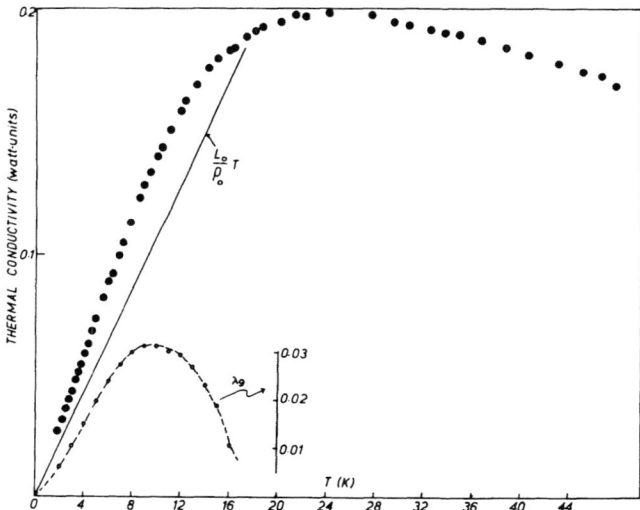

Fig.4: Thermal conductivity of Tb vs Temperature. Broken curve: Non-electronic contribution to thermal conductivity which can be attributed to lattice conductivity.

The observed W-F ratio for chromium[5] as a function of temperature is shown in fig.5.   This plot indicates a typical metallic
behaviour for chromium below 100K.   Attempts to explain the large
enhancement with a maximum of 4.2×10^{-8} watt units around 240K by
attributing it to lattice and band structure effects have been made.[6-8]
This anomalous behaviour in the W-F ratio is a consequence of the
antiferromagnetic ordering around 312K and the observed enhancement
over the Sommerfeld value persists in the region between 100 and 320K
where there exists a transverse spin-density-wave (SDW) magnetic
structure in chromium.   The peculiar band structure of chromium,
consisting of almost identical nearly octahedral-shaped electron and
hole jacks, stabilize a spin density wave having a $\vec{Q}$-vector that
measures the distance between one side of the electron-octahedron
and the same side of the hole-octahedron.   This is responsible for
the occurrence of the itinerant antiferromagnetic coupling of the
electron-hole surfaces provided by the Coulomb attraction of the
electrons and holes.   Such pairs do not respond to an external electric field and therefore give no contribution to the electrical
resistivity.   It is as if these pairs are locked-in and simply drop
out of the conduction band.   The effective number of conduction
electrons n' below the transition is then given by

$$n' = n\{1 - \alpha \, \Delta(T)/\Delta(o)\}. \tag{3}$$

Here, n is the conduction electron density in the paramagnetic state,
$\alpha \simeq 0.2$ is a constant related to the Fermi-energy, $\Delta(T)/\Delta(0)$ the

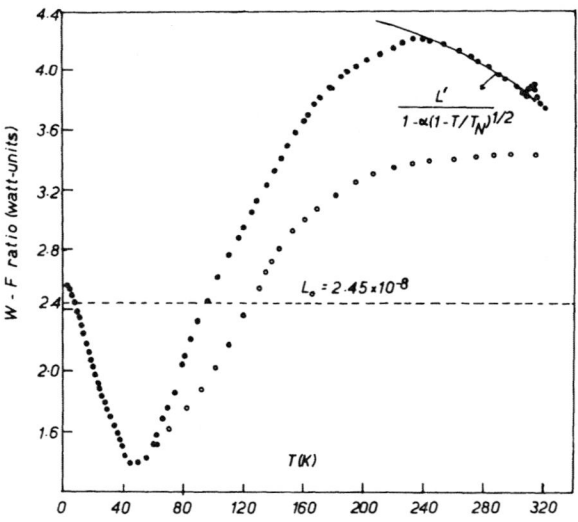

Fig.5: W-F ratio for chromium versus temperature.   The continuous curve is a fit to Equation 8 in the text.

SDW-energy gap ratio has a temperature dependence similar to that of the superconducting gap in the BCS theory

$$\Delta(T) = \Delta(0) \, (1-T/T_N)^{1/2} \qquad (4)$$

T_N, is the antiferromagnetic ordering temperature.   While these electron-hole pairs with an effective neutral charge do not respond to an external electric field they will certainly transport heat in a thermal gradient.   Thus there will be no change in the electron density as far as thermal conduction is concerned.   Now, it is well known from the classical electron theory that the thermal conductivity is related to the relaxation time τ, effective mass m^* of the electron, and the Boltzman constant, k_B, at any temperature, T, as

$$\lambda = \pi^2 k_B^2 T \cdot n \cdot \tau / \, 3m^* \qquad (5)$$

Similarly, the electrical resistivity is given by

$$\rho = m^* /n' \, e^2 \, \tau. \qquad (6)$$

In the paramagnetic state n'=n.   Below T_N, however, the W-F ratio becomes

$$L = \lambda\rho/T = (\pi^2 k_B^2/3e^2)(n/n') = L_0 \, n/n'. \qquad (7)$$

In the case of chromium, if we assume a W-F ratio L' just above the transition temperature, then below T_N due to the change in the electron density for the electrical conduction process Eqs.(4) and (8) predict the behaviour of L to be

$$L = L'\{1 - \alpha(1-T/T_N)^{1/2}\}^{-1} \qquad (8)$$

A remarkable agreement between Eq.8 and the experimental data for a temperature range of nearly 80K below the transition temperature is seen in fig. 5.   If one also incorporates the actual temperature dependence of the electrical resistivity below 310K in Eq.8, then it is clear that most of the enhancement observed in the W-F ratio for chromium is because the magnetic ordering has a profound effect on the electrical resistivity but little on thermal conduction.

Another feature of   some magnetic systems is the difference in the periodicity of the magnetic ordering vector and the lattice. This incommensurability introduces further magnetic superzone gaps which have a profound effect on the electrical resistivity, but not so on the thermal conduction.[9]   A plot of the W-F ratio therefore simply appears to reflect the effects of magnetic ordering on the electrical resistivity rather than indicate non-electronic contributions to thermal conduction.

In conclusion, we believe that, in magnetic systems, there are fundamental physical processes which effect the electrical resistance much more significantly than in thermal conduction.   The observed large values for the Wiedemann-Franz ratio and their complicated temperature dependencies are simply a manifestation of these mechanisms.

One of us, KVR, would like to thank Prof. J.O.Linde at the Royal Institute of Technology, Stockholm, for many a conversation on this.   This work is supported by a NSF-Grant.

REFERENCES:

1   K.V.Rao, H.U.Åström, and Ch. Johannesson
        Phys.Lett. 42A, 53 (1972)

2   K.T.Tee, K.V.Rao, and G.T.Meaden
        J. Less-Common Metals 31, 181 (1973)

3   K.T.Tee, M.S.Thesis, Dalhousie University, 1971.

4   K.V.Rao, G.T.Meaden, and H.Y.Loo
        Proc. 9th International Th. Cond. Conf.held at Iowa, 1969,
        US-ARC Conf. 69100, 181 (1969)

5   H.Y.Loo, M.S.Thesis, Dalhousie University 1969.

6   J.F.Goff, Phys. Rev. B2, 3606 (1970)

7   J.P.Moore, R.K.Williams, and D.L.McElroy
        Proc. 8th International Th.Cond. Conf. held at Purdue Univ.
        1968, Thermal Conductivity, ed.,C.Y.Ho and R.E.Taylor,
        Plenum Press, page 303, 1969

8   G.K.White, Proc. 10th International Th. Cond. Conf. held at
        Albuquerque, N.Mexico - extended abstracts, 1970.

9   See for example, D.W.Boys and S.Legvold,
        Phys. Rev. 174, 377 (1968).

THE THERMAL AND ELECTRICAL CONDUCTIVITY OF ALUMINUM[*]

J.G. Cook, J.P. Moore,[†] T. Matsumura, M.P. van der Meer

National Research Council
of Canada
Ottawa, Canada

[†]Metals and Ceramics Division
Oak Ridge National Laboratory
Oak Ridge, Tennessee   37830

The thermal conductivity, electrical resistivity, and absolute Seebeck coefficient of pure aluminum were determined from 80 to 400 K by measuring all three properties of samples with resistivity ratios of 11×10^3, 8.5×10^3, and 9.5×10^2 using three different techniques. Measurements were made on the purest sample down to 20 K. The thermal conductivity has a broad plateau from 180 to 400 K and a possible minimum of 0.25% which is insignificant compared to the experimental errors. The same properties were measured on an aluminum alloy with a resistivity ratio of 17. Measured values of the thermal conductivity of this alloy agreed to within ±1% with calculated values using parameters obtained from the pure aluminum.

INTRODUCTION

In their reviews of published data on the thermal conductivity, λ, of pure **aluminum, Powell et al.**[1] and Powell[2] have shown that λ at 300 K is known to approximately 2%. The uncertainty at 200 K is appreciably larger; in fact, some of the measurements indicate a deep minimum near 200 K, while others show no minimum at all. This has led Powell[2] to conclude that there is no doubt such a minimum exists, but that the temperature at which it occurs increases with sample purity.

---

[*] Part of research sponsored by the Energy Research and Development Administration under contract with Union Carbide Corporation.

Since we were inclined to doubt the validity of these data, and hence Powell's interpretation, we carried out a new investigation of λ, the electrical resistivity, ρ, and the Seebeck coefficient, S, of a number of Al specimens of different purities. At this conference we can present only a summary of our findings; a more detailed report by Cook et al.[3] is available upon request.

APPARATUS AND SAMPLES

Our samples and the three sets of apparatus used to measure their transport properties are listed in Table 1, where NRC refers to the National Research Council and ORNL refers to the Oak Ridge National Laboratory. Data obtained with these experimental systems have previously been intercompared.[4]

Table 1.   Apparatus and Specimens Employed in This Study

Apparatus		NRC-1[5]	NRC-2[6]	ORNL-3[7]
% Uncertainty(K;ρ)		(±0.8;±0.8)	(±0.6;±0.4)	(±1.2;±0.3)

Specimen Designation	RRR	$\rho(4.2K)$			
A	8500	2.8×10^{-4}	A-1	A-2	A-3
B	11000	2.1×10^{-4}	B-1	-	-
C	950	2.5×10^{-3}	-	-	C-3
D	17	0.14	-	-	D-3

RESULTS

Thermal Conductivity

Data for samples A, B and C are shown in Fig. 1. These data were first corrected for thermal expansion - as have all other data given here-and then for impurity scattering by subtracting $\rho(4.2K)/L_o T$ from $1/\lambda(L_o = 0.02445 \times 10^{-6} V^2 K^{-2})$. These impurity corrections were less than 0.1% for A-1, A-2, A-3 and B-1 above 100 K, leading us to conclude that Fig. 1 represents λ of ideally pure Al as corrected for thermal expansion. From 80 to 160 K, λ is very temperature sensitive, but becomes virtually independent of temperature from 160 to 400 K. The data in the latter range are shown on an expanded scale in the inset.

To obtain a mean curve of $\lambda(T)$, the data were fitted to polynomials.[3] In Fig. 2 the data for A, B and C are shown as deviations from the polynomial

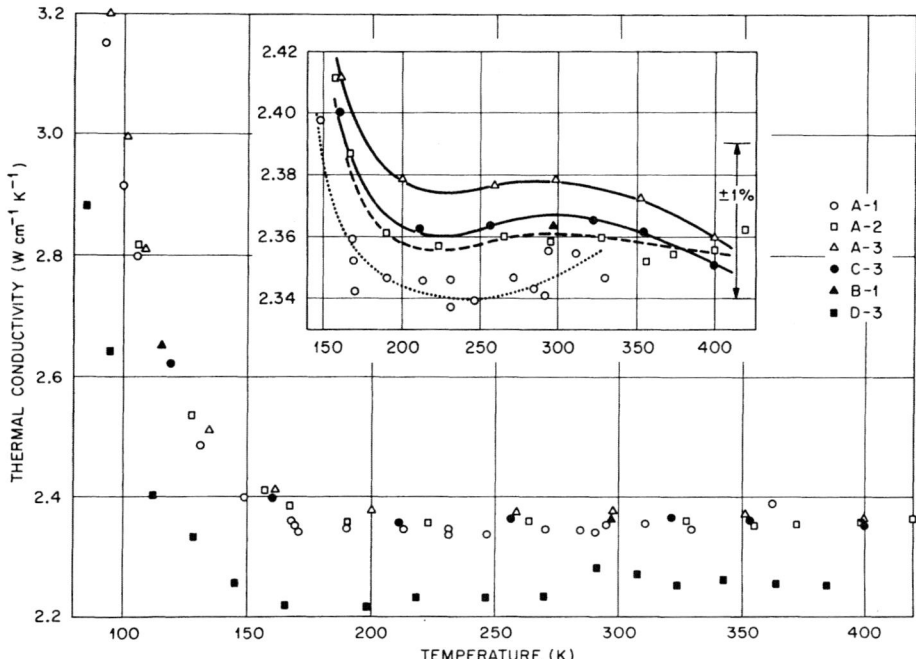

Fig. 1. Thermal conductivity from A-1, A-2, A-3, B-1, and C-3 as corrected for thermal expansion and for impurity resistance using Eq. 1. Results from D-3 were corrected only for thermal expansion.

$$\lambda = 0.779 \times 10^{8}T^{-4} - 26.27T^{-1} + 2.4507 - 0.1845 \times 10^{-6}T^{2} \quad (\text{W cm}^{-1}\text{K}^{-1})$$

in which the terms were chosen only to obtain a good fit to the data and have no theoretical significance. Although the deviations from this polynomial are generally less than 1%, the deviations do not appear to be entirely random. Therefore, the smooth line was drawn in the approximate middle of the experimental data from 80 to 400 K and values from this curve represent our best values for λ of pure Al as corrected for thermal expansion. This curve displays a minimum near 210 K which is only 0.3% deep and within our experimental uncertainty.

We have subtracted an estimated lattice conductivity ($\lambda_{\ell} = [T/17 + 5000/T^{2}]^{-1}$; ref. 3) to determine an electronic component of λ that has a minimum of 1.5% at 190 K. We do not wish to attribute much importance to this result, however, since it is only

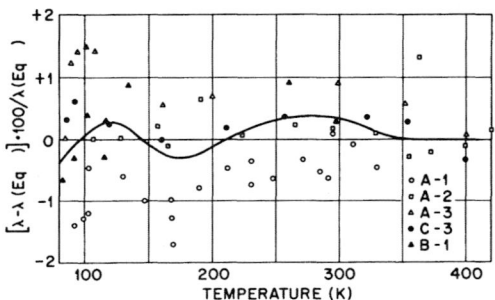

Fig. 2. Deviations of λ results from
           Eq. 2. The λ values from
A-1, A-2, A-3, B-1 and C-3 were
corrected for thermal expansion and
for impurity resistance. The solid
line represents $\bar{\lambda}$ which is the average
value for pure Al from these results.

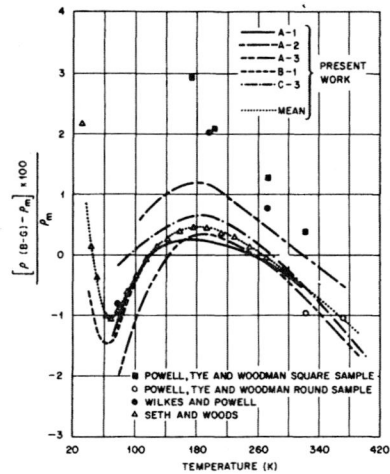

Fig. 3. Electrical resisti-
           vity of Al as devi-
ations from Eq. 3. The
dotted line represents $\bar{\rho}$
which is the average value
for pure Al from these
results.

slightly larger than our experimental uncertainty; and, furthermore,
the presence or absence of any minimum is simply related to the
relative strengths of the vertical and horizontal electron-phonon
scattering processes as noted by Ziman.[8] Values of $\bar{\lambda}$, λ_ℓ and the
electronic component, λ_e, are given in Table 2.

Electrical Resistivity

We have computed the ideal component of ρ simply by subtracting
$\rho(4.2\ K)$ and compared the results to the Bloch-Grüneisen theory in
a manner described by Cook et al.[3] We found a Debye temperature of
383.5 K, and the deviations shown in Fig. 3.

Seebeck Coefficient

Our results agree with literature values to within the experimental uncertainty.[3]

Table 2.   Thermal Conductivity and Electrical Resistivity of High
           Purity Aluminum from This Study, and Calculated Values of
           λ_ℓ, λ_e and L(T).*

T (K)	$\bar{\lambda}$ (W cm^{-1} K^{-1})	$\bar{\rho}$ ($\mu\Omega$ cm)	λ_ℓ (W cm^{-1} K^{-1})	λ_e (W cm^{-1} K^{-1})	L (T) $\times 10^8$ (V^2 K^{-2})
20	147.	0.0008	0.073	147	0.588
30	48.7	0.0043	0.137	48.6	0.696
40	21.2	0.0179	0.183	21.0	0.941
50	10.9	0.0472	0.202	10.7	1.010
60	7.05	0.0953	0.203	6.85	1.088
70	5.03	0.1618	0.195	4.84	1.118
80	4.007	0.2439	0.182	3.825	1.166
90	3.370	0.3377	0.169	3.201	1.201
100	2.968	0.4401	0.157	2.810	1.237
120	2.612	0.6601	0.135	2.476	1.362
140	2.465	0.8899	0.118	2.347	1.492
160	2.396	1.123	0.104	2.292	1.609
180	2.367	1.356	0.093	2.274	1.713
200	2.358	1.589	0.084	2.274	1.807
220	2.358	1.820	0.077	2.281	1.887
240	2.359	2.049	0.070	2.289	1.954
260	2.362	2.278	0.065	2.296	2.012
280	2.364	2.506	0.060	2.304	2.062
300	2.364	2.733	0.056	2.308	2.103
320	2.362	2.961	0.053	2.309	2.137
340	2.360	3.189	0.050	2.310	2.167
360	2.358	3.416	0.047	2.311	2.193
380	2.359	3.645	0.045	2.314	2.220
400	2.359	3.875	0.042	2.316	2.244

* $L(T) = (\bar{\lambda} - \lambda_\ell)\,\bar{\rho}/T$

Thermal Conductivity of D-3 and other Al Alloys

Our values for $\bar{\lambda}$ and $\bar{\rho}$, i.e. the mean values for λ and ρ of
pure Al from the results described above, permit a calculation of λ
of D-3 for comparison to the measured values. We first calculated
the Lorenz function of pure Al using $L(T) = \lambda_e\bar{\rho}/T$ and values of
L(T) are given in Table 2 at even T intervals. The values of
$\lambda(D-3)$ were then calculated using

$$\lambda(D-3) = (\rho_0/L_0 T + \bar{\rho}/L(T)T)^{-1} + (T/17 + 5000/T^2)^{-1} \qquad (1)$$

where ρ_0 is the impurity resistivity and λ_ℓ has been assumed independent of impurities for this material. For ρ_0 we may take either
$\rho(4.2\ K)$ or, taking deviations from Matthiessen's rule into account,

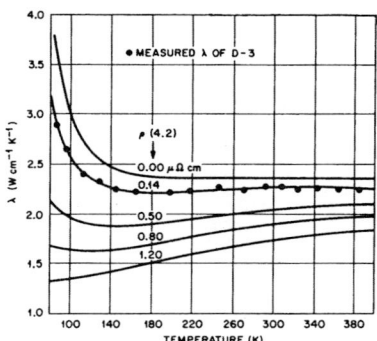

Fig. 4. The measured thermal conductivity of D-3 compared to $\bar{\lambda}$.
(Calculated curves for λ of D-3 and
less pure (hypothetical) samples are
also shown.)

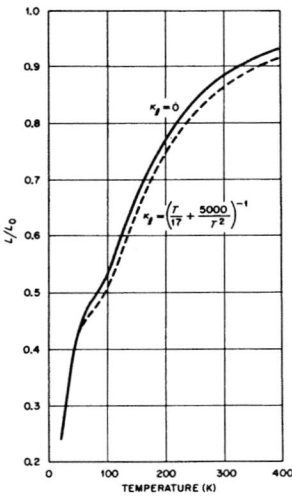

Fig. 5. The ratio of L(T)
to L_O with the
assumption that $\lambda_\ell = 0$ or
that λ_ℓ is given by Eq. 8.

ρ(D-3)-$\bar{\rho}$. We find the effect of such deviations to be small: above
110 K both curves lie within 1% of the actual λ(D-3) data, and both
diverge to about 4% at 80 K. This good agreement at the higher
temperatures has led us to extrapolate this procedure to larger ρ_O
values; the results are shown in Fig. 4. As ρ_O increases, a
minimum appears and becomes progressively deeper, with the temperature of the minimum occurring at progressively lower temperatures
(cf. Wilson Fig. IX 9).[9] In these alloys therefore the minimum
should behave as Powell[2] has described. For RRR values larger than
100 however, the minimum compared to that of ideally pure specimens
would be about 1% deep, a value comparable to the inaccuracy of
present state-of-the-art measurements.

The Thermal Conductivity at Low Temperatures

Upon combining our thermal resistivity data for sample B with
literature values, we found[3] that the low temperature thermal resistivity of Al does not vary quadratically with temperature. Thus,
we do not agree with the conclusions of Seeberg and Olsen.[10]

Lorenz Function

In addition to the $L(T)$ values listed in Table 2, the Lorenz Function was calculated with the assumption that $\lambda_\ell = 0$. That is, $L(T) = \bar{\lambda}\bar{\rho}/T$. Both results for $L(T)$ were normalized to L_0 and are compared in Fig. 5. The decrease of these functions with decreasing temperature is, of course, due to the increasingly inelastic nature of electron-phonon scattering; had we not corrected for impurity scattering these curves would have returned to unity at very low temperatures. A slight bump is visible in both curves near 80 K which is approximately one-fifth of the specific heat Debye temperature. Similar bumps may be found in the temperature range 0.14 θ – 0.20 θ in the alkali and noble metals. In each case the bump is found at those temperatures where the strengths of elastic and inelastic electron-phonon scattering are approximately equal.

REFERENCES

1. R.W. Powell, C.Y. Ho, and P.E. Liley, Thermal Conductivity of Selected Materials, NSRDS-NBS 8 (1966).

2. R.W. Powell, Contemp. Phys. 10, 579 (1969).

3. J.G. Cook, J.P. Moore, T. Matsumura, and M.P. van der Meer, ORNL-5079 (1975).

4. M.J. Laubitz and D.L. McElroy, Metrologia, 7(1), 1, 1971.

5. J.G. Cook, M.P. van der Meer, and M.J. Laubitz, Can. J. Phys., 50, 1386 (1972).

6. T. Matsumura and M.J. Laubitz, Can. J. Phys. 48, 1499 (1970).

7. The guarded-linear longitudinal shown in Fig. 6 of Laubitz and McElroy.

8. J.M. Ziman, Electron and Phonons (The Clarendon Press, Oxford, 1963).

9. A.H. Wilson, The Theory of Metals, 2nd ed., University Press, Cambridge, (1958).

10. P. Seeberg and T. Olsen, Phys. Norvegica 2, 197 (1967).

ACKNOWLEDGEMENT

Discussions with Dr. M.J. Laubitz on the transport properties of aluminium are gratefully acknowledged.

THERMAL AND ELECTRICAL TRANSPORT PROPERTIES OF HIGH PURITY SINGLE CRYSTAL BERYLLIUM[*]

W.E. Nelson and A.R. Hoffman[†]

University of Massachusetts, Department of Physics and

Astronomy, Amherst, Massachusetts 01002

Measurements of the electrical resistivity and thermal conductivity of a high purity single crystal of beryllium oriented along the c-axis have been made as a function of longitudinal magnetic field (up to 70 KG) and temperature (2 to 35 K). The residual resistivity ratio (RRR) as determined by the eddy-current method is about 2000. Completed results of the temperature dependent zero field thermal conductivity of this sample indicate a sharp peak of 143 watts/cm-K at a temperature of 13.5 K. The zero field electrical resistivity has a residual value of 1.85×10^{-9} ohm-cm, and Wiedmann Franz Lorenz (WFL) behavior typical for most metals is observed, with the residual plateau approaching the Sommerfeld value. A strong saturating magneto-thermal resistance is observed by 40 KG at selected temperatures both above and below the maximum in thermal conductivity. At 8.2 K, 14 K, 19 K and 25 K the thermal conductivity is observed to decrease by 73%, 65%, 58% and 51% respectively. Electrical resistivity is observed to increase by a similar factor, but the accuracy of the resistivity measurements was limited by excessive pickup in the higher field regions. Longitudinal D.C. electrical resistivity measurements give a RRR = 1750 in comparison to the eddy-current method. The same series of measurements are planned for a sample of RRR = 1000. Deviations from Matthiesson's rule, and the field dependence of the WFL ratio as a function of sample purity will be presented at a later date.

[*]Supported by the Atomic Energy Commission.

[†]A.P.S. Congressional Fellow on leave in Washington, D.C.

MEASUREMENTS OF THE TEMPERATURES AND MAGNETIC FIELD DEPENDENCE OF ELECTRICAL RESISTIVITY AND THERMAL CONDUCTIVITY IN OFHC COPPER [*]

W.E. Nelson and A.R. Hoffman[+]

Department of Physics and Astronomy

University of Massachusetts, Amherst, MA   01002

Experimental results between 3K and 35K are given for the thermal conductivity and electrical resistivity of a commercial grade OFHC[1] copper in a longitudinal magnetic field. The sample is characterized by a residual resistivity of 3.15×10^{-8} ohm-cm, and a thermal conductivity maximum of 13.7 Watts/cm.-Kelvin at 25K. The residual resistivity ratio (ρ_{273}/ρ_0) is 55. The Lorenz number is clearly observed to approach the Sommerfeld value below 10K, in contrast to other results for dilute copper alloys,[2] and to increase with field up to 15%, at 60 KG in the 30K region. The Wiedemann Franz ratio, and the thermal and electrical resistivities are compared with simplest theoretical descriptions, and with other experimental results. The predictability of thermal conductivity values for similar alloys is discussed.

I.   Introduction

The measurements described here were undertaken to verify and improve techniques used to perform identical measurements on small single crystal beryllium samples.[3] The scope of these results is thus intentionally limited to the geometries and regions of temperature and field associated with the beryllium work.

[*]Supported by the Atomic Energy Commission.

[+]A.P.S. Congressional Fellow on leave in Washington, D.C.

II. Apparatus and Sample Specifications

The experiments reported here were performed on   commercial grade OFHC copper from Admiral Brass and Copper Co.[1] The sample was unannealed except for soldering operations, which briefly reached a maximum temperature of 160°C. It was machined from a length of 0.5" dia. rod, with no other preparation involved. The total length was 2.5 cm, and the length (ℓ) between temperature sensing points was 1.13 cm to give a length to cross sectional area, ℓ/A, of 30.5. Though ℓ/A ratios of 100 or more are typically used on high conductivity samples to achieve larger temperature gradients during thermal conductivity measurements, this geometry was chosen to duplicate the dimensions of the beryllium samples.

During thermal conductivity measurements, temperature gradients along the sample, ΔT, were measured using matched pairs of cut-down (1/10 watt) Allen Bradley carbon resistors calibrated on the same cool-down against a commercial germanium resistance thermometer, and fit to a modified Clemment & Quinnell[4] resistance versus temperature expression. As a consistency check on thermometry, two independently matched pairs of carbons with different nominal resistances were monitored. Further, at selected temperatures, variations of a factor of two in the heater power used to establish the gradient, ΔT, were observed to result in essentially indistinguishable thermal conductivity values throughout the range of the measurements. The sample heater consisted of a manganin coil wound onto a 1/4" dia. by 1/4" length silver cap soldered to the upper end of the sample. The resistance of the heater was measured by voltage sensing leads so that power input to the sample was known to 0.1% at all times. The carbon resistances were monitored by A.C. Wheatstone bridge techniques at power levels of 10^{-8} watt using a phase sensitive detector. Based on the above techniques, it is estimated that the relative precision of the thermal conductivity values obtained is on the order of 1% for typical temperature gradients of 20 to 100 mK. The absolute accuracy of the results is limited by a systematic error due to the finite width of the sensor connections to the sample compared to their separation. Because of the small size of the sample, this error is about 5% for thermal conductivity values, and about 1.5% of electrical resistivity values.

A standard D.C. 4-probe technique was used to obtain the electrical resistivity, with the voltage leads attached to the sample using a low melting temperature Indium-Tin-Cadmium eutectic solder. Current polarity was reversed successively during measurements, and the voltage deflections averaged to allow for unknown offsets due to thermal emf's.

For all measurements performed in a field, H was parallel to the sample axis and the thermal or electrical currents. The carbon thermometers were calibrated in the field to account for non-uniform magnetoresistive drift which produced an apparent temperature difference between respective pairs. This drift was smooth, and always less than 10% of the observed <u>increase</u> in thermal gradient due to the application of the field, thus easily corrected for. The relative precision of the change in thermal conductivity with field is estimated to be about 4%.

III. Experimental Results and Discussion

The thermal conductivity, K, and electrical resistivity, ρ, are predicted by simple theoretical expressions[5] of the form:

$$\rho = \rho_o + \rho_i, \qquad (1)$$

$$\frac{1}{K_e} = W_e = W_o + W_i, \qquad (2)$$

where W_e is the electronic thermal resistivity, and the lattice component, K_g, is assumed much less than the electronic component, K_e. W_o and ρ_o are "residual" resistivities, dependent only on chemical impurities and imperfections. W_i and ρ_i are "intrinsic" resistivities due to the scattering of electrons by phonons, and assumed to be a characteristic of the metal. General expressions for the component resistivities are:

$$W_i = \alpha T^n; \ n \cong 2, \ T < 40K$$

$$\rho_i = C T^m; \ m \cong 4 \text{ to } 5 \qquad (3)$$

$$W_o = \beta/T.$$

β and ρ_o are constants depending on the impurities and physical imperfections in a particular sample, while α and C are constants characteristic of a given material. The separability of the components in Eq. (1) follows from Matthiesson's rule, and Eq. (2) is the electronic thermal analog. At low temperatures, dilute alloy resistivities are "impurity dominated", while for pure samples in which residual and intrinsic components may become comparable, a "deviation" term is often found in addition to the terms in (1) and (2), and strict separability of lattice and impurity effects no longer holds.

The experimental values for K (Fig. 1 and Table 1) are represented to within 1% from 3K to 35K by

$$K(T) = (3.882 \times 10^{-5} T^2 + 1.224/T)^{-1}. \qquad (4)$$

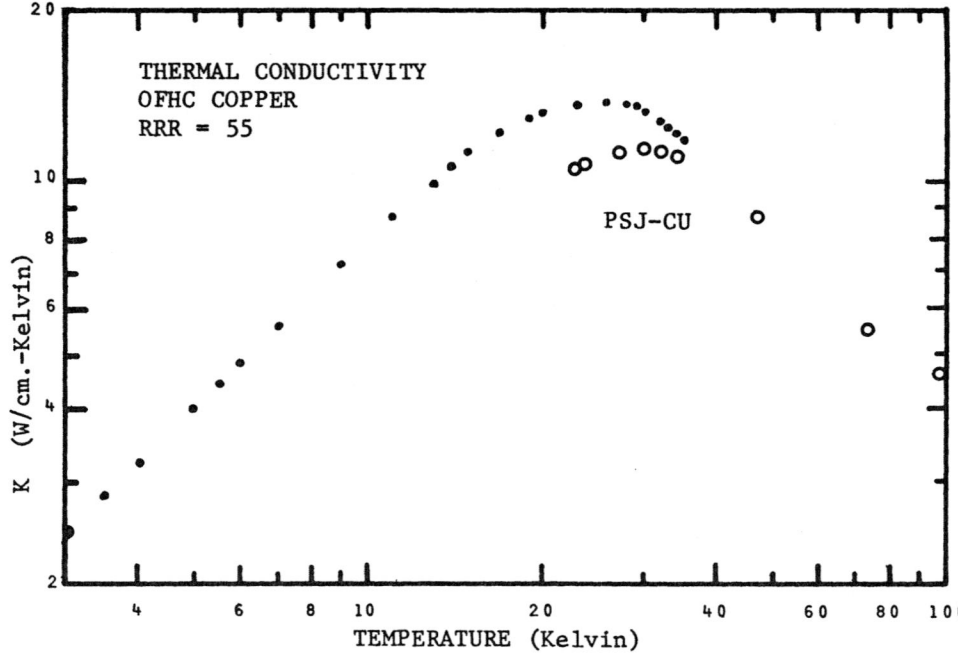

Fig. 1. Thermal Conductivity for OFHC Coppers. This work, •, described by the expression $K = (3.882 \times 10^{-5} T^2 + 1.224/T)^{-1}$, and Powers Schwartz and Johnston (1959),o.

The electrical resistivity data (Fig. 2 and Table 1), was fit to the expression $\rho = \rho_0 + CT^m$ by extracting ρ_0 graphically, subtracting it from the experimental values of ρ, and fitting the results to the ideal resistivity term, $\rho_i = CT^m$. The data is described to within 1% over the temperature range 7K to 35K, for $\rho_0 = 3.15 \times 10^{-8}$ ohm-cm by

$$\rho(T) = (3.15 + 2.28 \times 10^{-7} T^{4.42})$$

$$\times 10^{-8} \text{ ohm-cm.} \qquad (5)$$

Aside from extending the range of the temperature and field dependence of the Wiedemann Franz ratio (or Lorenz ratio, $L = K\rho/T$), a significant feature of this work is that L approaches the Sommerfeld value, $L_0 = 2.44 \times 10^{-8}$ V^2/K^2, in the residual plateau region (T < 10K). This is a generally observed characteristic of most pure metals, but other results for dilute copper alloys have indicated residual values of L 25% below L_0.[2] Because of the usefullness of the Lorenz ratio in predicting K, accurate knowledge of its temperature dependence for this class of alloys is important. Experimental results for L are shown in Fig.

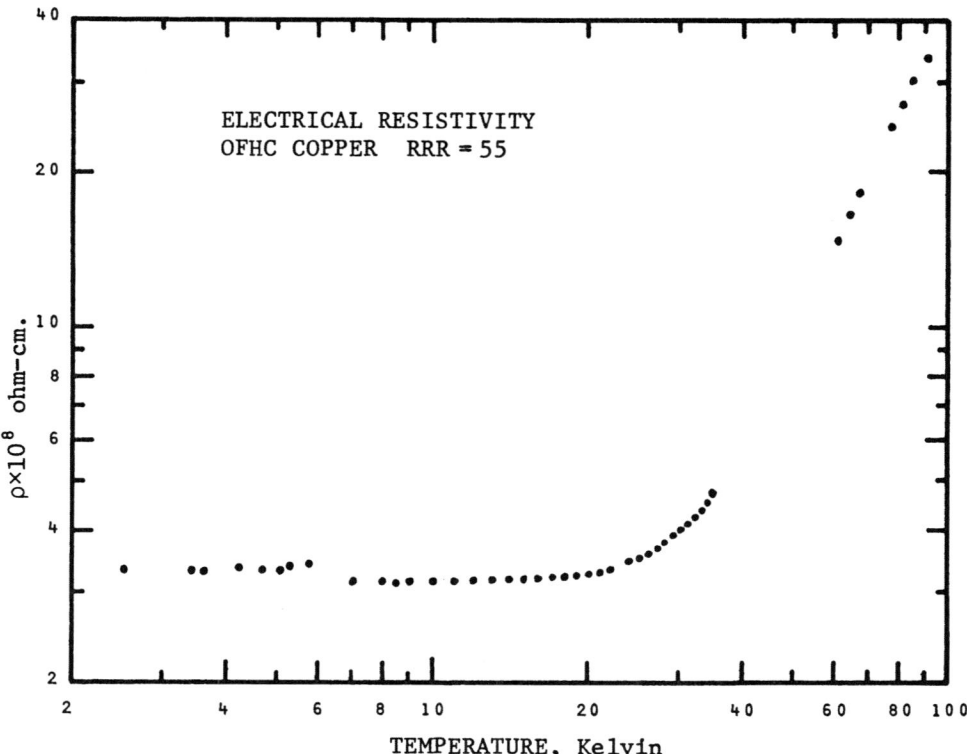

Fig. 2. Electrical Resistivity of OFHC Copper. From 7K to 35K the curve is given by $\rho = (0.0315 + 2.28 \times 10^{-7} T^{4.42})$ micro ohm-cm.

3, with numerical values given in Table 1.

The slight rise in ρ_0 (to 3.38×10^{-8} ohm-cm) below 6K is thought to be due to a superconducting transition in the solder attaching the voltage probes, and not an indication of a Kondo effect. Observations that 1 kG fields depressed the 2K to 6K value of ρ_0 to $3.12 \pm 0.04 \times 10^{-8}$ ohm-cm, coupled with the absence of magnetic impurities in the manufacturer's analysis support this hypothesis. Accordingly, values of the Lorenz number below 6K have about a 7% uncertainty associated with them. These are not shown, since it is most probable that the value $\rho_0 = 3.15 \times 10^{-8}$ ohm-cm correctly represents the sample behavior between 2K and 10K, and thus $L = 2.53 \times 10^{-8}$ V^2/K^2 in the plateau region (3.6% high).

The field variation of K, ρ, and thus L, was measured at 12 approximately equally spaced temperatures from 3K to 31K. Fig. 4

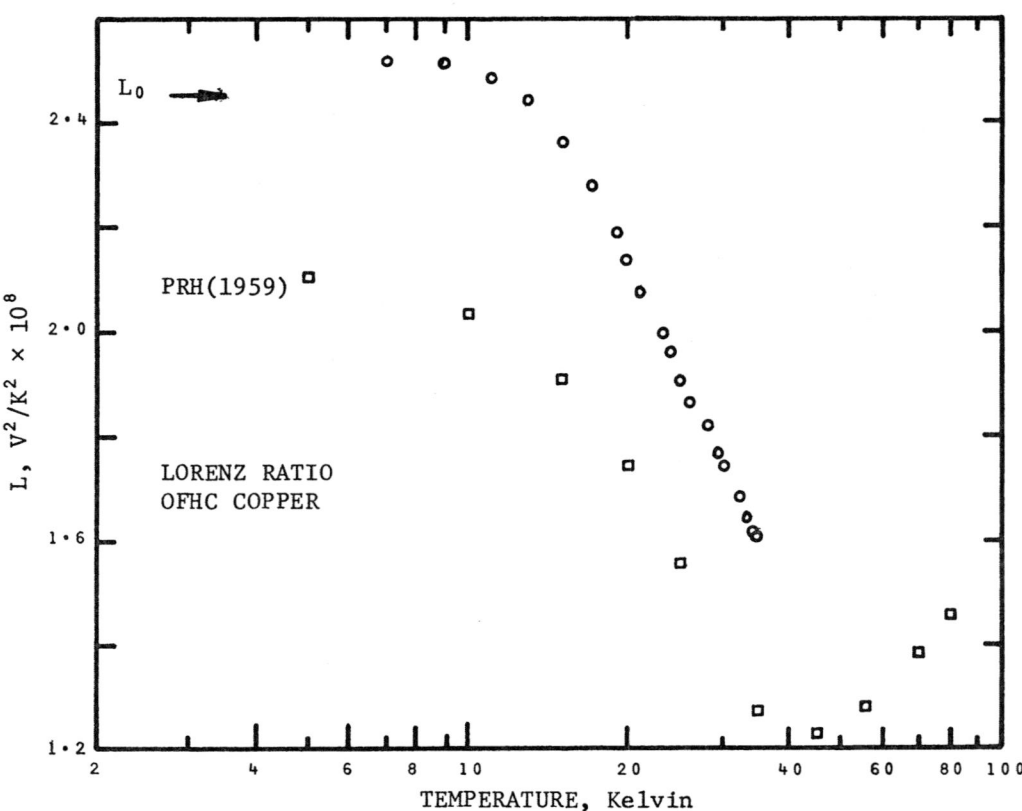

Fig. 3. The Lorenz Ratio of OFHC Copper. This work, o, Powell,
Roder and Hall, (1959) for a sample of RRR = 100, □ .

gives the % variation of thermal resistivity with field, $\Delta W/W$
(= $(W(H) - W(0))/W(0)$), where $W(0)$ is the zero field thermal
resistivity), at six representative temperatures. The analogous
change in electrical resistivity, $\Delta\rho/\rho$, exhibits very little temperature dependence in the 3K to 30K range, and the variation in
$\Delta\rho/\rho$ at 3K was almost indistinguishable from the 3K curve for
$\Delta W/W$. Because the entire temperature and field variation of $\Delta\rho/\rho$
is so closely approximated by the single curve $\Delta W/W$ at 3K (Fig. 4),
it is not graphed separately.

     To verify that the entire magnetoresistive effect occurs in the
residual resistance terms, the factor β was calculated at all twelve
values of $K(H = 60$ kG) using Eq. (4). The individual results were
all within 3% of the average result, $\beta(H = 60$ kG) = 1.626. Compared to the zero field value, $\beta(0)$ = 1.224, this is an increase
of 33% - consistent with experimental results for both $\Delta\rho/\rho$

Table 1. Some K, ρ and L Values for OFHC Copper (RRR = 55).

T (K)	K (W/cm·k)	ρ ($\times 10^8$ ohm cm.)	L ($\times 10^8$ V^2/K^2)
4.05	3.25	----	2.53[*]
7.0	5.61	3.15	2.52
9.0	7.16	3.15	2.51
11.0	8.62	3.17	2.48
13.0	9.95	3.19	2.44
15.0	11.1	3.19	2.36
17.0	12.1	3.21	2.28
19.0	12.9	3.25	2.21
20.0	13.1	3.28	2.15
21.1	13.2	3.31	2.07
23.0	13.6	3.38	2.0
24.0	13.7	3.44	1.96
26.0	13.6	3.57	1.87
28.1	13.5	3.76	1.81
30.0	13.2	3.95	1.74
32.11	12.8	4.21	1.68
33.1	12.5	4.36	1.65
34.9	12.1	4.66	1.62

[*]Assumes ρ_o = 3.15 x 10^{-8} ohm-cm.

and $\Delta W/W$ at 60 kG and 3K (where ρ_i and W_i are negligible).

The field variation of the Lorenz number may be written as

$$\Delta L(H)/L(0) = \frac{\Delta\rho/\rho(0) - \Delta W/W(0)}{\Delta W/W(0) + 1} . \qquad (6)$$

In the residual region, L is field independent, while at higher temperatures the Lorenz curve becomes shallower with field. Lorenz curves exhibit deeper minima for "purer" samples, so this behavior is consistent with the application of the field effectively "increasing" the impurity content.

The Lorenz behavior is presented for OFHC copper over part of the temperature region most sensitive to impurity (sample related) effects. Though evidence based on a single type of alloy is not conclusive, it is reasonable to expect the good agreement between the properties measured here and theoretical models to be a characteristic of the OFHC group and similar dilute alloys. This suggests that standard predictive schemes for K should be quite reliable for these alloys.

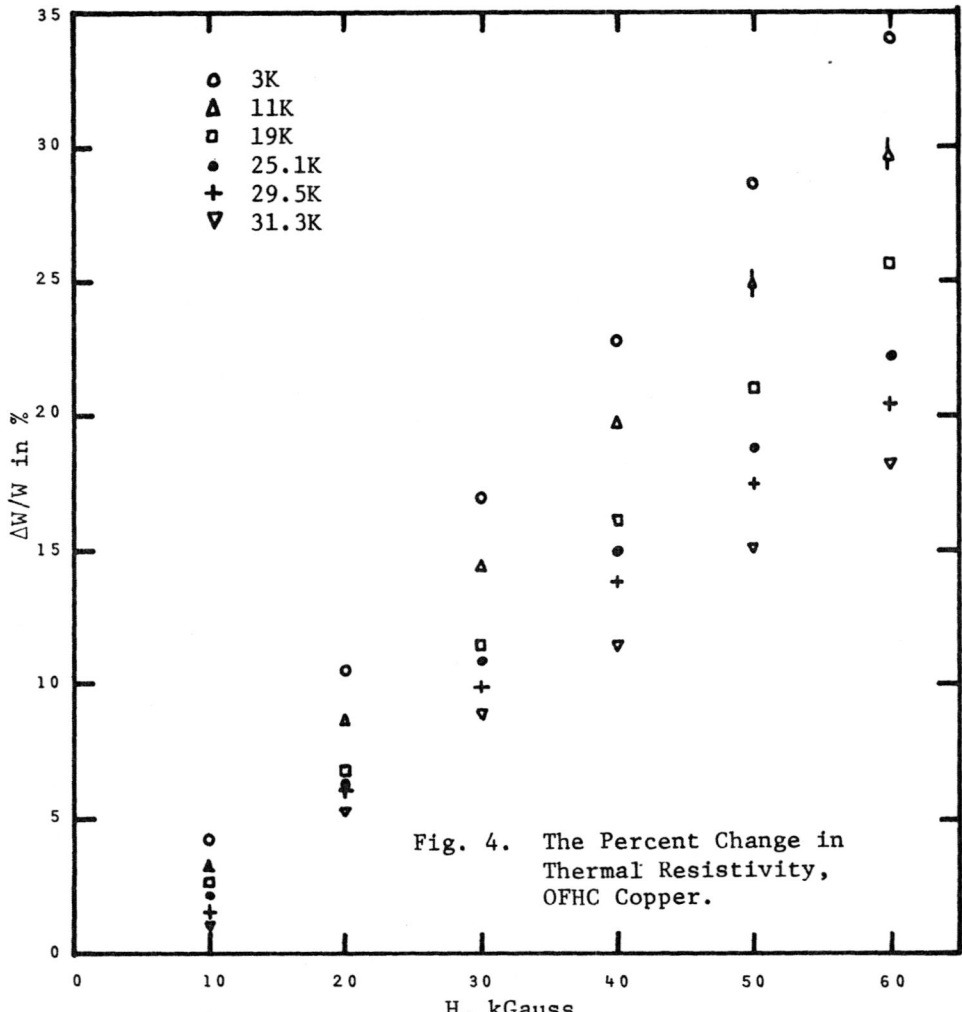

Fig. 4.   The Percent Change in
Thermal Resistivity,
OFHC Copper.

1) Oxygen Free High Conductivity: Alloy 101, Admiral Brass and
   Copper Co., 99.99% Cu with other impurities (by weight); 3ppm
   P, 1ppm Zn, 10ppm each Pb, O, Te and As, Bi, Mn, Sb, Sn
   totalling 40ppm.
2) R.L. Powell, H.M. Roder, W.J. Hall, Phys. Rev. 115 (1959)314;
   A. Feurier and D. Morize, Cryogenics (1973), for RRR of 100,
   and 60 respectively.
3) See Abstract this conference.
4) J.R. Clemment, E.H. Quinnell, Rev. Sci. Instr. 24 (1952) 213.
5) H.M. Rosenberg, Low Temperature Solid State Physics, (Oxford,
   1963).
6) R.W. Powers, D. Schwartz, H.L. Johnston, TR264-5, Cryogenics
   Lab, O.S.U., (1957).

THE LATTICE THERMAL CONDUCTIVITY OF COPPER AND ALUMINUM ALLOYS AT LOW TEMPERATURES*

A. C. Bouley, N. S. Mohan and D. H. Damon

Dept. of Physics and Institute of Materials Science

Univ. of Connecticut, Storrs, Connecticut 06268

ABSTRACT

Measurements of the lattice thermal conductivity of copper and aluminum alloys are reviewed. The lattice thermal resistivities due to electron-phonon scattering, phonon-phonon scattering, and phonon scattering by point defects are deduced from the measurements and compared to theoretical predictions. Good agreement between theory and experiment is found for the aluminum alloys. The results for the copper alloys are complicated by the presence of a defect which is the dominant phonon scatterer at very low temperatures.

INTRODUCTION

During the past eight years much of the work at this laboratory has been devoted to the measurement of the lattice thermal conductivity of aluminum and copper alloys at low temperatures. The objective of this work is the investigation of the properties of defects in alloys, particularly those defects introduced by cold work. The lattice thermal conductivity of a solid is sensitive to the type and concentration of the defects it contains since the dominant phonon wavelength varies with temperature from a value of the order of the lattice parameter at high temperatures to values of the order of hundreds of Angstroms at low temperatures. Therefore, the scattering of phonons and consequently the lattice thermal conductivity depends on the size and geometrical shapes of the lattice defects. The effect of cold work is first to introduce isolated dislocations. Then, because of the interactions between the dislocations and between dislocations and solute atoms, dislocation arrays and other defect aggregates may form. Both the formation and the dispersal of these extended defects are promoted by various heat treatments. In first approximation each type of defect adds a defect lattice thermal resistance to the total thermal resistance. The temperature dependence of the defect resistance characterizes the type of defect and its magnitude the defect concentration.

In order to identify the defect thermal resistivity one must not only subtract the electronic thermal conductivity from the measured total thermal conductivity [1], but also must separate the resulting lattice thermal resistivity into the sum of the total defect resistivity arising from cold work and other intrinsic lattice resistivities. In this paper we shall be concerned with these other lattice thermal resistivities, namely those arising from the scattering of phonons by electrons and other phonons in addition to one intrinsic defect resistivity, the scattering of phonons by solute atoms (point defects). The results of some of our measurements of the thermal conductivity of well annealed copper-aluminum, copper-nickel, copper-germanium and aluminum-magnesium alloys will be discussed together with the relevant theory. We shall be especially concerned with the intercomparison of the results for the copper and the aluminum alloys since the theory is often much more successful in predicting relative values than absolute values.

EXPERIMENTAL RESULTS

The separation of the measured thermal conductivity of aluminum alloys into lattice and electronic components was described in a previous contribution to these conferences [1]. The procedure is the same for all alloys and may be briefly reviewed. The total thermal conductivity is the sum of an electronic and a lattice conductivity,

$$\lambda = \lambda_e + \lambda_g. \tag{1}$$

The electronic thermal resistivity is the result of the scattering of conduction electrons by impurities and other defects and by the phonons so that,

$$\frac{1}{\lambda_e} = \phi_e = \phi_0 + \phi_i \tag{2}$$

where ϕ_0, the defect resistivity, is calculated from the Wiedemann-Franz law and measured values of the residual resistivity; ϕ_i, the phonon resitivity, is obtained from measurements made on the pure solvent metal. Figure 1 shows values of ϕ_i for aluminum and copper; these values were calculated from thermal conductivity data com-

piled by CINDAS [2]. The values of ϕ_o shown in the figure are
characteristic of the purest copper and aluminum specimens
(residual resistivities of the order of 6 x 10^{-10} ohm-cm) and show
that this procedure should yield reliable values of ϕ_i down to
temperatures of a few degrees K (the Debye temperature, θ, is 428K
for aluminum, and 343 K for copper).

Figure 2 shows the lattice thermal conductivity of some copper
and aluminum alloys. The maximum lattice thermal conductivity is
found at $T/\theta \simeq 0.1$ for all samples. At the lowest temperatures the
lattice thermal conductivity of the aluminum alloy varies as T^2.
As will be discussed below, this temperature dependence is the result of electron-phonon scattering if the electron mean free path

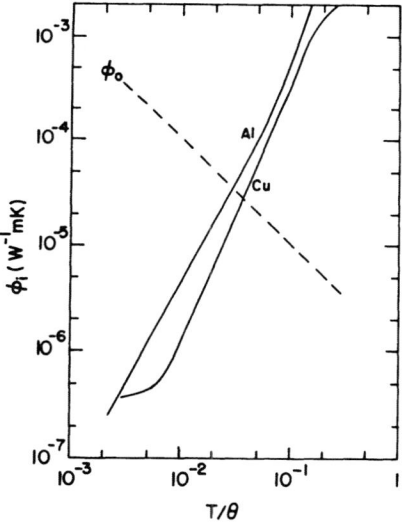
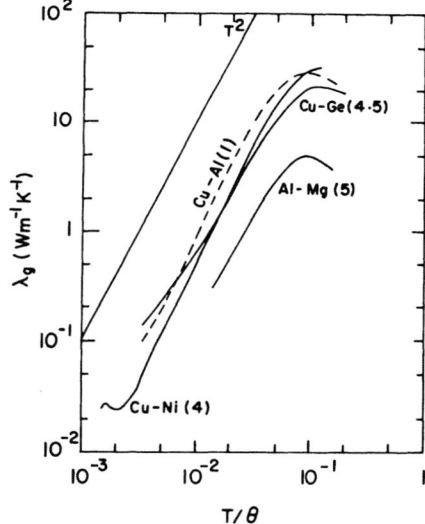

Fig. 1. Values of ϕ_i, the electronic thermal resistivity due to
electron-phonon scattering calculated from measured values of the
thermal conductivity of very pure
aluminum and copper samples vs
reduced temperature, T/θ. The
abrupt change in the slope of the
curve for Cu is almost certainly
due to experimental error ($\phi_o > \phi_i$
even for these pure metals).

Fig. 2. The lattice thermal
conductivity, λ_g, of some copper
and aluminum alloys vs reduced
temperature, T/θ. The solute
concentrations are given for
each alloy in the parentheses.

is not too short. None of the copper alloys show a T^2 dependence;
the lattice thermal conductivities of the Cu-Ni and Cu-Al alloys
vary faster than T^2 and the lattice conductivity of the Cu-Ge
alloy varies slower than T^2. The results for the Cu-Ge sample are
in agreement with those of Lindenfeld and Pennebaker [3] and can
be explained in terms of phonon-electron scattering if the electron mean free path is smaller than a typical phonon wavelength.
Further results for Cu-Ge alloys are presented elsewhere in this
conference [4]. The temperature dependence of the lattice thermal
conductivity of the Cu-Al and Cu-Ni samples is tentatively attributed to the effects of some cold work-induced defect that remains
even after annealing at temperatures near the melting point.

    At values of $T/\theta \simeq 4 \times 10^{-2}$ the effect of point defect scat-
tering can be observed. This is shown in Fig. 3 which displays the
lattice thermal conductivities of two aluminum alloys with different
magnesium concentrations. At the lowest and highest temperatures
the two lattice thermal conductivities are the same; the point defect scattering is relatively most important at temperatures just
below the maximum. For the copper alloys, Ni atoms provide the
weakest point defect scattering, Ge the strongest and Al is intermediate (note that the concentrations of Ge and Ni are almost the
same but the Al concentration is about 1/4 of that of the others).

           PHONON SCATTERING BY ELECTRONS AND OTHER PHONONS

    The theory of the lattice thermal conductivity of solids has
been described by Klemens [5]; the equations used here have been
taken from that article. If all lattice modes interact equally
strongly with the electrons and the electron mean free path is
longer than the phonon wavelength then the lattice thermal resistivity due to the scattering of the phonons by electrons, ϕ_{ge},
can be related to ϕ_i according to

$$\phi_{ge} = (\phi_i/313)(\theta/T)^4 N_a{}^{4/3} \qquad (3)$$

where N_a is the number of valence electrons. The values of ϕ_{ge}
calculated from this equation are shown in Fig. 4. The values of
ϕ_{ge} vary almost as T^{-2} which reflects the near T^2 variation of
ϕ_i. As described in a previous publication [6] Eq. (3) gives an
excellent account of the lattice thermal conductivity of aluminum-
magnesium alloys at lowest temperatures. (This agreement can be
seen by comparing the values shown in Fig.'s 2 and 4.) At $T/\theta =$
2×10^{-2} (which correspond to temperatures of 7 and 8.6 K respectively for copper and aluminum) this equation predicts that the
lattice thermal conductivity of aluminum should be about a factor
of ten smaller than that of copper. The experimental values of λ_g
at $T/\theta = 2 \times 10^{-2}$ show that the lattice thermal conductivity of

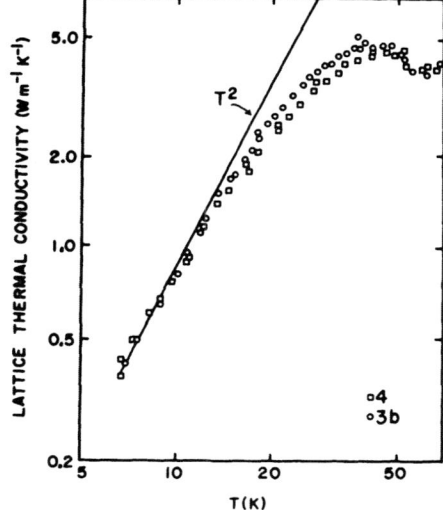

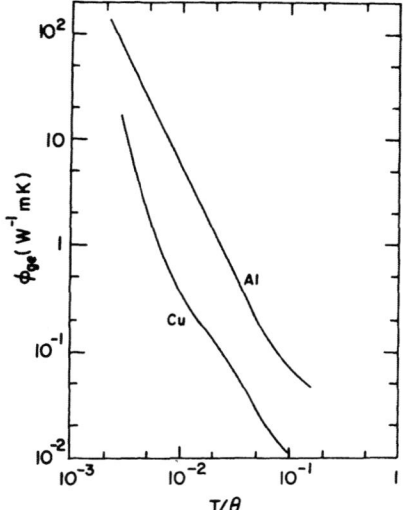

Fig. 3. The lattice thermal conductivity of two aluminum-magnesium alloys plotted against temperature. Sample 4 contains 7 at % magnesium and sample 3b, 5 at %.

Fig. 4. The lattice thermal resistivity due to the scattering of phonons by electrons vs reduced temperature for Cu and Al as calculated from Eq. (3) and the data given in Fig. 1.

copper is only 5 times that of aluminum. This discrepancy and the absence of a T^2 dependence for the copper alloys is attributed to defects remaining in these alloys after annealing. At very low temperatures the phonon wavelengths are longer than the electron mean free path. In this case Eq. (3) no longer describes ϕ_{ge}; an additional contribution, proportional to T, must be added to the equation. This should account for the lattice thermal conductivity of the copper-germanium alloy since the residual resistivity of this alloy is a factor two (or more) larger than any of the others and the mean free path of the electrons is correspondingly smaller.

At higher temperatures, near and above the maximum in the lattice thermal conductivity, phonon-phonon scattering becomes important. The lattice thermal resistivity due to three phonon processes, ϕ_{gu}, can be calculated in a low temperature approximation due to Klemens and a high temperature approximation due to Leibfried and Schloemann. The resulting formulae are

$$\phi_{gu} \quad = \quad \frac{16 \ \pi \ \hbar^3 \ \gamma^2 \ \theta}{M \ k^3 \ a \ J_3(\theta/T)} \ e^{-(\theta/\alpha T)} \qquad\qquad T<\theta \qquad\qquad (4)$$

$$\phi_{gu} \quad = \quad \frac{\gamma^2}{2.4 \ (4^{1/3})} \ (h/k)^3 \ (T/Ma\theta^3) \qquad T>\theta \qquad\qquad (5)$$

where γ is the Grüneisen constant, M is the atomic mass, $J_3(\theta/T)$ is a transport integral, a is the lattice constant and α is a numerical factor usually of the order of 2. The values given by these equations are shown in Fig. 5. As discussed in reference[6], Eq. (4) overestimates the magnitude of ϕ_{gu} for all substances sometimes by as much as two orders of magnitude; however, it correctly predicts the temperature dependence of ϕ_{gu} at low temperatures. The Leibfried-Schloemann formula correctly describes the temperature dependence of ϕ_{gu} at high temperatures and underestimates the magnitude of the lattice thermal resistivity by a factor of about 3 for a large number of dielectrics and metals including copper. According to Eq. (5) the lattice thermal conductivity of copper should be about a factor 2 larger than that of aluminum at the same value of θ/T. The results shown in Fig. 2 suggest that this factor is more like 6, and one concludes that the phonon-phonon scattering is unusually strong in aluminum.

SUMMARY

The lattice thermal conductivity of aluminum and copper alloys has been discussed in terms of lattice resistivities associated with a number of scattering mechanisms. It is well known that these resistivities cannot be simply added to give the total lattice resistivity. Instead, the scattering rates must first be expressed as functions of phonon frequency and then combined in a conductivity integral:

$$\lambda_g = \frac{k^3 \ T^2}{2 \ \pi^2 \ \hbar^2 \ v} \int_0^{\theta/T} \frac{e^x \ x^3}{(e^x - 1)^2} \left[\frac{1}{C + Ax^3(kT/\hbar)^3 + (E/T)e^{-(\theta/\alpha T)}} \right] dx. \quad (6)$$

In the integrand the constants C, A and E represent the strengths of the electron-phonon scattering, phonon-point defect scattering and phonon-phonon scattering mechanisms respectively, and v is the velocity of sound.

Using this equation, a successful account of the lattice

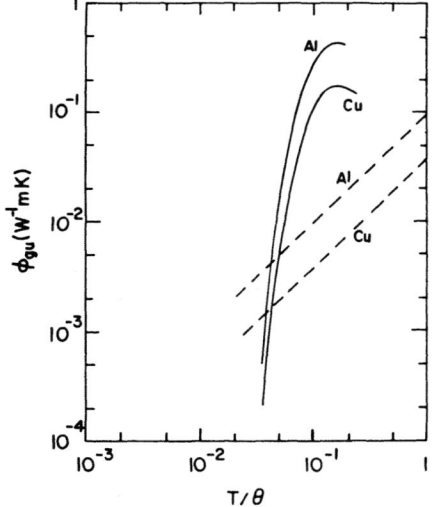

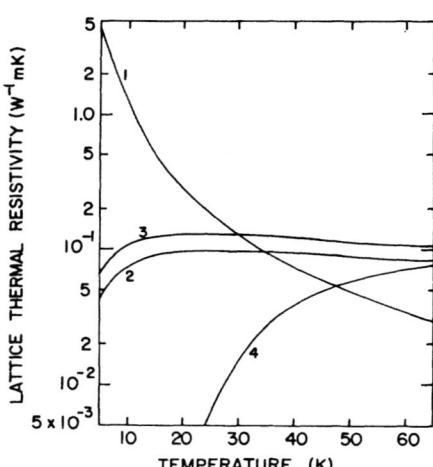

Fig. 5. The lattice thermal resistivity due to phonon-phonon
scattering for Al and Cu vs
reduced temperature. The solid
curves were calculated using
Eq. (4) and the dashed curves by
using Eq. (5).

Fig. 6. Lattice thermal resistivities for aluminum alloys vs
temperature. Curve 1 shows the
resistivity due to phonon-electron scattering, curves 2 and 3
show the resistivities due to
Mg atoms in concentrations of 5
and 7 at %, and curve 4 is the
resistivity due to phonon-phonon
scattering.

thermal conductivity of a number of aluminum alloys can be given
(see also [6]). The phonon-phonon interaction in aluminum is somewhat stronger than might be expected from other metals and insulators. The lattice thermal resistivities associated with the scattering processes in aluminum alloys are shown in Fig. 6. These
resistivities were calculated by first fitting Eq. (6) to the data
to obtain values of C,A and E and then calculating the resistivities due to individual scattering processes by using these constants one by one (for the point defect resistivity one calculates
the combined effect of phonon-electron and phonon-defect scattering
and then subtracts the calculated phonon-electron resistivity).
It is easily seen that only at the very lowest temperatures where
ϕ_{ge} dominates is it possible to attribute the lattice thermal resistivity to a single scattering mechanism.

The copper alloys have not yielded to this analysis. At low
temperatures where the electron-phonon scattering should be dominant the lattice thermal conductivity is smaller than predicted by
theory and is too strong a function of temperature. Certainly,
there is no reason to question the validity of Eq. (3) especially
for the case of copper. We, therefore, conclude that a defect,
introduced by cold work and not removable by annealing, even at
temperatures well above the recrystallization temperatures, is responsible for these results. Chu [7] has given an explanation of
some of these results in terms of the cellular structure formed by
the dislocations. Finally we note the minimum in the lattice
thermal conductivity of the Cu-Ni specimen at very low temperatures
shown in Fig. 2. This minimum has been found in the thermal conductivity of a number of other copper alloys and certainly can not
be explained as the result of the scattering of phonons by sessile
defects.

CITATIONS

* Supported by the United States Air Force Office of Scientific
Research under Contract #AFOSR 73-2418. Helium was supplied by
the Office of Naval Research under Contract #N00014-71-C-0249.

[1] R. W. Klaffky, N. S. Mohan and D. H. Damon, Proceedings of the
     Thirteenth International Conference on Thermal Conductivity
     (U. of Missouri Press, Rolla, Mo.), p. 7 (1973).
[2] Y. S. Touloukian, R. W. Powell, C. Y. Ho and P. G. Klemens,
     Thermal Conductivity, Metallic Elements and Alloys, Vol. 1
     (Plenum Press, N. Y.), p. 9 and p. 81 (1970).
[3] P. Lindenfeld and W. B. Pennebaker, Phys. Rev. 127, 1881 (1962).
[4] T. K. Chu and N. S. Mohan, Proceedings of this conference.
[5] P. G. Klemens, Solid State Physics 7, 1 (1958).
[6] R. W. Klaffky, N. S. Mohan and D. H. Damon, Phys. Rev. B11,
     1297 (1975).
[7] T. K. Chu, Accepted for publication in Jour. Appl. Phys.

LATTICE THERMAL CONDUCTIVITY AND COTTRELL ATMOSPHERE IN COPPER-

GERMANIUM ALLOYS*

T. K. Chu and N. S. Mohan[+]

Dept. of Physics and Institute of Materials Science

University of Connecticut, Storrs, Connecticut 06268

I. Introduction

The formation of solute atmospheres around dislocations was first proposed by Cottrell and has been shown to account for the yield point phenomena in a number of alloy systems. This atmosphere formation also affects the lattice thermal conductivity of alloys at low temperatures since it modifies the strain field around the dislocations. According to the theory of Klemens [1] and of Klemens and Ackerman [2], the scattering of phonons by solute atmosphere depends on two factors: the mass difference between the solute and the host atoms, and the lattice distortion introduced by the solute atoms. Furthermore, the temperature dependence of this scattering of phonons by the solute atmosphere is the same as that by bare dislocations. Therefore, lattice thermal conductivity determination at low temperatures should also yield information on the solute atmosphere.

A systematic approach to this problem is to make a series of measurements on a mechanically deformed specimen, and on the same specimen after successive partial anneals. The partial annealing should be done at relatively low temperatures (a fraction of the melting temperature) so that only the atmosphere, and not the dislocations would be annealed out.

In an earlier paper [3] , Mitchell et al reported on such experiments on a Cu+10a/oAl alloy, and demonstrated the existence of solute atmospheres in the Cu-Al system. A similar study by Friedman [4] for a Cu+7a/oGe sample, however, show a negative result. This non-existence of solute atmosphere in the Cu-Ge system is indeed quite puzzling, since the Ge atoms introduces large lattice distortion to the Cu lattice, and is expected to have a strong atmosphere effect.

The present investigation is aimed at resolving this inconsistency, by extending the measurements to two difference concentrations of Ge in Cu: 4.5a/o and 9a/o. The measurements are also carried out over a wider temperature range in order to examine the wavelength dependence of the phonon scattering by the solute stmosphere.

II. Samples

Samples for this investigation were made from two Cu-Ge ingots: 4.5 a/o and 9a/o Ge in Cu. The ingots were made from 5N pure materials melted in evacuated fused-quartz tubes. The ingots were then cast in quartz tubes of .3 in. I.D., and finally swaged down to rods of dia-meter .144 in. Two specimens for each alloy were cut from these rods. Measurements were carried out on sets of samples in the follow-ing states: well annealed, swaged, and partially annealed after swag--ing. Other informations on the samples are listed in Table I.

III. Results

For the purpose of this investigation, only the lattice compon-ent of the thermal conductivity will be examined. The lattice thermal conductivity, K_g, is extracted from the total measured thermal con-ductivity, K, by the following formula:

$$K_g = K - \frac{1}{L_o/(\rho_o T) + W_i} \tag{1}$$

where L_o is the ideal Lorenz number, ρ_o is the residual electrical resistivity, and T is the temperature. The ideal thermal resistivity

Table I

Alloy	Physical State	Electrical Resistivity ($10^{-8} \Omega$m)		
		273 K	77 K	4.2 K
Cu+4.5a/oGe	Annealed at 980 C − 24 hr	16.54	14.81	14.46
	Swaged	16.68	14.94	14.60
	Annealed at 125 C −24 hr	――	――	14.58
Cu+9a/oGe	Annealed at 890 C − 24 hr	26.54	24.36	23.86
	Swaged	28.16	25.98	25.51
	Annealed at 125 C − 24 hr	――	――	25.48

W_i of these alloys is assumed to be the same as that of pure copper and is obtained from the result of White [5]. The variation of K_g with temperature is displayed in Figure 1 for the Cu+4.5a/oGe alloy, and in Figure 2 for the Cu+9a/oGe alloy.

Well-annealed Specimens

As is seen in Figures 1 and 2, the lattice thermal conductivity of the well-annealed specimens follows a temperature variation that is slower than the T^2 dependence expected from phonon-electron inter- action. (The lines marked T^2 are drawn to illustrate the temperature variation; they are not drived from theory , nor from experimental results.) Furthermore, it is quite apparent that K_g for the higher concentration alloy departs more from the expected T^2 dependence than that of the lower concentration alloy. This behavior is attributed to two causes. (1) A strong point defect scattering of phonons in the Cu-Ge system. In a paper presented at this conference [6], Bouley et al show evidence that point defect scattering can have substantial contribution to the lattice thermal resistivity for alloys of moder- ate concentrations even at temperatures as low as 15 K. As a compar- ison, results of a Cu+9a/oAl [7] is also shown in Figure 2: the lower K_g for the Cu-Ge alloy from 10-30 K is interpreted as due to the stronger solute-phonon scattering in this alloy system. At tempera-

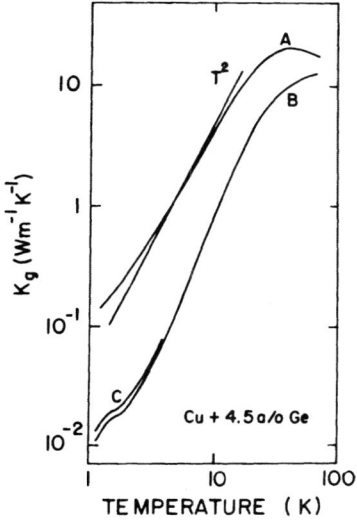

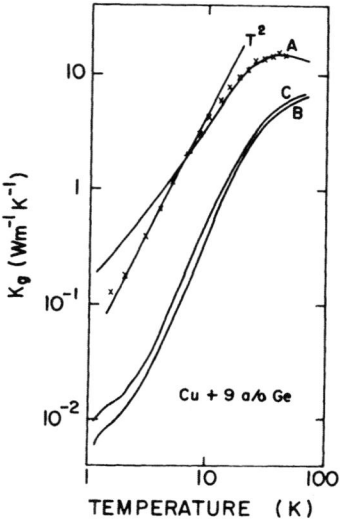

Figure 1. Plot of K_g versus T.
A: well-annealed , B: swaged,
C: partially annealed.

Figure 2. Plot of K_g versus T.
A: well-annealed, B: swaged,
C: partially annealed,
X: a well-annealed Cu+9a/oAl
specimen [7].

tures above the lattice thermal conductivity maximum , Umklapp-Processes are dominant and the two alloy systems behavior similarly. (2) The uncoupling of the longitudinal and the transverse phonon modes below the Pippard limit. Lindenfeld and Pennebaker [8] made calculations of the lattice thermal conductivity under such condition; and Figure 3 is a plot of the present results according to their scheme. It is seen that these results are consistent qualitatively with their calculations. They are also in good agreement with their experimental results (not shown). Results for the Cu+4.5a/oGe alloy at even lower temperatures indicate that the trend is continued down to $\sim$.5 K at a T/ρ_o value of $3 \cdot 10^6$ KΩ^{-1}m^{-1}[9]. The Pippard limit of these alloys is estimated to be at $T/\rho_o \cong 3 \cdot 10^7$ KΩ^{-1}m^{-1}.

Swaged and Partially Annealed Specimens

For the swaged and partially annealed specimens, dislocation-phonon interaction is dominant below the conductivity maximum: one expects, again, a T^2 dependence of K_g at low temperatures. The present results (Figures 1 and 2) show that K_g varies slightly faster than T^2 below $\sim$20 K and displays an inflection at 3 K. In order to illustrate more clearly the dislocation thermal resistivity, the quantity $W_{gd}T^2$ is plotted versus T in Figures 4 and 5; where W_{gd} is defined as the difference in thermal resistivity ($1/K_g$) between a swaged or partially annealed specimen and that of the well-annealed specimen. If dislocations are randonly distributed, the resultant lattice thermal resistivity would yield a horizontal straight line in such a plot. Instead, $W_{gd}T^2$ for the present specimens are only approximately constant above $\sim$15 K. They increase with decreasing temperature until they reach their maximun values at 3 K, and then

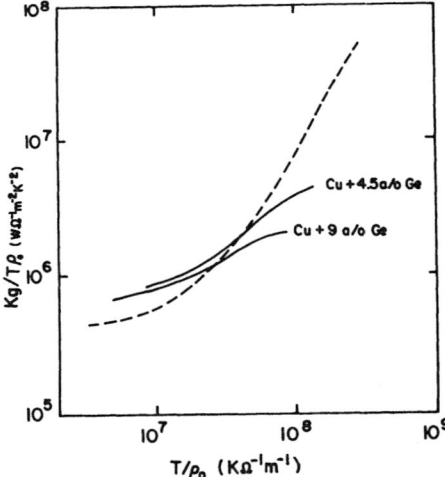

Figure 3. A universal plot to illustrate the Pippard limit. The dashed line is from the calculation of Lindenfeld and Pennebaker [8].

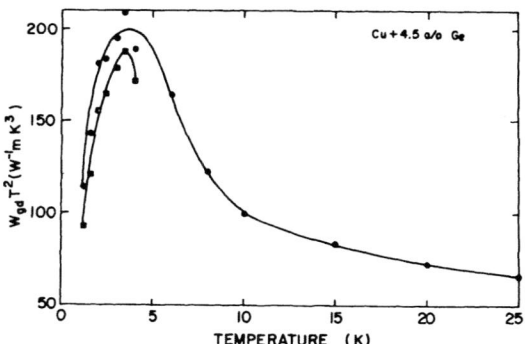

Figure 4. Plot of $W_{gd}T^2$ versus T.
⊙ : swaged, ⊡ : partially annealed.

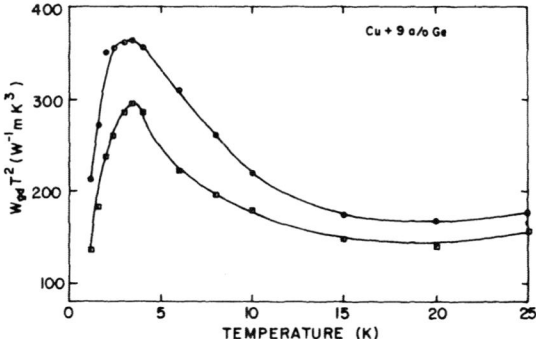

Figure 5. Plot of $W_{gd}T^2$ versus T.
⊙ : swaged, ⊡ : partially annealed.

decrease again at the lowest temperatures. This behvior is similar
to that observed in the Cu-Al system [10], and is probably due to
the wall-like (pile-ups and sub-boundaries) arrangement of dislocations common in copper and copper alloys.

When dislocations are non-randomly distributed and group together into pile-ups or some boundary structure, dislocations in
the groups can interfere constructively with each other in their
scattering of phonons, when the phonon wavelength becomes comparable
to the inter-dislocation distance as temperature decreases. At still
lower temperatures, when the phonon wavelength is longer even than
the spatial extent of these dislocation groups, the phonon scattering strength of these groups is weakened , and a decrease in W_{gd}
results. Chu has proposed a model of the dislocation structure in
copper alloys and has given a discussion of its phonon scattering
characteristics in Reference 9. Thus the gradual rise in $W_{gd}T^2$
below 15 K and the pronounced decrease below 3 K is interpreted as
due to dislocation groupings in these alloys also. The approximately
constant $W_{gd}T^2$ at higher temperatures is due to the scattering of
phonons by individual dislocations, when the phonon wavelength is
shorter than the inter-dislocation distance in the groups.

There are some qualitative differences between the present
specimens and the Cu-Al system. For Cu-Al, $W_{gd}T^2$ remains constant
above 15 K, whereas for the Cu+9a/oGe specimens, there is a slight
upturn at the higher temperatures. This upturn may be due to
stacking faults as the stacking fault energy of this alloy is very
low [11]. For the swaged Cu+4.5a/oGe specimen, $W_{gd}T^2$ seems to
continue decreasing as temperature increases. No satisfactory
explanation for this behavior can be found at the present time.

The effect of partial anneal can be examined only for the
Cu+9a/oGe alloy only, as measurements for the partially annealed
Cu+4.5a/oGe specimen are not completed yet. From Figure 5, it
appears that partial annealing causes a constant decrease of W_{gd} at
the higher temperatures. This constant decrease is consistent with
theoretical predications [2] if the partial anneal disperses part
of the solute atmosphere around the dislocations. The decrease in
W_{gd} at $T < 10$ K is larger and may be a result of possible dislocation
re-arrangement within the groupings. If this is true, the failure
to detect solute atmosphere in an earlier investigation [4] may be
attributed to the limited temperature range of those measurements.
Results below 4 K for the Cu+4.5a/oGe partially annealed specimen
indicates that dislocation re-arrangement also took place in this
alloy. Since the formation of solute atmosphere around groups of
dislocations has not been investigated, a separation of the effects
of dislocation rearrangement and the dispersion of Cottrell
atmosphere around individual dislocations has not been attempted.

Footnotes and References

* Supported by the U. S. Army Research Office, Durham; helium for this work was supplied by the U. S. Office of Naval Research under Contract #4500/NOO 14-71-0-0249.

+ Present address: Department of Mechanical Engineering, University of New Brunswick, Frederickton, New Brunswick, Canada.

1. P. G. Klemens, J. Appl. Phys. 39 5304 (1968), also J. Appl. 40 4696 (1969).
2. M. W. Ackerman and P. G. Klemens, J. Appl. Phys. 42 968 (1971).
3. M. A. Mitchell, P. G. Klemens and C. A. Reynolds, Phys. Rev. B3 1119 (1971).
4. A. J. Friedman, Phys. Rev. B7 663 (1973).
5. G. K. White, Aust. J. Phys. 6 397 (1960).
6. A. C. Bouley, N. S. Mohan and D. H. Damon, this Proceedings.
7. A. J. Friedman, T. K. Chu, P. G. Klemens and C. A. Reynolds, Phys. Rev. B6 356 (1972)
8. P. Lindenfeld and W. B. Pennebaker, Phys. Rev. 127 188 (1962).
9. A. C. Bouley, private communications.
10. T. K. Chu, to be published in the Journal of Applied Physics.
11. P. C. J. Gallagher, Met. Trans. 1 2429 (1970).

THE THERMAL CONDUCTIVITY OF HIGH PURITY VANADIUM

W.D. Jung and G.C. Danielson

Ames Lab. ERDA

Ames, Iowa

The thermal conductivity, electrical resistivity and Seebeck coefficient of high purity vanadium have been measured as functions of temperature from 5 K to 300 K. The samples were purified using the electrotransport technique and had resistance ratios ($R_{273K}/R_{4.2K}$) of 1524, 785, 81, and 38. The highest purity sample had a thermal conductivity maximum of 920 W/mK occurring at 9 K. The three highest purity samples had thermal conductivities which decreased to 35 W/mK at room temperature. The intrinsic thermal resistivity of vanadium varied as T^2 for temperatures less than 60 K owing to electron-phonon scattering. The residual electrical resistivity of the highest purity sample was 1.3×10^{-10} Ωm. The ideal resistivity of vanadium varied as $T^{3.4}$ at low temperatures. The Seebeck coefficient was positive from 5 K to 240 K and had a maximum at 60 K.

THE IMPORTANCE OF THERMAL CONDUCTIVITY IN INTERPRETING

THERMOPOWER DATA: ALLOYS WITH ATOMIC ORDER-DISORDER TRANSITIONS †

C.L. Foiles

Department of Physics, Michigan State University

East Lansing, Michigan 48824

Some earlier studies have interpreted the sign change in the total thermopower which accompanies atomic ordering in Cu_3Au as indirect evidence of changes in the Fermi surface. This paper presents an analysis combining thermopower and thermal conductivity data which contradicts such an interpretation; the individual contributions to diffusion thermopower do not change sign. The observed behavior is explained as a change in the balance of two competing terms.

There is a set of binary alloys, having constituent elements A and B, which occur with A_3B or AB_3 stoichiometry and display atomic order-disorder transitions. In the disordered state (DOS) these alloys possess an Al structure (FCC) and in the ordered state (OS) they possess a Ll_2 structure (SC with a basis of 4 atoms per unit cell). The introduction of atomic order has a marked influence upon the transport properties of these alloys. Unfortunately, even in an OS, the alloys exhibit significant residual scattering and study of these systems using Fermiology techniques is not feasible. Thus, identification of the specific features which accompany the introduction of atomic order and subsequently cause the changes in transport properties is difficult and interpretation must rely upon indirect methods.

A systematic comparison of alloy transport properties with the transport properties of pure copper has been one of the favorite indirect methods and a study of the first three rows in table I illustrates the reason for this popularity. In a DOS Cu_3Au should be quite similar to pure Cu and the entries in rows 1 and 2 confirm such an expectation. Even the anomalous positive sign for the

thermoelectric power appears to be reproduced. In an OS the introduction of additional band gaps is expected. Although no direct
evidence confirms this expectation, all the transport properties
appear to support such an interpretation; (i) the lattice contribution to electrical resistivity increases substantially without a
commensurate increase in the Debye temperature and thus suggests
band gaps that reduce the number of carriers, (ii) the sign of the
Hall effect changes from negative to positive and this can be
interpreted as the consequence of new band gaps which alter the
relaxation time anisotropy as well as the balance between electron
and hole contributions [1,2] and (iii) the sign of the thermoelectric power becomes negative and thus consistent with the sign
for polyvalent cubic metals.

Although the other systems listed in table 1 have not been
studied as comprehensively as Cu_3Au, there is sufficient data to
indicate that the general patterns of transport property changes
are not consistent with the changes observed in Cu_3Au. The
changes in lattice resistivity do appear to follow a uniform
pattern but the thermopower data fail to produce a pattern and
the more limited Hall effect data fail to display any additional
sign reversals. The fact that lattice resistivity, number of
valence electrons per unit volume and electronic contribution to
specific heat have similar values for all these alloys indicates
no complications arise from d-band considerations. This conclusion is supported by the fact that all of the alloys are diamag-
netic. Thus, the lack of general patterns despite the basic
generality of the arguments prompts a re-examination of the
apparent successes of the preceding interpretation.

In this paper we consider the thermoelectric behavior in
more detail and attempt to answer one specific question, "Is it
meaningful to consider the total thermopower when comparing the
behavior of these alloys with that of pure copper?" The manner
in which thermoelectric contributions add plays a crucial role in
answering this question. Generally, the total thermopower is the
weighted sum of diffusion components and a phonon drag component. [3]
For pure copper near room temperature the phonon drag contribution, S_g, is negligible and a single diffusion component from the
scattering of the electrons by the lattice, S_{th}^d , exists. For the
alloy systems listed in table I an entirely different situation
occurs. No definitive statement about phonon drag contributions
can be made for these systems but work on dilute alloys indicates
the phonon drag contribution is decreased by mass differences,
force constant changes and strain fields [3]. The decrease upon
alloying is usually rapid and we shall assume the alloys have no
significant phonon drag contributions over the temperature range
of present interest. The large residual scattering which remains
even in an OS, recall the earlier comment about the infeasibility
of Fermiology studies, supports this assumption but leads directly

Table I. Survey of selected properties. These data come from diverse sources and, for those instances where a range of values or conflicting values exist, no attempt was made to determine the "best" value. The listed values provide an accurate picture of the general variations. ρ_{th} is the electrical resistivity associated with phonon scattering of the electrons; Θ is the Debye temperature determined from resistivity measurements (sub-R) and calorimetric measurements (sub-D); S is the total thermopower; ΔS is the change which occurs in S near the ordering temperature (T_c); R_H is the Hall effect; n is the number of valence electrons per unit volume; γ is the electronic contribution to specific heat.

Material	ρ_{th} (R.T.) ($\mu\Omega$-cm)	Θ_R or (Θ_D) °K	S (~R.T.) (μV/°K)	ΔS near T_c (μV/°K)	R_H (R.T.) $10^{-13}\Omega$-cmG	n (10^{22} cm^{-3})	γ (mJ/mole°K^2)
Cu	1.55	333(320)	1.7		-5.4	8.46	0.69
Cu$_3$Au (DOS)	1.8	180(269)	0.2 to 0.45	-4	-6.4	7.62	0.68
Cu$_3$Au (OS)	2.9	200(285)	-1.2 to -1.65		1.7		0.65
Au$_3$Cu (DOS)	2.3		0.5	<-4		6.34	
Au$_3$Cu (OS)	3.7		-1.6				
Cu$_3$Pt (DOS)	1.8			only a change in slope		5.93	
Cu$_3$Pt (OS)	2.7		-5			7.91	
Cu$_3$Pd (DOS)		(323)	0<	8	-12.1	5.98	0.77
Cu$_3$Pd (OS) a			<-0.5		-16.7	7.98	
Cu$_{0.83}$Pd$_{0.17}$ (DOS)	1.4	(327)	1	10	-8.7	6.70	0.74
Cu$_{0.83}$Pd$_{0.17}$ (OS)					-2.7	8.08	

a Upon ordering a slight tetragonal distortion (c/a = 0.99) occurs.

to a serious problem. The residual scattering, if only by atten-
dant alteration of weighting factors, prevents the measured thermopower from being the lattice diffusion contribution. Moreover, if
any of the residual scattering processes possess an energy dependence then these processes can be expected to introduce additional
contributions to the diffusion thermopower. Therefore, a simple
and direct comparison of thermopower values at room temperature
contains several potential errors.

A reliable separation of S_{th}^{d} from the total thermopower is
needed. Electrical resistivities are often used in evaluating
weighting factors for different diffusion thermopower contributions; however, such a procedure is only an approximation. The
correct procedure is based upon the fact that the product of the
thermopower and the thermal resistivity is directly proportional
to the scattering cross-section. [4] Denoting the diffusion thermopower contribution of the ith scattering mechanism by S_{i}^{d} , the
thermal resistivity term by W_i and the total value for each of
these properties by the subscript t , this proportionality leads to

$$W_t \cdot S_t^{d} = \sum_i W_i \cdot S_i^{d} \equiv W_{th} \cdot S_{th}^{d} + \sum_j W_j \cdot S_j^{d} \qquad (1a,1b)$$

In eqn. (1b), the subscript "th" denotes the contribution associated with electrons being scattered by the lattice and the sum over
j includes all other electronic contributions.

One method of effecting a separation is to divide both sides
of eqn. (1b) by W_t and use temperature, T, or defect concentration,
X, as an implicit parameter. Unfortunately, this method requires
an "a priori" separation of the various W contributions and hence
additional information or assumptions. Instead, consider the general temperature dependence expected of eqn. (1b). From both practical and theoretical bases, S_{th}^{d} can be expected to vary linearly
with T and a similar variation is expected for each S_j^{d} . Near room
temperature, W_{th} for most pure metals is independent of T. Most
defects involve elastic scattering and W_j should vary inversely
with T. Therefore, the term $W_{th}S_{th}^{d}$ should display a linear T dependence and the term $\sum_j W_j S_j^{d}$ should be independent of T. A plot of
the product of experimental values for W_t and S_t^{d} as a function of T
permits a separation of the two terms. Without additional information it is not possible to determine the value for an individual
S_i^{d} but the signs of S_{th}^{d} and of the resultant defect contribution can
be uniquely determined. To the best of our knowledge only Cu_3Au
among the listed alloys has sufficient data to allow such a plot and
the data of Goff, etal [5] are plotted in figure 1. In both states of
order the anticipated linear variation in T is found over a substantial temperature range.

Two important features clearly emerge from figure 1. First and foremost is the slope of the linear T term which is related to S_{th}^d. In both the DOS and OS of Cu_3Au this slope is positive and hence S_{th}^d does not change sign as a function of order! In both states of order S_{th}^d for Cu_3Au is positive and has the same sign anomaly as pure copper. The naive use of total thermopower data employed in the second paragraph of this paper is not valid! Second, the sign of the resultant defect thermopower contribution is negative for both states of order.

Now that the signs of the contribution are clearly established we can introduce additional assumptions to estimate the magnitudes of the individual thermopower contributions:

(i)   Since $100<T<300°K$ is the range of the linear term, we assume the Wiedemann-Franz law is valid.

(ii) W_{th} is estimated from ρ_{th} near room temperature and the linear term is assumed to have the form $W_{th} \cdot AT$. The units of A are $\mu V/°K^2$.

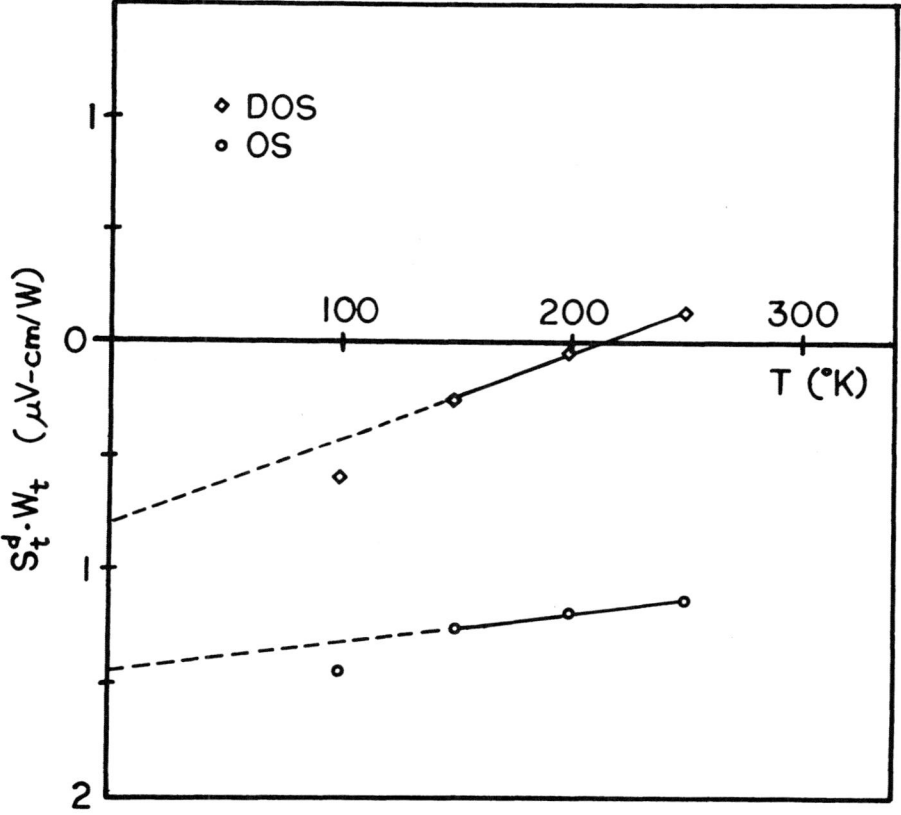

Figure 1. $W_t S_t^d$ versus T for Cu_3Au. The data of reference 5 are used.

(iii) A net defect term is estimated from the residual resistivity and the T=0 intercept with the assumption

$$\sum_j W_j S_j^d \equiv W_{defect} \cdot BT \qquad (2)$$

To gain some perspective of the accuracy and validity of these additional assumptions, a plot of data for pure copper, shown in figure 2, is analyzed in the same manner. The resultant values are given in table II.

The agreement between A values for pure copper is within 20%. Such agreement is not very satisfactory if precise values are the desired goal; however, in the present situation where only the correct sign and general values are sought, this agreement confirms the reliability of the separation process. Cu_3Au in the DOS has a S_{th}^d about twice that of pure copper while Cu_3Au in the OS has a S_{th}^d approximately 1/2 that of pure copper. In each state of order the resultant defect contribution is negative and appears to have a reasonable magnitude. The change in sign for the total thermopower arises from a change in the relative sizes of these two competing terms and not from a change in sign of any contribution.

One final point deserves consideration. Initially we assumed phonon drag contributions in Cu_3Au were negligible. If they were not, how would the preceding analysis differ? Consider figure 2 for pure copper where it is known that S_g is positive and vanishes near room temperature. In this figure the linear T relation is well obeyed at higher temperatures but displays deviations in a positive direction at lower temperatures. Since the plot involves the total thermopower, such a deviation is consistent with a

Table II. Values of thermopower contributions. All the values are given in units of $10^{-3} \mu V/°K^2$. B^a represents the range of values for non-transition metal impurities in copper. B^b represents some values from reference 7. In that study annealing was used to alter the antiphase domain boundaries in Au_3Cu and subsequent changes in ρ and S were measured. A linear fit over a 15% change in ρ permitted an extrapolation to determine the listed values. These values are not directly applicable but they do provide context for general magnitudes.

	Cu	Cu₃Au (OS)	Cu₃Au (DOS)
A (from figures)	6.3	2.8	14.0
A (accepted value)	5.5		
B (from figures)		-8.5	-2.1
Comparison	-0.7 to -6.0[a]		
Values for B		-15.7 to 19.4[b]	

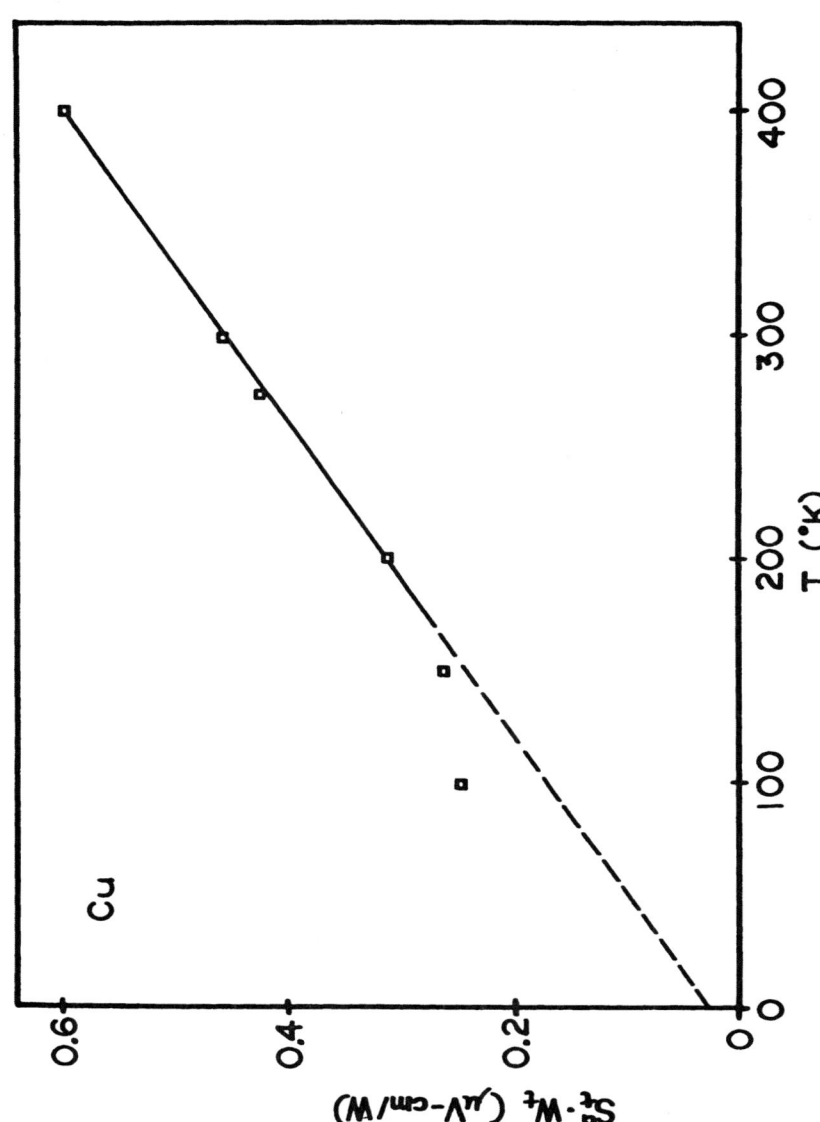

Figure 2. $W_t S_t^d$ versus T for Cu. The thermopower data are from N. Cusack and P. Kendall, Proc. Phys. Soc. 72, 898 (1958); the thermal resistivity data are the recommended values in Thermal Conductivity: Metallic Elements and Alloys, edited by Y.S. Touloukian, R.W. Powell, C.Y. Ho and P.G. Klemens (Plenum Press, New York and Washington, 1970).

positive S_g. Pearson [7] estimates S_g at 100°K as 0.5 μV/°K and this
value is consistent with the size of the deviation in figure 2.
When figure 1 for Cu_3Au is studied a related pattern occurs. For
each state of order a linear fit occurs at higher temperatures with
deviations at lower temperatures. However, for this system negative deviations occur and suggest negative phonon drag contributions.
A negative S_g is not only consistent with but actually suggested by
the low temperature data of Goff, etal [5].

   In closing we note that some earlier studies [8,9,10], which re-
lied upon the total thermopower, concluded that the introduction of
atomic order in Cu_3Au changed the sign of the lattice contribution
to diffusion thermopower. The present analysis of the thermopower
data in conjunction with thermal conductivity data contradicts that
interpretation. The lattice contribution to diffusion thermopower
does not change sign with the introduction of order. Instead, this
contribution always remains positive but its magnitude decreases by
approximately a factor of 5 with the introduction of order. The
present analysis also provides some consistent indications of the
presence of phonon drag. We suggest an analysis of the present type
is essential for understanding thermoelectric behavior in A_3B and
AB_3 alloys which display atomic order-disorder effects.

REFERENCES

[†] Supported in part by a NSF grant.
[1] M.H. Khan and A.S. Huglin, J. Mat. Sci. 1, 409 (1966) and
    Phys. Letters 10, 36 (1964).
[2] A.R. Von Neida and R.B. Gordon, Phil. Mag. 7, 1129 (1962).
[3] See, for instance, the review by R.P. Huebener in Solid State
    Physics Vol. 27, edited by Henry Ehrenreich, Frederick Seitz
    and David Turnbull (Academic Press, New York, 1972).
[4] M. Bailyn, Phys. Rev. 120, 381 (1960).
[5] J.F. Goff, J.J. Rhyne and P.G. Klemens in Intl. Conference on
    Phonon Scattering in Solids, edited by H. Albany (Service de
    Documentation, CEN-Saclay, 1972) p 199.
[6] K. van den Lee and A. van den Beukel, Scripta Met. 5, 901 (1971).
[7] W.B. Pearson, Phys. Rev. 119, 549 (1960).
[8] L.N. Petrova, Phys. Met. Metall. 13, 16 (1962).
[9] G. Airoldi, M. Asdente and E. Rimini, Phil. Mag. 10, 43 (1964).
[10] G. Airoldi and M. Asdente, Phil. Mag. 32, 691 (1969).

THE TRANSPORT PROPERTIES OF POTASSIUM AND ELECTRON-ELECTRON SCATTERING*

J.G. Cook and M.J. Laubitz

Division of Physics, National Research Council of Canada

Ottawa, Canada

The measured transport properties of Potassium display two unusual features at high temperatures: deviations from the Wiedemann-Franz relation, and deviations from linearity in the thermopower. We present here an analysis which shows that both these deviations can be consistently explained in terms of normal electron-electron scattering. The magnitude of this effect, in terms of the thermal resistivity, is $W_{ee}/T \sim 190 \times 10^{-6}$ cm/W.

INTRODUCTION

Some time ago (Cook et al. 1972), we have presented data on the transport properties of Sodium which displayed significant deviations from the Wiedemann-Franz relation at high temperatures. These deviations we have tentatively ascribed to normal electron-electron scattering. Since then, we have determined the properties of one specimen of pure Potassium. These results display not only the large deviations from the Wiedemann-Franz relation previously seen in Na, but also deviations from linearity in the high temperature thermopower. Suspecting that the two were related and due to electron-electron scattering, we have derived an approximate theoretical formula for the effect of such scattering on the thermopower. This we give below and, through comparison with experiment, show that indeed both observed deviations can be consistently ascribed to normal electron-electron scattering, with a magnitude of W_{ee}/T of approximately 190×10^{-6} (cm/W). Because of the shortness of space we concentrate here on the analysis, and do not

---

*NRC 14688

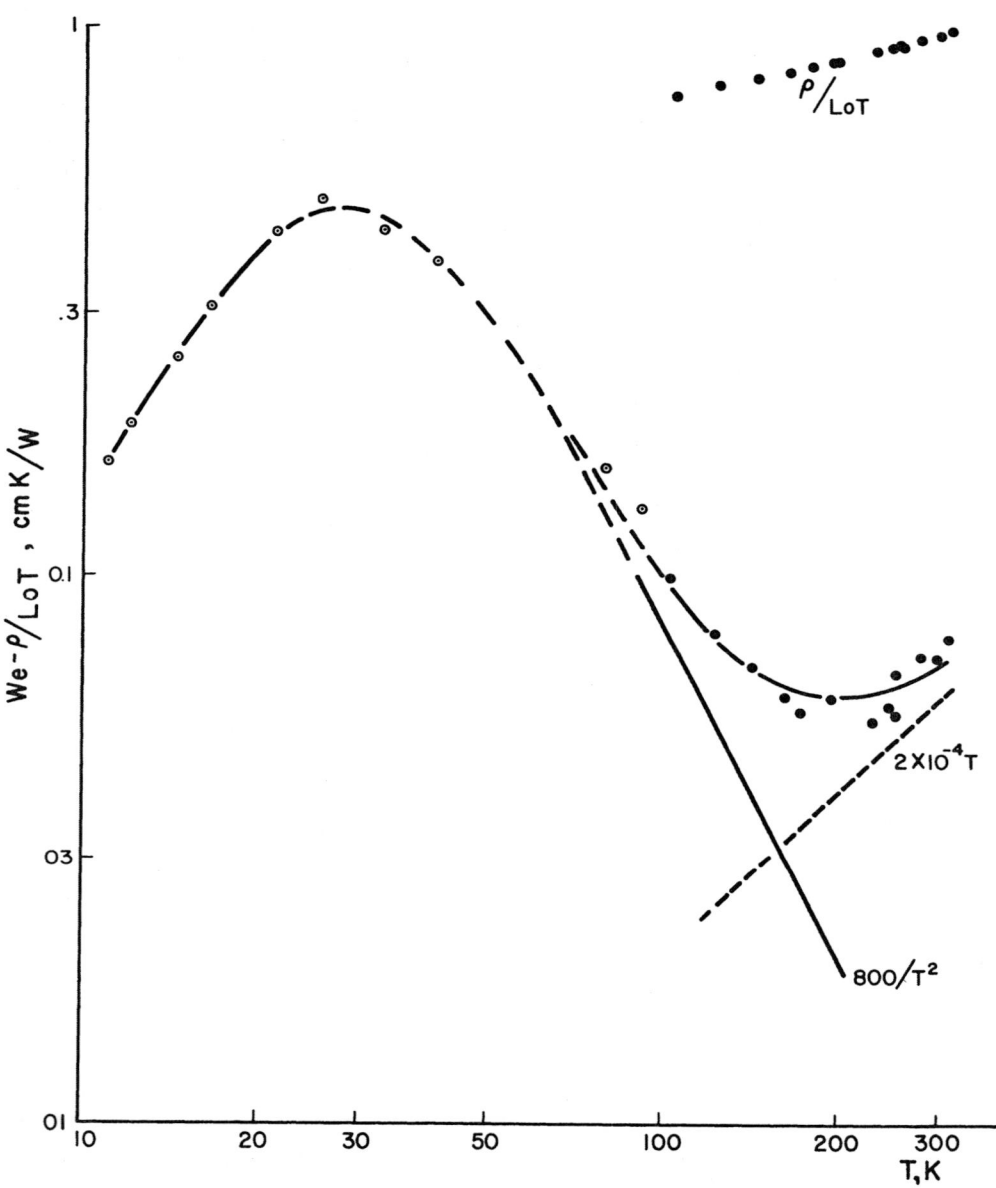

FIGURE 1. $W_e - \rho/L_0T$ as a function of T. Dots: our experimental
results; circles: results of MacDonald et al. (1956).
The solid line is the sum of the two terms, $800/T^2$ and
$2 \times 10^{-4}T$. For comparison, we give the experimental
values of ρ/L_0T.

present the original data for the transport properties of K; these can be obtained from the authors on request.

SAMPLE AND EXPERIMENTAL PROCEDURE

Our apparatus and method of sample preparation have been previously described in detail by Cook et al. (1972). Only two small differences from that description need be mentioned here. In the first place, the sample thermocouples consisted of Au(0.07% Fe)-chromel pairs, calibrated in situ against a standard resistance thermometer. The absolute thermopower of the sample was determined by comparing it to that of the Au(0.07% Fe) thermocouple lead, which in turn was compared to pure Pb. The values for Pb were taken from Christian et al. (1958). Secondly, the geometry of the very soft specimen was not determined directly, but through comparison of the ideal resistivities at 190K to the data reported by Dugdale and Gugan (1962). The residual resistivity ratio of the sample, $\rho(273\text{K})/\rho(4.2\text{K})$, was 141.

RESULTS AND ANALYSIS

In this section, we shall first derive the formulae for the thermal conductivity, λ, and the thermoelectric power, S, for mixed electron-phonon (e-p) and normal electron-electron (e-e) scattering, and then compare these with the experimental results. Our chief aim is to establish the effect that e-e scattering has on S, and, as this is a difficult problem, we use a number of approximations to derive a convenient formula. Not all of these approximations are necessary for the derivation of λ which, for mixed scattering, can be worked out exactly for spherical Fermi surfaces (cf. Jensen et al.,1969; Smith and Wilkins, 1969). However, even for this parameter we employ an approximate formula, since it seemed to us desirable to compare the e-e scattering effects in λ and S on the basis of a consistent formulation. In any case, the approximations employed are very reasonable.

If we define the deviation, ϕ, of the electronic distribution function, f_k, from its equilibrium value, f_k^o, through

(1) $f_k = f_k^o - \bar{v} \cdot (\phi_E e\bar{E} - \phi_T k_b \overline{\nabla T}) (\partial f_k^o / \partial \varepsilon)$

where $\bar{v}$ is the electron velocity, and $\bar{E}$ and $\overline{\nabla T}$ are the applied fields, then the transport coefficients are given by

(2) $\sigma = - e^2 \int (\bar{v} \cdot \bar{u})^2 \phi_E \, \partial f_k^o / \partial \varepsilon \, dk$

(3) $\lambda = -k_b^2 T \int (\bar{v} \cdot \bar{u})^2 \eta \phi_T \, \partial f_k^o / \partial \epsilon \, dk$

(4) $\sigma S = -k_b e \int (\bar{v} \cdot \bar{u})^2 \phi_T \, \partial f_k^o / \partial \epsilon \, dk$

where $\bar{u}$ is a unit vector in the direction of the applied fields, ϵ is the energy of the electron and ϵ_F the Fermi energy. $\eta = (\epsilon - \epsilon_F)/k_b T$, and σ is the electrical conductivity.

For temperatures in excess of the Debye temperature, Θ_D, e-p scattering can be considered elastic, and if it is dominant with just a small amount of e-e scattering added, then one can readily show that

(5) $\phi_T = \dfrac{\eta}{c + b(\pi^2 + \eta^2)}$

where c and b, both of which are functions of ϵ, pertain to details of e-p and e-e scattering respectively. This equation is not exact, but is a good approximation for monovalent metals under the conditions mentioned above. ϕ_E is unaffected by e-e scattering, at least for those metals for which c is not an excessively strong function of ϵ; this is certainly the case for K. Substitution of Eqn. (5) into (3), assumption of cubical symmetry and straightforward manipulation results in

(6) $W_e = W_{ep} + (3/5)(12 - \pi^2) W_{ee} = W_{ep} + 1.28 W_{ee}$

where W_e is the total thermal electronic resistivity and W_{ep} and W_{ee} are the thermal resistivities due to e-p and e-e scattering calculated as if these processes were acting in isolation. This equation is in very good agreement with the results of exact solutions of the Boltzmann equation which, for mixed elastic e-p and normal e-e scattering, yield $W_e = W_{ep} + 1.25 W_{ee}$ in the limit of $W_{ee}/W_{ep} \to 0$. Since $W_{ee} = BT$ and, at high temperatures, $W_{ep} = \rho/L_o T + A/T^2$ (Laubitz and Cook, 1972), where L_o is the standard Lorenz number and ρ the electrical resistivity, we have that

(7) $W_e - \rho/L_o T = A/T^2 + BT$

where A and B are constants.

To calculate the thermopower, we make a further approximation in that we ignore the energy variation of b (which, for instance, for Thomas-Fermi screening is proportional to ϵ^{-1}) in comparison with the energy dependence of the other factor due to e-e

scattering, $\pi^2 + \eta^2$. After some lengthy but straightforward manipulation, one can show that for a free-electron like metal

(8) $\sigma S = (e\lambda_{ep}/\varepsilon_F) [\xi - (3/5)(12 - \pi^2)(2\xi - 3/2) W_{ee}/W_{ep}]$

where ξ is the thermoelectric parameter (cf. Barnard 1972), and
which, with the assumption that $\lambda_{ep} = L_o\sigma T$, yields

(9) $S = (\pi^2 k_b^2 T \xi/3e\varepsilon_F) [1 - (3/5(12 - \pi^2)(2 - 3/2\xi) W_{ee}/W_{ep}]$

   The first term in the above is the usual "diffusion" thermopower due to elastic e-p scattering. Since $W_{ee} = BT$ and
$W_{ep} \sim$ constant, e-e scattering adds to S a contribution approximately quadratic in T, the magnitude of which is larger than could
be predicted from Kohler's relation (cf. Barnard, ibid) by the
factor $(2 - 3/2\xi)$*. It should be noted that this equation is
derived under the assumption of constant volume, and any comparison
with measured, constant pressure, values requires appropriate
corrections.

   Figure 1 illustrates the experimental value of $W_e - \rho/L_oT$,
where W_e was obtained from the measured λ by subtracting the contribution due to lattice conductivity, assumed to be 0.5/T (W/cmK),
one-half the value obtained from the Leibfried-Schloemann theory
(White, 1969). At high temperatures, this quantity can be readily
separated into contributions proportional to T^{-2} and to T. The
first yields an A of 800 (cmK^3/W), which compares very well with the
theoretically computed value of 840 using the Ashcroft pseudopoten-
tial (Laubitz and Cook, 1972). The second term yields a B of
160×10^{-6} (cm/W).

   Figure 2 shows the measured thermopower as deviations from an
equation of the form $S = aT + bT^2$, where a and b were obtained by
regression for T > 150K (this limit, approximately $2\theta_D$, was chosen
to ensure elasticity of the e-p scattering). The coefficient "a"
was used to derive ξ which, at 4.44, is significantly larger than
the value of 3.5 quoted in Barnard. The term in T^2 was corrected
for the effect of the volume expansion on the basis of the results
of Dugdale and Mundy (1961), and using the previously determined ξ,
yielded another value of B, 220×10^{-6} (cm/W). The two values of
B, obtained independently from the thermal resistivity and from the
thermopower, agree quite well with each other, and give, on aver-
aging, an experimental value of 190×10^{-6} (cm/W). As has been
shown by Kukkonen and Smith (1973), this value of W_{ee}/T, as well as

---

*Application of Kohler's relation yields $S = S_{ep}W_{ep}/W_T$, since
 $S_{ee} = 0$, as can be shown both by detailed and hand-waving arguments.

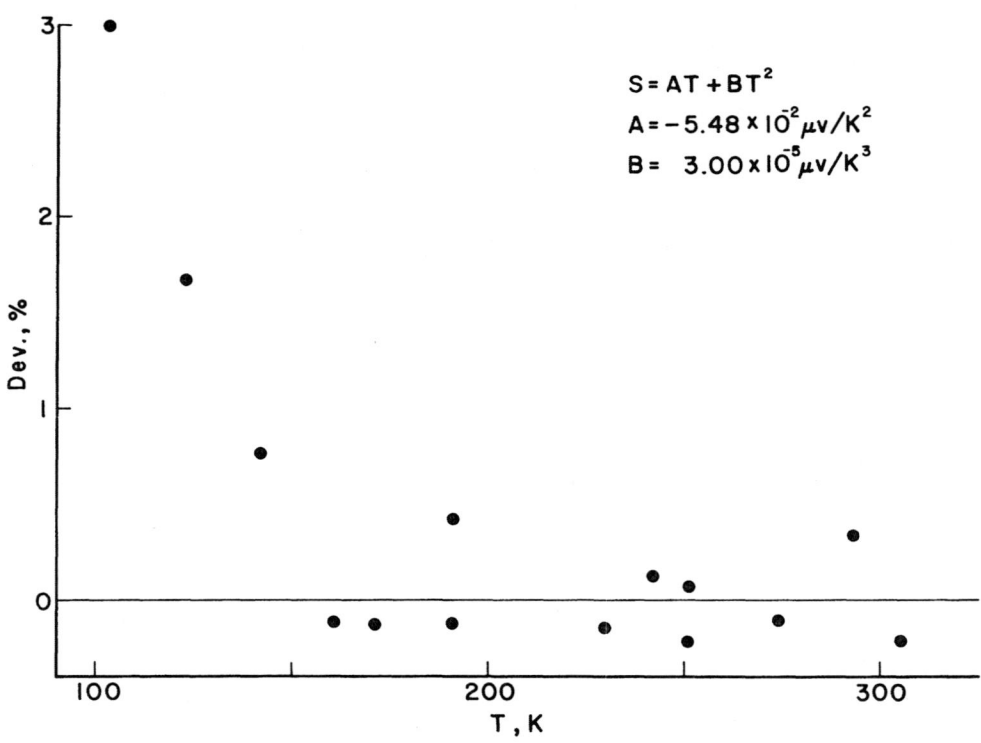

FIGURE 2. Deviations of the absolute thermopower, S, from
a quadratic equation fitted to the experimental
points above 150 K.

the one that we have previously obtained for Na, is substantially larger (by a factor of 3) than that expected from theory using Thomas-Fermi screening.

REFERENCES

1. R.D. Barnard, Thermoelectricity in Metals and Alloys (Taylor and Francis, London), (1972).

2. J.W. Christian, J.-P. Jan, W.B. Pearson, and I.M. Templeton, Proc. Roy. Soc. A245, 213, (1958).

3. J.G. Cook, M.P. Van der Meer, and M.J. Laubitz, Can. J. Phys. 50, 1386, (1972).

4. J.S. Dugdale, and D. Gugan, Proc. Roy. Soc. (London) A270, 186, (1962).

5. H.H. Jensen, H. Smith, and J.W. Wilkins, Phys. Rev. 185, 323, (1969).

6. C.A. Kukkonen, and H. Smith, Phys. Rev. B8, 4601, (1973).

7. M.J. Laubitz, and J.G. Cook, Phys. Rev. B6, 2083, (1972).

8. D.K.C. MacDonald, G.K. White, and S.B. Woods, Proc. Roy. Soc. A235, 358, (1956).

9. H. Smith, and J.W. Wilkins, Phys. Rev. 183, 624, (1969).

10. G.K. White, Proc. Eighth Thermal Conductivity Conference (Plenum Press, N.Y.), (1969).

ANISOTROPIC TRANSPORT DISTRIBUTION FUNCTIONS FOR POTASSIUM

C.R. Leavens and M.J. Laubitz

National Research Council of Canada

Division of Physics, Ottawa, Canada

Recent calculations of the electrical resistivity of potassium have shown that the effect of anisotropy in the electron distribution function, $\Phi_{\vec{k}}(T)$, becomes significant at low temperatures. However, the two approximate methods that have been used to calculate the resistivity give markedly different estimates as to the size of the effect. For example, at 3 K a variational calculation employing several different trial functions gave a correction that is less than one third that obtained using a one iteration approximation to the solution of the Boltzmann equation for $\Phi_{\vec{k}}(T)$. In an attempt to resolve this discrepancy, we have iterated the Boltzmann equation to convergence to obtain the correct distribution function at several temperatures. The corresponding corrections to the electrical resistivity due to any anisotropy in $\Phi_{\vec{k}}(T)$ are also presented and compared to the results obtained using the two approximate methods mentioned above. Finally, a qualitative discussion of the effect of phonon drag on our results is given.

THE SUPERCONDUCTING STATE AS A MAGNIFIER FOR THE STUDY OF PHONON INTERACTIONS*

D. A. Furst and P. Lindenfeld

Department of Physics, Rutgers University

New Brunswick, New Jersey 08903

In the superconducting state the scattering of phonons by electrons is reduced by a factor whose magnitude and temperature dependence are now well established. The effects of the scattering of phonons by electrons can therefore be separated from those which depend on the defect structure, and subtle effects in the lattice conductivity can be investigated more precisely than in metals which do not become superconducting.

We have measured the thermal conductivity of two lead-alloy specimens whose residual resistivity differed by a factor of twelve. They were measured in the normal and superconducting states, both annealed and deformed, between 1.2 K and the transition temperature. The measurements allow a determination of the parameters describing the scattering of phonons by electrons, by impurities, and by the defects introduced as a result of the deformation. They also give a value for the mean free path characteristic of the defect structure present in the annealed state.

Among our results are the following: the scattering by electrons is stronger in the more concentrated alloy; the effect of the deformation is compatible with that to be expected from sessile dislocations; and the residual mean free path is, surprisingly, of the same order of magnitude as that found in pure lead.

In metals the strong scattering of phonons by the conduction electrons usually masks the effect of other phonon interactions on the lattice thermal conductivity. In superconductors, however, the scattering by the electrons goes down rapidly as the temperature is reduced below the transition temperature, and the lattice conductivity increases accordingly. Earlier experiments[1] showed that the factor by which the electron scattering is reduced is, at least for weak-coupling superconductors, well described by the theoretical expressions of Bardeen, Rickayzen, and Tewordt (BRT).[2] It gets quite large, and is about 30 when $t=T/T_c$ is 0.5, and about 500 at $t=0.3$. The extent to which the actual ratio of the lattice conductivity in the superconducting state (K_{gs}) to that in the normal state (K_{gn}) departs from the calculated factor is then a measure of the other phonon interactions.

Our results for two lead-alloy specimens with nominal atomic percentages of 0.67% and 10% bismuth are shown in Fig. 1. The specimens (1/8 inch diameter rods) were annealed and measured in the superconducting state ($K_s^{(a)}$) and, with the application of a magnetic field, in the normal state (K_n). The specimens were then deformed by bending them back and forth many times, cooled rapidly, and remeasured. The results for the superconducting state are now given by the curves marked $K_s^{(d)}$. The normal state results, on the other hand, changed by less than 2%, and were not plotted again. The arrows marked T_c indicate the electrical transition.

To analyze the curves we first calculated the normal-state electronic conductivity from $K_{en}^{-1} = A/T + BT^2$ where the coefficient A is given by ρ_0/L, with $L = 2.45$ V^2/K^2 in accordance with the Wiedemann-Franz law, and the coefficient B is taken from Fig. 10 of the paper by Hilsch and Steglich.[3] The superconducting state electronic conductivity, K_{es}, was calculated using the BRT function[2] for K_{es}/K_{en} with an energy gap parameter equal to 4.3.[4] The lattice conductivities K_{gs} and K_{gn} could then be obtained by subtracting K_{es} and K_{en} from the measured conductivities.

To get an estimate of the contribution of the various scattering processes, we assume the lattice conductivities to be described by

$$K_g = PT^3 \int_0^\infty (\sum_i \tau_i^{-1}) \frac{x^4 e^x dx}{(e^x-1)^2}$$

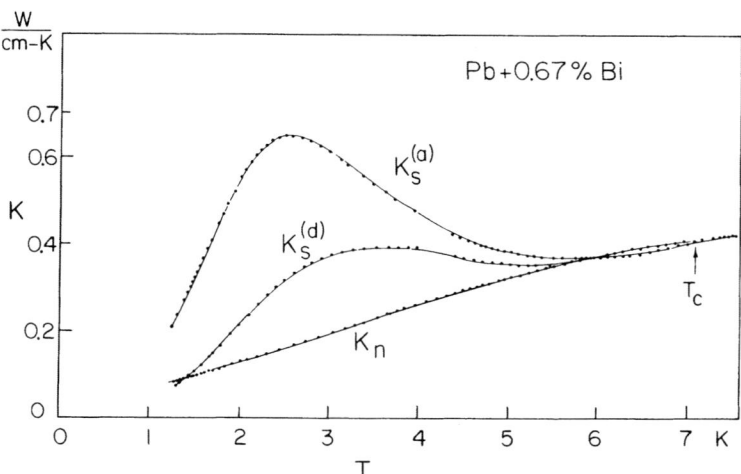

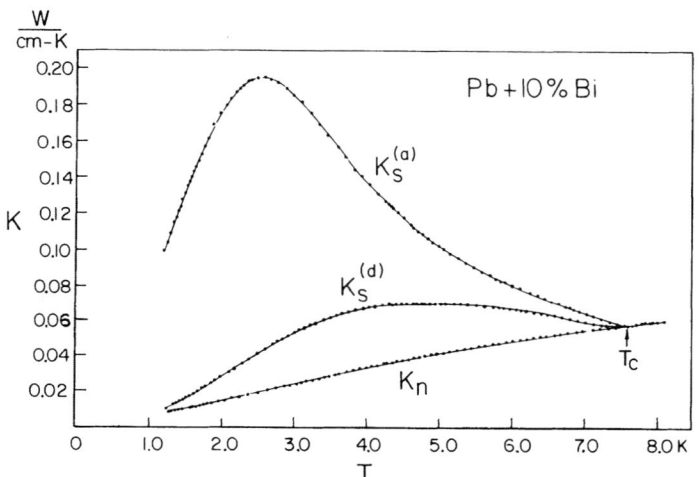

Fig. 1   The measured thermal conductivities.

where the τ_i are the relaxation times associated with the various scattering processes, and $x = \hbar\omega/kT$.

We used the following reciprocal relaxation times: Mx^4T^4 for point-defect scattering; $ExT\ g(x)$ for electron scattering, where $g(x)$ is equal to one for the normal state and is given by the BRT expressions for the super-conducting state; v/L for boundary scattering; DxT for dislocation scattering, and $\beta e^{-\theta/2T}x^2T^5$ for Umklapp scattering.[5]

The procedure was to fit $K_{gs}^{(a)}$ first without using the term DxT. K_{gn} was then calculated by letting $g(x)$ go to one, and $K_{gs}^{(d)}$ was calculated by including the dislocation term. Since each term has a different temperature dependence, the requirement that all three curves for each specimen be fit by this procedure is very severe. In fact we had to modify our procedure in the light of the results. In our first attempts we did not include Umklapp scattering, but found that the 10% Bi curves could not be fit near T_c.

For the 0.67% Bi specimen we could not get a good fit until we modified the function for K_{es}/K_{en}. This is not unexpected, since it is known that pure lead does not follow the BRT relations. The departure was not large, and we therefore used the same function $g(x)$ as for the 10% specimen. The fit for $K_{gs}^{(d)}$ for the 0.67% specimen can be slightly improved if, in going from $K_{gs}^{(a)}$ to $K_{gs}^{(d)}$, we not only include the dislocation term but also allow the mean free path L to vary.

Our best fits for the lattice conductivities are shown in Fig. 2, and the corresponding parameters in Table 1. Instead of the coefficient D we list N, the number of dislocations per cm^2 calculated by assuming that the dislocation thermal resistivity is equal to $3.2 \times 10^{-8}\ N/T^2$, as for copper alloys.[6] Fig. 3 shows the total conductivities as well as the separated lattice and electronic components, and illustrates vividly the magnification of the lattice conductivity which is achieved by going to the superconducting state.

To fit $K_{gs}^{(a)}$ none of the scattering mechanisms may be neglected in our temperature range. They can, however, be divided into two groups. The first consists of electronic and point-defect scattering which predominate above the maximum. They can be easily distinguished by the fact that only the electronic part changes in the

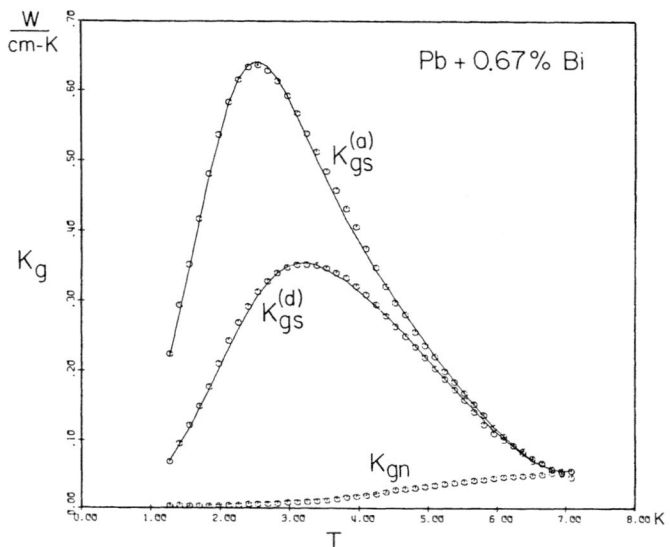

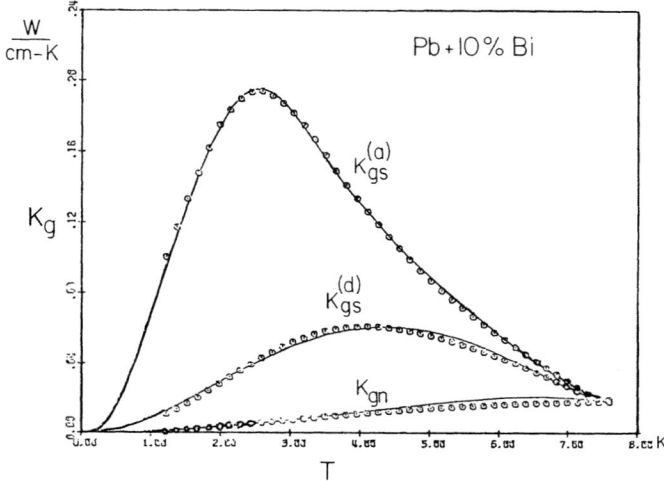

Fig. 2   The experimental lattice conductivities together
         with the computer fits. The points are taken
         from the smoothed curves resulting from the sub-
         traction of the electronic conductivities from
         the total measured conductivities. The lines
         are the computer fits corresponding to the
         parameters of Table 1. The figures were drawn
         directly on the Calcomp plotter attached to the
         computer.

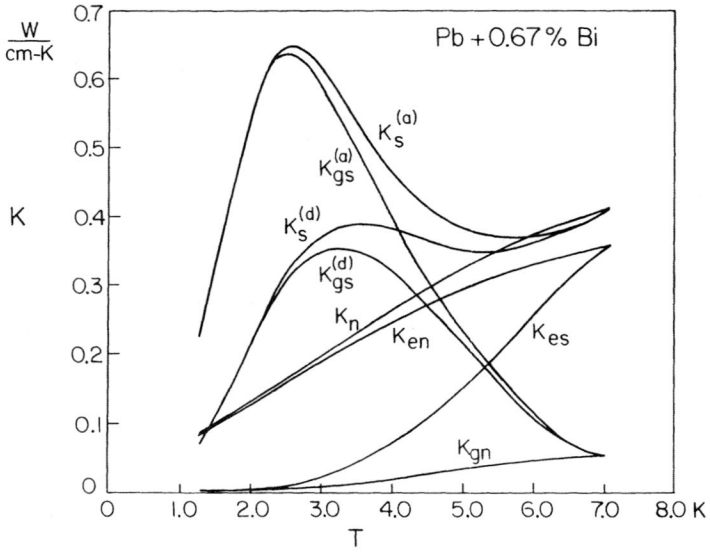

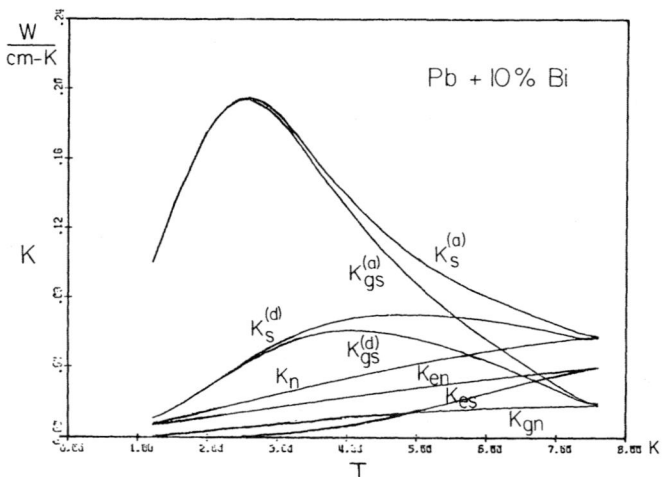

Fig. 3   The electronic, lattice, and total thermal con-
          ductivities in the normal and superconducting,
          annealed and deformed states.

transition to the normal state. Below the maximum the
main contribution is from dislocation and boundary
scattering. They can also be separated if we assume
that dislocations are effective only in the deformed
specimens. This is a dubious and even controversial
assumption, and our measurements have, so far, not decided its validity.

At still lower temperatures point-defect and electronic scattering will become negligible, and it may be
possible to gain unambiguous information on the remaining scattering processes. We plan to make such measurements, but it may be well to remember that for pure lead
the measurements of O'Hara and Anderson[7] between 0.05
and 0.4 K have raised more questions than they have an-
swered. The temperature dependence which they found does
not correspond to any of the relaxation times which we
have used; the magnitude corresponds to a phonon mean
free path between 0.03 and 0.06cm in the annealed state,
and changes only by a factor of about 2 when the specimen is deformed. They suggest that vibrating dislocations are the main scatterers even in the annealed state.

We note that the mean free paths determined by
O'Hara and Anderson are of the same order of magnitude
as the values of L which we observe, and also close to
those measured by Gupta and Wolf[8] in lead-indium alloys.
In each case they are far smaller than those which would
result from scattering by the external boundaries. It
seems that further work will be necessary to determine
the mechanism which is responsible.

Finally we see from Table 1 that electronic scattering is stronger in the more impure specimen. This is
in accord with the dependence on the electronic mean free
path first described by Pippard.[9,10] Recent work on
copper alloys[6] left some doubt on the existence of this
effect, but in normal metals the separation of the scattering by electrons and by dislocations is necessarily
ambiguous. This is another illustration of the advantage
which the superconducting state measurements provide.

We would like to acknowledge helpful conversations
with A. C. Anderson and G. Deutscher.

Table 1

	Pb + 0.67% Bi		Pb + 10% Bi	
	ann.	def.	ann.	def.
ρ_o ($\mu\Omega$-cm)	0.38	0.38	4.32	4.58
A	15.5	15.5	177	187
B	0.12	0.12	0.33	0.33
E	1.2×10^8	1.2×10^8	1.6×10^8	1.6×10^8
M	5.5×10^2	5.5×10^2	9.8×10^3	9.3×10^3
L(cm)	.023	.0078	.022	.022
N	-	10^8	-	3×10^9
β	-	-	7.5×10^6	7.5×10^6

Specimen parameters

REFERENCES

*   Supported by the National Science Foundation and
      the Rutgers University Research Council.

1.   See for example L. G. Radosevich and W. S. Williams,
     Phys. Rev. 188, 770 (1969).
2.   J. Bardeen, G. Rickayzen, and L. Tewordt, Phys. Rev.
     113, 982 (1959).
3.   R. Hilsch and F. Steglich, Z. Physik 226, 182 (1969).
4.   B. J. Mrstik and D. M. Ginsberg, Phys. Rev. B5, 1817
     (1972).
5.   See for example J. Callaway, Phys. Rev. 122, 787
     (1961).
6.   R. J. Linz, T. K. Chu, A. C. Bouley, F. P. Lipschultz,
     and P. G. Klemens, Phys. Rev. B10, 4869 (1974).
7.   S. G. O'Hara and A. C. Anderson, Phys. Rev. B10, 574
     (1974).
8.   A. K. Gupta and S. Wolf, Phys. Rev. B6, 2595 (1972).
9.   A. B. Pippard, Phil. Mag. 46, 1104 (1955).
10. P. Lindenfeld and W. B. Pennebaker, Phys. Rev. 127,
     1881 (1962).

ON THE POSSIBILITY OF DETECTING PHONON DRAG DUE TO AN ELECTRONIC HEAT CURRENT*

R. FLETCHER

PHYSICS DEPT., QUEEN'S UNIVERSITY

KINGSTON, ONTARIO, CANADA

The influence of the non-equilibrium distribution of the electrons on the thermal conductivity of the lattice λ_g has been theoretically investigated in the literature. It was found that the effect is very slight and leads to corrections to the calculated λ_g of order kT/μ, where μ is the chemical potential of the electrons, and thus was thought to be unobservable. A simple physical picture is presented showing how, in favourable circumstances, the Righi-Leduc effect may be used to magnify the non-equilibrium effects by factors of the order of 10^4 which should lead to observable changes of λ_g. Experimental results on tungsten and cadmium are briefly reported.

The concept of phonon drag as applied to the calculation of electrical resistance and thermal conductivity is over four decades old; the early developments have been summarized by Klemens.[1] In its simplest form the idea applies to N-process scattering between the electron and phonon systems with the result that any net shift of either system from equilibrium (in reciprocal space) tends to cause a shift in the other by an equivalent amount. The situation is made considerably more complex in real metals by the existence of electron-phonon and phonon-phonon U-processes but these problems need not concern us here. The application of an electric current J will displace the electron system by some (essentially) constant vector parallel to the resultant electric field $\tilde{E}$ and this will in turn be reflected in the phonon system, causing both a heat current to be carried by the phonons, i.e. a thermolectric effect, and a decrease in the part of the resistivity arising from electron-phonon

scattering. Such an effect is of first order at low temperatures
and its inclusion is essential in any realistic theory of thermopower and resistivity.

However a quite distinct type of phonon drag has also been considered which involves the interaction between an electronic *heat*
current and the phonons. Here the displacement of the electronic
system from equilibrium is far more subtle and does not correspond
to a uniform shift in reciprocal space. Indeed the net centre of
gravity, so to speak, of the electronic system remains unchanged and
and the net result is that there are equal numbers of 'hot' and
'cold' electrons travelling in opposite directions with no net J.
In this case the effect on the phonon system is very small. It has
been estimated[1] that if one calculates the lattice conductivity λ_g
in the two cases of the electronic system carrying or not carrying
any heat, then the difference between the two results for λ_g is of
order kT/μ, where μ is the chemical potential of the electrons, k
the Boltzmann constant and T the temperature. For a degeneracy temperature of perhaps 10^4 K and for T ~ 4K, this correction is thus $\lesssim$
10^{-3} and appears to be negligibly small in all normal cases. However the purpose of this paper is to examine a very abnormal case
in which there is an effective magnification of possibly $10^3 - 10^4$
so that one might hope to detect the effect. The magnifier is the
Righi-Leduc effect which is the thermal analogue of the Hall effect;
we refer the reader elsewhere for more details of this coefficient.[2]

If a heat current density $\tilde{U}$ is present, then there exists a
temperature gradient ∇T which is related to $\tilde{U}$ via the thermal conductivity tensor $\bar{\lambda}$ i.e., $\tilde{U} = -\bar{\lambda}\nabla T$. In fact there are also thermoelectric terms due to the presence of either an electric field or
electric current but these are very small and can be neglected for
our purposes. If the material is isotropic in the x-y plane and the
heat flow is restricted to be along x (with the magnetic field B
along z) then we have:

$$U_x = -\lambda_{11}(\partial T/\partial x) + \lambda_{21}(\partial T/\partial y)$$
$$U_y = 0 = -\lambda_{21}(\partial T/\partial x) - \lambda_{11}(\partial T/\partial y)$$
(1)

If we further restrict ourselves to the case of high magnetic
fields, and no open orbits, then theory[3] gives

$$\lambda_{21} = \lambda_{21}^e + \lambda_{21}^h = L_0|e|T\{N_e - N_h\}/B + \text{small terms}$$

where the superscripts or subscripts e and h refer to electrons and
holes, L_0 is the ideal Lorenz number, $|e|$ is the magnitude of the
electronic charge and N refers to the number densities of electrons
or holes. For our purposes the most interesting case is that of com-

pensated metals with $N_e = N_h$ and for which we thus find λ_{21} to be very small and typically a factor 10^4 smaller than the magnitude of the individual terms, e.g. $L_o |e| TN_e/B$. Thus $\partial T/\partial y \approx 0$ for this case (though not actually zero in practice except perhaps coincidentally). However, even though $\partial T/\partial y \approx 0$ and $U_y = 0$, the individual hole and electron heat currents along y are not small but are in fact extremely large being almost exactly equal to each other but having opposite directions.

Thus for example the hole heat current U_y^h is

$$U_y^h = \lambda_{21}^h (\partial T/\partial x) + \text{ very small term in } (\partial T/\partial y).$$

Roughly speaking at fields of a few Tesla with high purity materials, it is quite easy to produce a U_y^h of about 10^4 times larger than that which would exist if the hole current were due solely to the actual $(\partial T/\partial y)$, (and similarly U_y^e). Another way of looking at the result is that the electronic distribution (both electron and hole) is greatly perturbed from equilibrium along y corresponding to the large heat flows, but is only weakly perturbed along x. Thus if we can measure λ_g along both x and y we might hope to detect a difference due to the influence of phonon drag along y. Notice that the phonon drag may increase or decrease the apparent λ_g along y, say λ_{gy}, and the effect would be zero if the holes and electrons had precisely the same effect on the phonons.

To determine λ_g along x, say λ_{gx}, is straightforward in compensated metals since

$$\lambda_{11} = \lambda_{11}^{e+h} + \lambda_{gx}$$

where λ_{11}^{e+h} is due to the electronic system which theory[3] predicts to be of the form $A(T)/B^2$ at high B, $A(T)$ being a constant which does not depend on B but does generally depend on T. Thus one simply measures λ_{11} and plots against B^{-2}, the intercept giving λ_{gx}. For compensated metals one finds from Eqn.1 that the measured thermal resistivity $\gamma_{11}^m = -(\partial T/\partial x)/U_x$ is almost precisely λ_{11}^{-1} (since $|\lambda_{21}| \ll \lambda_{11}$) so that λ_{gx} is easily measured.

Fortunately the Righi-Leduc effect itself allows a determination of λ_{gy} to be made. Thus one can show[4] that the measured Righi-Leduc thermal resistivity $\gamma_{21}^m = -(\partial T/\partial y/U_x)$ is given by

$$\gamma_{21}^m = \gamma_{21}^e /(1 + \lambda_{gx}/\lambda_{11}^{e+h})(1 + \lambda_{gy}/\lambda_{11}^{e+h}) \qquad (2)$$

where γ_{21}^e is the value γ_{21}^e would have had if the lattice did not transport heat. To obtain λ_{gy} we must have some information about γ_{21}^e and we assume that γ_{21}^e and the Hall resistivity ρ_{21} (i.e. E_y/J_x) have the same field dependence (this is a sufficient requirement) i.e. $\rho_{21} = L_2'T\gamma_{21}^e$ (3)

where L_2' is a Lorenz number that may be T dependent. In the elastic
scattering limit $L_2' = L_o$ and Eqn. 3 should be perfectly obeyed[3]; thus
we should examine compensated metals of high Debye temperature to
minimize phonon scattering (Eqn. 3 may well be accurate for phonon
scattering but theory is silent on this point) and we chose W and
Cd for our initial experiments.

If we now make the assumption that $\lambda_{gy} = s\lambda_{gx}$, where s is a
constant that is not too B dependent (the validity of this assumption
is by no means obvious), then the denominator on the right hand side
of Eqn. 2 can be accurately approximated (to about 1% for the pre-
sent case) by $(1 + t\lambda_{gx}/\lambda_{11}^{e+h})^2$ where t is another constant related
to s. Thus Eqn. 2 can then be rewritten

$$(\rho_{21}/\gamma_{21}^m L_2'T)^{\frac{1}{2}} = 1 + t\lambda_{gx}B^2/A(T)$$

so that a plot of $(\rho_{21}/\gamma_{21}^m)^{\frac{1}{2}}$ against B^2 should be a straight line
with ratio of slope to intercept of $t\lambda_{gx}/A(T)$, a result which can be
compared with the directly measured ratio of $\lambda_{gx}/A(T)$ from the previously mentioned experiment. Clearly if $t \neq 1$ then $\lambda_{gx} \neq \lambda_{gy}$.

Space does not permit a detailed presentation of our data and
techniques and these will be presented elsewhere. Data was taken on
a Cd crystal (B//[0001]) of residual resistivity ratio $(\rho_{293K}/\rho_{1.2K})$
of 91,000 and on a W crystal (B//[001]) with $(\rho_{293K}/\rho_{1.2K})$ of 29,000.
Fields were provided by a Nb-Ti solenoid and data was taken over the
range of 0.5 - 3.5T in the ^{4}He range. The accuracy is limited mainly
by the small magnitude of the transverse temperature differences
(typically 1mK).

Fig.1 shows typical data used in the evaluation of λ_{gx}, in this
case for W but the Cd results are similar. Fig. 2 gives a plot of
$(\rho_{21}/\gamma_{21}^m L_oT)^{\frac{1}{2}}$ against B^2 for Cd.

For various reasons it was found that Cd was not an ideal material for the study and the analysis could be made only in the narrow temperature range of 1.6 - 2K. We find that t = 1.04 ± 0.10 for
this case i.e. the result is negative. However W is a much more convenient material and useful data over the whole ^{4}He range could be
obtained. Fig. 3 shows the measured variation of t with T and it is
seen that t is much higher than unity and appears to be decreasing
as T decreases, as theory would lead us to expect. Obviously one must
question the accuracy of the data when one obtains such an unusual
result; there are a number of checks that one can make, the most significant being the observation that L_2' tends to L_o (within 2%) as T
tends to zero. This result gives us a strong measure of confidence
in the experimental values of t.

One might enquire why Cd does not show an observable effect.
This may be due to the fact that the electronic states in Cd are

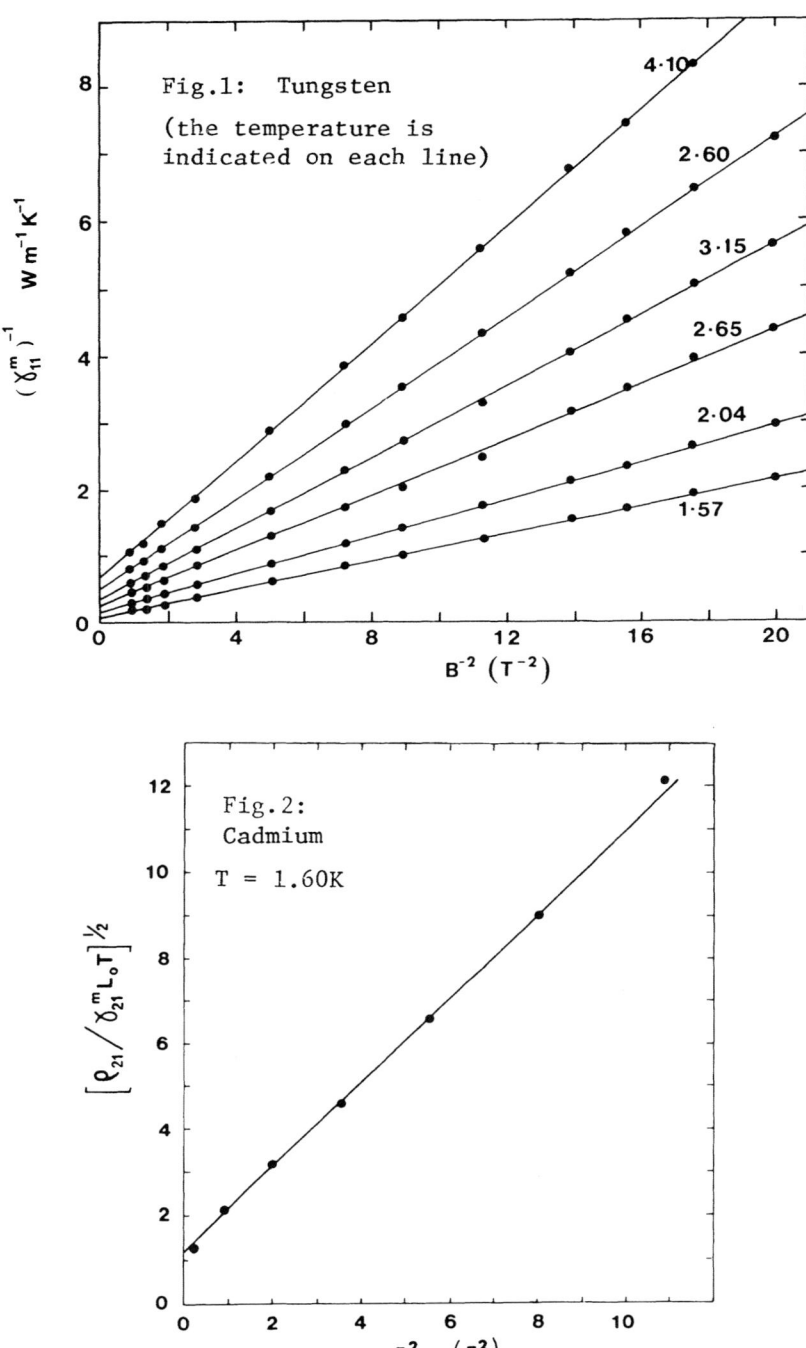

Fig.1: Tungsten

(the temperature is
indicated on each line)

Fig.2:
Cadmium

T = 1.60K

basically single O.P.W. (except near the zone boundaries) for both
the electron and hole surfaces and, bearing in mind that the effect
we are looking for would be zero if the electrons and holes had exactly the same interaction with the phonons, then the result for Cd
may not be unreasonable. On the other hand, W has a much more complex electronic structure and the same comments are not applicable
to that metal.

    The arguments we have used are clearly somewhat speculative and
require theoretical justification. If it happens that they are incorrect, the rather unexpected experimental results will still require an explanation. We are continuing the investigation with
other metals of both simple and complex electronic structure.

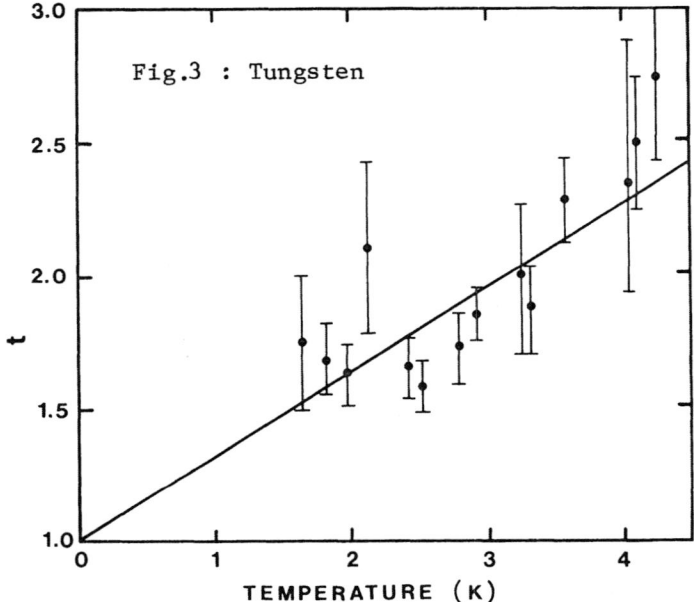

REFERENCES

*Work supported by N.R.C.

1)   P.G. Klemens H. der Physik $\underline{14}$, 199, (1956)
                   See page 255 et seq.

2)   J.-P. Jan Solid St. Phys. $\underline{5}$, 1 (1957)

3)   M. Ia Azbel', M.I. Kaganov and I.M. Lifshitz Sov. Phys. JETP
                   $\underline{25}$, 967, (1957)

4)   The methods are indicated in the paper:
                   R. Fletcher J. Phys.(F) $\underline{4}$, 1155 (1974)

EVIDENCE FOR SUBTHERMAL PHONONS CONTRIBUTION TO THE THERMAL

CONDUCTIVITY OF A RHOMBOHEDRAL CRYSTAL

J-P. Issi, J-P. Michenaud & J. Heremans

Université Catholique de Louvain - Laboratoire PCES

1, Place Croix du Sud , B-1348 Louvain-la-Neuve,Belgique

Measurements performed on two tuning fork bismuth samples of different arms ratios show a size-effect persisting well above the Casimir range. The effect is ascribed to the low energy subthermal phonons contribution to the lattice thermal conductivity at intermediate temperatures. The experimental results are consistent with Herring's predictions for this class of solids.

INTRODUCTION

Heat is transported in bismuth by phonons and charge carriers. In the lowest temperature range the electronic thermal conductivity is small compared to the measured total conductivity, which exhibits the typical dielectric behaviour with a maximum around 4K [1]. These data on the lattice thermal conductivity were interpreted so far in terms of the Debye formula, which implies that at a given temperature only the thermal phonons give a significant contribution.

Previous measurements on the thermal conductivity of bismuth at low temperatures showed that a size-effect could still persist a little above 10K [2]. Intermediate temperatures size-effects if detected experimentally would reveal a contribution of the low energy phonons to the thermal conductivity. This was first pointed out by Herring [3] and experimentally verified for a cubic material by Geballe and Hull [4]. Such studies might help in the interpretation of phonon drag observations in the thermopower of bismuth [2,5], which remains an arduous job because of the many unknown parameters concerning the low energy phonons.

The present work aims at extending the size-effect measurements on bismuth [2] at higher temperatures, mainly above 20K. Since
at these temperatures the effect was expected to be small, a sensitive comparison method, which avoids a precise measurement of
the geometrical dimensions of samples was used.

EXPERIMENTAL

The method used here is a modified version of that originally
described by Geballe and Hull [4]. For a given temperature of the
heat sink, the current I_1 in the heater H_1 on the top of the thick
arm of the tuning fork (Fig.1) is adjusted in order to establish
the desired temperature difference; i.e. approximately 2K above 77K
to reach a few 10^{-2} K in the vicinity of the liquid helium range.
This temperature difference is measured by means of the differential
thermocouple ΔT_1 . The current I_2 is then adjusted in the heater H_2

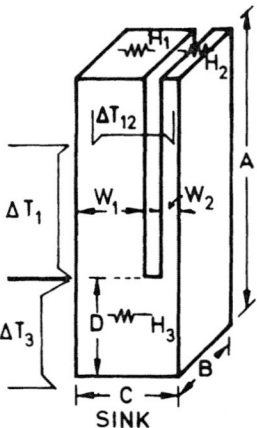

FIG.1.   Tuning fork sample with heaters and thermocouples. H_1 and
H_2 are the heaters used to establish the gradient on the thick and
thin arms respectively, while H_3 is a dummy heater. The differential thermocouple ΔT_1 measures the temperature difference across the
thick arm, ΔT_3 between the lower part of the sample and the heat
sink, while ΔT_{12} measures the temperature difference between the
upper parts of the two arms, and is used to adjust the currents in
heaters H_1 and H_2 . All thermocouples are electrically insulated
from the sample. The dimensions A, B, C, D, W_1 and W_2 of the two
tuning fork samples investigated TF01 and TF02 are given in Table I.
Note that the sample represented here is mounted upside down in the
cryostat.

at the top of the thin arm to give a reading for the differential thermocouple ΔT_{12} the nearest possible to $\Delta T_{12}(0)$, which is the e.m.f. read on this thermocouple when there is no imposed temperature gradient along the sample ($I_1 = I_2 = 0$). The main problem is to keep this stray e.m.f. $\Delta T_{12}(0)$ very small (here less than 10^{-7}v) and, above all, constant during the experiment (no more than 10^{-8}v) variation. Then, with I_1 fixed, we chose about ten values of I_2 which yield values of ΔT_{12} slightly above and below that of $\Delta T_{12}(0)$. By linear interpolation we determine the ratio $R = I_1^2 / I_2^2$ which corresponds to the $\Delta T_{12}(0)$ at the temperature considered. In the lowest temperature range, where the thermal contact resistance sample-sink is large compared to that of the arms of the sample, we used the dummy heater [6] at the bottom of the sample (H_3) to determine $\Delta T_{12}(0)$ by adjusting the current I_3 such that ΔT_3 (when $I_1 = I_2 = 0$) indicates the same values $\Delta T_3 + \Delta T_1$ as when H_1 and H_2 were on.

TABLE I : APPROXIMATE DIMENSIONS OF SAMPLES

(cfr. Fig.1)

sample	A	B	C	D	W_1	R_0 [(+)]	W_2 [(*)]
	cm	cm	cm	cm	cm	—	cm
TFO1 [(o)]	4.3	0.3	0.6	0.65	0.280	2.352	0.119
TFO2	4.6	1.0	1.5	1.33	0.905	4.933	0.183

(+) R_0 is the ratio of the thermal conductances of the two arms as determined between 77 and 100K; where they were found to be constant for the two samples within 0.2%

(*) W_2 was determined from $W_2 = W_1 / R_0$

(o) Measurements below 10K were previously reported for this tuning fork sample (cfr. Ref.2)

Two tuning fork samples were studied and their characteristics
are given in Table I. They were initially of rectangular cross
section with their axis oriented along the binary direction. They
consisted of 6N purity bismuth. A single spark cut along the length
transformed them into tuning forks of different arms thicknesses as
shown in Table I. The samples were then etched to remove surface
damage and mounted in a vertical liquid helium cryostat. Special
care was taken when designing the differential thermocouple ΔT_{12}
in order to reduce the stray thermal e.m.f. to a value around
$10^{-7}V$ at any temperature. The heat flux was along the sample axis.
The temperature sensors were Au + 0.03% Fe v/s p-chromel Johnson
and Matthey thermocouples and e.m.f.'s were measured by means of
a Leeds and Northrup K-5 potentiometer and a sensitive null detector
(L & N 9838-1). A resolution of better than 10^{-8} was achieved in
the measuring circuits. The temperature of the sink was regulated
by means of a Harwell temperature controller and the currents in
the heaters H_1, H_2 and H_3 were determined by measuring their voltage
drops across standard resistors by means of a four digit voltmeter.

RESULTS AND DISCUSSION

Fig.2 shows the temperature dependence of the size-effects
observed on the two tuning fork samples as a function of temperature. $\Delta K/K$ is the percentage increase in the lattice thermal conductivity $(K_1 - K_2) / K_2$, where the indices 1 and 2 refer to the
thick and thin arms respectively. This relative increase is equal
to $(R/R_o - 1)$, R being the ratio of the thermal conductances of the
two arms measured at the temperature considered, and R_o the same
ratio measured in the region 77 to 100K, where it was found to be
temperature insensitive for both tuning fork samples within 0.2%.
It is clear from Fig.2 that for both samples TFO1 and TFO2 a detectable size-effect persists well above 20K. By extrapolating to
higher temperatures the values of the mean free paths of the thermal phonons for phonon-phonon interactions determined at low temperature [1], one can roughly estimate the ratio of the thermal conductivities in the two arms of both samples due to a Casimir type
boundary scattering. For sample TFO2 the size-effect for this type
of scattering amounts to 6% at 6K, 1% at 8K and becomes negligibly
small above 10K compared to the observed size-effect. This indicates, as expected, that the boundary scattering we observe in the
highest temperature range could not be ascribed to the thermal
phonons. However, the subthermal low energy phonons have much
larger mean free paths than the thermal ones.

Following Herring's procedure [3] we divide the phonons into
two classes:

1. The low energy subthermal phonons with wave number $q < q_c$,
where the energy corresponding to q_c is approximately equal to
0.1 kT at temperatures below the Debye temperature (θ_D = 120K).

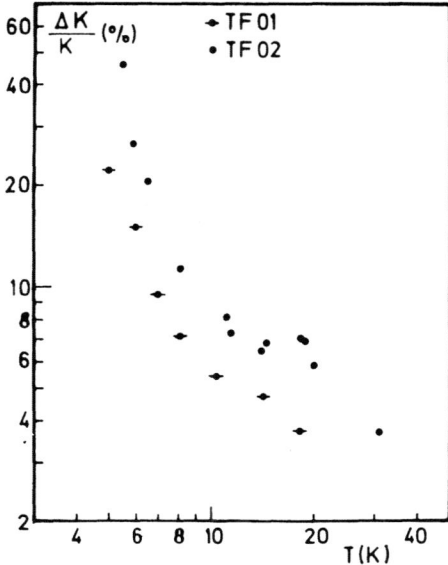

FIG.2.   Percentage increase of the lattice thermal conductivities
of the two arms of samples TF01 and TF02 as a function of temperature. $\Delta K/K$ is equal to the ratio of the difference in conductivities between the thick and thin arms $(K_1 - K_2)$, over that of the
thin arm, K_2.

To these phonons, we associate a single mode relaxation time [7] $\tau(q)$
given by the asymptotic relation [3]

$$\tau^{-1}(q) = A \; q^a \; T^{5-a} \tag{1}$$

2.   The higher energy phonons which are those with wave number
$q > q_c$ and thermal conductivity $K(q > q_c)$

    Thus, the total lattice thermal conductivity will be
given by [3]

$$K = K(q > q_c) + \frac{1}{6 \; \pi^2} \; kv^2 \; \int_o^{q_c} \tau(q) \; q^2 \; dq \tag{2}$$

where v is the velocity of sound that we suppose independent of
energy and k the Boltzmann constant.

    According to Herring for longitudinal phonons in a material
like bismuth a=3. Using this value and introducing $\tau(q)$ from (1)

into (2) we find for the two arms of thicknesses W_1 and W_2 of the tuning fork, supposing $K(q > q_c)$ unaffected by size (cfr. APPENDIX)

$$\Delta K = K_1 - K_2 = \frac{kv^2}{18 \; \pi^2 \; AT^2} \; \ln \frac{W_1}{W_2} \qquad (3)$$

Thus, if the size-effect is to be ascribed to the subthermal low energy phonons we should find a T^{-2} variation of ΔK. This is what we observe in the intermediate temperature range for both samples (Fig.3).   Also, from the observed T^{-2} curves, and taking $v = 2.55 \times 10^5$ cm.sec^{-1} in the binary direction, we find an experimental value for A of 39×10^{-17} cm^3 .sec^{-1}. K^{-2} for both samples TF01 and TF02.

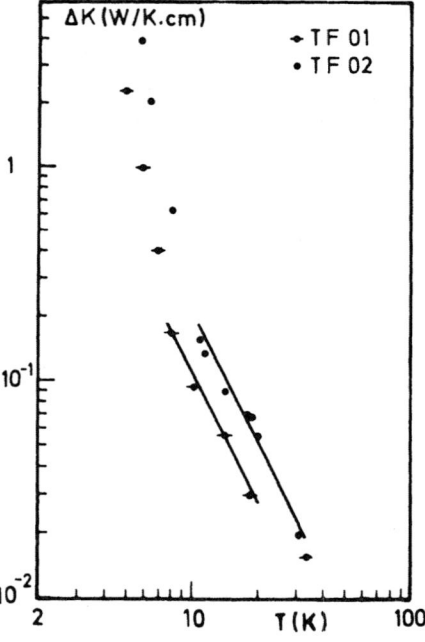

FIG.3.   Difference of thermal conductivities between the two arms of the tuning fork samples plotted versus temperature. The two straight lines represent a T^{-2} variation.

Further, the fact that the T^{-2} law is only verified in a narrow
temperature range may be attributed to the assumptions underlying
Herring's theory; i.e., that it should only hold above the Casimir
range and below the Debye temperature.

It is interesting to note that thermopower measurements on
these samples showed a size-effect exceeding 15% at 77K which
seems to persist at much higher temperature. This tends to favour
the idea of a phonon-drag contribution to the thermopower of this
material at high temperatures [8]. These data will be reported
elsewhere.

The authors are indebted to Dr. S.H. Koenig and Dr. R.D. Brown
of IBM for kindly supplying the single crystals and to P. Coopmans
for his skilful technical help.

REFERENCES

1. G.K. WHITE & S.B. WOODS, Phil. Mag. $\underline{3}$, 342 (1958)

2. J-P. ISSI & J.H. MANGEZ, Phys. Rev. $B\underline{6}$, 4429 (1972)

3. C. HERRING, Phys. Rev. $\underline{95}$, 954 (1954); $\underline{96}$, 1163 (1954)

4. T.H. GEBALLE & G.W. HULL, Conférence de Physique des basses
   températures p.460 (Institut International du Froid, Paris,1955)

5. I.Ya. KORENBLIT, M.E. KUZNETSOV & S.S. SHALYT, Sov. Phys. JETP,
   $\underline{29}$, 4 (1969)

6. See, for example, R. BERMAN & D.J. HUNTLEY, Cryogenics, $\underline{3}$, 70
   (1963)

7. P.G. KLEMENS, Proc. Roy. Soc. (London), $\underline{A208}$, 108 (1951)

8. J-P. ISSI, Proceedings of the International Conference on the
   Physics of Semimetals and Narrow Gap Semiconductors, Nice –
   Cardiff, September, 1973 (to be published)

APPENDIX

The relaxation time of subthermal phonons is supposed to be
limited by phonon-phonon scattering, which characteristic single
mode relaxation time τ_{pp} is given by (1) with $a=3$, and by boundary
scattering with a relaxation time $\tau_i = W_i / v$ (i = 1 or 2) where
we assume that the boundary scattering length equals the thickness
of one of the tuning fork arms W_i. Using Matthiessen's rule

$$\tau^{-1}(q < q_c) = \tau_{pp}^{-1} + \tau_i^{-1} \qquad (A1)$$

and since the integral in (2) becomes

$$\int_o^{q_c} \frac{q^2\, dq}{A\, q^3\, T^2 + \tau_i^{-1}} = \frac{1}{3\, A\, T^2} \ln \left| \frac{q_i^3 + q_c^3}{q_i^3} \right| \tag{A2}$$

introducing (A2) in (2) we find

$$\Delta K = \frac{k\, v^2}{18\, \pi^2\, A\, T^2} \ln \left| \frac{q_1^3 + q_c^3}{q_2^3 + q_c^3} \cdot \frac{q_2^3}{q_1^3} \right| \tag{A3}$$

In the range where q_1 and q_2 are sufficiently smaller than q_c

$$\Delta K = \frac{k\, v^2}{18\, \pi^2\, A\, T^2} \ln \frac{q_2^3}{q_1^3} \tag{A4}$$

which leads to (3).

Chapter II

Solids at High Temperatures

CONDUCTION PROPERTIES AND THERMAL EXPANSION*

P. G. Klemens

Dept. of Physics and Institute of Materials Science

Univ. of Connecticut, Storrs, Connecticut 06268

I. INTRODUCTION

In the theory of solids the thermal expansion and the conduction properties are topics which impinge on each other. A major component of the thermal expansion can be attributed to the anharmonic components of the interatomic forces. This anharmonicity gives rise also to the interchange of energy amongst the normal modes. The strength of this interaction enters explicitly into the conduction processes involving lattice waves, and some of the information needed to estimate anharmonic interaction rates is obtained from thermal expansion.

Another area of common interest is the dependence of various conduction properties on dilatation or pressure. The carriers of interest can be either electrons or lattice waves; both carriers are scattered by lattice waves, but these lattice waves are affected by dilatation. It is possible, in spite of some uncertainties in the theory, to describe at least qualitatively the effects of anharmonicity on the pressure dependence of the conduction properties.

Finally, one can combine the dependence of various conduction properties with the thermal expansion to reduce these conductivities, observed at high temperatures, to constant volume. In many cases one expects the various carrier mean free paths to vary inversely with temperature at high temperatures, provided the volume is kept constant. To compare theory and observations, it is thus necessary to reduce the observed conductivities to constant volume.

II. INTERACTIONS BETWEEN NORMAL MODES

A lattice wave can be specified by its wave-vector $\underline{q}$ and its polarization. Its displacement field is

$$\underline{u}(\underline{x}) \propto e^{i(\underline{q}\cdot\underline{x} - \omega t)} \tag{1}$$

when ω is the frequency, $\underline{x}$ a lattice site. Keeping $\underline{q}$ fixed and subjecting the solid to a uniform strain, ω will vary, so that

$$\omega = \omega_o \left(1 - \gamma_i \varepsilon_i\right) \quad and \quad \omega = \omega_o \left(1 - \gamma \Delta\right) \tag{2}$$

where ε_i is a component of strain and Δ the dilatation. This defines the anharmonic coefficients γ_i and γ. The latter, defined for each mode, is $\gamma(\underline{q})$, the Grueneisen parameter. It enters into the expression for thermal expansion

$$\frac{d\Delta}{dT} = \frac{1}{\lambda} \sum_{q,j} C_j(\underline{q})\, \gamma(\underline{q}) \tag{3}$$

where λ is the bulk modulus and $C_j(\underline{q})$ the specific heat of a normal mode of wave-vector $\underline{q}$ and polarization j.

Owing to cubic anharmonicities, energy is exchanged between triples of modes $\underline{q}$, $\underline{q}'$, $\underline{q}''$. The strength of the interaction involves the interaction Hamiltonian and the energies of the three modes. The unperturbed Hamiltonian of the structurally perfect harmonic lattice contains terms which are simply the squares of the amplitudes of individual modes. The perturbation Hamiltonian due to cubic anharmonicities has terms of the form

$$H' = \sum c_3(\underline{q}, \underline{q}', \underline{q}'')\, \alpha(\underline{q})\, \alpha(\underline{q}')\, \alpha^*(\underline{q}'') \tag{4}$$

where the a's are time-dependent displacement amplitudes of the individual modes. The rate of energy transfer between a triplet of modes – from $\underline{q}$ and $\underline{q}'$ into $\underline{q}''$ for the typical term in (4) – contains the factor $C_3{}^2$, and a knowledge of the coefficients C_3 is basic to the calculations.[1]

One can calculate $C_3(\underline{q}, \underline{q}', \underline{q}'')$ from a detailed model of the interatomic linkages, since it is related to the $(\underline{q} + \underline{q}' - \underline{q}'')$'th Fourier coefficient of the spatial distribution of the anharmonicity. The latter has the periodicity of the lattice, so that all allowed interactions satisfy

$$\underline{q} + \underline{q}' - \underline{q}'' = \underline{b} \tag{5}$$

where $\underline{b}$ is an inverse lattice vector. Interactions for which $\underline{b} = 0$ are normal processes; they are the only ones allowed in an anharmonic elastic continuum. Cases of non-zero $\underline{b}$ are Umklapp processes.

From a detailed model one could deduce the coefficients C_3, and all
the $(\underline{q})$'s as special cases when q' is longitudinal and $\bar{q}'$ small.
The dilatation then contains q' $a(\bar{q}')$, and

$$\lim_{q' \to 0} C_3 \, a(\underline{q}') \, a(\underline{q}) \, a^*(\underline{q}+\underline{q}') = 2M\omega^2 \gamma(\underline{q}) \, q' \, a(\underline{q}') \, a(\underline{q}) \, a^*(\underline{q}) \quad (6)$$

Unfortunately this detailed knowledge is not available, so
that one usually reverses the procedure, and searchs for an expres-
sion for $C_3(\underline{q},q',\underline{q}'')$ which is consistent with known values of $\gamma(\underline{q})$
through (6). This procedure is not unique. Also, we do not know
individual $\gamma(\underline{q})$'s from thermal expansion, but only weighted
averages. Finally, we know $\gamma(\underline{q})$ only for dilatational strains,
so that (6) applies only if q' is longitudinal. For transverse q'
one needs to know $\gamma_i(\underline{q})$ for shear strains, but these coefficients
cannot be inferred from macroscopic observations. The individual
$\gamma_i(\underline{q})$ do not vanish - they vanish only if q is along a symmetry
direction - but the modes can be grouped in pairs with the same
frequency but with opposite sign of γ_i, so that these anharmonicities do not lead to macroscopic changes of shape.

For phonon-phonon interactions, and for the related problem
of scattering of phonons by inhomogeneous strain fields, the interaction depends on an average value of $\gamma^2(q)$, while the thermal expansion depends on an average of $\gamma(\underline{q})$. Attenuation of ultrasonic
waves in the Akhieser limit[3] - ultrasonic waves much longer than
the mean free path of thermal phonons-depends on the average of
$(\gamma(q)-\langle\gamma\rangle)^2$, where $\langle\gamma\rangle$ is the average of $\gamma(\underline{q})$.

If one assumes C_3 to be independent of the polarization of the
participating modes and averages over polarizations, one obtains[1]

$$C_3(\underline{q},q',\underline{q}'') = \frac{2M}{\sqrt{3G}} \gamma \, \omega \, \omega' \, \omega'' / v \quad (7)$$

where v is the phonon velocity and G the number of atoms in the
crystal. In (7) it is implied that at least one of the modes is
of low frequency; when all three modes are of high frequency, (7)
cannot be trusted. For example, in the decay of an optical mode
into two acoustic modes, each of half the frequency, expression (7)
would overestimate the interaction, while an estimate of C_3 from a
model of individual linkages gives better results.[4] Use of (7)
gives fair quantitative agreement for the relaxation times of low-
frequency phonons, including the attenuation of ultrasonic waves in
lithium fluoride of high frequency in the direct-interaction
limit,[5] the relaxation time for normal three-phonon processes
deduced from the isotope effect,[6] and the scattering of phonons
by static dislocations.[7] On the other hand, there are uncertainties in the relaxation times of high-frequency phonons due to
Umklapp processes, with discrepancies between different theories[8,9]
and discrepancies in the theory of thermal resistance at high tem-

peratures from observations.[1] Some of these uncertainties arise
from the values for C_3.

III. CHANGE OF CONDUCTIVITIES WITH DILATATION

The various conduction properties vary with dilatation and
thus with pressure. Some of that variation arises from a change
in the normal modes, which scatter the carriers and thus cause resistance. The pressure dependence of the electrical resistivity ρ
has been discussed in these terms already by Mott and Jones.[10]
Compression increases the normal mode frequencies - see (2) - and
ρ varies as $1/\theta^2$, so that

$$\frac{1}{\rho} \frac{d\rho}{d\Delta} = - \frac{d}{d\Delta} \ln \theta^2 = 2\gamma \tag{8}$$

This gives a qualitative fit to the data of simple metals, but not
an exact fit, because the electronic band structure also changes
with dilatation. This has been discussed by later workers, particularly Dugdale,[11] who was concerned particularly with the elec-
tronically simple but very compressible alkali metals. Lacking a
detailed understanding of the effects of band structure on the
electrical resistivity of real metals, it is not surprising that we
do not fully understand the pressure dependence either.

The pressure dependence of the thermal conductivity by lattice
waves is similarly not well understood; furthermore, experimental
data is still scarce. Here the phonons are both carriers and also
the source of resistance. Differentiating the expression for the
intrinsic thermal conductivity[8] with respect to dilatation, using
(2), and including the dependence of γ on Δ, Mooney and Steg[12]
obtained

$$\frac{d}{d\Delta} \ln \varkappa = - \left(3\gamma + \frac{2}{3} + 2\gamma'/\gamma \right) \tag{9}$$

where $\varkappa$ is the thermal conductivity, and $\gamma' = d\gamma/d\Delta$ the second
Grueneisen coefficient.

Both (8) and (9) apply to the case of high temperature and intrinsic resistivity. If ρ or $1/\varkappa$ is due to imperfections, the
theory must be modified. This was done for strong point-defect
scattering of phonons.[12]

IV. EFFECTS OF THERMAL EXPANSION ON THE CONDUCTION PROPERTIES

At sufficiently high temperatures the mean free path of both
electrons and phonons, when it is limited by scattering with thermal
vibrations, should simply vary inversely with T, i.e.

$$1/\ell \propto \sum_q E(\underline{q}) \propto \overline{\varepsilon^2} \propto T \tag{10}$$

where $E(\underline{q})$ is the energy content of a lattice mode $\underline{q}$ and $\overline{\varepsilon^2}$ is the mean thermal strain. This assumes that the dominant interaction is the scattering of the carrier (electron or phonon) by one phonon at a time. Higher order interactions (scattering of electrons by two phonons, or scattering of phonons by two phonons, i.e. four-phonon processes) should be reduced by a factor which is also of order $\overline{\varepsilon^2}$. Now

$$\overline{\varepsilon^2} \simeq \frac{kT}{Mv^2} = \frac{T}{T_0} \tag{11}$$

where M is the atomic mass and v the velocity of shear waves, i.e. Mv^2 is equal to μa^3, where μ is the shear modulus and a^3 the atomic volume. The characteristic temperature T_0 thus defined is typically around 50,000 K. Higher order processes (scattering of carriers by two phonons) should therefore provide only a small fraction of the total resistivity even at the melting point.

The actual intrinsic electrical resistivity of even the simple metals varies faster than T. If expressed as

$$\rho(T) = aT + bT^2 \tag{12}$$

then the quadratic term is perhaps 30 to 50% of the linear term at the melting point, so that bT/a is much larger than T/T_0. This was already attributed by Mott and Jones[10] to the effects of thermal expansion and the dependence of ρ on dilatation. Let $\rho_0(T)$ be the electrical resistivity as function of temperature if the specimen volume is kept constant. The actual resistivity $\rho(T)$ becomes

$$\rho(T) = \rho_0(T)[1 + p\Delta T] \tag{13}$$

where $p = (1/\rho)\, d\rho/d\Delta$, and $\Delta(T)$ is the thermal expansion dilatation. Using (3) for $\Delta(T)$ and (8) for p, Mott and Jones obtained for (13)

$$\rho(T) = \rho_0(T)[1 + 2\gamma^2 CT/\lambda] \tag{14}$$

where C is the specific heat per unit volume. If $\rho_0(T) = aT$, as demanded by simple theory, $\rho(T)$ has the form (12) with bT/a very roughly of the observed magnitude.

One can improve over (14) by using observed values for p and $\Delta(T)$, since (3) and (8) are only rough approximations. From observed values of $\rho(T)$ one should be able to derive $\rho_0(T)$, the electrical resistivity corrected to constant volume. One would expect $\rho_0(T)$ to vary linearly with T; corrections due to two-phonon electron scattering processes should be typically only about 2% at 1000 K. There may also be a small correction due to the next term in the expansion (13), i.e. $(1/2\rho)(d^2\rho/d\Delta^2)\Delta^2(T)$, which would also appear as a T^2 term in $\rho_0(T)$.

For many metals, the interval between θ and the melting temperature T_M is not wide enough to adequately test the linearity of $\rho_0(T)$. However, gold is a very suitable case, since θ is around 170 K while $T_M = 1336$ K. Values of p at room temperature are available from high-pressure data. This data has been reviewed by Dugdale;[11] according to him p = 5.5 for gold.

Dr. R. A. Matula, of CINDAS, Purdue University, has kindly supplied me with a set of electrical resistivity values of gold, derived as a composite from various sources, and at present only of provisional significance, as well as recent values of $\Delta(T)$. We have calculated $\rho_0(T)$ from these data and equation (13). These values are shown in Table I; it can be seen that ρ_0/T is fairly constant above 300 K, and more so than values derived by Mott and Jones[10] from (14).

We have also calculated values of $\rho_0'(T)$ derived from (13) with p replaced by $(p-1/3)$. This would be more appropriate if the observed resistivity values are not corrected for the change in specimen geometry on heating, so that the temperature dependence is simply that of the specimen resistance. The values of $\rho(T)$ are a composite from various sources. Each author covered only part of the temperature range, and in the best measurements geometrical uncertainties are the greatest source of error. It is thus difficult to know to what extent the values of $\rho(T)$ in Table I incorporates changes of specimen geometry with temperature. Table I gives both $\rho_0(T)$ and $\rho_0'(T)$. The differences are small: it seems that ρ_0/T approaches a constant value a little better than ρ_0'/T.

At highest temperatures, values of ρ_0/T and ρ_0'/T increase slightly. This may be due to the formation of thermally activated defects, probably vacancies, which have been studied by Simmons and Balluffi.[13] Due to uncertainties in the present analysis, we cannot reliably deduce the magnitude or activation energy of that resistivity component.

There could also be a term proportional to Δ^2 due to the next term in the expansion, i.e. $\frac{1}{2}p'\Delta^2(T)$, where $p' = (1/\rho)d^2\rho/d\Delta^2$. We have no prior knowledge of the magnitude of p'. If one assumes p' to be comparable in magnitude to p, this correction term would be unimportant. If one regards (13) as the leading terms of an expansion in $p\Delta$, the next term, of order $\frac{1}{2}p^2\Delta^2$, would be about 20% of the term in $p\Delta$ around 1000 K. This is equivalent to assuming $p' \simeq p^2$ instead of p. With such a term one could account

TABLE I

Electrical Resistivity of Gold

T	$\Delta(T)$	$\rho(T)$	$\rho_o'(T)$	$\rho_o(T)$	ρ_o'/T	ρ_o/T
(°K)		..... $\mu\Omega$- cm			$10^{-3}\mu\Omega$–cm/K	
100	0.002	0.620	0.613	0.613	6.13	6.13
200	0.0059	1.44	1.40	1.40	7.00	7.00
300	0.0105	2.24	2.12	2.12	7.08	7.07
400	0.0145	3.07	2.85	2.84	7.13	7.10
500	0.0190	3.92	3.57	3.55	7.14	7.10
600	0.0237	4.80	4.27	4.25	7.12	7.08
700	0.0285	5.75	5.01	4.97	7.16	7.10
800	0.0335	6.70	5.71	5.66	7.14	7.08
900	0.0387	7.76	6.46	6.40	7.18	7.11
1,000	0.0442	8.90	7.24	7.16	7.24	7.16
1,100	0.0500	10.02	7.95	7.86	7.23	7.15
1,200	0.0560	11.30	8.75	8.64	7.29	7.20
1,300	0.0626	12.7	9.58	9.45	7.37	7.27

for the apparent increase of ρ_o/T at the highest temperatures, but only at the expense of the apparent constancy of ρ_o/T or ρ_o'/T at the middle temperatures.[14] A fit of the form

$$\rho(T) = \rho_o(T) \, exp(\rho\Delta) \qquad (15)$$

had in fact been suggested by Misek and Polak.[15] In my opinion there is no theoretical justification for this form.

While there is still some uncertainty about the finer details of the fit, and while it is not yet possible to draw meaningful conclusion about the defects in thermal equilibrium from electrical resistivity data alone, it is clear that the thermal expansion explains the major part of the deviation of $\rho(T)$ from linearity remarkably well, if observed values of p and $\Delta(T)$ are used to correct $\rho(T)$ to constant volume.

The thermal conductivity of dielectric crystals should vary as $1/T$ at constant volume. Ecsedy[16] has discussed at this meeting how the thermal conductivity, which generally varies more rapidly than $1/T$, can be brought into better agreement with theory through a similar correction to constant volume.

Footnotes and References

*   Supported by the U. S. Army Research Office, Durham.
(1)   See, for example, P. G. Klemens in Solid State Physics, Vol. 7, p. 1, Academic Press, New York, 1958.
(2)   R. E. Peierls, Ann. Physik 3, 1055 (1929).
(3)   A. Akhieser, J. Phys. (U.S.S.R.) 1, 277 (1939).
(4)   P. G. Klemens, Phys. Rev. 148, 845 (1966); Phys. Rev. B11, 3206 (1975).
(5)   J. de Klerk and P. G. Klemens, Phys. Rev. 147, 585 (1966).
(6)   R. Berman and J. C. F. Brock, Proc. Roy. Soc. (London) A289, 46 (1965).
(7)   M. W. Ackerman, Phys. Rev. B5, 2751 (1972).
(8)   G. Leibfried and E. Schloemann, Nachr. Ges. Wiss. Goett. Math.-Phys. Kl. 2 (4), 71 (1954).
(9)   M. Roufosse and P. G. Klemens, Phys. Rev. B7, 5379 (1973).
(10)   N. F. Mott and H. Jones "Theory of the Properties of Metals and Alloys" Oxford U. Press, London, 1936.
(11)   J. S. Dugdale, Science 134, 77 (1961); J. S. Dugdale and D. Gugan, Proc. Roy. Soc. (London) A270, 186 (1962).
(12)   D. L. Mooney and R. G. Steg, High Temperatures-High Pressures 1, 237 (1969).
(13)   R. O. Simmons and R. W. Balluffi, Phys. Rev. 125, 862 (1962).
(14)   I am indebted to Prof. H. James for drawing my attention to this term.
(15)   K. Misek and J. Polak, J. Phys. Soc. Japan 18, Suppl. II, 179 (1963).
(16)   D. J. Ecsedy, in these Conference Proceedings.

MOLECULAR DYNAMICAL CALCULATIONS OF THE THERMAL DIFFUSIVITY OF A PERFECT LATTICE*

R. A. MacDonald and D. H. Tsai

National Bureau of Standards

Washington, D.C. 20234

We have used the method of molecular dynamics to make
a detailed study of thermal diffusivity in a perfect
monatomic lattice. The interatomic potential is that
appropriate for iron. We limit the atomic motions to
two dimensions in order to shorten the computation.
We maintain one end of the lattice at a given kinetic
temperature and obtain the temperature profile in the
lattice as a function of time. The total energy added
to the system is recorded. We fit diffusive curves to
the temperature profiles and thus obtain the thermal
diffusivity of the lattice. Its value is 4×10^{-6} $m^2 sec^{-1}$
at a mean lattice temperature of 75K.

1. INTRODUCTION

We have used the method of molecular dynamics to study the
transport of thermal energy in a perfect monatomic lattice. The
complexity and immensity of such calculations make it desirable to
relate our results to physical reality in as many ways as possible.
To this end, we obtained a rough estimate of the lattice thermal

---
*This work was performed with partial support from the U.S. Energy
Research and Development Administration and from the U.S. Army
Research Office, Durham, N.C.

conductivity from an earlier calculation of heat pulse propagation
in a three-dimensional lattice[1]. This led to the present, more
detailed study of thermal diffusivity in a lattice. We have limited the atomic motions to two dimensions in order to shorten the
computation time, otherwise, this restriction is insignificant.

We shall first describe our model, then in section 3 we shall
treat the problem using conventional theory of heat transfer by
diffusion in a continuum. In section 4 we present the results of
our molecular dynamical calculations and compare them with the
continuum theory. In the final section we discuss our results.

2. MODEL

Our two-dimensional system is made up of bcc unit cells forming
a ribbon-like filament 25 units wide, 250 units long, and 1 unit
thick. This ribbon is placed in Cartesian coordinates with x denoting the width, z the length and y the thickness. Two-dimensional motion is obtained by limiting the motion in the y direction to
zero. Mirror boundary conditions are imposed on the transverse
boundaries in the $\pm$ x directions and at the heated end, z = 0. The
interatomic potential is that constructed by Chang[2] for α-iron.
Initially, this lattice is in thermal equilibrium at a temperature
T_i. Then, the first ten lattice (x-y) planes are heated quickly to
an average temperature T_h and maintained at that level for the rest
of the calculations. The amount of energy added to these ten planes
is recorded. The classical equations of motion for the lattice
atoms are solved by numerical integration, yielding the position
and velocity of each atom. The average kinetic energy, potential
energy, stress components and density of lattice planes are obtained as functions of time. These results show the transport of
energy into the lattice.

We use the rigid mirror boundary condition in the transverse
direction to prevent lateral displacement and distortion of the
filament. This was necessary because we found that when periodic
boundary conditions were used, as in earlier calculations, the
filament would drift laterally from the equilibrium position, due
to accumulation of small errors in the position and velocity of the
atoms, and perhaps also due to the procedure of numerical integration. Since this motion was not uniform along the length of the
filament, and since the undisturbed end was constrained to stay in
place, the filament became laterally distorted and strain energy
developed. Over the duration of a run (400 time units) the strain
energy was such that a two fold increase in the total energy was
observed. This strain energy is eliminated when the mirror boundary condition is applied.

3. CONTINUUM THEORY

Since the lattice is at a moderate initial temperature ($\sim 0.1\ \Theta_D$, where Θ_D = Debye temperature) the mechanism of heat transfer can be expected to be diffusive. It is therefore appropriate for us to test our results against those of the continuum theory of thermal diffusion[3]. We proceed as follows:

When a time-dependent boundary temperature $T_b(t)$ is imposed on a system initially at a uniform temperature T_i, the temperature profile is given by (see ref. 3 pp. 273-4)

$$T(z,\tau) = T_i\ \text{erf}(u) + (2/\sqrt{\pi})\int_u^\infty T_b\,(\tau - z^2/4\alpha u^2)\exp(-u^2)\,du \qquad (1)$$

where $u^2 = z^2/4\alpha(\tau - t)$, $\text{erf}(u) = (2/\sqrt{\pi})\int_o^u \exp(-t^2)\,dt$,

α is the thermal diffusivity and z is the distance in the direction of heat flow. Although we maintain the average value of the kinetic energy in the first ten planes of the lattice at a constant value, the local kinetic temperature of the tenth (boundary) plane varies considerably on an atomic scale. For the diffusive calculation we need the average temperature of the boundary plane. In the present calculation, this may be approximated by three linear segments (see Fig. 1):

$$T_b(\tau) = T_0 + C_1\tau, \qquad 0 < \tau \leqslant \tau_1$$

$$T_b(\tau) = T_1 + C_2\tau, \qquad \tau_1 < \tau \leqslant \tau_2 \qquad\qquad (2)$$

$$T_b(\tau) = T_2, \qquad\qquad \tau_2 < \tau,$$

Substitution into Eq. 1 yields the following expression for the temperature profile:

$$T(z,\tau) = T_i + [T_0 - T_i + C_1\tau]F(u_o) - [T_0 - T_i + (C_1 - C_2)\tau]F(u_1)$$

$$+ [T_2 - T_1 - C_2\tau]F(u_2)$$

$$+ 2\,C_1\tau[u_o F(u_o^2) - u_o \exp(-u_o^2)/\sqrt{\pi}]$$

$$- 2(C_1 - C_2)(\tau - \tau_1)[u_1^2 F(u_1) - u_1 \exp(-u_1^2)/\sqrt{\pi}]$$

$$- 2C_2(\tau - \tau_2)[u_2^2 F(u_2) - u_2\exp(-u_2^2)/\sqrt{\pi}]. \qquad (3)$$

where $F(u) = 1 - \text{erf}(u)$, $u_o = z/2\sqrt{\alpha\tau}$,
$u_1 = z/2\sqrt{\alpha(\tau - \tau_1)}$, $u_2 = z/2\sqrt{\alpha(\tau - \tau_2)}$.

$T(z,\tau)$ can be fitted to our computed temperature profiles at one value of τ. Since the diffusive profile is given in terms of $u(=z/2\sqrt{\alpha\tau})$, the distance scale is arbitrary when α is unknown. This arbitrariness is removed by requiring that the area under the diffusive curve be equal to the total energy added to the system up to the time τ. With the value of α thus obtained, the diffusive curves at other times are obtained by appropriately scaling the distance. The area under these other diffusive curves must also be in agreement with the total energy added to the system at their respective times. If this is so, it can be concluded that the heat transfer in our system is indeed diffusive.

4. RESULTS

In Fig. 1, we show the boundary temperature as a function of time, $T_b(\tau)$. Even though we have averaged the kinetic temperature of planes 6 to 15 to obtain $T_b(\tau)$, there is still considerable fluctuation in the boundary temperature and we approximate it by the dashed curve C, as given in Eq. 2. The constants in these equations have the following values:

$T_o = 59$ K, $T_1 = 102$ K, $T_2 = 115$ K,

$\tau_1 = 20$, $\tau_2 = 208$, $C_1 = 2.214$ K and $C_2 = 0.0628$ K.

The time τ is measured in units of lattice plane spacing d divided by longitudinal sound velocity (equal to 0.264×10^{-13}s). The initial temperature of the lattice, T_i, is 44 K. These values are used in Eq. 3 to calculate the diffusive temperature profiles.

In Fig. 2 we show the kinetic temperature, density and longitudinal stress profiles in the lattice at time $\tau = 300$. Each point is the average of 10 neighboring planes. The kinetic temperature

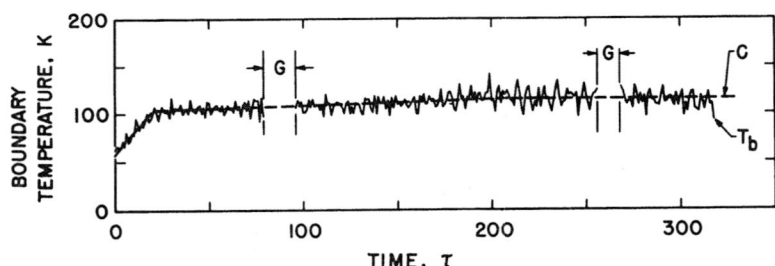

Fig.1. Boundary temperature (Kelvin) as a function of time τ (unit of $\tau = 0.264 \times 10^{-13}$s). T_b is the average kinetic temperature of planes 6 to 15. The boundary is at plane 10. C labels the approximate value of boundary temperature used in the continuum theory. G denotes a gap in computer output.

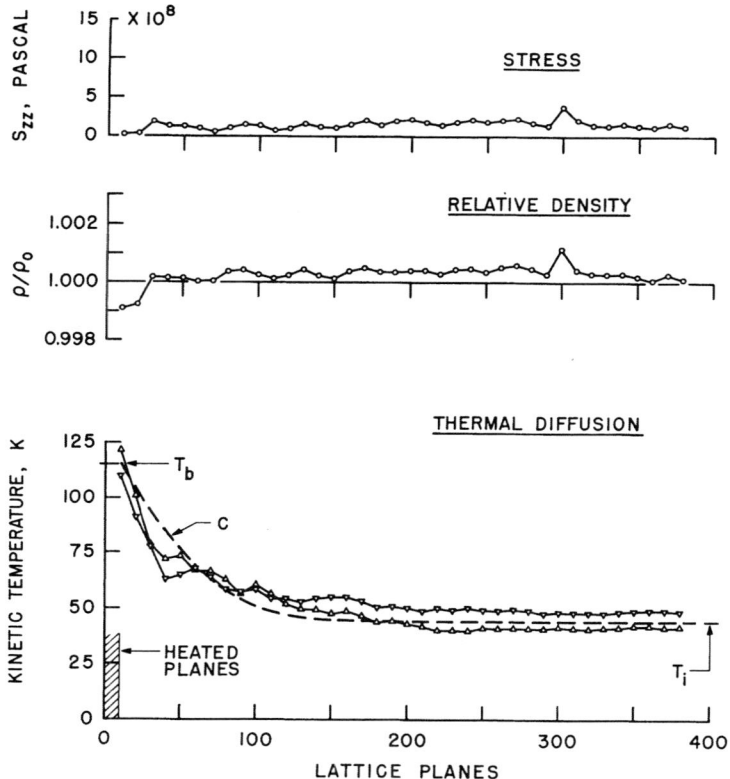

Fig.2. Kinetic temperature (Kelvin), relative density and longitudinal stress (Pascal) versus lattice plane number. Each point is the average over 10 neighboring planes. For the density ρ/ρ_0 and stress, S_{zz}, $\tau = 300$; for the kinetic temperature, $\tau = 298$ ($\triangle$) and $\tau = 303$ ($\triangledown$). Continuum theory diffusive profile, C, is at $\tau = 300$ (---). $T_i = 44$ K is the uniform initial temperature of the lattice.

of a plane is defined as $m\langle V^2\rangle/2k_B$ where m is the mass of an atom, V its velocity, $\langle \cdots \rangle$ indicates the average over all atoms in the plane, and k_B is Boltzmann's constant. We plot two temperature profiles at slightly different times, $\tau = 298$ and $\tau = 303$, to show how much local fluctuation there is in the temperature. The diffusive curve, C, is calculated from Eq. 3 at $\tau = 300$. The prominent pulses in the stress and density profiles near lattice plane 300 are due to the stress and strain in the heated planes (and their image across the mirror) generated during the initial rapid heating. These pulses propagate with the longitudinal velocity of sound, as may be determined from their trajectories in a series of profiles. Although there is a slight drop in pressure near the heated

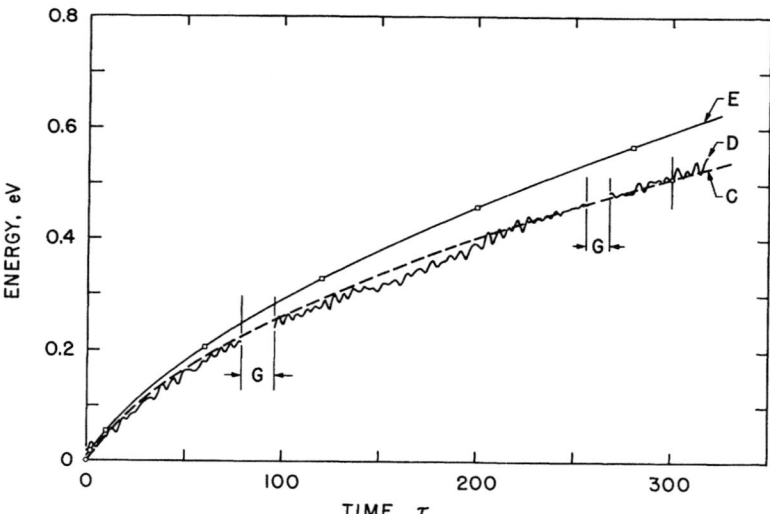

Fig.3. Energy in lattice (eV) as a function of time τ. D is the
        energy added to the heated planes, per filament of unit cell
        cross section. G denotes a gap in computer output. C is
        energy added to the filament as calculated by continuum
        theory of diffusion with $\alpha = 4\times10^{-6}$ m^2 s^{-1}. E is total
        energy in filament minus its initial energy content.

boundary at this particular time (τ = 300), our data show that both
the time average and the spatial average of the stress have a
constant value of 1.84×10^8 Pascal (1.84 kilobar) behind the initial
pulse. This indicates that, as expected, the stresses equilibrate
with the velocity of sound. The density profile shows that the
density near the heated end is definitely lower than the average
density. This is clearly due to the higher temperature at the
heated end and is, in our view, responsible for the observed fall
of the computed temperature profile below the diffusive curve C
near the heated end of the lattice. These results therefore show
quite clearly the thermoelastic coupling in the lattice, a feature
that is usually ignored in the continuum theory of thermal diffusion.

     In Fig. 3, curve D is the energy added to the lattice at the
heated end per filament of unit cell cross-section as a function of
time. We calculate the diffusivity of the lattice from the diffusive curve C at τ = 300 in Fig. 2, by requiring that the area under
that curve be equal to the value of curve D at τ = 300 in Fig. 3,
as described in section 3. We obtain the following diffusivity,

$$\alpha = 4.0\times10^{-6} \ m^2 \ s^{-1},$$

or conductivity

$$\kappa = 9.4 \ \mathrm{W \ m^{-1} \ K^{-1}}$$

where we have used the relation

$$\kappa = \alpha \rho C_p = \alpha k_B / 2d^3.$$

We use this value of α to obtain diffusive curves at other times from Eq. 3. The areas under these curves are plotted as curve C in Fig. 3, in good agreement with the computed curve D.

The curve E shows the total energy that is actually in a filament of the lattice minus the energy that was in the lattice at uniform temperature T_i before heating. The difference between curves D and E shows the lack of energy conservation in the system due to truncation error in the numerical procedure. By the end of our run, this error amounts to about 3% of the total energy of the system.

5. DISCUSSION

The present calculation is an outgrowth of our earlier study of the propagation of a heat pulse in a lattice. One of our major objectives here is to seek additional confirmation, beyond what has been established in previous calculations, that our model is indeed realistic. Toward this end, we are able to show that the heat flow in our model is diffusive, as we expected. There is no experimental value of the lattice thermal conductivity of pure crystalline iron with which to compare our value of κ, but for the alloy Fe (99.5%) Ni (0.5%) there is a recent estimate viz., $\kappa = 28.2 \ \mathrm{W \ m^{-1} \ K^{-1}}$ at 75 K[4]. Considering the fact that the effective interatomic potential for iron is not known at all accurately, we feel that the agreement obtained here is really quite satisfactory. In addition, our results provide a wealth of detail: for example, we see that there is thermoelastic coupling in the process of heating, and there is energy non-conservation due to the unexpected strain energy in the lattice caused by the cumulative errors in the numerical procedure. In this connection, we are able to show that when mirror boundary conditions are imposed the residual numerical error is indeed small, a few percent. There is also evidence of second sound, albeit heavily damped, propagating at this moderately high temperature. However, we shall not pursue this intriguing problem here. Further tests of the model may be devised. For example, it would be useful to investigate the effect of temperature on the lattice thermal conductivity. If the result should also prove to be reasonable, it would then be of interest to investigate thermal conductivity under conditions not easily accessible by present-day theory or experiment, e.g., under conditions of extreme pressure. Our experience indicates, however, that the calculation of thermal diffusivity by the method of molecular dynamics is so lengthy as to be too costly and too inefficient. If the transient nature of the heat flow is not under study, then we believe that a model for steady heat flow, from which the thermal conductivity may be obtained directly[5], would be considerably more economical and more convenient to investigate than the diffusive model employed here.

REFERENCES

1. D.H. Tsai and R.A. MacDonald, Solid State Comm. $\underline{14}$, 1269 (1974).
2. R. Chang, Phil. Mag. $\underline{16}$, 1021 (1968).
3. P. J. Schneider, "Conduction Heat Transfer," (Addison-Wesley Publishing Company, Inc., Cambridge, 1955) pp. 272-6.
4. C.Y. Ho, M.W. Ackerman, K.Y. Wu, S.G. Oh, and T.N. Havill, "Thermal Conductivity of 10 selected binary alloys," CINDAS, (Purdue University, 1975) T.P.R.C. Report #30.
5. D.N. Payton, M. Rich and W.M. Visscher, Phys. Rev. $\underline{160}$, 706 (1967).

LATTICE THERMAL CONDUCTIVITY OF PbTe: INFLUENCE OF A DISPERSION ANOMALY

J. F. Goff and Bland Houston

U. S. Naval Surface Weapons Center, White Oak

Silver Spring, Maryland 20910

and

P. G. Klemens

Department of Physics and Institute of Materials Science

University of Connecticut

Storrs, Connecticut 06268

While each of the measurements of three different N-type PbTe samples that have been reported by Shalyt and co-workers appear to be understandable in terms of conventional theories of thermal conductivity, they seem to be inconsistent with respect to each other. We suggest that this apparent inconsistency is the result of the unusual dispersion found in the normal modes of PbTe and adduce qualitative arguments to explain it. A model has been constructed which incorporates the necessary modifications of the usual Debye sphere; there are conducting hemispheres at the six (100) points of the Debye sphere which have an activation temperature of approximately 44 K. Preliminary calculations indicate that it will be possible to explain the data once certain questions concerning relaxation times are resolved.

I.   INTRODUCTION

Shalyt, Muzhdaba, and Galetskaya[1] have measured the thermal
conductivity of three N-type samples of PbTe, obtained the electronic component by studying the magneto-resistance, and thus
deduced the lattice thermal conductivity K_g for the temperature
range between 20 K and 100 K. The electronic Lorenz No., which
can be obtained from these measurements, is somewhat less than the
Sommerfeld value L_0 as would be expected but extrapolates to L_0 at
14 K. Thus, the data shown in Fig. 1 represent their values of K_g
for temperatures greater than 20 K and values derived from the total
thermal conductivity by use of L_0 or interpolated values at lower
temperatures.

At the lowest temperatures K_g varies almost inversely with
carrier concentrations - as is shown in Fig. 2. Since the carriers
in N-type PbTe arise from Te vacancies, this variation with carrier
concentration indicates that the thermal resistivity arises either
from an electron-phonon interaction or from the scattering of phonons
by some defect which correlates with the carrier concentration or
some combination of these.

At temperatures above the thermal conductivity maximum, K_g
appears to be influenced by strong point defect scattering. Sample
#8, which has the largest carrier concentration ($N = 1.3 \times 10^{19}$ cm-3),
exhibits a maximum that is rounded or blunted. This rounded shape
is often found to be due to strong point defects. However sample #7,
which is only slightly more pure ($N = 5.0 \times 10^{18}$ cm-3), displays the
concavity which is usually associated with phonon-phonon Umklapp
processes at low temperatures. Finally #1, the purest sample
($N = 7 \times 10^{17}$ cm-3) and the one with the highest thermal conductivity,
fails to show this concavity; K_g varies as $1/T$ above about 20 K as
is usually observed for phonon-phonon interactions at intermediate
and high temperatures. Thus, while each curve by itself can be interpreted in terms of conventional theory, the curves in combination are
not consistent. One usually finds that the purest sample shows the
concavity associated with the exponential dependence of Umklapp
processes, while the addition of point defects not only lowers the
curve but obliterates the concavity and progressively broadens the
maximum. It is the purpose of this paper to explain this behavior -
at least qualitatively - in terms of the unusual phonon dispersion
of PbTe.

II.   THE DISPERSION

The dispersion curves for PbTe have been measured at room
temperature by Cochran and co-workers.[2] Their data (Fig. 1, ref. 2),
presented in four panels, is in an extended zone format. These
panels, reading from left to right represent the following Brillouin

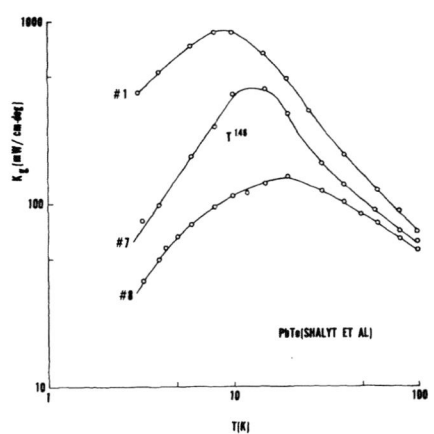

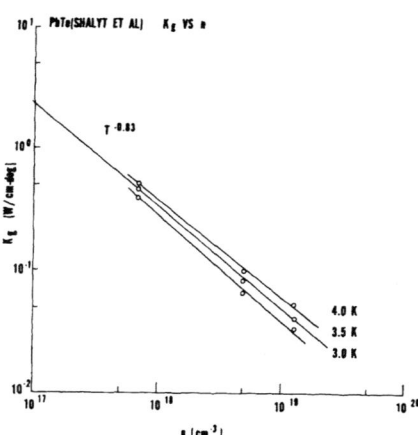

Figure 1. The lattice thermal conductivity of PbTe, which was measured by Shalyt and co-workers.

Figure 2. The lattice thermal conductivity versus carrier concentration.

zone directions (see Fig. 45, ref. 3): $\Gamma \rightarrow X$, $X \rightarrow U \rightarrow \Gamma$, $\Gamma \rightarrow L$, $L \rightarrow K$. The acoustic branches show little dispersion in the (111) direction but extreme dispersion in the (100) direction. In that direction ν (longitudinal branch) increases linearly with q in the normal manner, reaches a maximum of approximately 2.0×10^{12} Hz at $q_m/2$, and then decreases more or less linearly to a value of 1.0×10^{12} Hz at the zone boundary q_m.

Thus the dispersion curve has a negative group velocity for a large group of modes; but since the thermal conductivity for a cubic crystal is given by

$$K_g = \frac{1}{3} \sum_{q,j} S(q,j) v_g^2(q,j) \tau(q,j) \tag{1}$$

where $S(q,j)$ is the heat capacity per unit volume for mode q and polarization j, v_g is the group velocity, and τ is the relaxation time, a section of the dispersion curve with a negative group velocity will still contribute additively to the conductivity. However, the modes of this group do not act in the usual manner; the minimum frequency ν_m (q_m) $\sim 1 \times 10^{12}$ Hz causes them to contribute to the conduction with an activation energy that corresponds to a temperature of about 44 K. The corresponding contribution to the conductivity increases exponentially below that temperature. Therefore this term may well reduce or cancel the exponential decrease in conductivity that would be expected in curve #1 (Fig. 1) in the same temperature range as a result of Umklapp processes.

More importantly, we want to explain the curves of the specimens with higher carrier concentrations. We note that the ν_m are located at the X points of the Brillouin zone, which are the midpoints of the (100) faces. The q vector connecting two such points is exactly the same as the intervally separation of the electron surfaces at the L points of the hexagonal faces of the zone. Thus if there are carriers, intervally scattering could well involve precisely those phonons with the negative group velocity that are contributing exponentially to the thermal conductivity. If such scattering would quench this exponentially increasing conductivity in sample #7, then one would expect the remaining exponentially decreasing Umklapp-limited conduction to be revealed. Further addition of carriers in sample #8 would eliminate the concavity in the usual manner as a result of the increased point defect scattering.

III. THE MODEL

We have made two approximations in constructing a model based upon these dispersion curves. Firstly we ignore the optical branch at the Γ point. This branch has an activation energy of approximately the same value at room temperature as does the acoustic branch at the X-point. However it is temperature dependent[4] and so has been temporarily neglected to simplify the problem. Secondly it can be seen in Fig. 1 of reference 2 that the polarization of any branch is maintained in the extended zone so that if one starts at Γ on a branch of one polarization then he returns to Γ on a branch of the opposite polarization. The branches cross on the X face. We ignore this crossing. Finally, since the thermal conductivity integral is weighted by the square of the group velocity v_g, we neglect the transverse branch in the (100) direction when $\bar{v}_g$ becomes approximately zero.

The model shown in Fig. 3 approximates the Brillouin zone by a sphere. This sphere has three distinct sections. The interior sphere ($q \leq q_m/2$) is assumed to be dispersionless and isotropic with one acoustic and two transverse branches. At the X-point the sphere is pushed in by Debye hemispheres with an activation temperature of 44.2 K, and a maximum hemisphere radius of $q_m/2$. The longitudinal hemispheres are assumed to conduct; the transverse ones do not. The third part of the model is the remnant sphere which is the portion of the sphere for $q_m/2 < q \leq q_m$ left after the hemispheres are subtracted. The volumes of these sections in terms of the sphere volume V_0 are given in Table I along with other parameters of the model.

Table I. The parameters of the modified Debye sphere. Note that the "ω" constants must be obtained from "Θ" with corresponding subscripts by a conversion of units.

Sphere	Branch	Volume (V_o)	Θ_U (K)	Θ_L (K)	Θ_o (K)	$v_g \times 10^{-5}$ (cm sec^{-1})	$v_p(\omega)/v_g$	$(2\pi^2 v_g^3) \times G(\omega)$ (sec^{-2})
Inner	L	1/8	97.4	0	-	2.95	1	ω^2
	T	1/4	47.6	0	-	1.34	1	$2\omega^2$
Hemi	L	3/8	97.4	44.2	-	1.32	1	$3(\omega-\omega_L)^2$
	T	3/4	-	-	-	0	-	-
Remnant	L	1/2	135.7	97.4	27.8	1.47	$\omega/(\omega-\omega_o)$	$(\omega-\omega_o)^2 F_L(\omega)$
	T	1	75.6	47.6	36.5	6.10	$\omega/(\omega-\omega_o)$	$2(\omega-\omega_o)^2 F_T(\omega)$

Each branch of the model will be evaluated by an integral of the form

$$K_g = (1/3) \int_{\omega_L}^{\omega_U} v_g^2 \, \tau(\omega) \, S(\omega) \, G(\omega) \, d\omega$$

where v_g is the group velocity, ω is the angular frequency, $\tau(\omega)$ is the combined relaxation time, $G(\omega)$ is the density-of-states per unit volume per unit angular frequency, and

$$S(\omega) = k(\hbar\omega/kT)^2 \exp(\hbar\omega/kT)/[\exp(\hbar\omega/kT) - 1]^2$$

is the heat capacity per unit mode. The value of v_g, ω_U, and ω_L and the function $G(\omega)$ can be obtained from Table I for each branch. ω_U and ω_L are expressed as temperatures θ_U and θ_L.

The interior sphere is Debye-like; that is, there is no dispersion in either the acoustic or transverse branches. The Debye velocity v calculated from the dispersion curves is 1.64×10^5 cm/sec as compared with 1.78×10^5 cm/sec calculated from the Debye temperature of 130 K. The density-of-states is given in Table I.

The conducting hemispheres are also Debye-like but with an activation energy. The phase and group velocities are equal and calculated with respect to this activation energy at the X-point. Since there are six such hemispheres they are calculated as three spheres. The density-of-states is given in Table I.

The remnant spheres are the most complicated part of the model because they contain dispersive portions of the phonon modes and have sections deleted by the hemispheres. The dispersion is approximated by assuming that the branch has a constant group velocity between the radius of the interior sphere and the exterior. The phase velocity is given in Table I where ω_0 is the angular frequency of the intercept of the extrapolation of the group velocity to the Γ point and is also in Table I expressed as the temperature θ_0. Because the states within the hemispheres must be subtracted out, the density-of-states is reduced by the fraction

$$F_B(\omega) = 3 \left\{ \frac{3 + 4[(\omega - \omega_0)/(\omega_U - \omega_0)]^2}{8[(\omega - \omega_0)/(\omega_U - \omega_0)]} \right\} - 2.$$

The constants are determined by the branch B.

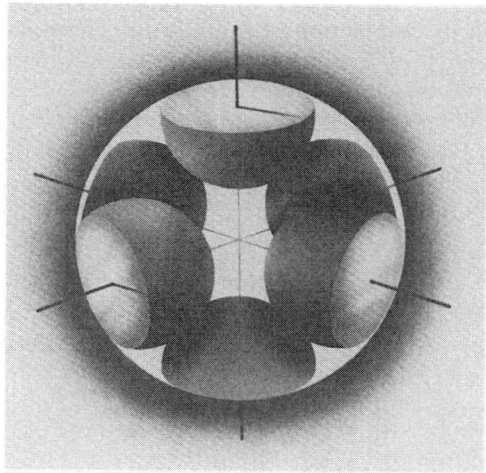

Figure 3. The model of the Debye
sphere for PbTe as modified by the
dispersion of the normal modes.

IV. PRELIMINARY CONCLUSIONS

There remain questions about several relaxation times before
meaningful calculations can be carried out. The first of these
concerns the effect of the actual Fermi surface and normal mode
system on the electron-phonon interaction. In materials with less
than about 1/4 of an electron per atom, one expects[5] this interaction
to be described by a Ziman type relaxation time.[6] However, this
interaction allows K_g to vary faster than T^3 at low temperatures
while the present data appears to vary at $T^{1.5}$, a dependence that
is less than that found for even a normal electron-phonon interaction. Such a temperature dependence has been seen in very impure
P-type Ge[7,8,9] and so could be a general effect of an unknown
mechanism. At the present time we are considering the possibility
that it results simply from the real modal and electronic structure
of the material.

The second question concerns the effect of the anisotropy of
the model in Fig. 3 upon the phonon-phonon relaxation times. Clearly,
the different sections of the model will differ in the form and
strength of the phonon-phonon interaction. We have begun to work
on this problem. Finally, it would be helpful to have a formulation
of the electron-phonon intervally scattering.

Preliminary calculations in which these considerations have been
ignored indicate that it will be possible to explain the data in a

consistent manner. The inner sphere contributes greatly to the thermal conduction up to temperatures somewhat above the thermal conductivity maximum. At the higher temperatures the contributions of the remnant spheres and the hemispheres depend greatly on the strength of their phonon-phonon relaxation times. It appears that it is possible to explain the data quantitatively by changing the relative contributions of these two sections of the model.

References

1. S. S. Shalyt, V. M. Muzhdaba, and A. D. Galetskaya; Sov. Phys. - Sol. St. 10, 1018 (1968).

2. W. Cochran, R. A. Cowley, G. Dolling, and M. M. Elcombe; Proc. Roy. Soc. A. 293, 433 (1966).

3. H. Jones, "The Theory of Brillouin Zones and Electronic States in Crystals," North Holland-Interscience (New York, 1962).

4. H. A. Alperin, S. J. Pickart, J. J. Rhyne, and V. J. Minkiewicz; Phys. Letters 40A, 295 (1972).

5. A. H. Wilson, The Theory of Metals, University Press (Cambridge, 1954), p. 264, section 9.351.

6. J. M. Ziman, Phil. Mag. 1, 191 (1956). See also "Corregendum" Phil. Mag. 2.

7. J. F. Goff and N. Pearlman, Unpublished data.

8. J. A. Carruthers, T. H. Geballe, H. M. Rosenberg, and J. M. Ziman; Proc. Roy. Soc. A, 238, 502 (1957).

9. J. A. Carruthers, J. F. Cochran, and K. Mendelssohn; Cryogenics 2, 1 (1962).

THERMAL TRANSPORT IN REFRACTORY CARBIDES

R. E. Taylor and E. K. Storms*

School of Mechanical Engineering
Purdue University
West Lafayette, Indiana 47906

Thermal energy transport in zirconium carbide samples with various degrees of carbon vacancies was investigated. Thermal diffusivities and electrical resistivities were measured from 100 to 800 K and some measurements were extended to 2000 K. The results confirmed a remarkable increase in the electron and phonon contributions to the thermal conductivity as the carbon vacancy concentration was decreased near the stoichiometric composition. The temperature dependency of the thermal conductivity and its magnitude could be explained in terms of a substantial residual resistivity and a Wiedemann-Franz-Lorenz ratio which increased with increasing temperature and was somewhat dependent upon the carbon vacancy concentration.

INTRODUCTION

Energy transport studies in refractory carbides began in the early 1950's[1,2]. Contradictory results concerning both the magnitude of the thermal conductivities and the sign of the temperature coefficient were reported[3]. At the present time, the general behavior of the conductivity can now be considered to be established and the peculiarities of the observed thermal conductivity can be qualitatively explained as the result of the characteristics of the carbides; namely, high point defect concentration, few conduction electrons, high density of states and polar modes. In particular, the importance of carbon vacancies, postulated by Méndez-Peñalosa and Taylor[4] was dramatically shown by the work of Storms and Wagner[5], which revealed a sharp increase in thermal conductivity as x was decreased from 0.99 to 0.95 in ZrC_x. The present work is an attempt to

---

* University of California, Los Alamos Scientific Laboratory, Los Alamos, New Mexico 87544.

further the quantitative understanding of the transport mechanism in refractory carbides.

EXPERIMENTAL RESULTS

Sample Characterization

Samples of ZrC_x with x varying from 0.682 to 0.976 were prepared by Dr. E. K. Storms. The samples were fabricated as 1.27 cm diameter x 0.25 cm thick discs by hot pressing powdered carbide under pressure of 100 MN/cm^2 at 3000 K using a graphite die. The billets were ground to the desired diameter, sliced into discs using a wire saw, and ground to the desired thicknesses. They were then purified by vacuum heat treatment. The samples used in the studies are characterized in Table 1.

Thermal Diffusivity Results

The diffusivity of the zirconium carbide samples were measured from 150 to 1000 K and in several cases were measured from 100 to 2000 K. Smooth curves for all the samples are included in Figure 1. In general, the reproducibility of the diffusivity results was within ± 2% of the smooth curves and the smooth curves are believed accurate within the same percentage. The results of other investigators are also shown on Figure 1.

The present data for the $ZrC_{0.682}$, $ZrC_{0.870}$, and $ZrC_{0.897}$ samples agree with the general temperature dependencies reported by Morrison and Sturgess[6], Chafik[7], Wilkes[8], and Fridlender and Neshpor[9]. However, the temperature dependencies of the $ZrC_{0.976}$ and $ZrC_{0.965}$ samples above 800 K are negative rather than positive, as noted for the other samples. Also, the present results differ with previous results in

Table 1. Sample Characterization

C/Zr	Oxygen ppm	a_0 Å	Measured Density g/cm^3	% Porosity	Theoretical Density g/cm^3
0.965	1000	4.6992	6.03	8.6	6.60
1.002*	525	4.6988	4.99	24.4	6.60
0.897	148	4.7014	5.71	12.4	6.52
0.870	640	4.7015	5.23	19.5	6.50
0.682	842	4.6972	6.23	2.2	6.37

* $ZrC_{0.976}$ + free carbon.

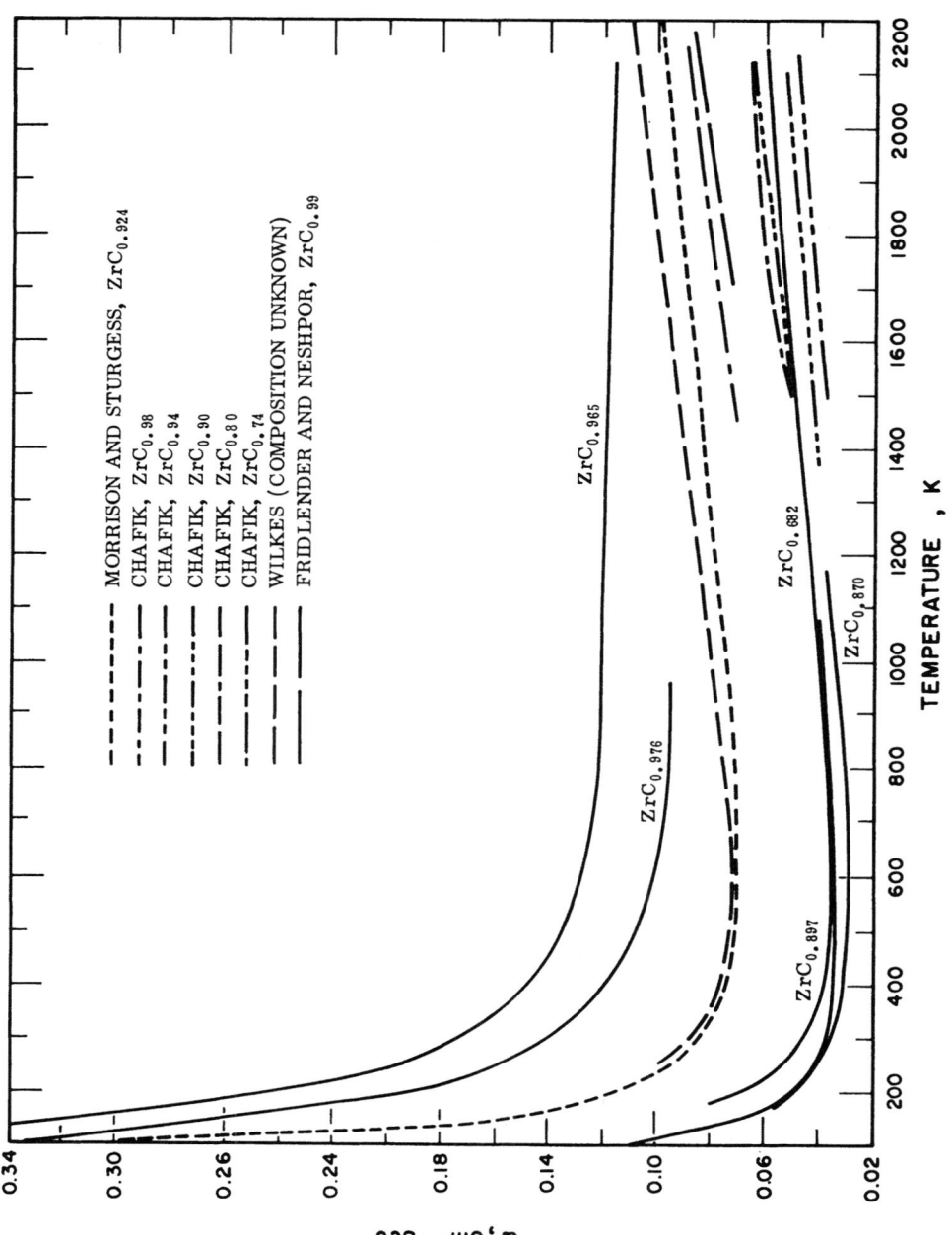

Figure 1. Thermal Diffusivity of ZrC

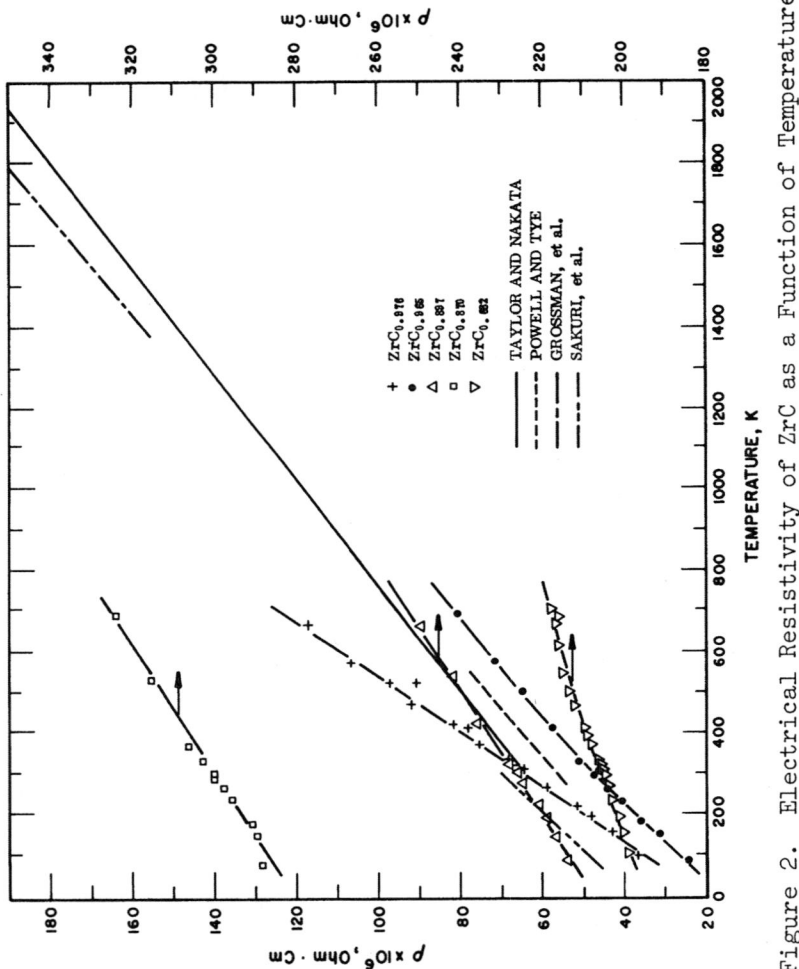

Figure 2. Electrical Resistivity of ZrC as a Function of Temperature

the dependency of the magnitude upon the carbon content. The present results are substantiated by diffusivity measurements[5] made on the same material at Los Alamos Scientific Laboratory.

Electrical Resistivity Results

Electrical resistivities of the diffusivity samples were made using a previously described apparatus[10]. The electrical resistivity results as a function of temperature are shown in Figure 2. The resistivity values were generally reproducible to ± 2 $\mu\Omega$ cm and the curves representing the data are believed accurate to about half this uncertainty. The data for the $ZrC_{0.965}$, $ZrC_{0.897}$, and $ZrC_{0.870}$ are reasonably parallel and follow the temperature dependency reported by Taylor and Nakata[11], Powell and Tye[12], Grossman[13], and Sakuri, et al. [14]. However, the data for $ZrC_{0.682}$ and $ZrC_{0.976}$ do not follow the general trend. The temperature dependency of the resistivity of $ZrC_{0.682}$ is much less and the temperature dependency of the resistivity of $ZrC_{0.976}$ is much greater than this general trend. Since the $ZrC_{0.976}$ sample also contains free carbon and is very porous, its behavior may not be too surprising. However, the results for the $ZrC_{0.682}$ sample indicate that Matthiessen's rule is not obeyed for this degree of carbon vacancies.

The electrical resistivities at 300 K of ZrC as a function of carbon-to-zirconium atom ratio are given in Figure 3. These resistivity values have been corrected to zero porosity using two different fits; namely, $(1-1.2P)$ and $(1-0.3P-5.5P^2)$ where P is the porosity[5]. The present data exhibit a maximum in ρ about $ZrC_{0.82}$. Also shown in Figure 3 are data of Taylor and Nakata[11] and Grossman[13] (both corrected for porosity). The data of Neshpor, et al. [15], also shown in Figure 3, are generally much lower and do not exhibit a maximum. Not shown in Figure 3 are the data of Fridlender and Neshpor[9] and Avgustinik, et al. [16] which generally are much higher than the present data. No attempt has been made to include all the resistivity data. Unfortunately, much of the data in the literature were not very useful as composition or density information was not included. Also, the data in the literature, even from the same group of authors, are often contradictory. It should be noted that the data of Taylor and Nakata[11] and Grossman[13] may not be in serious disagreement with the present results when one considers the uncertainty in the determinations of C/Zr (not including free carbon). Wu and Taylor results[17] for TiC + VC are very similar to Figure 3.

Thermal Conductivity Results

Thermal conductivity values (λ) were computed from thermal diffusivity (α), specific heat (C_p), and density (d) values using $\lambda = \alpha\, C_p\, d$. The value for bulk densities (Table 1) and for thermal diffusivities were corrected for expansion.

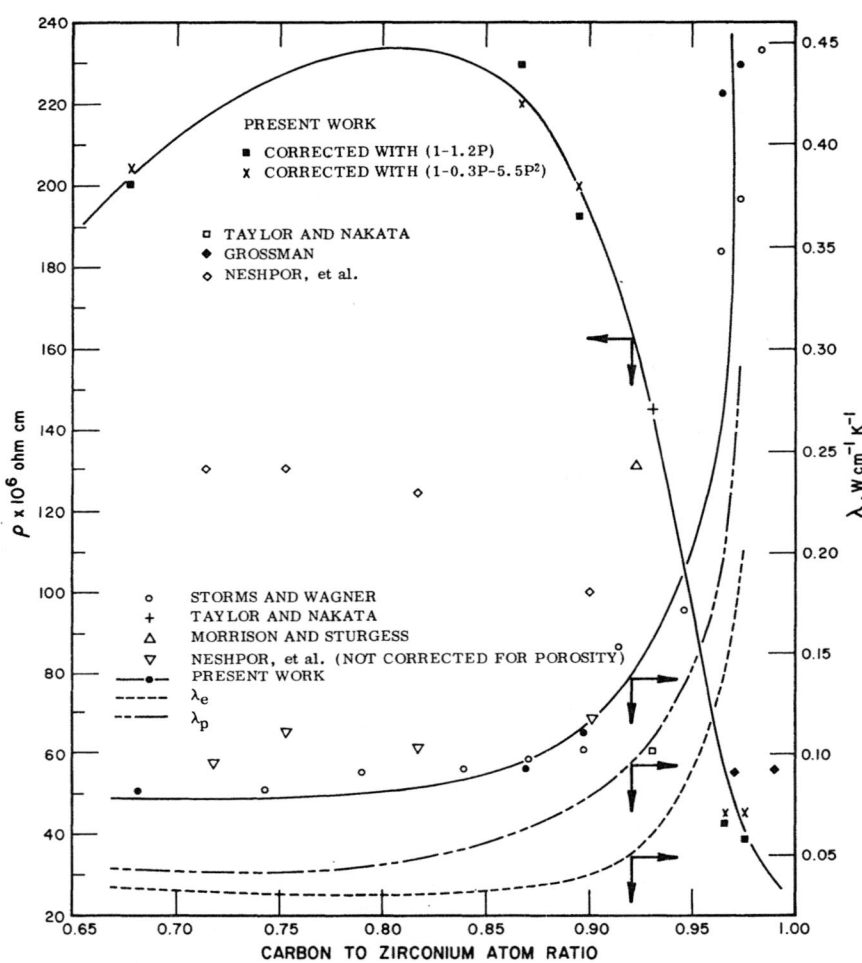

Figure 3. Electrical Resistivity and Thermal Conductivity of ZrC
          as a Function of Carbon-to-Zirconium Ratio

Specific heat values for $ZrC_{0.95}$ are based on the results of Westrum and Feick[18] and Mezaki, et al. [19]. The values were corrected for carbon content using the Neumann-Kopp relation.

Thermal conductivity values at different temperatures for the $ZrC_{0.976}$ and $ZrC_{0.965}$ samples are shown in Figure 4. Also shown in Figure 4 are the electron and phonon components of the total conductivity calculated using the assumption that L equals the classical value L_0 in the relation $\lambda_e = L_0 T/\rho$ where λ_e is the thermal conductivity due to electrons, ρ is the electrical resistivity, and T is the absolute temperature. Based on these results, a more realistic assumption is presented under Discussion and Conclusions.

The thermal conductivity results for $ZrC_{0.897}$ and $ZrC_{0.682}$ are shown in Figure 5 along with calculated electronic and phonon components based on L equal to L_0. The calculated ratio of the electronic component for the $ZrC_{0.682}$ sample slightly exceeds the value for the total conductivity above 1400 K. This may indicate that the total conductivity values are several percent high, possibly due to the uncertainty in the specific heat values for this composition. The latter values have a larger uncertainty than those for the other compositions because the correction factor for carbon vacancies is much greater.

The present conductivity results (corrected to 100% dense materials) are compared over a wide temperature range to each other in Figure 6. From this figure, it can be seen that the present results show that increasing the carbon-to-metal ratio increases the thermal conductivity. Also shown in Figure 6 are some of the data reported in the literature. These results have not been corrected to 100% density. The data of Radosevich and Williams[20] are the only data below 100 K available and their values are probably greater than the present results for about the same non-stoichiometry at 100 K. The results of Taylor and Nakata[11], Grossman[13], Paderno, et al. [21], Morrison and Sturgess[6], and others not shown on Figure 6, show that conductivity increases with increasing temperature above 1000 K. However, from the present results, it would appear that the non-stoichiometry of the Grossman[13], Paderno, et al. [21], and other data reported in the literature are understated, while the non-stoichiometry of the Taylor and Nakata[11] sample is overstated. There is a maximum in conductivity curves between 50 and 300 K, depending on the composition. Thus, in the temperature range 100 to 700 K, the temperature coefficient of the conductivity may be either positive or negative.

The thermal conductivity values at 300 K (corrected for porosity) are included in Figure 3 along with the data of Storms and Wagner[5], Taylor and Nakata[11], and Neshpor, et al. [15]. The results of Storms and Wagner[5] showing the remarkable increase in conductivity as the carbon to zirconium is increased above 0.93 are confirmed by the present study. When the Wu and Taylor data[17] for TiC and TiC + VC are plotted versus

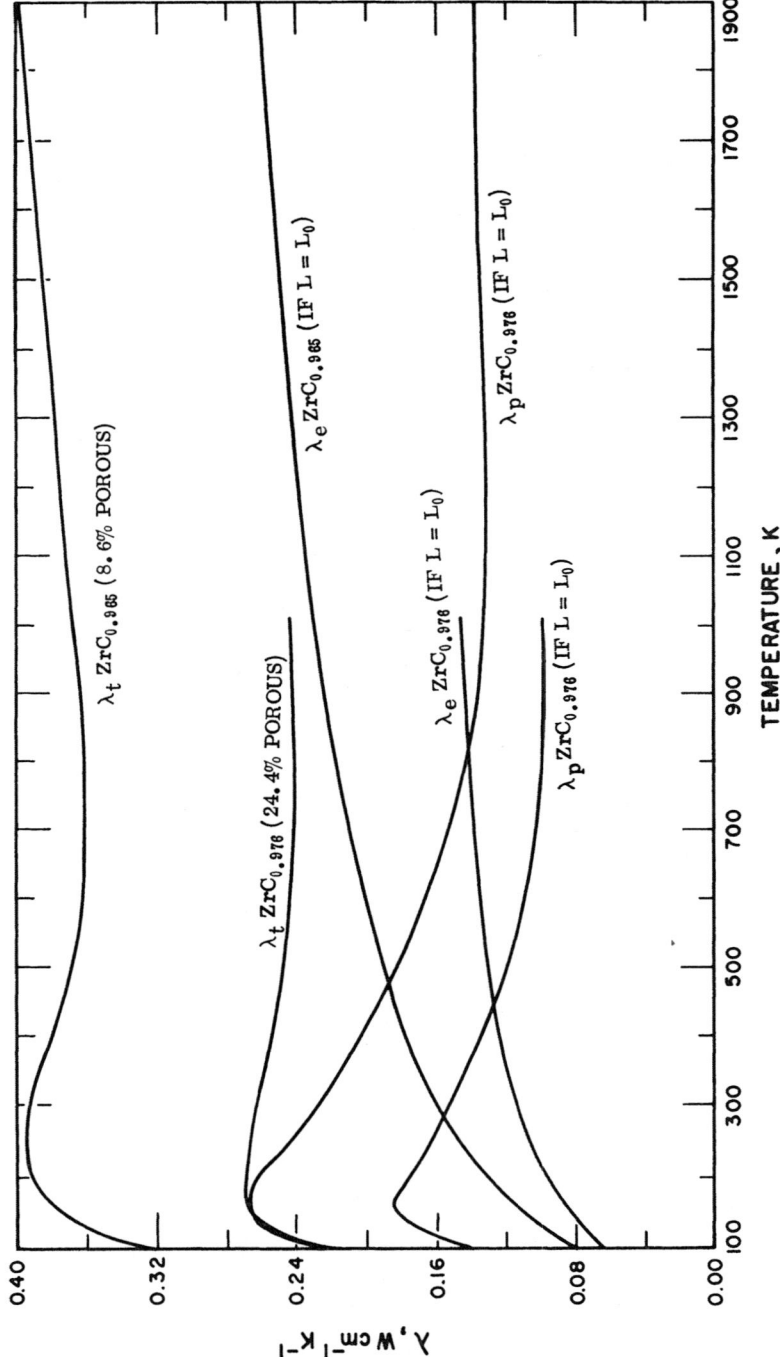

Figure 4. Thermal Conductivity of $ZrC_{0.976}$ and $ZrC_{0.965}$

metal/carbon ratio, a similar graph is obtained. Also shown in Figure 3 are the relative magnitude of the electronic (λ_e) and phonon (λ_g) contributions to the total conductivity deduced from the present work, assuming L equals L_0 at 300 K. The phonon contribution was calculated from the relation $\lambda_g = \lambda_t - \lambda_e$ where λ_t is the total (observed) conductivity. The relative magnitude of λ_e and λ_p depends on temperature, but it is obvious from Figure 3 that both components increase rapidly as the carbon sublattice approaches completeness.

DISCUSSION AND CONCLUSIONS

The resistivity curves for the $ZrC_{0.965}$, $ZrC_{0.897}$, and $ZrC_{0.870}$ samples are nearly parallel (Figure 2) with increasing temperature. However, the temperature coefficient of the resistivity for the $ZrC_{0.976}$ (plus free carbon) is greater than the temperature coefficient of the $ZrC_{0.682}$ sample and is less than that for the $ZrC_{0.965}$ to $ZrC_{0.870}$ composition range. This indicates that Matthiessen's rule is valid over a limited range of non-stoichiometry. The resistivity of ZrC_x decreases rapidly as x increases from 0.92 towards 1.0 (Figure 3).

The thermal conductivity of ZrC_x increases rapidly as x increases from 0.92 towards 1.0 (Figure 3). Both the electronic and phonon contribution to the total conductivity contribute to the marked dependence on x over this range. The phonon contribution predominates below about 300 K and the electronic contribution predominates above about 500 K for all compositions. From a consideration of the data shown in Figure 5, it is clear that in the composition range $ZrC_{0.682}$ to $ZrC_{0.897}$, L probably does not exceed the classical value by more than a few percent. At lower temperatures there is expected to be a region where L is less than L_0 due to small angle inelastic electron-phonon scattering. Above 300 K, the phonon contribution for the $ZrC_{0.682}$ to $ZrC_{0.897}$ composition range decreases as T^{-y} where y is nearly 1.0 if $L = L_0$. If L becomes greater than L_0, y increases above 1. Since a value of y of about 0.5 is expected for a combination of three-phonon interactions and strong point defect scattering[22,23], it is unlikely that L increases significantly above L_0. On the other hand, to force y to decrease to 0.5 requires that L be only about 0.8 L_0 at 300 K, which is conceivable.

In the composition range of $ZrC_{0.92}$ to $ZrC_{0.97}$, setting $L = L_0$ results in a constant λ_p above 900 K, which is unlikely. The results can be explained by allowing L to increase above the classical value at higher temperatures. This permits λ_p to continue to follow the $T^{-1/2}$ relationship which it closely follows between 300 and 900 K based on $L = L_0$. An increase in L of 18% above the classical value at 2000 K is required in the case of the $ZrC_{0.965}$ specimen to maintain the $T^{-1/2}$ relationship. This increase in L is in line with the results obtained on TiC using the alloying method[17]. Thus, the value of L apparently depends upon the carbon vacancy concentration and the temperature. For the composition range $ZrC_{0.68}$ to $ZrC_{0.90}$, L is limited to values between 0.8 L_0 and L_0 above

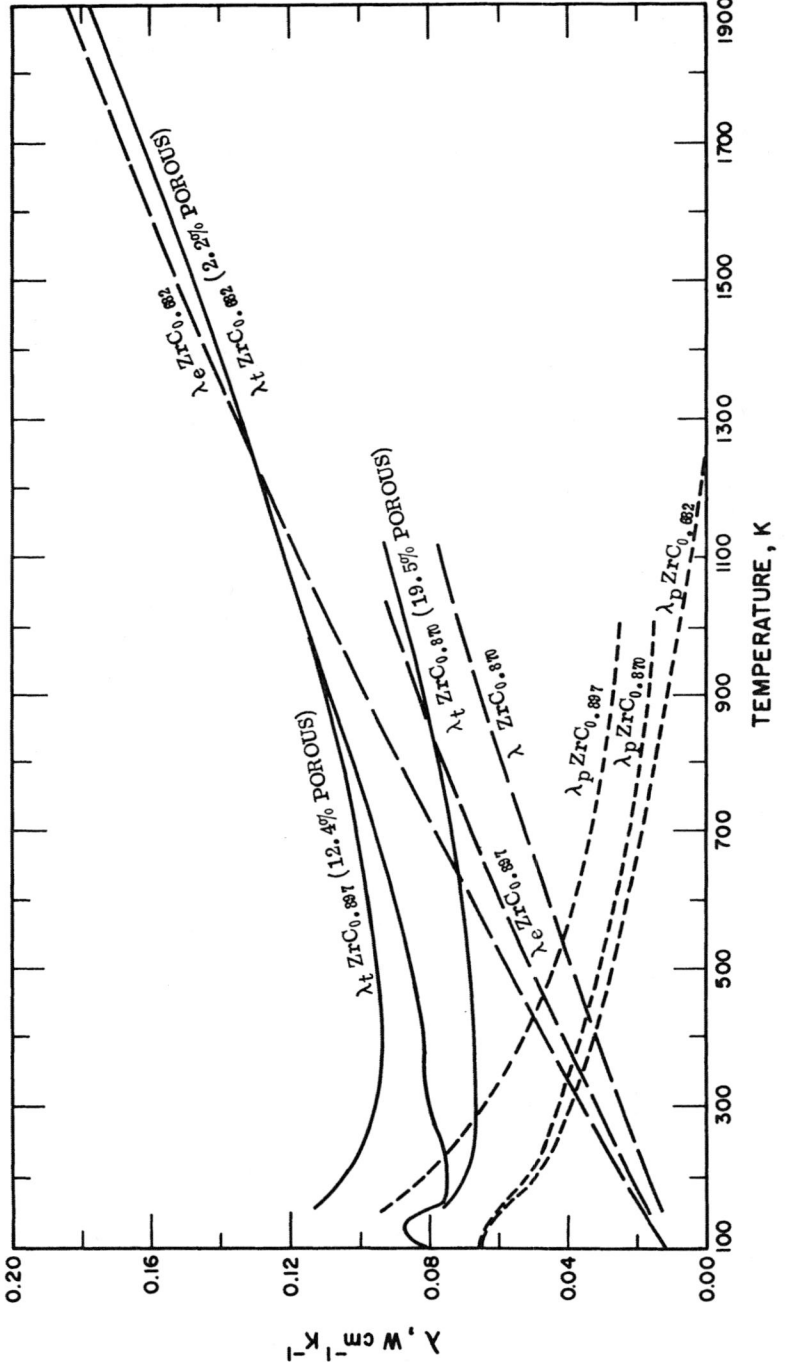

Figure 5.   Thermal Conductivity of $ZrC_{0.897}$, $ZrC_{0.870}$, and $ZrC_{0.682}$

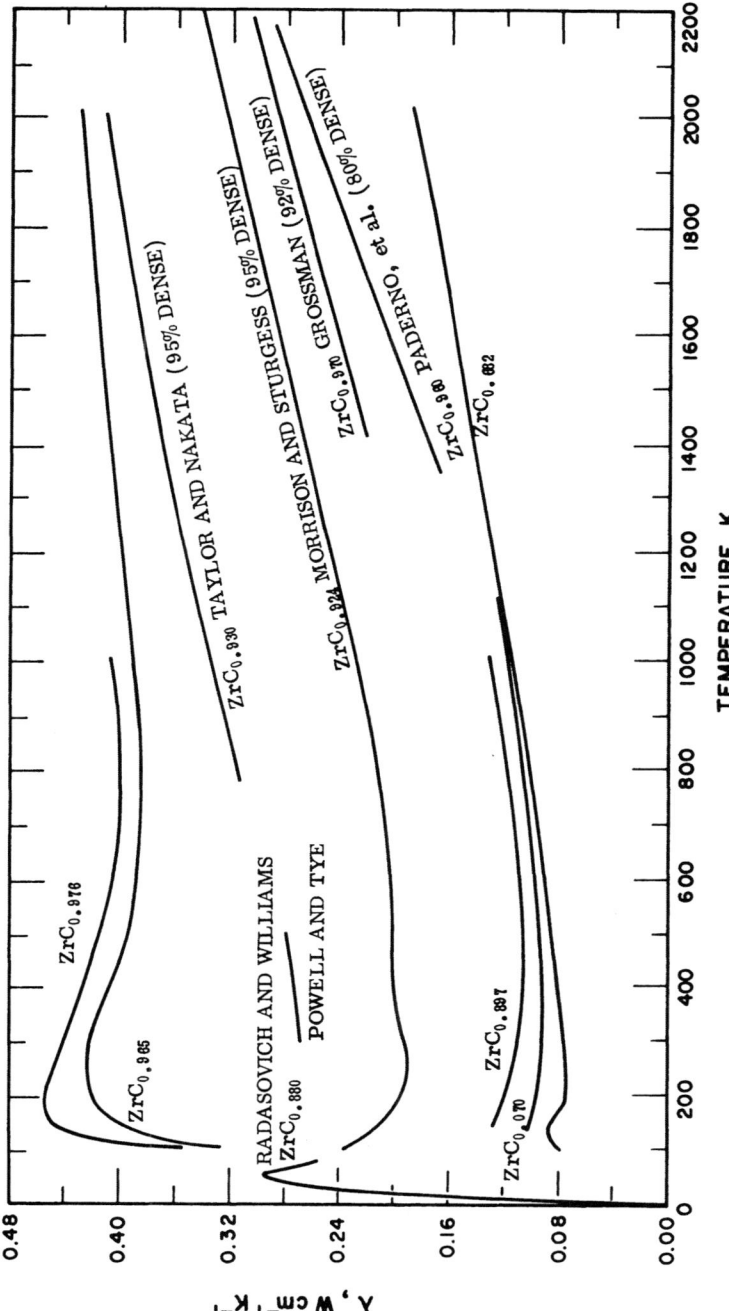

Figure 6.   Thermal Conductivity of ZrC_x as a Function of Temperature

300 K. For compositions in the range $ZrC_{0.92}$ to $ZrC_{0.97}$, L is probably close to L_0 at 300 K and increases to about $1.2 L_0$ at 2000 K. It should be noted that one reason λ increases with increasing temperature is the large value of ρ_0, i.e., in the equations $\rho = \rho_0 + \rho_i T$ and $\lambda_e = LT/\rho$ if ρ_0 is comparable to $\rho_i T$, then λ_e must increase as T increases, even if L remains constant. Since, at higher temperatures, λ_e is much greater than λ_p, the increase in λ_e with increasing temperature more than offsets the decrease in λ_p, causing the total conductivity to increase with increasing temperature. On this basis, one would expect the total conductivity of samples of ZrC_x with x approaching 1 (and therefore having smaller values of ρ_0) to have smaller values of temperature coefficients of thermal conductivity and this is observed in the present work (Figure 6). However, this tendency may be partially offset by the value of L increasing more rapidly with increasing temperature as x approaches 1.

The value of λ_p for stoichiometric ZrC was calculated from the Leibfried-Schlömann equation

$$\lambda_p \sim \frac{5 \times 10^{-6} \; \bar{A} \; a \; \theta^3}{\gamma^{\frac{2}{2}} \; T}$$

where $\bar{A}$ is the mean atomic weight, a is the lattice parameter, θ is the Debye temperature, and γ is the Grüneisen constant. The calculated value 1.4 W cm^{-1} K^{-1} is about three times the value obtained by extrapolating the present results (Figure 3). Often, the calculated values of λ_p, using this equation, are about a factor of two too high, so the present results are not unreasonable, especially considering the uncertainties in θ and γ. The dependence of λ_p on x (Figure 3) approximates the behavior predicted by strong defect scattering, but tends to rise more sharply as x approaches 1.

The results to date show for zirconium carbide, titanium carbide, and niobium carbide that there is a marked dependence of thermal and electrical conductivities on the carbon vacancy concentration for low concentrations. However, this behavior has not been reported in uranium carbide, which has been extensively studied. There has not been sufficient research to determine whether this behavior also occurs in the nitrides and diborides, although it seems likely that it would occur at least with nitrides analogous to the carbides which exhibit this phenomenon. Obviously, a careful study of other carbides, nitrides, and diborides is required before the general pattern is fully established.

As for the understanding of the temperature dependency of the thermal conductivity and its magnitude for transition metal carbides, the situation has improved markedly. The total conductivity can be explained in terms of the addition of the electronic and phonon contributions with allowances for a substantial value of ρ_0 and for an L which is somewhat dependent on temperature and degree of non-stoichiometry.

ACKNOWLEDGMENT

The work was performed under the Air Force Office of Scientific Research sponsorship, under the cognizance of Dr. J. F. Masi and is described in detail in AFOSR-72-2375 (July 1974).

REFERENCES

1 G. R. Finlay, Chem. Can., 4, 41 (1952).

2 F. W. Glaser, W. Arbiter, W. Ivanick, V. Tabola, W. Pierazio, and L. Alter, Progress Report, p. 81 (1951). [AD 201 475]

3 R. E. Taylor, J. Amer. Ceram. Soc., 44(10), 525 (1961).

4 R. Méndez-Peñalosa, Thermal Conductivity Conference, National Physical Laboratory, Teddington, England (July 1964).

5 E. K. Storms and P. Wagner, High Temp. Sci., 5, 454 (1973).

6 B. H. Morrison and L. L. Sturgess, Rev. Int. Hautes Temper. Réfract., 7(4), 351 (1970).

7 E. M. Chafik, University of Stuttgart, Institute for Nuclear Energy, Ph.D. Thesis (1970).

8 K. Wilkes, private communications (1974).

9 B. A. Fridlender and V. S. Neshpor, Teplofiz. Vys. Temp. (USSR), 8(4), 795 (1970); English translation: High Temp., 8(4), 750 (1970).

10 R. E. Taylor and H. Groot, High Temp.-High Pressures (in press).

11 R. E. Taylor and M. M. Nakata, U.S. Air Force Rept. WADD-TR-60-581, Part IV, p. 109 (1963). [AD 428 669, AD 441 079]

12 R. W. Powell and R. P. Tye, Spec. Ceram. Proc. Symp. Brit. Ceram. (Academic Press, London and N.Y., 1965), p. 243.

13 L. N. Grossman, J. Amer. Ceram. Soc., 48(5), 236 (1965).

14 J. Sakuri, J. Hayashi, M. Mekata, T. Tsuchida, and H. Takagi, Jap. Inst. Met. J., 25, 289 (1961).

15 V. S. Neshpor, S. S. Ordanyan, A. N. Avgustinik, and M. B. Khusiaman, Zh. Prikl. Khim., 37(11), 2375 (1964).

16 A. N. Avgustinik, O. A. Golikova, G. M. Klimashin, V. S. Neshpor, S. S. Ordanijav, and V. A. Snetovova, Neorg. Mater., 2(8), 1439 (1966).

[17] K. Y. Wu and R. E. Taylor, Advances in Thermal Conductivity (R. L. Reisbig and J. Sauer, Editors), 57 (1974).

[18] E. F. Westrum, Jr. and G. Feick, J. Chem. Eng. Data, 8(2), 176 (1963).

[19] R. Mezaki, E. W. Tiileux, T. F. Jambois, and J. L. Margrave, Advances in Thermophysical Properties at Extreme Temperatures and Pressures (Third ASME Symposium on Thermophysical Properties, Purdue University, March 1965).

[20] L. G. Radosevich and W. S. Williams, Phys. Rev., 181(3), 1110 (1969).

[21] Yu. B. Paderno, I. G. Barantseva, and V. L. Yupko, Vys. Neorg. Soedin., Akad. Nauk. Ukr. SSR, 199 (1965).

[22] P. G. Klemens, Phys. Rev., 119(2), 507 (1960).

[23] P. G. Klemens, G. W. White, and R. J. Tainsh, Phil. Mag., 7, 1323 (1962).

INFLUENCE OF PRESSURE ON THE THERMAL CONDUCTIVITY OF POLYMER GLASSES

R. S. Frost, R. Y. S. Chen and R. E. Barker, Jr.

University of Virginia, Department of Materials Science

Charlottesville, Virginia 22901 USA

The influence of pressure on the thermal conductivity K of four vitreous poly(alkyl methacrylate) polymers has been measured by steady state techniques. The measurements were made under pressures up to 2Kbars and over a temperature range between 173 and 300°K. For each member of the homologous series K was found to increase with applied pressure. Shifts in thermal conductivity transition temperatures (attributed to glass transition phenomena) of 25, 26 and 16°C per Kbar of applied pressure were observed for PMMA, PEMA and PnBMA. A quantitative model for the pressure dependence of K is presented. By extending the hole theory of liquids to temperatures below the glass transition, an equation of state is obtained and an expression for the pressure dependence of K is then formulated from a consideration of the anharmonicity associated with the segmental vibrations. The resulting equation $\Delta K/K = \gamma_G f_o \{1 - \exp(-P/B^*)\}$, where γ_G is the Grüneisen parameter, f_o is the fractional free volume and $B^* = KTg/V_o$, is compared with experimental results for the poly (alkyl methacrylates).

INTRODUCTION

The existing theories of thermal conductivity in polymeric glasses[1-8] tend to be based on either liquid[5,7] or solid state models[6]; the difference being that the former assumes energy transfer to occur as uncorrelated events while the latter consider the collective motions of repeating units. Neither of these models has been formulated to explicitly include the effects of pressure however.

175

Experimentally there have been many investigations of the temperature dependence[9-12] of the thermal conductivity of polymers, but
relatively little attention has been paid to the influence of pressure
on the conductivity. Except for our work, the only quantitative data
reported are those of Lohe[13] (at 2 pressures only) and those of
Andersson[19,20] and his coworkers in Sweden. The present work had as
its goal the correlation of the pressure dependence of the thermal
conductivity with the chemical structure of the poly(alkyl methacrylates).

The thermal conductivity was measured using a steady state
technique which has been described elsewhere.[14] Essentially polymer
films between one and two hundred microns thick were sandwiched in
conductivity cells, to which pressure was applied through pistons
located at the top and bottom of the cell. Polymer films of poly-
methyl methacrylate (PMMA), poly ethyl methacrylate (PEMA), poly
normal butyl methacrylate (PnBMA) and poly iso butyl methacrylate
(PiBMA) were prepared from benzene-polymer solution cast on clean
glass at room temperature.

In Figures 1-4 the conductivity of each of the methacrylates
is shown as a function of temperatures and pressure. As has been
indicated the conductivity of PMMA appears to pass through a maximum
at a temperature characteristic of each pressure. Moreover the
position of the maximum seems to be systematically shifted by an
amount estimated to be 25°C per Kbar of applied pressure. The value
is comparable in magnitude to that reported in the literature for
the shift of the glass temperature as a function of pressure for
this polymer.[18] Wada[17] and his coworkers have reported a mechanical
loss maximum near -53°C at 1 bar pressure (at 10^7 Hz) for PMMA and
suggest that it is due to motions of the α-methyl side groups, which
we believe to be responsible for the conductivity maxima as well.
The results for PEMA indicated that K initially increases linearly
with T but eventually passes through a maximum. The shift in the K
maximum was estimated to be 26°C/Kbar in this case. Measurements
on PnBMA showed it to undergo a transition that appears as a discontinuity in the temperature coefficient of K. The transition
temperature, believed due to a subglass transition[15,16] is in this
case shifted by 16°C/Kbar. PiBMA was unique in that its conductivity
decreased over the entire temperature range of the investiation, but
as with the other acrylics, K increased with an application of external pressure.

Viewed collectively, these data suggest that as the number of
carbon atoms attached to the side chain increases the magnitude of
the thermal conductivity decreases. Presumably this effect would
be due to the less efficient chain packing among the higher members
of the series.

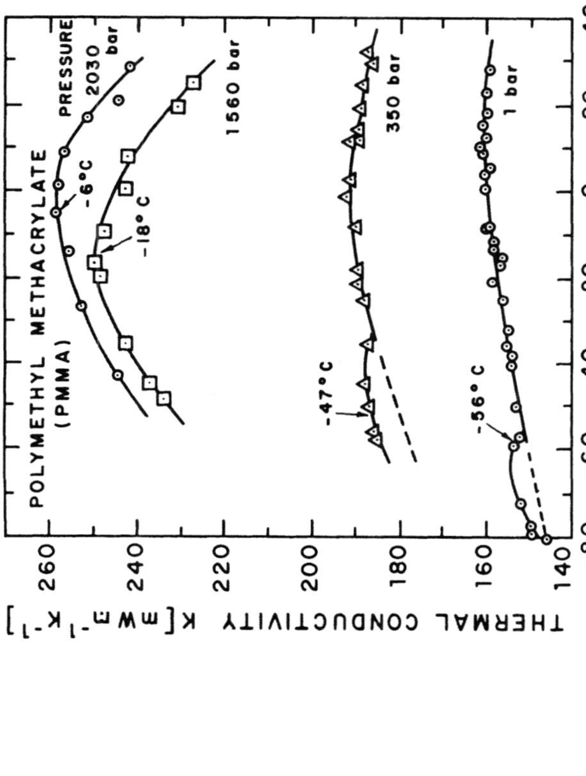

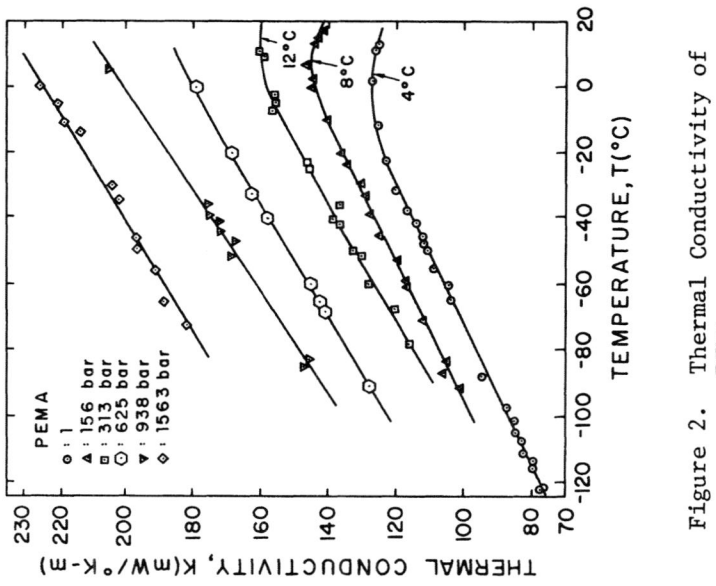

Figure 2.   Thermal Conductivity of PEMA

Figure 1.   Thermal Conductivity of PMMA

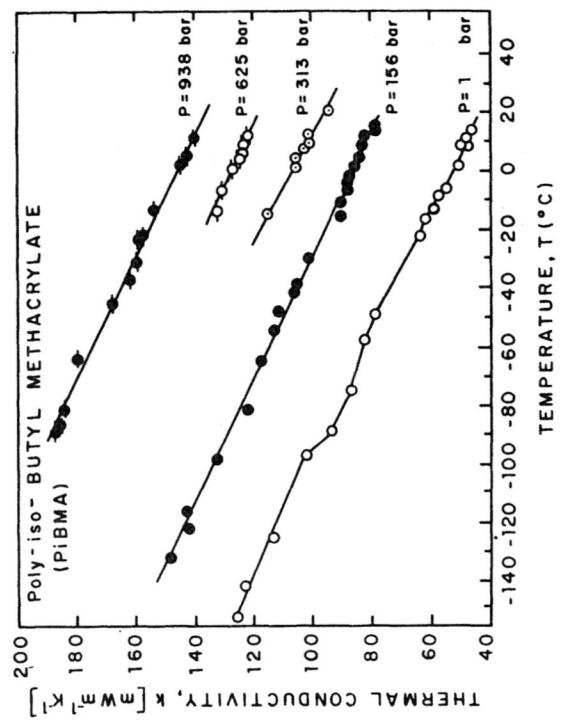

Figure 4.   Thermal Conductivity
             of PiBMA

Figure 3.   Thermal Conductivity
             of PnBMA

MODEL

These observations can be quantified. First we derive an
equation of state for glassy polymers, then an expression for the
pressure dependence of the thermal conductivity is formulated by
considering the anharmonicity of segmental vibrations.

As in the hole theory[21] of polymer solutions, consider a
polymer melt to consist of a volume divided into identical cells,
some of which are occupied by chain segments while the rest remain
vacant. If the total number of ways to place the molecules on the
lattice is represented by $W(n_o)$ where n_o is the number of lattice
vacancies, the hole formation partition function is

$$Z = \sum_{n_o} W(n_o) \exp(-n_o E_o/kT) \tag{1}$$

Here E_o is the vacancy formation energy, and the summation is over
the number of vacancies in the lattice. Upon evaluating the partition function and substituting into the expression for the Helmholtz
free energy, $F = -kT \ln Z$, we have a relation that can be used to evaluate the fractional free volume f_o, that is the ratio of unoccupied
to available lattice sites, as a function of pressure and temperature.
According to the iso-free volume hypothesis, f_o remains frozen below
the glass temperature and we derive

$$f_o(p,T_g) = f_o(0,T_g) \exp(-pV_o/kTg) \tag{2}$$

Here V_o is the volume of a lattice cell and p denotes pressure.

The quasilattice model may be extended to describe thermal
conductivity. Energy is assumed to be transported down a thermal
gradient when molecules oscillating about their equilibrium positions
collide. Thus, for the conductivity one can write

$$K = \sigma P C_v \nu. \tag{3}$$

Here P is the probability that energy is transferred on each collision, ν is a characteristic vibrational frequency, C_v is the

heat capacity and σ is a geometrical constant. Assuming that the equipartition limit applies for C_v and that only the frequency changes appreciably with decreasing volume, differentiation yields

$$-d\ln K/d\ln V = \gamma_G \qquad\qquad (4)$$

Here $\gamma_G \equiv -\partial\ln\nu/\partial\ln V$ is the interchain Grüneisen parameter. Assuming that the dilation $\Delta V/V$ may be represented by differences in free volume, we may combine Equations (2) and (4) to obtain an expression for the fractional change in K

$$\frac{\Delta K}{K(p)} = \gamma_G f_o(0,T_g)\{1-\exp(-pV_o/KT_g)\} \qquad\qquad (5)$$

which states that $\Delta K/K(p)$ is directly proportional to the change in free volume.

DISCUSSION

When the experimental data was fit to Equation 5, good agreement was found by taking $\gamma_G = 4.3$ and using the values of f_o and v_o indicated in the following table.

Polymer	$f_o(273^{\circ}K)$	Density	T_g	kT_g/V_o	$\sigma_{T.S.}$
PMMA	0.097	1170 Kg/m³	376°K	635 bars	620 bars
PEMA	0.110	1125	337	596	345
PnBMA	0.152	1053	291	187	69
PiBMA	0.158	1041	327	189	234

The densest polymer, PMMA, had the smallest free volume and the other members had values consistent with their relative densities. The values of f_o are in close agreement with those reported recently by Aharoni.[22] The term kT_g/V_o has the dimensions of an internal pressure and its relative magnitude for the polymers is consistent with such an interpretation; moreover, it appears to be correlated with the tensile strength, $\sigma_{T.S.}$.

In conclusion, we note that as the success of our model can only be judged in terms of the qualitative reasonableness of the parameters derived by fitting the theoretical equations to the experimental data, these results are encouraging in view of the enormously complicated systems described.

ACKNOWLEDGMENTS

This work was supported in part by grants from the National Institute of Dental Research (US PHS DE 0211) and by a John Lloyd Newcomb Fellowship (RSF).

REFERENCES

1. R. E. Barker, Jr., J. Appl. Phys., 38, 4234 (1967).
2. R. E. Barker, Jr., R. Chen, J. Chem. Phys. 53, 2616 (1970).
3. P. G. Klemens, Proc. Roy. Soc. (London) A68, 1113 (1955).
4. P. Carruthers, Rev. Mod. Phys., 33, 92 (1961).
5. K. Eiermann, J. Polym. Sci., Pt. C6, 157 (1964).
6. G. K. Chang and R. E. Jones, Phys. Rev. 126, 2055 (1962).
7. W. Reese, J. Appl. Phys. 37, 3959 (1966).
8. D. Hansen and B. D. Washo, Kolloid-Z. 210, 111 (1966).
9. R. Berman, Proc. Roy. Soc. (London) A208, 90 (1951).
10. R. Koulouch and R. Brown, J. Appl. Phys. 39, 3999 (1968).
11. W. Reese and J. E. Tucker, J. Chem. Phys. 43, 105 (1965).
12. K. Eiermann, Kunstoffe 55, 355 (1965).
13. P. Lohe, Kolloid-Z. Z. Polym 203, 115 (1965).
14. R. Chen and R. E. Barker, Jr., J. Biomed. Mat. Res. 6, 147 (1972).
15. J. Powles, B. Hunt and D. Sandiford, Polymer 5, 505 (1964).
16. T. J. Kawai, Phys. Soc. Japan 16 1220 (1961).
17. J. Hirose and Y. Wada, Rep. Polym. Phys. IX, 287 (1966).
18. A. Zosel, Kolloid-Z. 199, 133 (1964).
19. P. Andersson and G. Backstrom, High Temp.-High Press. 4, 101 (1972); J. Appl. Phys. 44, 2601 (1973).
20. P. Andersson and B. Sundqvist, J. Polym. Sci 13, 243 (1975).
21. P. J. Flory, Principles of Polymer Chemistry, Cornell Univ. Press, Ithaca, New York, 1953.
22. S. M. Aharoni, J. Polym. Sci. 42, 795 (1973).

TRANSPORT PROPERTIES OF SOME TRANSITION METALS AT HIGH TEMPERATURES

M. J. Laubitz and P. J. Kelly

National Research Council of Canada

Division of Physics, Ottawa

The measured high-temperature transport properties of the paramagnetic, cubic, transition metals of Group VIII cannot be described in terms of pure electron-phonon scattering: to achieve consistency among the measured properties, strong electron-electron scattering must be included in the analysis. Furthermore, the old postulate of Mott, that the relaxation time of the electrons is inversely proportional to the density of states, does not appear to be universally true. We briefly discuss the results and the analysis that leads to the above conclusions.

ANISOTROPIC LATTICE THERMAL CONDUCTIVITY OF α-QUARTZ AS A FUNCTION

OF PRESSURE AND TEMPERATURE*

D.M. Darbha and H.H. Schloessin

Dept. of Geophysics, University of Western Ontario

London, Canada

The lattice thermal conductivity of α-quartz was measured parallel and perpendicular to the optic axis as well as perpendicular to the rhombohedral plane in the pressure range of 20-56 kbars and temperatures up to 350°C. The pressure and temperature variations of the conductivity show that the conductivity anisotropy increases with increasing pressure and decreases with increasing temperature. The values perpendicular to the rhombohedral plane appear to lie almost on the conductivity ellipse formed by the conductivities along the two principal axes.

At ordinary pressure and temperature α-quartz exhibits a pronounced anisotropic lattice thermal conductivity, with the value in $\langle 0001 \rangle$ direction being about twice the value in $\langle 10\bar{1}0 \rangle$ direction. This was first established by Lees [1] who applied Forbe's compound bar method to measurements of anisotropic heat conduction. He found the thermal conductivitiy coefficient to be 29.9×10^{-3} cal $deg^{-1}cm^{-1}sec^{-1}$ parallel to the optic axis and 15.8×10^{-3} cal $deg^{-1}cm^{-1}sec^{-1}$ perpendicular to it. These values have been confirmed and upheld by later measurements (Birch & Clark [2], Kanamori et.al. [3]). Variations in absolute values, at ordinary T and P, change from 24.6 to 33.3 cal $deg^{-1}cm^{-1}sec^{-1}$ and from 14.8 to 15.5 cal $deg^{-1}cm^{-1}sec^{-1}$ for c - and a - direction respectively. They can probably be accounted for by differences in impurity concentration and structural imperfections between samples of different origin.

The conductivity coefficients λ_{ij} in general relate two polar vectors according to the relation

$$h_i = - \lambda_{ij} \partial T/\partial x_j \tag{1}$$

183

where, h_i is the heat flux and $\partial T/\partial x_j$ the temperature gradient. The
coefficients λ_{ij} form a second-rank polar tensor. If the heat flux
vectors coincide with principal axes (1) becomes

$$h_i = \lambda_i \; \partial T/\partial x_i \; . \tag{2}$$

α-quartz belongs to the trigonal - trapezohedral class 32. Three
SiO_4 molecules form a unit triangular prismatic cell with a = 4.89 Å
and c = 5.375 Å . The tensor components for this structure are
determined by two constants $\lambda_{11} = \lambda_{22}$ (λ_a, $\lambda_{10\bar{1}0}$) and λ_{33}, (λ_c, λ_{0001}).

The present measurements are concerned with the variations of
the anisotropic lattice thermal conductivity with pressure and temperature in the range from 20 to 56 Kb and 50 to 300°C. Several
experiments were performed for each of three orientations of the
heat flux vector in directions parallel and perpendicular to the
optic axis and perpendicular to the rhombohedral plane (10$\bar{1}$1). Each
experiment involved stacks of four cylindrical discs of small and
different thicknesses cut from single crystals of optically clear
and twin-free Madagaskar quartz. These stacks were accommodated in
5/8" or 1¼" cubes made of pyrophyllite the thermal conductivity of
which varies between ~7 x 10^{-3} and 15.0 x 10^{-3} cal $cm^{-1}sec^{-1}deg^{-1}$
with a temperature coefficient of ~1.7 x 10^{-5} cal $deg^{-2}cm^{-1}sec^{-1}$.

A 0.051 cm Pt wire was placed in the centre of the stack to
serve as a heat source. Two 0.025 cm Pt/Pt 10% Rh thermocouples
were placed at the far sides of the two discs adjacent to the heat
source. The discs measured 0.80 cm in diameter and their thickness
varied between 0.1 and 0.3 cm. Thus the geometical arrangement
satisfied the condition for approximately linear heat flow which
requires that the dimension along the heat flux vector is small compared to the cross sectional dimensions. The heat generation at the
source was determined by measurements of the a.c. current and voltage
applied to the Pt wire.

Thermal conductivity for heat flux vectors perpendicular to
(0001), (10$\bar{1}$0) and (10$\bar{1}$1) planes are shown in Fig. 1. As one would
expect the coefficients increase with increasing pressure and decrease with increasing temperature. However, the conductivity in
a-direction is considerably less pressure and temperature sensitive
than in c-direction; the conductivity perpendicular to (10$\bar{1}$1) is
intermediate and rather temperature sensitive. In particular,
values of ($\Delta\lambda/\Delta T$) at constant pressure and ($\Delta\lambda/\Delta P$) at constant
temperature are as follows:

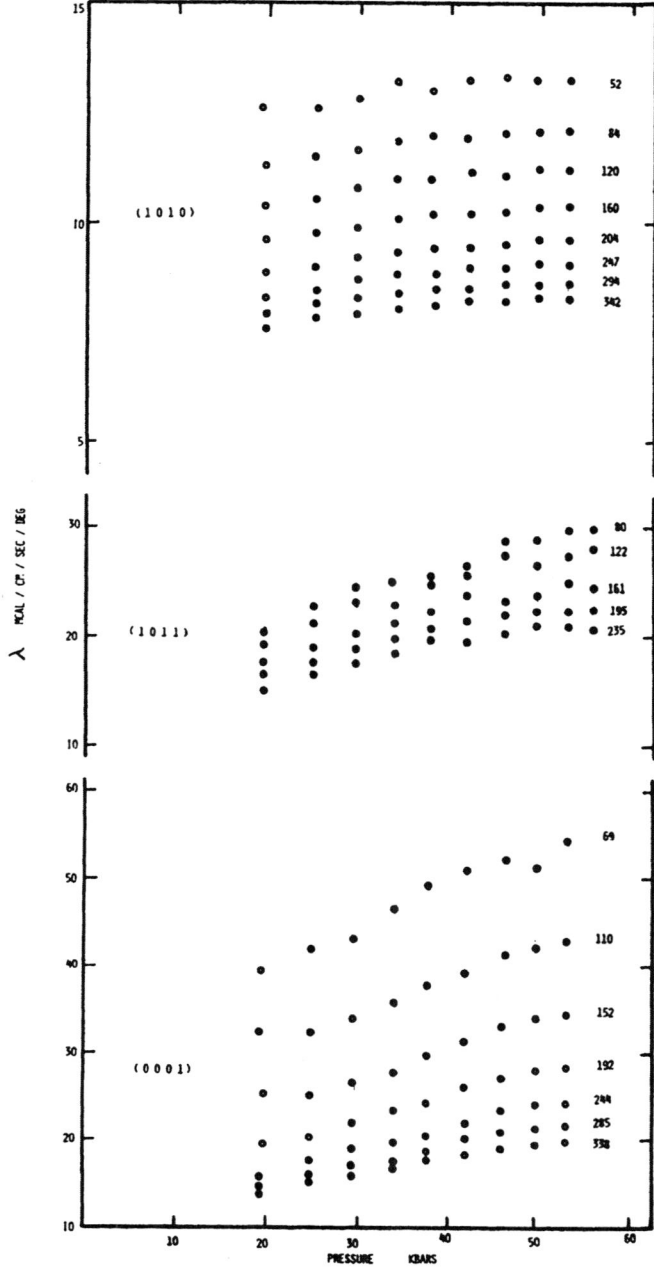

Figure 1   Lattice conductivity perpendicular to (0001), (10$\bar{1}$0), and (10$\bar{1}$1) planes versus pressure at various mean sample temperatures (°C).

		$(\partial\lambda/\partial P)_T$		$(\partial\lambda/\partial T)_P$
		(mcal deg^{-1}cm^{-1}sec^{-1}Kb^{-1})		(mcal deg^{-2}cm^{-1}sec^{-1})
(0001)	400K	3.92 x 10^{-1}	25Kb	1.80 x 10^{-1}
	450K	3.36 x 10^{-1}	50Kb	1.88 x 10^{-1}
(10$\bar{1}$0)	400K	2.60 x 10^{-2}	25Kb	2.20 x 10^{-2}
	450K	2.32 x 10^{-2}	50Kb	2.34 x 10^{-2}
(10$\bar{1}$1)	400K	2.5 x 10^{-1}	25Kb	4.2 x 10^{-2}
	450K	2.1 x 10^{-1}	50Kb	5.6 x 10^{-2}

As variations of λ with pressure are approximately linear, it is
easily possible to extrapolate the values to zero pressure. A
comparison of the extrapolated values with the dependence of λ on
temperature as observed by Birch and Clark[2] is shown in Fig. 2.
From this it would appear that the general behaviour of the temperature dependence found in the present study, especially for the
⟨10$\bar{1}$0⟩ direction, is in very close agreement with the one previously
established by room pressure experiments. The differences in absolute value can be accounted for by ambiguities in linear extrapolation, uncertainties in temperature value, and, possibly, significant
differences in the concentrations of impurities and imperfections
of the samples investigated in the two studies. Comparison of the
conductivities in a and c-direction for different pressure (figure
3) indicates that the degree of anisotropy diminishes with increasing   temperature and increases with increasing pressure. If we
define the ratio λ_a/λ_c as a measure of the anisotropy its value
with varying temperature follows the curves shown in Fig. 3.

   Variations of λ with the inverse temperature are shown in Fig.
4. In a-direction the dependence at the two pressures is reasonably
linear as would be expected from lattice theory (Roufosse and Klemens
[4]),whereas in c-direction and perpendicular to (10$\bar{1}$1) there is
a pronounced non-linearity. This could arise from a slight but
noticeable radiative conductivity component, the contribution of
which would increase with increasing temperature. As the radiative
component is proportional to the square of the refractive index,
such a contribution could become more pronounced in c-direction than
in a-direction, especially because of the higher absorption coefficient in a-direction.

   The representation quadric(Nye[5])for the lattice conductivity
of α-quartz is an ellipsoid. In rectangular coordinates with axes
coinciding with the crystallographic axes, its semi-major axes are
determined by $(\lambda_{11})^{-\frac{1}{2}} = (\lambda_{22})^{-\frac{1}{2}} = (\lambda_a)^{-\frac{1}{2}}$ and $(\lambda_{33})^{-\frac{1}{2}} = (\lambda_c)^{-\frac{1}{2}}$. The
length of the radius vectors defined by the conductivity surface
represents the magnitude and direction of the temperature gradient
required for constant heat flux in the direction of the surface
normal. Isothermal and isobaric deformations of the conductivity

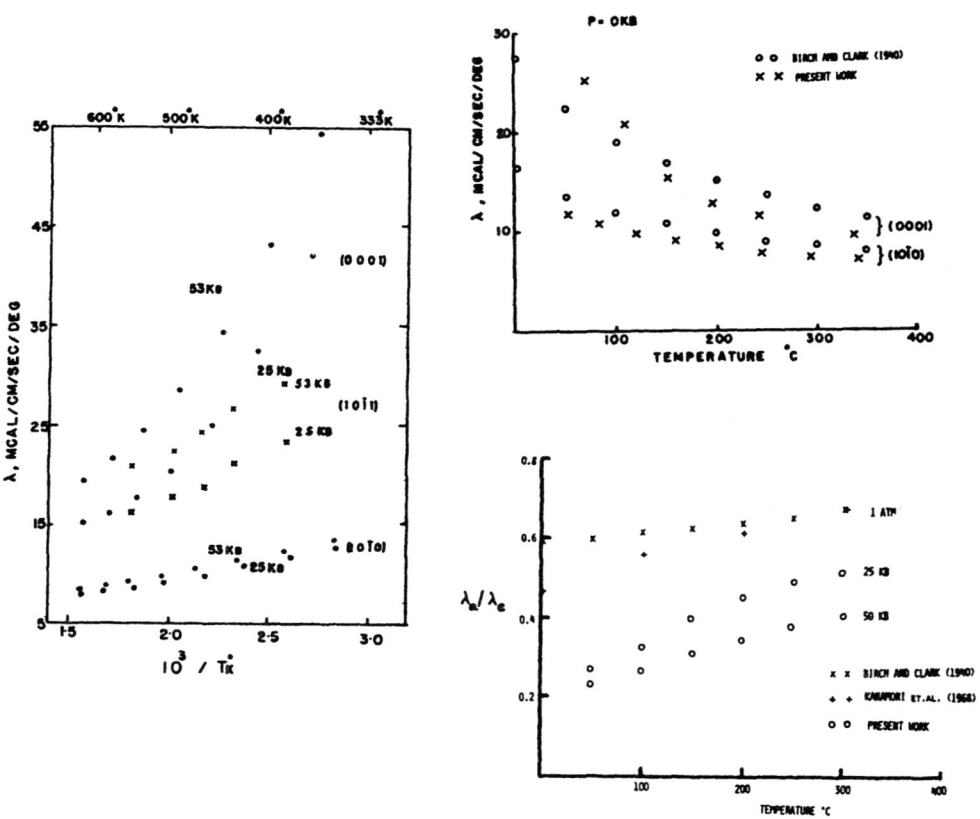

Figure 2 (on right top) Lattice conductivity values from high pressure experiments extrapolated to zero pressure as a function of temperature. Conductivity variations with temperature is determined by Birch and Clark[2] as shown for comparison.

Figure 3 (on right bottom) The ratio λ_a/λ_c versus temperature at 25 and 50 Kb.

Figure 4 (on left) Lattice conductivity perpendicular to (0001), (10$\bar{1}$0), and (10$\bar{1}$1) planes versus the inverse temperature at 25 and 53 Kb.

surface with pressure and temperature are illustrated in 4 sections shown in Fig. 5.

   As can be seen the values obtained for heat flux normal to (10$\bar{1}$1) planes at the higher temperatures (>100°C) do not fall exactly on the conductivity ellipsoid as defined by the heat flux in a-and c-directions. Apart from experimental error and the possibility of changes due to plastic deformation, some of the discrepant results could have arisen from the fact that the (10$\bar{1}$1) samples were of different origin.

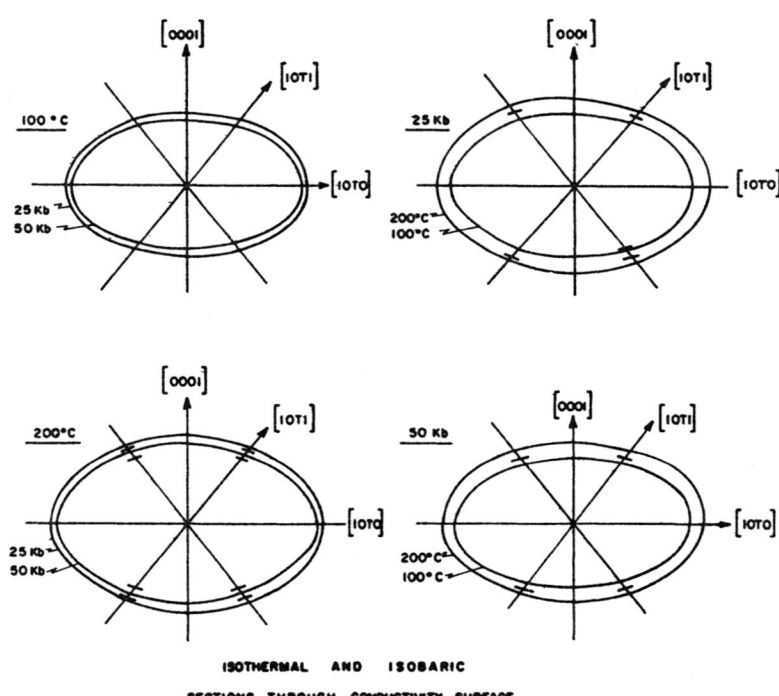

Figure 5   Isothermal and Isobaric sections of the conductivity surface of α-quartz as determined from the conductivities measured perpendicular to (10$\bar{1}$0) and (0001) planes. The dashes on the radius vector normal to (10$\bar{1}$1) planes mark the actual $(\lambda_{10\bar{1}1})^{-\frac{1}{2}}$ values calculated from the measured conductivities.

Considering thermal conductivity in the range from 300 to 600K to be dominated by umklapp processes (Klemens[6]), the anisotropic conductivity coefficients may be expressed as

$$\lambda_i = \frac{1}{3} \bar{c}_{ij} \bar{v}_{ij} \bar{\Lambda}_{ij} , \qquad (3)$$

where $\bar{c}_{ij}$ denotes the phonon heat capacity (0.48 cal cm^{-3}degr^{-1} at room temperature), $\bar{v}_{ij}$ phonon velocity and Λ_{ij} the phonon mean free path (~40Å at room temperature), all averaged over the full range of vibrational frequencies, and where the ij refer to directions of phonon transport resolved onto principal axes. If we assume the v_{ij} to be identical with sound velocities, their values will be determined by the elastic constants which, in turn, can be related to first order force constants. The phonon mean free path Λ_{ij} depends on the cross section for umklapp collisions between phonons and on the magnitude of the reciprocal lattice vector. It would seem, therefore, that with the choice of suitable potentials for the Si-0, 0-0, and Si-Si interactions, the pressure and temperature dependence of the phonon conduction can ultimately be reduced to pressure and temperature dependence of interatomic distances or lattice constants.

The pressure coefficient $(\partial\lambda/\partial P)_T$ is positive and decreases with increasing temperature and the temperature coefficient $(\partial\lambda/\partial T)_P$ is negative and increases with increasing pressure. Therefore, to maintain constant λ for a given P,T pair we have

$$(\partial\lambda/\partial P)_T \Delta P + (\partial\lambda/\partial T)_P \Delta T = 0 . \qquad (4)$$

On the basis of (4) certain $(\Delta T/\Delta P)_{\lambda,P,T}^{a(c)}$ values can be determined for definite paired-off values of pressure and temperature.

From the results $(\Delta T/\Delta P)_{\lambda,P,T}$ values which satisfy (4) are for example:

	$(\Delta T/\Delta P)_{a,\lambda}$			$(\Delta T/\Delta P)_{c,\lambda}$
25Kb 357 K	1.8 K Kb^{-1}	25 Kb 342K	3.1 K Kb^{-1}
53Kb 615 K	1.4 K Kb^{-1}	53 Kb 611K	1.2 K Kb^{-1}

The immediate significance of these $(\Delta T/\Delta P)_{\lambda,P,T}^{a(c)}$ values in terms of other parameters such as linear compressibility and thermal expansion coefficients and their pressure and temperature dependence (Bridgman[7]) in thermodynamic applications similar to isothermal, isentropic or isenthalpic $(\Delta T/\Delta P)$ values is difficult to establish.

It would appear that a more correct interpretation of the $(\Delta T/\Delta P)_\lambda$ values can be given in terms of first and second order force constants, on the basis of suitable interatomic repulsion and attraction potentials, possibly restricted to 0-0.

*Support of this work by the National Research Council of Canada is gratefully acknowledged.

[1] C.H. Lees, Mem and Proc. Manchester Philos. & Lit. Soc, 4, 17 (1890-91).
[2] F. Birch & H.C. Clark, Am. J. Sci., 238-529, (1940).
[3] H. Kanamori, N. Fujii, and H. Mizutani, J. Geophy. Res., 73, 2, 595 (1968)
[4] M. Roufosse, and P.G. Klemens, J. Geophys. Res., 79, 5, 703 (1974).
[5] J.F. Nye, Physical Properties of crystals (Oxford, London, 1969) pp. 195.
[6] P.G. Klemens, Solid state physics, edited by F. Seitz and D. Turnbull, 7, 1 (Academic Press, New York, 1958).
[7] P.W. Bridgman, Collected experimental papers, V. 1-7 (Harvard, 1964).

PRESSURE DEPENDENCE OF THE THERMAL CONDUCTIVITY OF COMPLEX DIELECTRIC SOLIDS[*]

Micheline C. Roufosse

Department of Orthodontics, University of Connecticut

Farmington, Connecticut 06032

ABSTRACT

The pressure dependence of the thermal conductivity at high temperatures is determined in the case of complex dielectric crystals. As in the theory of Mooney and Steg, the thermal conductivity is expanded in a Taylor series as function of dilatation, but the changes in the position of the zone boundaries are considered. The results are similar to that of the previous theory which applied to simple lattices. The changes in thermal conductivity depend upon the Grüneisen parameter and its first derivative with respect to dilatation or pressure. The theory is applied to minerals at high temperatures and high pressures.

I. INTRODUCTION

The pressure dependence of the thermal conductivity of dielectric solids at high temperature is important in geophysics. The variation of the thermal conductivity as a function of depth in the earth's mantle must be known in order to predict the temperature gradient in the earth's interior. The pressure dependence of the lattice thermal conductivity has been investigated by Mooney and Steg[1] for the case of simple crystals. The present note aims to extend their result to the case of complex dielectric crystals which are more relevant to geophysical problems.

II. HIGH TEMPERATURE THERMAL CONDUCTIVITY FOR COMPLEX CRYSTALS

The expression for the thermal conductivity at high tempera-

ture has first been derived by Leibfried and Schlömann[2] in the
case of simple crystals

$$K_1 = \frac{24}{10} \, 4^{1/3} \, \frac{1}{\gamma^2} \left(\frac{k}{\hbar}\right)^3 M a \, \theta^3 / T \qquad (1)$$

where K_1 is the thermal conductivity at normal pressure for lattices
containing one atom per unit cell, M is the mass of the unit cell,
γ is the Grüneisen parameter, θ the Debye temperature and a the
interatomic spacing.

Their result has been extended to the case of complex lattices
by Slack[3] and by Roufosse and Klemens.[4] Slack considered the case
of a complex crystal where the extra zone boundaries introduce
large gaps into the phonon dispersion curve so that the phonon group
velocity of all branches is greatly reduced except that of the
acoustic branches. Neglecting the contribution of all higher, or
optical, branches to the conductivity, he defined a reduced Debye
temperature appropriate to the extent of the acoustic branches in
the momentum space and smaller than the ordinary Debye temperature
by a factor $N^{1/3}$ where N is the number of atoms per unit cell.
His expression for the thermal conductivity is

$$K_N' = K_1 / N^{1/3} . \qquad (2)$$

He did not treat the effect of the crystal structure on the relaxation time.

Roufosse and Klemens[4] extended this model to treat the relaxation time in detail. For that purpose, they neglected changes in
the dispersion curve by zone boundaries since this does not change
the geometry of the phonon interactions materially. They found
that the relaxation time is essentially unaffected by the additional zone boundaries owing to the cancellation of two factors.
The summation over additional zone boundaries increases the relaxation rate for U-processes but the anharmonic matrix element is
reduced because the repeat distance of the unit cell is enlarged
so that coherent reinforcement of the participating waves is reduced. Their expression for the thermal conductivity becomes

$$K_N = \frac{3}{4} K_1' . \qquad (3)$$

The factor 3/4 arises from a factor

$$\tan^{-1}(A \omega_D a/v)/(A \omega_D a/v)$$

where A is a constant introduced by additional zone boundaries, ω_D
is the Debye frequency, v the sound velocity and K_1' is their expression for the thermal conductivity of solids containing one atom per
unit cell. It differs from that of Leibfried and Schlömann[2] by a
numerical factor but this should not influence the relation between
K_N and K_1 given by equation 3.

Now, one can make a further assumption, previously made by Slack,[3] that only the acoustic modes contribute to the conductivity. The additional zone boundaries may well affect the dispersion curve and lower the group velocity in all higher branches without essentially changing the interactions, so that the neglect of the higher branches is not inconsistent with the approximation made to determine the relaxation time. With this approximation, the expression for the thermal conductivity becomes

$$K_N'' = \frac{3}{4} K_1' / N^{1/3} \tag{4}$$

III. PRESSURE DEPENDENCE

The thermal conductivity given by equation 3 is written in terms of the sound velocity v, the interatomic spacing a and the Grüneisen parameter γ, which are functions of the dilatation $\Delta = (V-V_0)/V_0$. These parameters can be expanded in terms of the dilatation as

$$v = v_0 (1 - \gamma \Delta) \tag{5}$$

$$a = V_0^{1/3} (1 + \Delta/3) \tag{6}$$

$$\gamma = \gamma_0 + \gamma_0' \Delta \tag{7}$$

where γ_0 and γ_0' are respectively the Grüneisen parameter at normal pressure and its first derivative with respect to dilatation. These can be replaced into the expression for the thermal conductivity. If we limit ourselves to first order terms in dilatation and use the relation

$$a^3 \omega_D^3 / v^3 = 6 \pi^2 \tag{8}$$

we can see that the argument of $\tan^{-1}$ appearing in the expression for K_N is independent of pressure. The other factors vary with pressure as in the paper by Mooney and Steg. The following expression is obtained

$$K = K_0 \left[1 - \left(3\gamma_0 + \frac{2}{3} + 2\gamma_0'/\gamma_0 \right) \Delta \right] \tag{9}$$

where K_0 represents the thermal conductivity of any dielectric, complex or simple, at ordinary pressure. Similarly, if we use equation 4, we are led to an identical dependence upon dilatation, since the application of pressure does not change the number of atoms per unit cell as long as one does not cross a phase boundary. If there is a change of phase, the change in the factor $N^{1/3}$ would have to be considered.

IV. CONCLUSIONS

If we write the dilatation in terms of pressure, $\Delta = -\alpha\beta$ we can see that in the case of simple crystals as well as in the case of complex crystals, the thermal conductivity

$$\kappa = \kappa_o \left[1 + \alpha \left(3\gamma_o + \frac{2}{3} + 2\gamma_o'/\gamma_o \right) p \right] \qquad (10)$$

increases linearly with pressure and depends upon the Grüneisen parameter and its first derivative, provided there is no phase change with application of pressure.

It has been pointed out by Ecsedy,[5] that at high temperature, the thermal expansion will cause the lattice thermal conductivity to decrease faster than $1/T$.

In the conditions of high temperature and high pressure prevailing in the earth's mantle, we can conclude that thermal expansion on one hand and pressure on another hand will tend to compensate each other. As a consequence, the $1/T$ law for the lattice component of the thermal conductivity should hold throughout the earth's mantle.

ACKNOWLEDGMENT

I would like to express my gratitude to Dr. Klemens for his advice during this research.

[*]Work supported by the United States Army Research Office – Durham, and the University of Connecticut Research Foundation.

1. D. L. Mooney and R. G. Steg, High Temperatures. High Pressures 1, 237 (1969).
2. G. Leibfried and E. Schlömann, Nachr. Ges. Wiss. Goett. Math-Phys. K1.2 (4), 71 (1954).
3. G. A. Slack, Phys. Rev. 139, A 507 (1965); Phys. Rev. B4, 592 (1971).
4. M. C. Roufosse and P. G. Klemens, Phys. Rev. B7, 5379 (1973).
5. D. J. Ecsedy, Doctoral Thesis, University of Connecticut (1975). See also his contribution in these proceedings.

FOUR-PHONON PROCESSES AND THE THERMAL EXPANSION EFFECTS IN THE THERMAL RESISTIVITY OF CRYSTALS AT HIGH TEMPERATURES[*]

David J. Ecsedy

Dept. of Physics and Institute of Materials Science

Univ. of Connecticut, Storrs, Connecticut 06268

The temperature dependence of the high temperature thermal resistivity for many crystals is observed to contain a T^2 component in addition to the linear component expected from three-phonon processes. Four-phonon processes yield a T^2 resistivity component. The four-phonon inverse relaxation time is estimated in terms of the Grueneisen parameter and its strain derivative, but its magnitude is found to be much smaller than that needed to account for the experimental T^2 resistivity term. A formula for the pressure dependence of the thermal conductivity can be used to estimate the effect of the thermal expansion on the three-phonon resistivity. It is found that the thermal expansion effect can account for the T^2 resistivity component in KCl. The high Debye temperatures of Al_2O_3, BeO, and MgO allow other effects to mask the T^2 component at most temperatures. An inspection of lower Debye temperature materials, RbBr, Ge, and Si reveals that the thermal expansion effect accounts for only a fraction of the T^2 component. The behavior of the low frequency longitudinal phonons may be the reason why the resistivity, corrected to constant volume, still varies faster than T.

I.-INTRODUCTION

Available high temperature thermal conductivity data from dielectric solids allows for the testing of phonon scattering theories. Even a cursory examination of experimental results shows a wide variety of temperature dependences, while convincing derivations,[1,2] have demonstrated that the dominant phonon scattering mechanism, the three-phonon U-process should yield a T^{-1} dependence. It is most disturbing that the thermal resistivity is often seen to increase faster than T. This leads one to postulate the existence of a process which allows for a T^2 resistivity component.

II.-FOUR-PHONON PROCESSES

Traditionally four-phonon processes, which for many years have been known to contribute a T^2 term to the resistivity, have been regarded as the reason why the observed temperature dependence of the thermal conductivity is faster than T^{-1}. Several authors[3] have attempted to fit high temperature data to the form $aT+bT^2+c$. The four-phonon contribution can be estimated by deriving the acoustic mode relaxation time. The perturbation Hamiltonian, derived by writing the frequency changes resulting from strains, is proportional to the square of the dilation. As in the case of the well known three-phonon interaction relaxation time, this perturbation Hamiltonian gives a relaxation time which is the product of three terms: (i) An anharmonic coefficient, written in terms of the Grueneisen parameter and its strain derivatives. From multiphonon interaction theory this coefficient can be found to be $(\gamma^2 - 2\gamma')^2$, where γ is the usual Grueneisen parameter and where $\gamma' = d\gamma/d\Delta]_{\Delta=0}$. (ii) An explicit frequency and temperature dependent part which is found to be $\omega^2 T^2$ by invoking the high temperature approximation ($KT \gg \hbar\omega$). This is in contrast to $\omega^2 T$ for the three-phonon case. (iii) An assortment of constants representing the physical properties of the material. The combination of these terms yields

$$\left(\frac{1}{\tau}\right)_4 = (\gamma^2 - 2\gamma')^2 \frac{\pi^2}{18} \frac{K_B^2 a}{M^2 v^5} \omega^2 T^2 \tag{1}$$

This expression can be compared directly with a three-phonon relaxation time such as that of Roufosse and Klemens.[2] The ratio of the scattering strengths is given by

$$\left(\frac{1}{\tau}\right)_4 / \left(\frac{1}{\tau}\right)_3 = \left[\frac{(\gamma^2-2\gamma')^2}{\gamma^2}\right]\left[\frac{K_B T}{Mv^2}\frac{\pi}{72}\right] \tag{2}$$

At 1000K the second term in brackets typically has a value of 10^{-3}. Furthermore the first term is generally less than 1.0 in value, since according to the work of Thomsen[4] γ and γ' have the same sign. In addition one often finds that $\gamma' \approx \gamma \approx 2$, in which case the first term and therefore the entire ratio approximately vanishes. Consequently even at the highest measurement temperatures, four-phonon processes are probably not the origin of the T^2 resistivity component.

III.-THE EFFECT OF THERMAL EXPANSION

There is, however, the effect of the thermal expansion on the three-phonon processes at high temperatures, which may be responsible for the conductivity varying faster than T^{-1}. Mooney and Steg[5] obtained an expression for the thermal conductivity of compressed material to explain the pressure dependence of the conductivity of NaCl:

$$K = K_0 \left(1 - \left(3\gamma + \frac{2}{3} + 2\frac{\gamma'}{\gamma} \right) \Delta \right) \qquad (3)$$

It is now proposed to use

$$K_0 = \frac{K_{exp}}{1 - \left(3\gamma + \frac{2}{3} + 2\frac{\gamma'}{\gamma} \right)\Delta} = \frac{K_{exp}}{1 - \beta(T)} \qquad (4)$$

to correct the experimental results for the effect of thermal expansion. If the thermal expansion effect alone is responsible for the T^2 component, then the temperature dependence of K_0 should be very close to T^{-1}, while if the thermal expansion were responsible for only a part of the T^2 component then the conductivity corrected to constant volume should at least be closer to a T^{-1} dependence than the uncorrected.

Before proceeding to test the above equation on experimental results it should be noted that this procedure is an approximation in at least three senses: (i) The second and higher order terms in Δ have been disregarded. It is clear that if Δ becomes an appreciable fraction of 1.0, then Δ^2 and perhaps higher order terms become important. (ii) Eq. (4) is not strictly correct at temperatures less than θ_D. At temperatures considerably less than θ_D the specific heat is still increasing rapidly and the three-phonon U-process relaxation time assumes exponential form. (iii) Eq. (4) holds only if the three-phonon U-process is by far the dominant process. If point defects and/or extended defects are important then the addition of their relaxation rates in the thermal conductivity integral leads to a different and generally slower temperature dependence.

Finally at the high temperatures the experimental results clearly show radiation contributions to the thermal conductivity, and these should be corrected for. The actual increase with temperature of the thermal conductivity at the highest temperatures allows one to determine at least a major portion of the radiation contribution at lower temperatures by assuming a T^3 dependence.

The values of KT for KCl, where K is the experimental value of the conductivity with the suspected radiation contribution subtracted out, is found to decrease markedly with temperature (see Table I). When, however, Eq. (4) is used, K_oT is very nearly a constant over a wide temperature range. Because the Debye temperature is relatively low, there is in fact a wide temperature range over which the high temperature three-phonon U-process form is applicable and yet the radiation is still sufficiently small.

If one plots KT against T for BeO and Al_2O_3 the effect of radiation is clearly seen, as the experimental results turn up sharply at the highest temperatures. The corrected K_oT does not begin to approach a constant value until relatively high temperature (approximately 800K). In fact the Debye temperatures of BeO and Al_2O_3 are so high that it is almost impossible to separate the low temperature U-process, defect, and radiation contributions. It would therefore be helpful to consider other materials with low Debye temperatures. RbBr (θ_D = 150K) when corrected for both radiation and expansion does not show K_oT to be constant; it still decreases with increasing temperature indicating perhaps the existence of yet another process whose associated resistivity increases faster than T. The same effect is seen in even more striking fashion for Ge and Si; here it is clear that radiation and expansion can account for only a fraction of the resistivity component which varies faster than T.

IV.-THE EFFECT OF LOW FREQUENCY LONGITUDINAL PHONONS

The four-phonon process as a mechanism contributing a T^2 resistivity term has been ruled out as the dominant interaction causing the faster than T dependence because it is almost vanishingly small compared to the three-phonon U-process. The low frequency longitudinal phonons, however, are severely restricted, due to the conservation conditions and the dispersion relations, as to the interactions in which they can participate.[6] The Herring rule gives

$$\left(\frac{1}{\tau}\right)_{3N} = A\omega^4 T \tag{5}$$

for three-phonon N-processes.[7] The four-phonon relaxation time has the form

$$\left(\frac{1}{\tau}\right)_4 = B\omega^2 T^2 \tag{6}$$

If one assumes that longitudinal phonons of frequency lower than ω_o undergo these interactions, and one takes ω_o as the frequency at which the two relaxation times are equal, one finds a contribution of the longitudinal phonons to the conductivity which

TABLE I.

KT, where K is the experimental thermal conductivity, and K_0T, where K_0 is the thermal conductivity corrected for radiation and thermal expansion, at selected temperatures.

	RbBr			KCl	
T	KT [8]	K_0T		KT [9]	K_0T
100	12.89	13.49			
160	12.08	13.04			
200	11.38	12.52		22.2	24.4
240				21.0	24.0
260	10.37	12.05			
300	9.69	11.75		19.6	23.7
360	8.78	11.38			
400	8.20	11.08		18.2	24.6
460				17.0	24.8

	Al_2O_3			BeO	
T	KT [10]	K_0T		KT [11]	K_0T
400				88.8	92.0
480	975	1020			
600				66.6	72.1
740	740	821			
800				50.1	56.9
900	693	799			
1000				44.7	54.2
1200	663	792			
1320				41.7	53.8

	Ge			Si	
T	KT [12]	K_0T		KT	K_0T
150	195	196.2			
200	190	192.3			
250	183	186.5			
300	180	184.8			
400	176	183.5		420	425.5
500	168.5	179.1		400	407.2
600	160.2	173.2		384	393
700				364	375
800				344	355.5
1000				310	322.5

varies as $T^{-3/2}$.

V.–OTHER EFFECTS AND SUMMARY

Another possible contributing factor in determining the thermal conductivity temperature dependence at high and intermediate temperatures may be the interaction of acoustic and optical modes. The precise form of the resulting temperature dependence is not clear, but the magnitude of such scattering appears to be too small to account for the observed deviations from T^{-1}. However, more detailed calculations should be made for the above mentioned materials.

In summary, it is found that there are deviations from the expected T^{-1} dependence in the high temperature thermal conductivity. Four-phonon processes per se appear to be unimportant. The discrepancies are reduced, if not eliminated, if the thermal conductivities are corrected to constant volume. The remaining discrepancies may possibly arise from the low-frequency longitudinal phonons.

[*]Research supported by U.S. Army Research Office, Durham.
[1]G. Leibfried and E. Schloemann, Nachr. Akad. Wiss. Goett. Math.-phys. Kl. IIa, 71 (1954).
[2]M. Roufosse and P. G. Klemens, Phys. Rev. B7, 5379 (1973).
[3]D. Billard and F. Cabannes, High Temp. - High Press. 1, 201 (1971).
[4]L. Thomsen, J. Phys. Chem. Solids 31, 2003 (1970).
[5]D. L. Mooney and R. G. Steg, High Temp. - High Press. 1, 237 (1969).
[6]P. G. Klemens, J. Appl. Phys. 38, 4573 (1967).
[7]C. Herring, Phys. Rev. 95, 954 (1954).
[8]J. P. Moore, R. K. Williams, and R. S. Graves, Phys. Rev. B11, 3107 (1975).
[9]E. D. Devyatkova and I. A. Smirnov, Soviet Phys. Solid State 4, 1445 (1963).
[10]M. McQuarrie, J. Am. Ceram. Soc. 37, 91 (1954).
[11]W. D. Kingery, J. Am. Ceram. Soc. 37, 107 (1954).
[12]C. J. Glassbrenner and G. A. Slack, Phys. Rev. 134, A1058 (1964).

THE EFFECT OF A DELTA-PHASE STABILIZER ON THE THERMAL DIFFUSIVITY OF PLUTONIUM

Homer D. Lewis, Jerry F. Kerrisk, and Karl W. R. Johnson

Los Alamos Scientific Laboratory

P. O. Box 1663, Los Alamos, New Mexico 87545

Thermal diffusivities of high purity plutonium and a delta-phase alloy containing 1.0 weight percent gallium were measured over the temperature range 20°C to 600°C. Measured diffusivity values of the high purity plutonium increased from about 0.011 cm^2/s at 20°C to 0.057 cm^2/s at 550°C, while the diffusivity for the delta-plutonium alloy increased from about 0.040 cm^2/s at 20°C to 0.065 cm^2/s at 400°C.

Thermal conductivities were calculated from measured diffusivities for each type of sample using available specific heat and thermal expansion data. Thermal conductivities of the high purity plutonium samples showed a slight linear increase with temperature for each of the six allotropic crystalline forms. Values ranged from 3.2 W/m·deg C in the alpha temperature range to over 14 W/m deg C in the delta temperature range, and are consistently lower than predicted for a free electron metal.

Conductivities calculated for the delta-phase alloy ranged from 8.6 W/m deg at 25°C to 16.8 W/m deg at 400°C.

The typical variation in L/L_0 over the temperature range was 0.66 to 0.89 for the high purity plutonium and 1.18 to 1.10 for the delta-phase alloy, where L_0 is the Lorenz number from the free electron model, and L is calculated from the experimental conductivities.

This work was performed under the auspices of the United States Energy Research and Development Administration.

INTRODUCTION

The experiments to be discussed are a part of a more general investigation into the effect of delta-phase stabilizers on the thermal conductivity of plutonium above 300°K. Several elements, including aluminum, americium, gallium, and cerium form solid solutions with face-centered cubic delta plutonium, resulting in a stable delta-phase to below room temperature. The results of thermal diffusivity measurements on three types of samples will be considered.

a) High purity "crackfree" plutonium.
b) High purity plutonium, heat treated to produce microcracks.
c) 1.0 weight percent (3.35 atom percent) gallium-plutonium alloy.

The complexity of the crystallographic structure of un-alloyed plutonium is illustrated by the abbreviated properties summary shown in Table I. The un-alloyed metal exhibits six crystalline forms below the liquidus. Several investigations have shown that the α, β and γ transformation temperatures are strongly dependent on heating or cooling rate. For example the α-β transition temperature reported by Adams[2] et al. was 126 ± 2°C on heating at 1°C/min. and 142 ± 5°C at 4°C/min. The metal shows approximately a twenty percent volume increase at about 315°C, and the close packed fcc delta-plutonium exhibits the lowest density of the six crystalline forms.

The 1.0 weight percent gallium-plutonium alloy is single phase from room temperature to approximately 500°C. The upper limit of the two-phase ($\delta + \epsilon$) region at this composition occurs at about 570°C.[3] The volumetric thermal expansion is about 0.5 percent between room temperature and 400°C[4] as compared to 20 percent for the unalloyed Pu.

EXPERIMENTAL PROCEDURE

The diffusivity measurement method used is similar to the flash technique described by Parker.[5] A commercial high temperature furnace mounted inside an inert atmosphere glovebox was used to control absolute sample temperature. The furnace atmosphere was high purity argon at slightly below ambient glovebox pressure. The normal glovebox atmosphere is high purity argon containing less than 7 ppm H_2O and 5 ppm O_2. Both absolute sample temperature and the back surface temperature were monitored by means of a "Platinel 2" thermocouple, tack welded to the sample. The back surface temperature history was recorded on film from a commercial oscilloscope operating in single sweep, memory mode, and

TABLE I

SOME PROPERTIES OF PLUTONIUM[1]

Phase	Stability Range C	Lattice	Structure Atoms Per Unit Cell	Density g/cm^3	Expansion $\Delta l/l \times 10^2$
α	Below ~115	Simple Monoclinic	16	19.86	~0.50
β	~115 - ~200	Body-centered Monoclinic	34	17.70	~3.90
γ	~200-310	Face Centered Orthorhombic	8	17.14	~5.40
δ	310-452	Face Centered Cubic	4	15.92	~7.70 - ~7.30
δ	452-480	Body-centered Tetragonal	2	16.00	~7.1
ϵ	480-640	Body-centered Cubic	2	16.51	~6.6

the absolute sample temperature was determined using a Rubicon-B potentiometer. The input energy pulse was supplied, through a simple optical system, to the sample by a commercial neodymium-glass pulsed laser. The apparatus was "checked out" before each run by determining the diffusivity of a known molybdenum sample. The diffusivity of the molybdenum sample was determined after initial calibration of the equipment against a sample of Round-Robin Armco Iron.

The samples used in this investigation were 10 mm in diameter. Diffusivity determinations were made on sample pairs for each type of material considered except for the "heat treated" unalloyed plutonium. Sample thicknesses for each pair were selected so that the measured half times differed by about a factor of two. The impurity level of the unalloyed plutonium was less than 200 ppm total Fe, Si, C, W, Ta, Al, Ni, Mo and trace impurities. The "crack free" samples were machined from chill cast and annealed rod.

The "heat treated" or micro-cracked sample was prepared by heating a crack free sample to 140°C at about 2°C/min and furnace cooling in the diffusivity apparatus furnace. Density of the crack free samples was 19.77 g/cm^3, and the "heat treated" sample density was 19.73 g/cm^3.

Three pairs of cast Pu-Ga alloy samples were considered. Each pair was subjected to a different homogenization treatment. The impurity levels of the three pairs were respectively less than 1,500 ppm, less than 500 ppm and less than 200 ppm. Densities of these six samples ranged from 15.72 to 15.84 g/cm^3.

Thermal conductivity was calculated from the diffusivities determined for each sample using specific heat and electrical resistivity data from the literature. The specific heat data of Pu and Pu-Ga alloy respectively by Kay and Loasby[6] and Rose[7] et al. are shown in Figure 1. Resistivity data by Sandenaw and Gibney[8] and Andrew[9] are shown in Figure 2. The transition temperatures indicated along the abscissa of Figures 1 and 2 are for unalloyed plutonium.

RESULTS AND DISCUSSION

The measured diffusivities of the unalloyed plutonium samples are summarized in Figure 3. The values have been corrected for thermal expansion and heat loss.[10] Discontinuities in the general data trend reflect the various transition temperatures. It was expected that the "microcracked" sample would show lower values than the "crack free" samples in the temperature range 300° to 450°C, but the behaviour in the range 200° to 300°C is not understood. Diffusivity values for the Pu-Ga alloy samples are shown in Figure 4. The maximum variability among the values determined for the six samples is about ± 2% up to 400°C. The increased variability between 450 and 500°C is attributed to micro-inhomogeneity of the alloy. Variability in the ϵ phase is also believed to

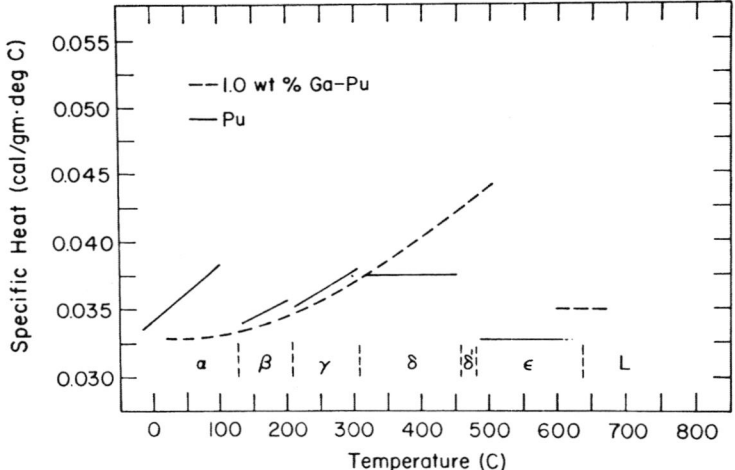

Figure 1. The specific heat of plutonium[6] and plutonium-1.wt% gallium.[7]

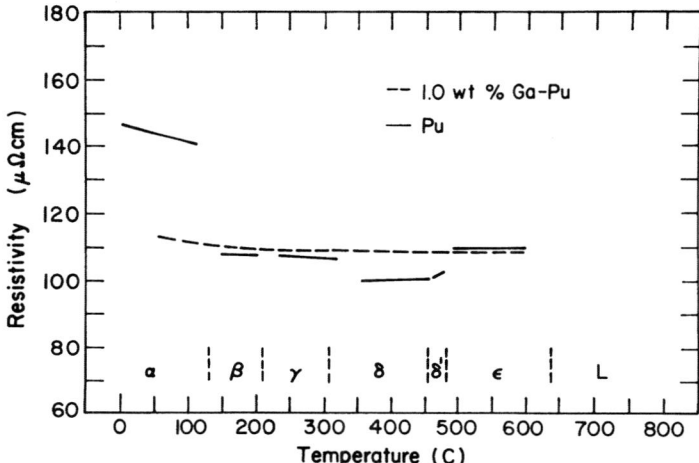

Figure 2.   The electrical resistivity of plutonium[8] and plutonium 1.wt% gallium.[9]

be the result of inhomogeneity.

Calculated conductivity values are shown in Figure 5. The lines between the large black dots represent the thermal conductivity calculated from the free electron model for unalloyed Pu. The line between the X's is the free electron model conductivity calculated for the Pu-Ga alloy.

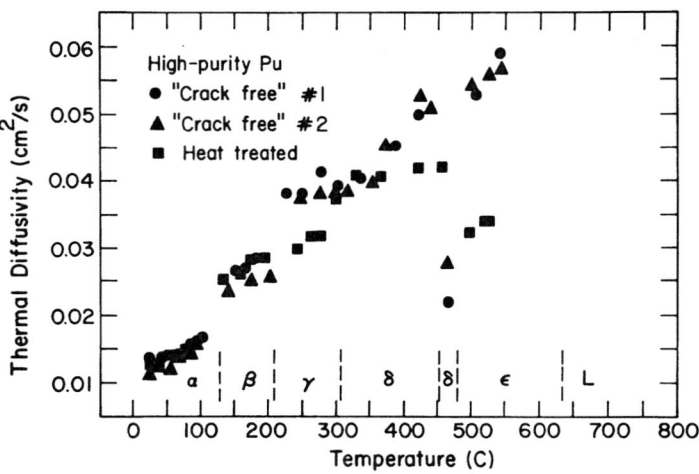

Figure 3.   Measured diffusivity for high-purity plutonium.

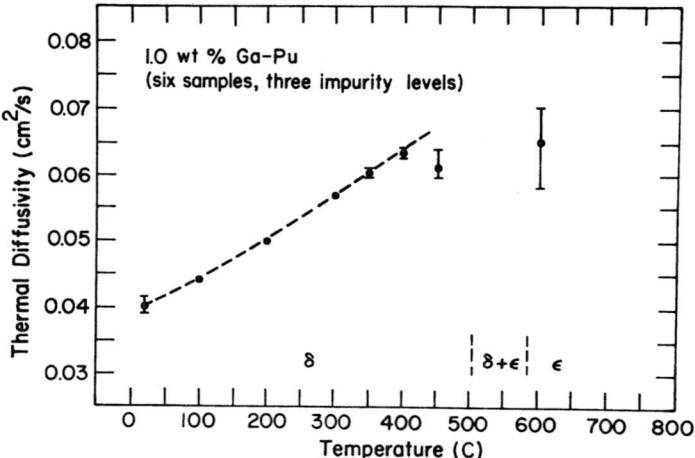

Figure 4.   Measured diffusivity for plutonium-1.wt% gallium alloy.

Conductivity values calculated from the diffusivity of the "crack free", unalloyed samples are shown by the unenclosed dots. Values were calculated using the specific heat data of Kay and Loasby[6] and density corrections from expansion data from Wick.[1] The variability between the two samples is less than about ± 7%. The data seem to show the same discontinuities in conductivity at transition temperatures as calculated from the theoretical Lorenz number and the resistivity except at the β-γ transition. If data near this transition temperature were ignored, the conductivity could be considered to increase linearly from about 6 W/m·deg at the α-β transition to about 10.2 W/m·deg at the γ-δ transition. Conductivity of the "heat treated" plutonium sample is shown by the circled dots. It was expected that conductivity would be lower than that for the "crack free" material, however, as stated for the diffusivity values, the rather sudden decrease at the β-γ transition is not understood.

The conductivity values calculated from diffusivity for the Pu-Ga alloy samples are shown in Figure 5 as vertical bars representing the variability among the six samples. The conductivity is higher than that calculated from the free electron model, and increases smoothly with temperature up to about 450°C, or just below the (δ + ϵ) transition temperatures. The values just above 600°C show ordering dependent on impurity level with the highest impurity level material showing the highest conductivity.

The ratio of $^L/Lo$ for the "crack free" high–purity plutonium ranged from about 0.66 to 0.89, Lo being the Lorenz number from the free electron model, and L the Lorenz number calculated from the experimental thermal conductivity. The largest values were obtained just below the δ–δ' transition. Wittenberg's[11] conductivity for unalloyed plutonium from 250 to 450°C was consistently higher, and from 500 to 600°C lower than determined in this investigation. His data give $^L/Lo$ values which range from 2.18 in the γ temperature range to 0.45 in the ε temperature range. Andrew's[12] measurements of thermal conductivity of high purity plutonium between 50 and 300°K gave 6 W/m·deg and 1.27 for conductivity and $^L/Lo$ respectively at 300°K. Extrapolation of Andrew's conductivity curve shows about 8 W/m·deg at the α–β transition and approximately 12 W/m·deg at the β–γ transition.

The $^L/Lo$ values calculated for the Pu-Ga alloy ranged from about 1.18 at 60° to 1.10 at 400°C. The conductivity at 25°C agrees well with the value 9 W/m·deg at 300°K reported by Andrew[13] from thermal conductivity measurements on a 1.0 wt.% Ga-Pu alloy from 75 to 300°K.

ACKNOWLEDGEMENTS

The authors wish to thank D. R. Harbur for sample preparation and K. A. Johnson for metallographic examination of samples used in this work.

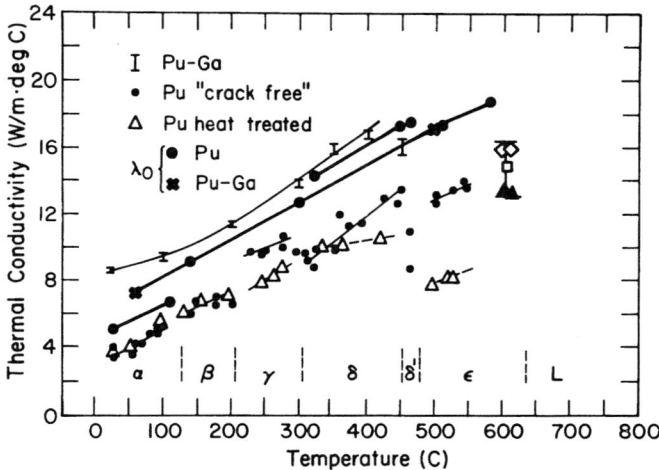

Figure 5. Thermal conductivity values for plutonium and plutonium-1 wt% gallium, calculated from the diffusivity.

REFERENCES

1.   Wick, O. J., (Ed.) Plutonium Handbook Vol. 1, pp. 33-41, Gordon and Breach, N.Y., (1967).

2.   Adams, E. T., Mardon, P. G., North, H. M., Pearce, J. H., The Metal Plutonium, Univ. Chicago Press, p. 139, (1961).

3.   Ellinger, F. H., Miner, W. N., Boyle, D. R., Schonfeld, F. W., Constitution of Plutonium Alloys, LA-3870, Los Alamos Scientific Laboratory, (1968).

4.   Reuscher, J. A., Schmidt, T. R., Harbur, D. R., Properties of Pu Materials Under Conditions of Rapid Fission Heating, (U), SC-DR-720740.

5.   Parker, W. J., Jenkens, R. J., Butler, C. P., and Abbot, G. L., J. Appl. Phys., 32, 9, 1679, (1961).

6.   Wick, O. J., op. cit., p. 38, and Kay, A. E., and Loasby, R. G., Phil. Mag., 9, 43, (1964).

7.   Rose, R. L., Robbins, J. L., and Massalski, T. B., J. Nucl. Mater., 99-107, (1970).

8.   Wick, O. J., op. cit., p. 40, and Sandenaw, T. A. and Gibney, R.B., Phys. Chem. Solids, 6, 81-88, (1958).

9.   Andrew, J. F., Resistivity of 1.0 wt% Ga-Pu, (unpublished data).

10.   Cowan, R. D., J. Appl. Phys. 34, 4, 926-927, (1961).

11.   Wittenberg, L. J., Thermochem. Acta, 7, 13-23 (1973).

12.   Andrew, J. F., Phys. Chem. Solids Vol. 28, 577-580, (1967).

13.   Andrew, J. F., J. Nucl. Mater. 30, 343-345 (1969).

EXPERIMENTAL DETERMINATION OF THERMAL PROPERTIES

OF FUSED QUARTZ, POLY- AND MONOCRYSTALLINE Al$_2$O$_3$

A.A. El-Sharkawy[*], R.P. Yourchack, S.R. Atalla[*]

MOSCOW STATE UNIVERSITY

Using the apparatus described in previous work[1], the thermal capacity, diffusivity and conductivity coefficients of fused quartz and poly - and monocrystalline Al$_2$O$_3$ in the temperature range 1200-2000°K have been measured. The role of radiative energy transfer was investigated. The absorption coefficient of fused quartz was calculated. The thermal properties of monocrystalline Al$_2$O$_3$ were measured along the three axes of symmetry of the lattice.

INTRODUCTION

Wikks and Block[2] have measured the thermal capacity of fused quartz in the temperature range 300-1800°K. Experimental data on the thermal conductivity of quartz have been reported by Kingery and Lee[3] in the temperature range from 300°K to 1300°K, and by Romashin[4] in the range 300-1200°K. No experimental data for the thermal diffusivity, to our knowledge, have been reported.

For the thermal capacity of polycrystalline Al$_2$O$_3$, Berman[5], Chekovskoi[6], and Gronow[7] have reported experimental data in the temperature range 1200-2200°K. Luliev[8] and McQuerrie[9] have reported thermal conductivity data in the temperature range 1000-2000°K. Thermal diffusivity data have been reported by Plummer[10] for temperatures up to 1300°K and by Mavashev[11] from 1250 to 2000°K.

[*]Present address: Al-Azhar University, Faculty of Science, Department of Physics, Nasr-City, Cairo-EGYPT.

For monocrystalline Al_2O_3 the thermal capacity was measured from 300 to 1000°K both by Lietz[12] and Chekovskoi[6]. In this work the thermal capacity, conductivity and diffusivity coefficients of the mentioned substances have been measured simultaneously for the same specimens.

EXPERIMENTAL

Specimens in the form of circular discs of diameter 10 mm and thicknesses 0.5-2.5 mm were used. The mean temperature of the specimen was achieved by means of electron-bombardment of the metallized side surface. One face of the specimen have been subjected to a periodically changing (on-off) thermal flux produced by an infrared radiation from a powerful incandescent lamp using a shutter. The temperature oscillations of the second surface were detected by means of a photo-multiplier. Processing the obtained curves, the amplitude and phase of temperature oscillations can be found. Hence, the thermal capacity, diffusivity and con-ductivity coefficients can be computed. The necessary corrections for the role of radiation heat transfer at different frequencies have been taken into consideration.

RESULTS

Fused Quartz. Specimens of density 2.22 gm/cm³ and melting point 1728°C have been investigated. Measurements have been carried out using discs of diameter 10 mm and thicknesses 2.07 and 0.5 mm. The obtained data for the thermal capacity are shown in Fig. (1), for thermal conductivity in Fig. (2) and for thermal diffusivity in Fig. (3).

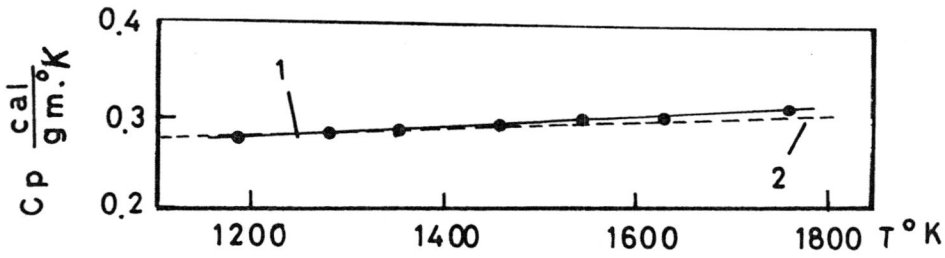

Fig. (1)   Thermal capacity of fused quartz
1-   Present Data
2-   Wikks K., Block F.   2

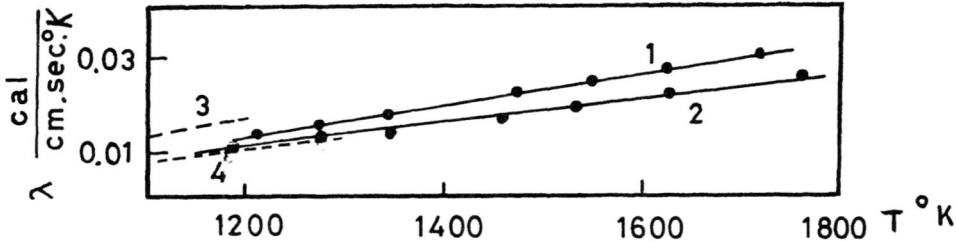

Fig. (2)   Thermal conductivity of fused quartz
          1-   Thickness 2.07 mm
          2-   Thickness 0.5 mm
          3-   Romashin A. /4/
          4-   Kingery W.and Lee D. /3/

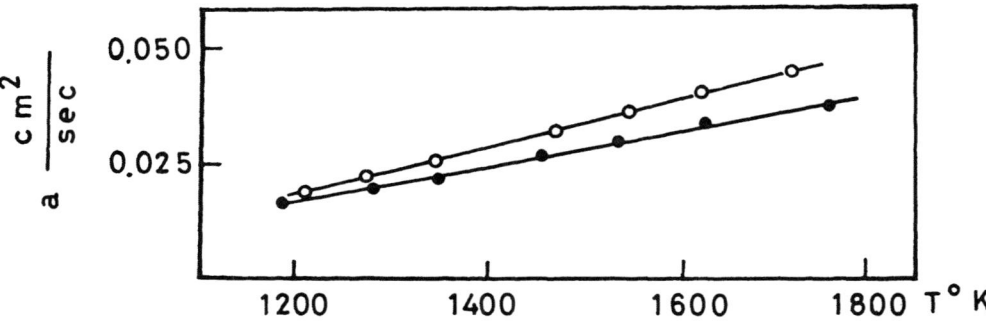

Fig. (3)   Thermal diffusivity of fused quartz
          •-   Thickness 2.07 mm.
          o-   Thickness 0.5 mm.

Al_2O_3. Monocrystalline specimens of Al_2O_3 of density 3.98 gm/cm^3 have been cut in three different planes $(10\bar{1}0)$, $(\bar{1}000)$ and $(1\bar{1}20)$ to find the influence of anisotropy on the thermal properties. The obtained results for the thermal capacity are given in Fig. (4),for thermal diffusivity in Fig.(5), and for thermal conductivity in Table(1).The same data for polycrystalline specimens were given in the previous work[1].

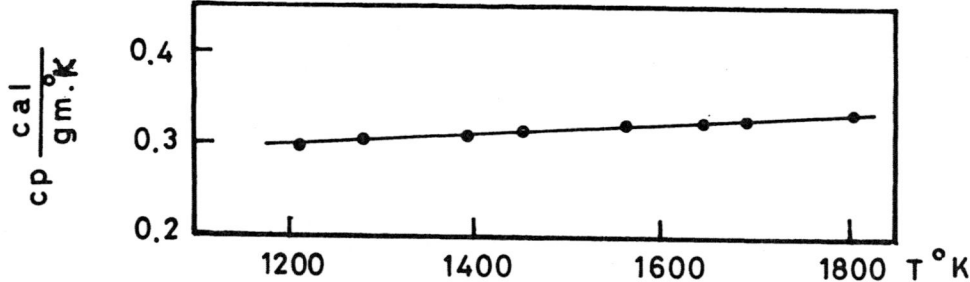

Fig. (4)   Thermal capacity of monocrystalline Al_2O_3.

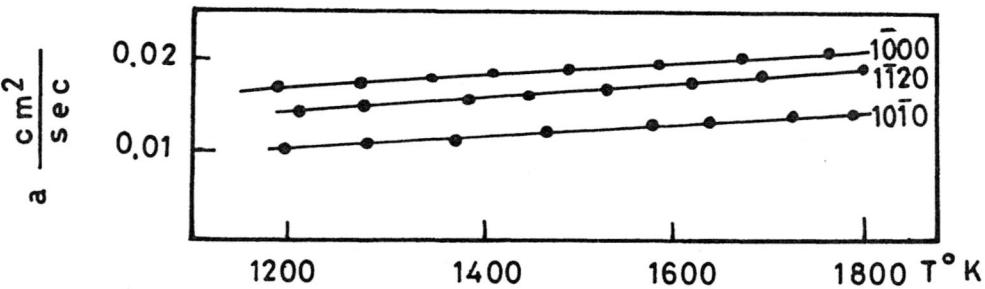

Fig. (5) Thermal diffusivity of monocrystalline Al_2O_3

TABLE 1

Thermal conductivity of monocrystalline Al_2O_3

T^oK	Thermal conductivity Cal/cm.sec.deg.		
	$(10\bar{1}0)$	$(\bar{1}000)$	$(1\bar{1}20)$
1200	0.0126	0.0170	0.0169
1300	0.0131	0.0215	0.0182
1400	0.0143	0.0223	0.0148
1500	0.0150	0.0236	0.0208
1600	0.0160	0.0253	0.0218
1700	0.0169	0.0262	0.0235
1800	0.0191	0.0279	0.0253

DISCUSSION

For fused quartz, as seen in Fig.(1), the thermal capacity does not depend upon temperature, while the thermal conductivity increases with temperature as shown in Fig.(2). The maximum difference in the thermal diffusivity from the experimental values for the thickn esses 2.07 and 0.5 mm are 11% at 1200°K and 20% at 1750°K. This indicates that the mechanism of energy transfer in fused quartz has conductive and radiative components.

The radiative thermal conductivity can be expressed as follows[13];

$$\lambda_{photon} = \frac{4 \ \sigma \ n^2 \ T^3 L}{1 + \ 3/4 \ \alpha \ L}$$

where; σ = Stefan-Boltzman constant
         n = refractive index
         α = absorption coefficient
         L = thickness of the specimens

From the thermal conductivity values for the two different thicknesses, it is possible to calculate the absorption coefficient α using the previous formula. The calculated results for the absorption coefficient are given in Table (2).

TABLE 2

The absorption coefficient of fused quartz

$T°K$	cm^{-1}
1200	4.90
1300	5.29
1400	5.75
1500	5.76
1600	5.80
1700	6.00

For Al_2O_3, unlike the obtained results for the thermal diffusivity of the polycrystalline specimens, the diffusivity coefficient increases with temperature. The obtained values of the thermal conductivity for (1000) plane is 1.5 times greater than those for the perpendicular plane (1010). This is in a qualitative agreement with the obtained results for the other thermal characteristics of anisotropic insulators at lower temperatures[14,15,16]. The increase in the thermal conductivity

214

value from 0.017, at 1200°K, to 0.0279 Cal/cm.gm.sec, at 1800°K, cannot be explained by the lattice conductivity alone because the thermal conductivity of polycrystalline specimens was 0.0167 at 2000°K. As the obtained values of the thermal conductivity are the same for two polycrystalline specimens with different thicknesses, 2 and 2.5 mm, it can be deduced that there cannot be any radiative energy transfer. However, for the monocrystalline specimens, the existence of the radiative compon-ent of energy transfer is confirmed by the difference in thermal conductivity values for specimens with different thicknesses.

CONCLUSION

For fused quartz there exists an additional radiative mechanism of energy transfer in the investigated range of temperature. In case of polycrystalline Al_2O_3 radiative energy transfer can be neglected. The monocrystalline Al_2O_3 is highly anisotropic material with three different planes of symmetry $(10\bar{1}0)$, and $(\bar{1}000)$ and $(1\bar{1}20)$.

REFERENCES

[1] A.A. El Sharkawy et al, The 4th European conference on the thermal properties of solids at high temperature, September 1974, Orleans, France.

[2] K.E. Wikks, and F.E. Block, Thermodynamical Properties of 65 elements, their oxides, carbides and nitrides. Metallurgy, Moscow (1965) (in Russian).

[3] W.D. Kingery and W.D. Lee, Radiation energy transfer and thermal conductivity of ceramic oxides, J.Am.Cer.Soc. 43, (1959).

[4] A.G. Ramashin, Teplofizika Vysokikh Temp. 4 (1969).

[5] R. Berman. The thermal conductivity of some dielectric Solids, Proc. Roy. Soc. p. 90 (1961).

[6] V.R. Luliev et al. Zavodskaja laboratoria No. 8 (1957).

[7] H.E. Gronow, and H.E. Schwiete Die Spesifisohen Warmen von Cao, Al_2O_3 Von 20 dis 1500°C. Ztschr anorg chem. 216 (1933).

[8] A.N. Luliev et al. Zavodskaja laboraotoria No. 8 (1957).

[9] M.G. McQuerrie et al. High Temperature method and results for alumina, magnesia, beryllia from 1000 to 18000°C. J. Am. Ceram. Soc. 37, N2 (1954).

[10] W.A. Plummer et al. Method of measurement of thermal diffusivity to 1000°C. J.Am. Ceram. Soc. 1, N7 (1962).

[11] U.Z. Mavashev: Ph.D. Thesis, Tashkent (1971).

[12] J. Lietz: The specific heat of futile and anatase Hamburger Beitr angew, Mineral u. Kristallphys. 1, 229-238 (1956).

[13] J. Schatz and G. Simmons. Methods of Simultaneous measurement of radiative and lattice thermal conductivity. J. Applied Physics, 43, N6 (1972).

[14] T. Birch and N. Clark. The thermal conductivity of rocks and its dependence upon temperature and composition. Am. J. of Science 233, 1940.

[15] A. Goldsmith et. al., Handbook of thermophysical properties of solid materials, Pergamon Press, Oxford, London, New York, Paris (1963).

[16] A.P. Dimitriev et al. Physical Properties of rocks at high temperatures, Nedra, 1969.

THE THERMAL CONDUCTIVITY AND ELECTRICAL RESISTIVITY OF COPPER AND SOME COPPER ALLOYS IN THE MOLTEN STATE +

R. P. Tye

Dynatech R/D Co.

Cambridge, Massachusetts

Measurements of the thermal and electrical conductivity of two different pure copper materials and ten commercial copper alloys have been undertaken in vacuum over the approximate temperature range 1300-1500K. The alloys ranged from simple low alloy to complex high alloy materials presenting a representative range of commercial products in wide use.

The thermal conductivity measurements were carried out by guarded axial rod longitudinal heat flow technique based upon that used by Powell and Tye for other materials at lower temperatures. The samples were placed in a thin-walled high-purity molybdenum cylinder between cylindrical reference sections of the same material. The heat flow in the test sample was determined from a knowledge of the thermal conductivity of the molybdenum which had been evaluated in a prior experiment. The electrical resistivity measurements were carried out using a standard four-probe technique in a high-purity alumina system. The estimated accuracy of the thermal conductivity experiment is ± 5% and ± 2% for the electrical resistivity.

Results are presented and discussed both in comparison to the relatively few other experimental results available and in relation to the emperical equation proposed by Powell and Tye for molten metal systems. The values for the Lorenz Number of the two pure materials are some 5 to 10% above the theoretical value and does not show a tendency to decrease rapidly with temperature as has been indicated recently for some other molten metals. The Lorenz Number for the copper alloys appears to be consistently higher than the value for the pure material which would indicate that there is still a significant lattice component of thermal conductivity for the alloy system.

+ Work carried out on behalf of the International Copper Research Association, New York.

Chapter III

Reference and Technical Materials

STANDARD REFERENCE MATERIALS FOR

THERMAL CONDUCTIVITY AND ELECTRICAL RESISTIVITY*

J. G. Hust

Cryogenics Division, Institute for Basic Standards

National Bureau of Standards, Boulder, Colorado

ABSTRACT

A brief historical review is presented of thermal conductivity reference material research. Recent reference material reseach by the National Bureau of Standards is described. Critically evaluated thermal conductivity and electrical resistivity data are summarized for the following three Standard Reference Materials now available from the Office of Standard Reference Materials, National Bureau of Standards, Washington, D. C.: electrolytic iron (4 to 1000 K), austenitic stainless steel (4 to 1200 K), and tungsten (4 to 3000 K).

Key Words: Austenitic stainless steel; electrical resistivity; high temperature; iron; low temperature; standard reference materials; thermal conductivity, tungsten.

INTRODUCTION

Design and development engineers require thermal and electrical property data of technically important materials. Often these data are not in the published literature and immediate measurements must be performed. Since only a few laboratories have the proven experimental capability to make such measurements, usually they are performed by inexperienced personnel using unproven apparatus. The results, as can be seen from the literature, exhibit excessive scatter; 50% differences are common. In such situations, Standard Reference Materials (SRM's) are invaluable to ascertain the accuracy of the engineering measurements. Currently, an inaccuracy of 10% is allowable for most engineering thermal property data; therefore,

---

*Contribution of the National Bureau of Standards, not subject to copyright.

221

SRM data for engineering applications need to be established with an uncertainty no larger than about 5%.

A few research laboratories performing thermal and electrical measurements are obtaining data with uncertainties at the state-of-the-art level, 1% for thermal conductivity and lower for electrical resistivity. SRM's for use at such laboratories must be correspondingly more accurate and may indeed be possible but have not yet been established.

Considerable effort has been directed toward the development of suitable thermophysical SRM's*, over a period of many years, with limited success. This lack of success may be due, in part, to the tacit assumption that SRM data must be accurate to state-of-the-measurement-art to be useful. There are several reasons why the achievement of thermal and electrical property SRM's with certified inaccuracies of less than 1% is extremely difficult. The principal reason is that material variability, generally, causes property variations of greater than 1% even with the most up-to-date production control techniques. The effects of material variability lead to the consideration of three categories of calibration materials and three concomitant certification inaccuracies: (1) A characterized type of material, e.g., copper, gold, iron, etc. Based on past experience it appears that inaccuracies of 5-10% can be expected. (2) A characterized specific lot of a given type of material. Data uncertainties of one percent appear to be near the lower limit of current production control techniques. (3) Characterized specimens of material. At first glance, it may be thought that the latter SRM's would be invariant; but it is known that the thermal and electrical properties of some specimens change spontaneously with time (aging effects) and depend on thermal and mechanical histories. These effects are especially significant at low temperatures, especially for highly purified materials. Appropriately chosen well-characterized specimens, handled with care to avoid physical and chemical changes, and frequently reexamined to detect changes, presently represent the only means to achieve accuracies in the state-of-the-measurement-art range. This is the basis of round-robin measurements used by standardizing laboratories for state-of-the-art apparatus intercomparisons (see, for example, Laubitz and McElroy [1]). Category (2) is considered to be the most cost-effective to satisfy engineering needs, and, to

---

*The term SRM is used in this paragraph and the following historical review in a broad sense to denote any material or specimen that is to serve as a calibration standard. The term, as coined by the National Bureau of Standards, Office of Standard Reference Materials, generally implies a specific lot of material prepared under strict control and subsequently characterized for chemical composition and homogeneity.

a lesser extent, the needs of standards laboratories. It is also
the philosophical basis of the NBS Office of Standard Reference
Materials.

This report is a result of a program to establish several
thermal and electrical conductivity metal SRM's with conductivities
ranging from pure metals (high conductivity) to structural materials
(low conductivity). Plans are being formulated to extend this
program to insulating materials and dielectric solids as well.

This paper reviews the historical development of thermal
conductivity SRM's. A listing is given of selection criteria for
SRM's and a justification is presented for the establishment of
both engineering and standards laboratory SRM's. Data are compiled
and best values are selected to establish iron, steel, and tungsten
as electrical resistivity and thermal conductivity SRM's. These
materials appear to have the qualities of excellent SRM's. An
adequate supply exists to insure measurement compatibility among
laboratories for about ten years.

The following historical review of SRM efforts is presented to
indicate the relatively large amount of research that has been conducted, and the few thermophysical SRM's that have been established.
It is this divergence between expended efforts and concrete results
that has prompted us to establish potentially useful SRM's, at what
may seem to some as a premature phase of the work. Based on past
experience, it appears that if this is not done, a vast amount of
research is lost. Not because the data are lost, but rather because
the stock of material on which the research was performed is lost.
This consideration also points out the significance of continuity
in SRM projects.

HISTORICAL REVIEW

Early Efforts

Thermophysical property reference material investigations began,
for all practical purposes in the 1930's with the work of Powell at
the National Physical Laboratory (NPL), Teddington, England [2] on
iron and Van Dusen and Shelton at NBS [3] on lead. These successful
efforts resulted in frequently used thermal conductivity reference
materials. Powell's work resulted in the establishment of ingot
iron* (category 1) as a standard which is still being used. Lucks [4]

---

*The ingot iron used for this purpose is Armco iron produced by
Armco Steel Corporation. The use of trade names of specific products is essential to the proper understanding of the work
presented. Their use in no way implies any approval, endorsement,
or recommendations by NBS.

recently reviewed the massive amount of work which has been done on
this material and recommended the continued use of ingot iron as a
reference material. Van Dusen and Shelton's work resulted in an
unofficial lead standard based on a well-characterized lot of pure
lead (category 2) distributed by NBS as a freezing point standard.

Iron

Since the 1930's reference material investigations have been
sporadic with notable efforts by researchers from the NBS (National
Bureau of Standards, U.S.), NPL (National Physical Laboratory,
England), ORNL (Oak Ridge National Laboratory, Tennessee), BMI
(Battelle Memorial Institute, Ohio), and AFML (Air Force Materials
Laboratory, Ohio). The material which has been the subject of
the most extensive investigations is ingot iron. Renewed interest
in this material was spurred by the round-robin experiments
initiated by Lucks of Battelle Memorial Institute during 1959.
Twenty-four laboratories requested and received the round-robin
material for measurements. Data from eight laboratires on a single
lot of ingot iron were ultimately reported and compiled by Lucks [4].
The literature (see Lucks), however, contains data on a total of
eleven distinct lots of ingot iron. Lucks [4] has shown that ingot
iron is an acceptable reference material at temperatures from about
100 K to 1000 K. In this range, material variability affects thermal
conductivity and electrical resistivity by about 5%. At higher
temperatures reported variations increase. At lower temperatures,
especially at liquid helium temperatures, variations of 10% have
been reported on a single 30 cm long rod by Hust et al. [5,6].
Electrolytic iron, SRM 734, was established as a low-temperature
standard by Hust and Sparks [7] becuase it exhibits relatively
small low-temperature variability. Based on their high temperature
study of ingot iron and a high purity iron, Fulkerson et al. [8]
also concluded that high purity iron is a more homogeneous and
stable SRM then ingot iron.

NBS (Washington) Efforts

Flynn of NBS, Washington began a study of potential thermal
conductivity SRM's during the early 1960's. He examined several
ceramics* and alloys†. None of these materials has achieved the
status of an SRM. Descriptions of these efforts appear in the
unpublished proceedings of the early thermal conductivity conferences.
Laubitz and Cotnam [9] reported that Inconel 702 exhibits transformation effects of several percent in thermal conductivity and
recommended against its use as a reference material.

---

*Pyroceram 9606 and Pyrex 7740 (trade names of Corning Glass Works).

†Inconel 702 (trade name of International Nickel Company, Inc.),
 lead, and 60% platinum - 40% rhodium alloy.

At the 1963 thermal conductivity conference, Robinson and
Flynn [10] presented the results of a survey of thermal conductivity
SRM needs. SRM's with a data uncertainty of 3-5% were in greatest
demand. The intended use of SRM's, most often stated, was to check
and calibrate apparatus. Needs were indicated for SRM's of conductivities from 0.01 W/mK to 500 W/mK at temperatures from 4 to 3300 K.

NBS (Boulder) Efforts

Powell of NBS (Boulder) initiated a low-temperature SRM project
during the early 60's. This project has been continued by the author
since that time. Materials studied include ingot iron, electrolytic
iron, gold, tungsten, graphite, and stainless steel. As a result of
these studies, electrolytic iron, stainless steel, and tungsten have
been established as SRM's of electrical resistivity and thermal
conductivity from 4 to 3000 K. It is anticipated that this project
will continue until a sufficiently wide range of conductivities and
temperatures are included to satisfy existing demands for thermophysical SRM's.

AFML-AGARD Project

Minges [5th Thermal Conductivity Conference, 1965] reported
the initiation of an AFML sponsored high-temperature reference
materials program. This program was divided into two phases.
Phase I included the preliminary selection and characterization of
materials as potential reference materials. Selection criteria
were established, dozens of materials screened, and about 15 were
chosen for experimental evaluation. Phase II included further
measurements on those materials selected from Phase I studies.
Arthur D. Little Corp. contracted with AFML to perform this study.
The results were reported in reference [11]. The materials of
particular interest in Phase II of this program were aluminum oxide,
thorium oxide, tungsten, and graphite.

After partial completion of the AFML program an international
program, principally high-temperature, was initiated under the
auspices of the Advisory Group for Aerospace Research and Development, NATO (AGARD). Fitzer of Karlsruhe University, Germany,
directed this program in close cooperation with the AFML program.
The establishment, progress, and results of this program are
described in a series of report by Fitzer [12]. Minges has also
summarized some of the results on AFML-AGARD programs [13]. The
materials, internationally distributed and measured by numerous
laboratories, are: platinum, gold, copper, austenitic steel,
tungsten (both sintered and arc-cast), tantalum - 10% tungsten
alloy, alumina, and graphite.

SRM SELECTION CRITERIA

The criteria for screening and selecting potentially useful materials for physical property SRM's are generally well understood and accepted. These criteria are not met absolutely by any material, but serve as a guide to determine which materials are most suitable. Some of the more significant factors are:

1. The material should be homogeneous* and isotropic through- out a lot. The lot should be large enough to be adequate for at least a decade and renewable with a minimum of effort.

2. Thermophysical properties should not vary with time and should be relatively unaffected by the environment of the measurement apparatus. The material should have chemical stability, thermal shock resistance, low vapor pressure, and insensitivity to stress.

3. The material should be readily available, machinable, be relatively inexpensive, and have sufficient strength to be handled without suffering damage.

4. The material should have characteristics similar to the material to be measured.

5. The material should be useful over a wide temperature range.

The materials described in this report satisfy these criterial reasonably well.

The following sections contain a summarization of the charac- terization and recommended values of thermal conductivity and elec- trical resistivity for these materials. The detailed description of this research has been presented by Hust and Giarratano [14, 15, 16].

---

*The term homogeneous refers here to the uniformity of the thermophysical property in question. Homogeneity of a thermo- physical SRM implies not only chemical homogeneity, as in chemical composition SRM's, but also homogeneity of physical characteristics of the material. The parameters affecting physical property homogeneity are so numerous that detailed characterization of each is prohibitive. Instead, one often reverts to aggregate characterization methods, such as by electrical resistivity as discussed later.

MATERIAL CHARACTERIZATION

Steel

The steel selected for low conductivity SRM's is an austenitic stainless steel produced by a German facility containing 20 wt. % nickel, 16 wt. % chromium, about 1 wt. % of other elements, and the balance iron. Its average grain size is 0.05 mm, hardness is Rockwell B-46, and density is 8.006 g/cm^3. Electrical resistivity measurements were performed on about three dozen specimens at 4 K and 273 K after various heat treatments. These data indicated a material inhomogeneity of less than 1%. Thermal conductivity and other thermophysical properties were measured on several specimens of this lot of steel. These measurements were consistent with the remarkable homogeneity indicated by the resistivity measurements.

Iron

The iron selected for medium-to-high conductivity SRM's is somewhat more pure (99.98%) than ingot iron (99.9%). Its residual resistivity ratio, referenced to the ice-point, is 23.5; hardness is Rockwell B-24; average grain size is 0.05 mm; and density is 7.867 g/cm^3. Numerous electrical resistivity measurements at 4 K and 273 K indicated the lot inhomogeneity to be less than 1%. Thermal conductivity and electrical resistivity measurements were performed on a single specimen at temperatures below 300 K. At higher temperatures data were obtained from literature data on similar iron.

Tungsten

Two lots of tungsten, one sintered and the other arc cast, were selected for SRM use. After annealing at 2000°C these two are nearly the same in their thermophysical properties. However, the residual resistivity variations in both of these high conductivity tungsten lots are significantly larger than in the iron lot. As explained in reference [16], the effect of these variations can be accounted for through a residual resistivity measurement on each SRM specimen. The mean residual resistivity ratio is 75; density is 19.23 g/cm^3; hardness is 405 DPH; and the purity is 99.98 wt. % tungsten.

Results

The recommended thermal conductivity and electrical resistivity values are based on several sources. The low temperature data (below 300 K) for each of these materials are based on NBS (Boulder) measurements. The high temperature values for steel and tungsten are based upon the results of the AFML-AGARD reference material

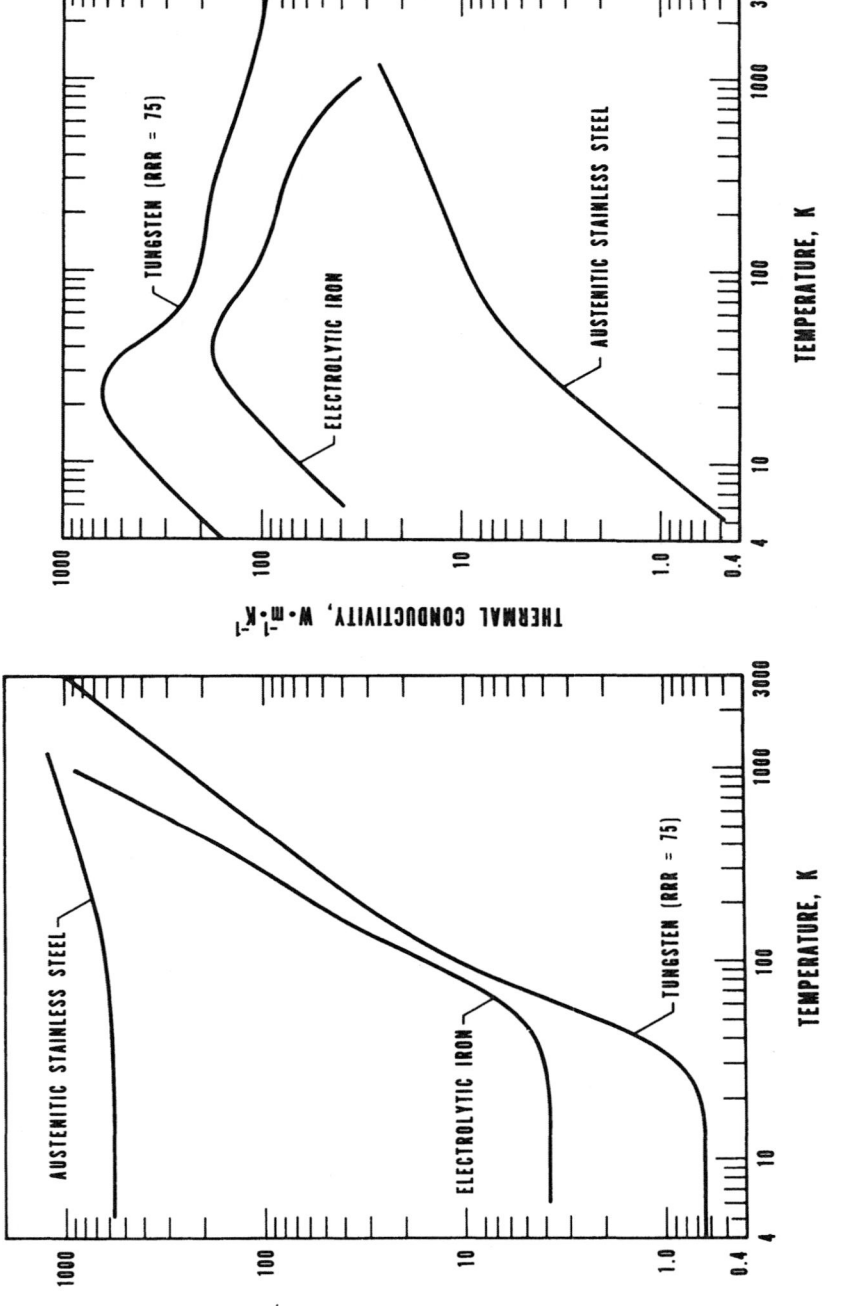

Figure 1. Electrical resistivity of iron, steel, and tungsten SRM's.

Figure 2. Thermal conductivity of iron, steel and tungsten SRM's.

program [12,13] (see also [14,16]). The high temperature data for electrolytic iron are based on measurements on a nearly identical iron [8] and on the numerous literature data on ingot iron [4].

The recommended thermal conductivity and electrical resistivity values for these three materials are illustrated in figures 1 and 2. Closely spaced tabular values are available in references [14, 16]. The uncertainties of these data are 2-5% for thermal conductivity and 1-2% for electrical resistivity. These SRM's are for sale from the Office of Standard Reference, National Bureau of Standards, Washington, D. C. Various sizes are available up to approximately 3 cm in diameter and lengths below about 30 cm.

ACKNOWLEDGMENTS

This work has been in progress for several years and over this period of time many people have assisted in various ways. R. L. Powell was instrumental in suggesting the importance of the work and in formulating the preliminary program. Larry L. Sparks assisted in the measurement of thermal conductivity with the multi-property apparatus. Thanks are given to R. E. Michaelis of NBS, OSRM for supporting this work and for many helpful discussions.

REFERENCES

1.   M. J. Laubitz and D. L. McElroy, Precise Measurements of Thermal Conductivity at High Temperatures (100 - 1200 K), Metrologia 7, No. 1, 1-15 (1971).

2.   R. W. Powell, The Thermal and Electrical Conductivities of Metals and Alloys: Part I, Iron from 0° to 800°C, Proc. Phys. Soc. 46, 659-679 (1934).

3.   M. S. Van Dusen and S. M. Shelton, Apparatus for Measuring Thermal Conductivity of Metals up to 600°C, Bur. Stand. J. Res. (USA) 12, 429-440 (1934).

4.   C. F. Lucks, Armco Iron: New Concept and Broad-Data Base Justify Its Use as a Thermal Conductivity Refernece Material, J. of Testing and Evaluation 1, No. 5, 422-431 (1973).

5.   J. G. Hust, Thermal Conductivity Standard Reference Materials from 4 to 300 K. I. Armco Iron, Proc. of the 9th Thermal Conductivity Conference, Iowa Sate Univ., Ames, Oct. 6-8, 1969.

6.   J. G. Hust, R. L. Powell, and D. H. Weitzel, Thermal
     Conductivity Standard Reference Materials from 4 to 300 K.
     I. Armco Iron: Including Apparatus Description and Error
     Analysis, J. Res. Nat. Bur. Stand. 74A, No. 5, 673-690 (1970).

7.   J. G. Hust and L. L. Sparks, Thermal Conductivity of Elec-
     trolytic Iron, SRM 734, from 4 to 300 K, Nat. Bur. Stand.
     Special Publication 260-31 (1971).

8.   W. Fulkerson, J. P. Moore, and D. L. McElroy, Comparison of
     the Thermal Conductivity, Electrical Resistivity, and Seebeck
     Coefficient of a High-purity iron and Armco iron to 1000°C,
     J. Appl. Phys. 37, No. 7, 2639-2653 (1966).

9.   M. J. Laubitz and K. D. Cotnam, Thermal and Electrical
     Properties of Inconel 702 at High Temperatures, Can. J.
     Phys. 42, 131-152 (1964).

10. H. E. Robinson and D. R. Flynn, The Current Status of Thermal
     Conductivity Reference Standards at the National Bureau of
     Standards, unpublished proceedings of the 3rd Thermal Conduc-
     tivity Conference, Oak Ridge National Laboratory, Oct. 16-18
     (1963).

11. A. D. Little, Inc., Development of High Temperature Thermal
     Conductivity Standards, Technical Reports AFML-TR-66-415
     (1967); AFML-TR-69-2 (1969), Wright-Patterson Air Force
     Base, Ohio.

12. E. Fitzer, Thermophysical Properties of Solid Materials,
     Advisory Report 12 (1967), Advisory Report 38 (1972);
     Report 606 (1972), AGARD, NATO, France.

13. M. L. Minges, Evaluation of Selected Refractories as High
     Temperature Thermophysical Property Calibration Materials,
     AFML Technical Report TR-73-278 (1974); Int'l. J. Heat. and
     Mass Transfer, Pergamon Press, London, (to be published in
     1974).

14. J. G. Hust and P. J. Giarratano, Thermal Conductivity and
     Electrical Resistivity Standard Reference Materials:
     Austenitic Stainless Steel, SRM's 735 and 798, from 4 to
     1200 K, Nat. Bur. Stand. Special Publication 260-46 (1975).

15. J. G. Hust and P. J. Giarratano, Thermal Conductivity and
     Electrical Resistivity Standard REference Materials:
     Electrolytic Iron, SRM's 734 and 797, from 4 to 1000 K,
     Nat. Bur. Stand. Special Publication 260-50 (1975).

16. J. G. Hust and P. J. Giarratano, Thermal Conductivity and
     Electrical Resistivity Standard Reference Materials:
     Tungsten, SRM's 730 and 799, from 4 to 2000 K, Nat. Bur.
     Stand. Special Publication 260-52 (1975).

THE NEED FOR REFERENCE MATERIALS OF LOW THERMAL CONDUCTIVITY

R. P. Tye

Dynatech R/D Co.

Cambridge, Massachusetts

The overall emphasis of energy conservation combined with the continued increase in the development and installation of thermal insulation materials and systems often upon a guaranteed performance basis make it imperative that we must be able to measure reliable performance values. This applies to whatever the method or technique used to simulate real conditions the type of material or system and the temperature and environmental conditions under which the evaluation is to take place. In order to help accomplish more reliable evaluation, reference samples or materials are needed to cover a range of thermal conductivity over a wide temperature range and an environmental gas and pressure range.

The whole subject of thermal conductivity reference materials will be reviewed briefly to indicate where problem areas exist to confront those working in this field of measurement of heat transmission in thermal insulations or of the thermal conductivity of poor thermal conductors. Examples will be given to illustrate how poor, inadequate or insufficient measurements produce problems when information obtained from systems in actual use are compared with the experimentally obtained values.

Details will be given of the current progress that is being made in the subject both in the definition of the materials required and the establishment of the criteria for their evaluation. The international and national aspects of the problem and its solution will also be emphasized.

THERMAL CONDUCTIVITY OF COPPER-NICKEL AND

SILVER-PALLADIUM ALLOY SYSTEMS

C. Y. Ho, M. W. Ackerman, and K. Y. Wu

Center for Information and Numerical Data Analysis and Synthesis
Purdue University
West Lafayette, Indiana 47906

The recommended thermal conductivity values for copper-nickel and silver-palladium alloy systems are presented which cover the full range of alloy composition and temperature. The recommended values are based on both the critically evaluated experimental thermal conductivity data and the values obtained from methods developed for the calculation of the electronic and lattice thermal conductivities. The electronic thermal conductivity of an alloy is calculated from the electrical resistivity and thermoelectric power of the alloy and the electrical resistivity and thermal conductivity of the pure constituent elements. The lattice thermal conductivity is calculated semi-theoretically based upon the Klemens-Callaway theory.

I. INTRODUCTION

Due to the difficulties in accurate measurement of the thermal conductivity of solids and in exact characterization of test specimens, the available experimental data on thermal conductivity extracted from various research documents are in many cases widely divergent and subject to large uncertainty. It is, therefore, very important to critically evaluate the validity and reliability of the available data and related information, to resolve and reconcile the disagreements in conflicting data, and to generate recommended reference values. For the thermal conductivity of alloys, furthermore, there are serious gaps in the experimental data for either the compositional or the temperature dependence or both. Hence, in addition to data evaluation and analysis, estimated thermal conductivity values are needed for filling the gaps in data.

In order to check the validity, consistency, and reliability of experimental data and to fill the data gaps, methods for the calculation of the electronic and lattice thermal conductivities of alloys were developed. The reliability of these methods has been sufficiently tested with selected key sets of reliable experimental data on alloys in various binary alloy systems. From the critically evaluated experimental data and the values calculated from the developed methods, recommended thermal conductivity values for copper-nickel and silver-palladium alloy systems have been generated. Methods and procedures for data evaluation and analysis and for the generation of recommended values are discussed. The resulting recommended values for copper-nickel and silver-palladium alloy systems are presented.

II. DATA EVALUATION AND GENERATION OF RECOMMENDED VALUES

In the critical evaluation of the validity and reliability of a particular set of thermal conductivity data, the temperature dependence of the data is examined and any unusual dependence or anomaly carefully investigated, the experimental technique reviewed to see whether the actual boundary conditions in the measurement agreed with those assumed in the theory and whether all the stray heat flows and losses were prevented or minimized and accounted for, the reduction of data examined to see whether all the necessary corrections were appropriately applied, and the estimation of uncertainties checked to ensure that all the possible sources of errors were considered.

Experimental data can be judged to be reliable only if all sources of systematic error have been eliminated or minimized and accounted for. Major sources of systematic error include unsuitable experimental method, poor experimental technique, poor instrumentation and poor sensitivity of measuring devices, sensors, or circuits, specimen and/or thermocouple contamination, unaccounted for stray heat flows, incorrect form factor, and perhaps most important, the mismatch between actual experimental boundary conditions and those assumed in the analytical model used to derive the value of thermal conductivity. These and other possible sources of errors are carefully considered in critical evaluation of experimental data.

In the process of critical evaluation of experimental data outlined above, erroneous data are eliminated. The remaining data are then subjected to further critical analysis. For those alloys for which experimental data on both the thermal conductivity and the electrical resistivity are reported, the electrical resistivity data are used for the calculation of the electronic thermal conductivity values using the equation:

$$k_e = \frac{1}{\dfrac{\rho - \rho_i}{(L_0 - S^2)\,T} + W_{ei}} \tag{1}$$

where ρ and S are the electrical resistivity and the thermoelectric power of the alloy, L_0 is the classical theoretical Lorenz number and has a value of 2.443×10^{-8} $V^2\,K^{-2}$, T is the temperature, and ρ_i and W_{ei} are the intrinsic electrical and thermal resistivities of the "virtual" crystal obtained by interpolating between the values for the elements, linearly for alloys of ordinary metals and according to Mott's theory [1,2] for alloys containing transition elements such as those of the present two alloy systems. The intrinsic electronic thermal resistivity of an element is calculated from the expression:

$$W_{ei} = \frac{1}{k - k_g} - \frac{\beta}{T} \tag{2}$$

where k and k_g are the total and lattice thermal conductivities of the element and β is the impurity-imperfection parameter of the element.

The lattice thermal conductivity values are derived as the differences of the experimental thermal conductivity data and the calculated electronic thermal conductivity values from Eq. (1). These "experimental" lattice thermal conductivity values derived from different sets of experimental thermal conductivity data are then intercompared as well as compared with the calculated lattice thermal conductivity values from the expression [3]:

$$k_g = k_u(T') \frac{I_2(\theta/T) + I_4^2(\theta/T)/[I_6(\theta/T) + I_8(\theta/T)/x_0^2]}{J_2(\theta/T') + J_4^2(\theta/T')/J_6(\theta/T')} \tag{3}$$

where $I_n(\theta/T)$ is the modified transport integral given by

$$I_n(\theta/T) = \int_0^{\theta/T} \frac{x^n\,e^x\,dx}{(e^x - 1)^2 \left[1 + \dfrac{x^2}{x_0^2\,(1+\alpha)}\right]} \tag{4}$$

$J_n(\theta/T')$ is the standard transport integral given by

$$J_n(\theta/T') = \int_0^{\theta/T'} \frac{x^n\,e^x\,dx}{(e^x - 1)^2} \tag{5}$$

$k_u(T')$ is the lattice thermal conductivity of the "virtual" crystal at some fixed temperature T', θ is the Debye temperature, α is the temperature-independent ratio of reciprocal relaxation times for three-phonon N- and U-processes, $x = \hbar\omega/\kappa T$ where ω is the frequency and $\hbar$ and κ are the reduced Planck constant and the Boltzmann constant, and $x_0 = \hbar\omega_0/\kappa T$ where ω_0 is the frequency at which the reciprocal relaxation times for point-defect scattering and U-processes are equal. It is important to note that Eq. (3) is applicable only to disordered solid-solution alloys and only for moderate and high temperatures. At low temperatures the lattice thermal conductivity values are obtained only as the difference of the experimental total thermal conductivity and the calculated electronic thermal conductivity.

In the comparison of the "experimental" lattice thermal conductivity values with one another and with the calculated lattice thermal conductivity values from Eq. (3), the validity, consistency, and reliability of the available experimental data can further be judged. The evaluated "experimental" lattice thermal conductivity values are then graphically smoothed and synthesized to obtain the values for alloys of the selected compositions. At moderate and high temperatures, the "experimental" lattice thermal conductivity values are used to check the calculated values from Eq. (3). If there are disagreements and the "experimental" lattice thermal conductivity values are considered more reliable, the values of the lattice thermal conductivity of the "virtual" crystals, k_u, used in Eq. (3) would be adjusted so that the calculated lattice thermal conductivity values are in agreement with the "experimental" values.

The total thermal conductivity values are thus obtained as the sum of the electronic thermal conductivity values calculated from Eq. (1) and the lattice thermal conductivity values derived from the "experimental" lattice thermal conductivity values or calculated from Eq. (3), which might have been adjusted to fit the "experimental" lattice thermal conductivity values, if such values are available and reliable.

In graphical smoothing and synthesis of data, cross-plotting from conductivity versus temperature to conductivity versus composition and vice versa are often used. Smooth curves are drawn which approximate the best fit to the conductivity data versus temperature, and points from the smoothed curves are used to construct conductivity versus composition curves for a convenient set of selected temperatures. In the conductivity versus composition graph, the families of isotherms are similar and any required smoothing of the data can be done more easily and with greater confidence than when working directly with the conductivity-temperature curves. The points from the smoothed curves are then used to construct conductivity-temperature curves for the selected compositions, and these curves are further smoothed. In the graphical smoothing process it is extremely important that the alloy phase diagrams be constantly consulted and the phase boundaries between solid solutions and/or mechanical mixtures and the boundaries of magnetic transitions be kept in mind, so as to be aware of any possible discontinuity or sudden change of slope in the thermal conductivity curves.

III. RESULTS AND DISCUSSIONS

Both copper-nickel and silver-palladium alloy systems exhibit complete solid solubility and are free of all transformations except that of ferromagnetism in the copper-nickel alloy system.

The experimental thermal conductivity data on the copper-nickel alloy system are shown in Figures 1 and 2. There are 153 sets of experimental data available for the thermal conductivity of this alloy system. However, of the 104 data sets available for Cu + Ni alloys shown in Figure 1, 27 sets are merely single data points and 25 sets cover only a narrow temperature range from around room temperature to about 500 K. Of the 49 data sets for Ni + Cu alloys shown in Figure 2, 23 sets are single data points. Furthermore, many sets of data show large discrepancies.

The recommended electrical resistivity values for the copper-nickel alloy system are presented in Figures 3 and 4, which are used in Eq. (1) for the calculation of the electronic thermal conductivity values. These values cover the full range of temperature and alloy composition and are generated from both the electrical resistivity data reported for the test specimens on which thermal conductivity measurements were made and those for all other alloys of the copper-nickel system in the literature. As shown in Figure 4, the electrical resistivity versus temperature curves for Ni + Cu alloys change slope abruptly at the Curie temperature of the alloys. The Curie temperature decreases as the concentration of copper in the alloys increases. The ferromagnetism disappears and the Curie temperature drops to zero as the concentration of copper reaches 61. 88% (60 At.%). The insert in Figure 4 shows the Curie temperature as a function of percent copper in nickel, which is linear for the atomic percent of copper. This straight-line relationship was determined from the electrical resistivity data shown in Figure 4. The behavior of the electrical resistivity of these alloys has a direct bearing on the behavior of the thermal conductivity, and therefore the knowledge of the former is important to the understanding of the latter.

The recommended thermoelectric power values for the copper-nickel alloy system are shown in Figures 5 and 6, which are likewise used in Eq. (1) for the calculation of the electronic thermal conductivity values. Figure 6 shows also the Curie temperature of each alloy as the point at which the slope of the curve changes abruptly.

Using the methods and procedures outlined in the last Section, the recommended thermal conductivity values for the copper-nickel alloy system are generated, as presented in Figures 7 and 8, which cover the full range of temperature and alloy composition. These values are for well-annealed alloys. It is interesting to observe the variations of the slope of the curves at both the low- and high-temperature ends and of the individual curves as a function of temperature. The abrupt changes of the curves at the Curie temperature are also shown in Figure 8.

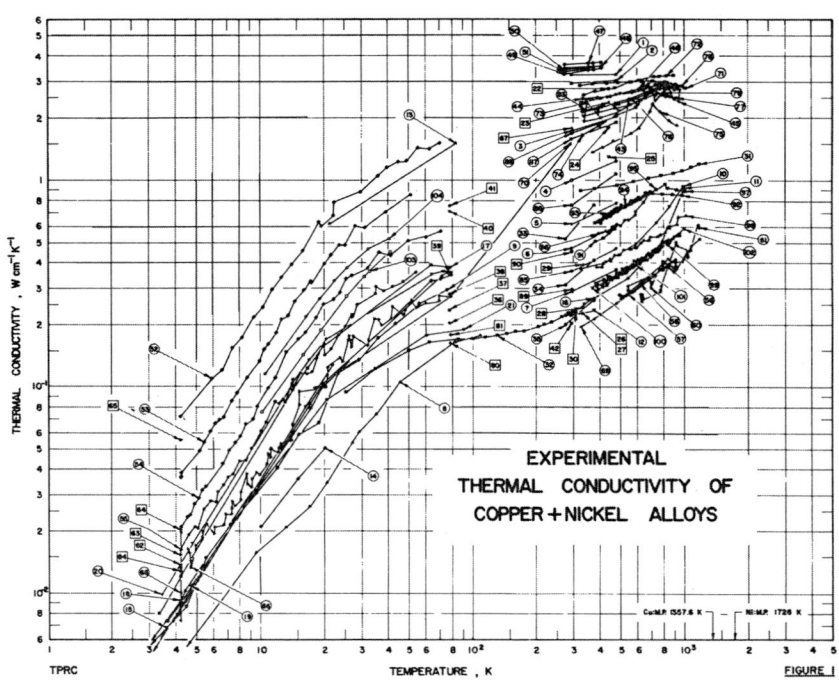

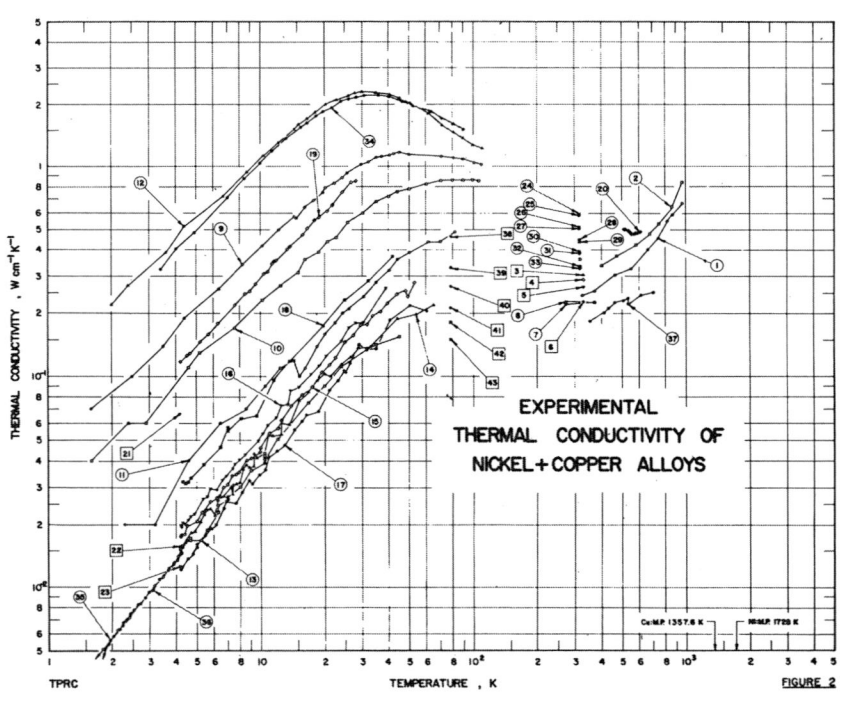

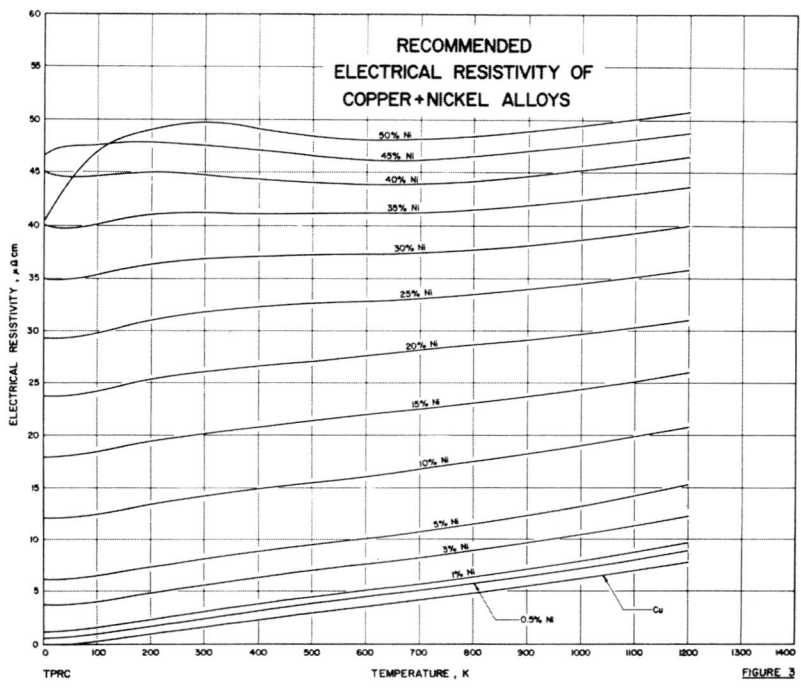

RECOMMENDED
ELECTRICAL RESISTIVITY OF
COPPER + NICKEL ALLOYS

TPRC   TEMPERATURE , K   FIGURE 3

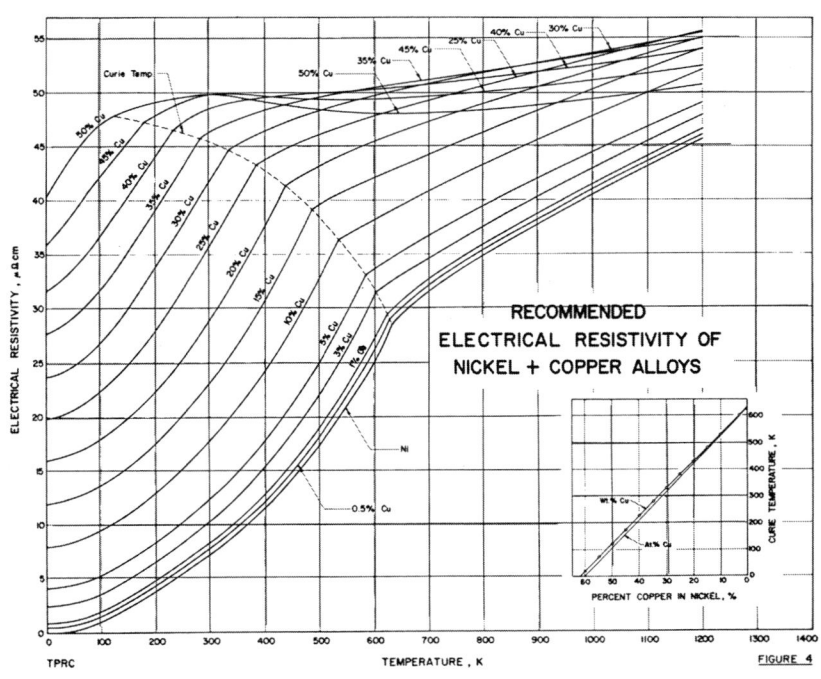

RECOMMENDED
ELECTRICAL RESISTIVITY OF
NICKEL + COPPER ALLOYS

TPRC   TEMPERATURE , K   FIGURE 4

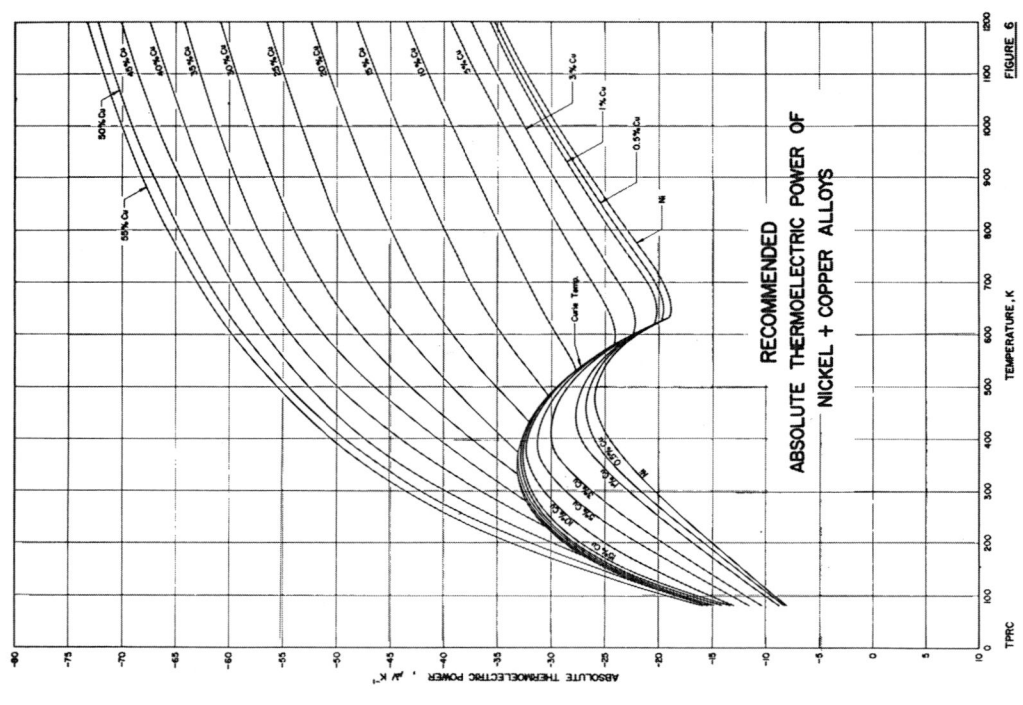

RECOMMENDED
ABSOLUTE THERMOELECTRIC POWER OF
NICKEL + COPPER ALLOYS

FIGURE 6

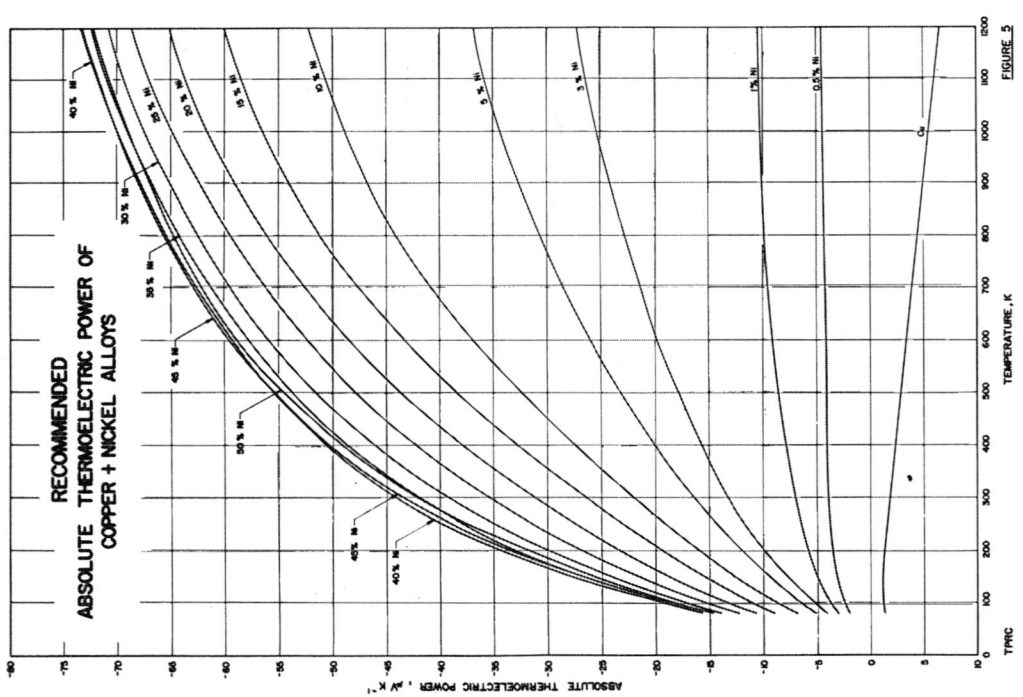

RECOMMENDED
ABSOLUTE THERMOELECTRIC POWER OF
COPPER + NICKEL ALLOYS

FIGURE 5

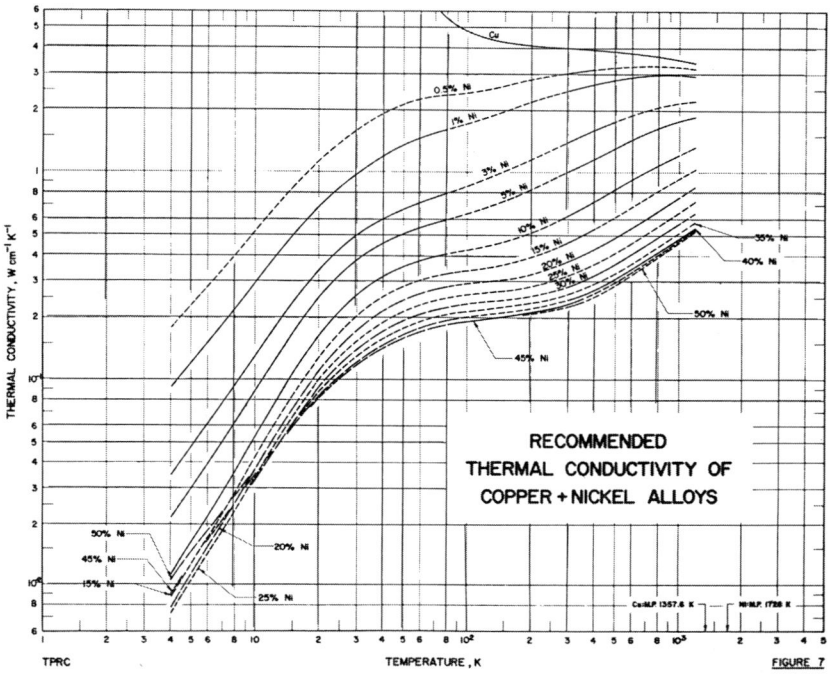

RECOMMENDED
THERMAL CONDUCTIVITY OF
COPPER + NICKEL ALLOYS

TPRC   TEMPERATURE , K   FIGURE 7

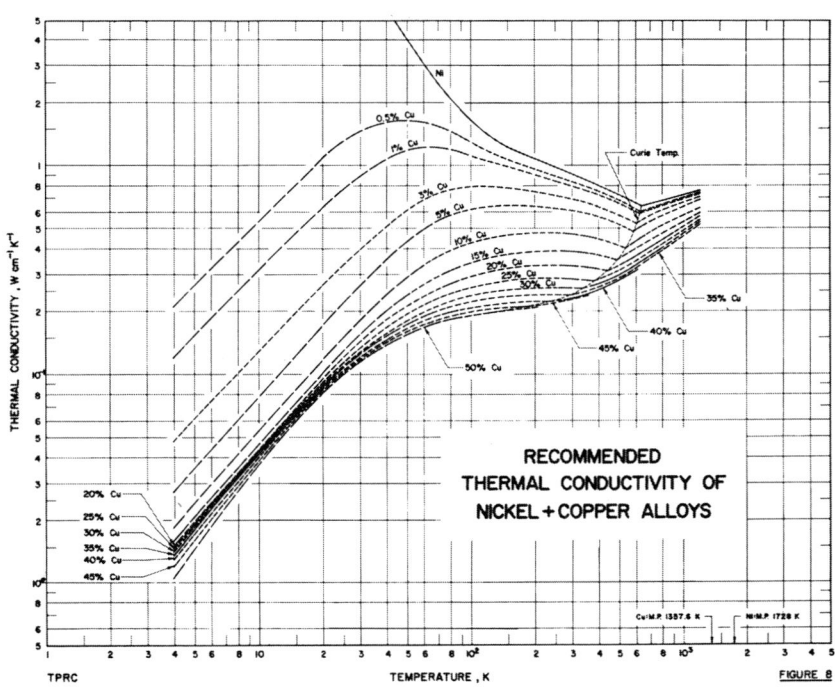

RECOMMENDED
THERMAL CONDUCTIVITY OF
NICKEL + COPPER ALLOYS

TPRC   TEMPERATURE , K   FIGURE 8

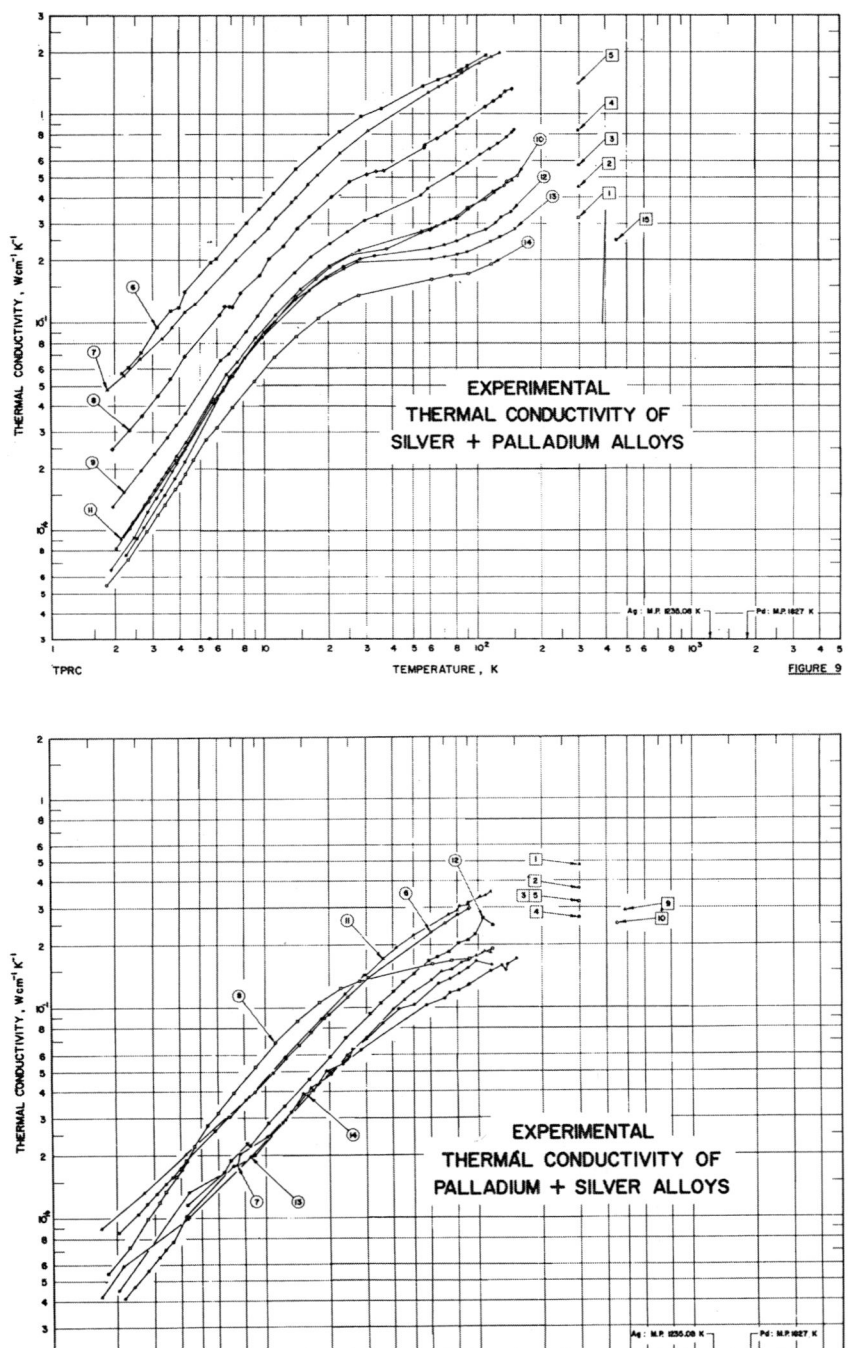

FIGURE 9

FIGURE 10

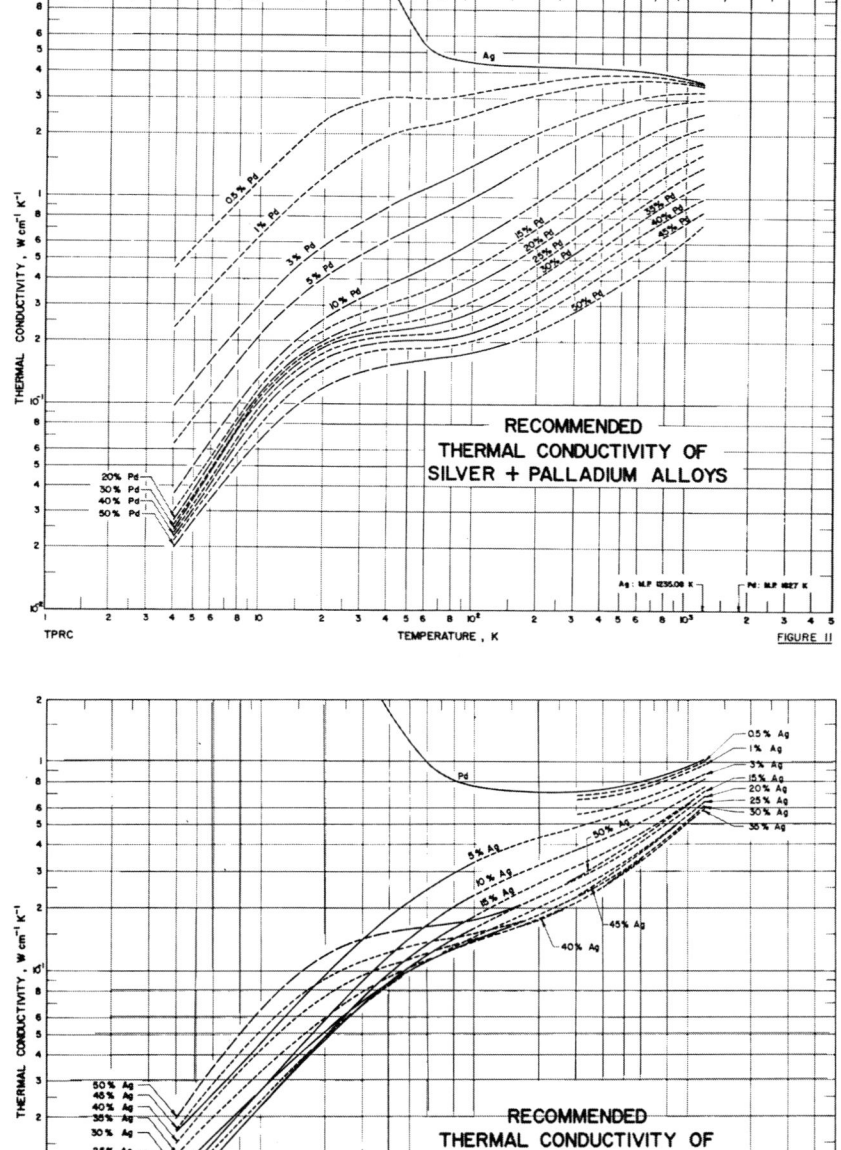

RECOMMENDED
THERMAL CONDUCTIVITY OF
SILVER + PALLADIUM ALLOYS

FIGURE 11

RECOMMENDED
THERMAL CONDUCTIVITY OF
PALLADIUM + SILVER ALLOYS

FIGURE 12

Similarly the experimental thermal conductivity data on the silver-palladium alloy system are shown in Figures 9 and 10. There are 32 sets of experimental data available for the thermal conductivity of this alloy system. However, of the 18 data sets available for Ag + Pd alloys shown in Figure 9, six sets are merely single data points, and of the 14 sets for Pd + Ag alloys shown in Figure 10, seven sets are single data points. Figures 11 and 12 present the recommended thermal conductivity values for the silver-palladium alloy system, which are generated using the same methods and procedures. These values are likewise for well-annealed alloys.

The uncertainty of the recommended thermal conductivity values for both alloy systems is of the order of ±5% to ±10% at moderate and high temperatures and of ±10% to ±15% at low temperatures.

IV. REFERENCES

1. Mott, N.F., Proc. Phys. Soc. (London), 47, 571 (1935).

2. Mott, N.F. and Jones, H., The Theory of the Properties of Metals and Alloys, Clarendon Press, Oxford (1936); Dover Publications, New York, 196-8 (1958).

3. See the article by M. W. Ackerman, K. Y. Wu, and C. Y. Ho in this Proceedings.

ACKNOWLEDGMENT

This work was supported by the Office of Standard Reference Data, National Bureau of Standards, U. S. Department of Commerce, Washington, D.C. under Contract No. 3-35707.

LATTICE THERMAL CONDUCTIVITY AND LORENZ FUNCTION OF

COPPER-NICKEL AND SILVER-PALLADIUM ALLOY SYSTEMS

M. W. Ackerman, K. Y. Wu, and C. Y. Ho

Center for Information and Numerical Data Analysis and Synthesis
Purdue University
West Lafayette, Indiana 47906

Following a brief discussion of the separation of
the electronic and lattice components of thermal conductivity at low temperatures, a method for calculating the lattice
thermal conductivity of binary alloys at temperatures above
that corresponding to the lattice thermal conductivity maximum is described. The method is based on the Klemens-
Callaway theory. Curves of the lattice thermal conductivity
of copper-nickel and silver-palladium alloy systems are
presented which cover the full range of alloy composition
and temperature. Curves of the electronic and total Lorenz
functions of these alloy systems are also presented which
likewise cover the full range of alloy composition and temperature. The electronic Lorenz function is derived from
the electronic thermal conductivity and the electrical resistivity, whereas the total Lorenz function is derived from
the total thermal conductivity and the electrical resistivity.

I. INTRODUCTION

At the lowest temperatures the lattice thermal conductivity of an
alloy is limited by the phonon-electron interaction and phonon scattering
by residual dislocations anchored in place by solute atoms; both of these
resistive mechanisms result in approximately a T^2 temperature dependence. At somewhat higher temperatures point-defect scattering and
scattering by dislocation cores cause the lattice conductivity to depart
from its T^2 temperature dependence, and at still higher temperatures
the combination of three-phonon anharmonic interactions and point-defect
scattering cause the conductivity to decrease approximately as $T^{-1/2}$.
On the basis of the Klemens-Callaway theory, a method has been developed

245

for the calculation of the lattice thermal conductivity of well-annealed
disordered binary solid-solution alloys at temperatures above the region
of the maximum of lattice thermal conductivity. At temperatures in the
region of lattice conductivity maximum and below, however, there is no
adequate method available for the calculation of the lattice thermal conductivity because we lack knowledge of both the phonon-electron coupling
constant and the residual dislocation densities, and at present we must
depend on experimental data to obtain the lattice thermal conductivity by
subtracting the electronic component from the total. This situation may
improve presently because below about 1 K the dominant phonon wavelengths are larger than the range of the dislocation strain field so that
scattering by dislocations should be greatly reduced, and this may make
it possible to improve our knowledge of the effect of alloying on the phonon-
electron interaction and the residual dislocation densities.

This paper describes the method developed for the calculation of
the lattice thermal conductivity at temperatures above the region of the
lattice conductivity maximum and presents the calculated results for
copper-nickel and silver-palladium alloy systems together with their
lattice conductivity values at low temperatures derived from the experimental thermal conductivity data and values of the calculated electronic
component. These values as a whole thus cover the full range of temperature as well as alloy composition. In addition, this paper presents also
the electronic and total Lorenz functions of these alloy systems, which
likewise cover the full range of temperature and alloy composition.

II. METHODS OF CALCULATION

The low temperature values of the lattice conductivity were
obtained by subtracting the values of the calculated electronic component
from the measured data on the total thermal conductivity. For this purpose the electronic thermal conductivity was calculated at the lowest
temperatures from the Wiedemann-Franz-Lorenz law and above 20 K
from

$$k_e = \frac{1}{\dfrac{\rho - \rho_i}{(L_0 - S^2)\, T} + W_{ei}} \tag{1}$$

where ρ_i and W_{ei} are the intrinsic electrical and thermal resistivities of
the "virtual" crystal obtained by interpolating between the values for the
elements, linearly for alloys of ordinary metals and according to Mott's
theory [1,2] for alloys containing transition elements such as those of
the present two alloy systems, ρ is the electrical resistivity, L_0 is the
Sommerfeld value of the theoretical Lorenz number, S is the thermoelectric power, and T is the temperature. This expression for k_e is derived
on the assumption that the deviations from Matthiessen's rule and its

thermal analog are related by the Wiedemann-Franz-Lorenz law, which is approximately true provided that only one band of electrons contributes significantly to the electrical and thermal conduction.

At temperatures above that of the maximum of the lattice component one can obtain the strength of the three-phonon U-processes from measurements of the thermal conductivity of very dilute alloys and can calculate the lattice thermal conductivity as limited by these processes and point-defect scattering on the basis of a theory developed by Klemens [3,4] and Callaway [5]. The principal problem in this region was how to incorporate the three-phonon N-processes, which do not contribute directly to the thermal resistivity but do so indirectly by redistributing energy from the low frequency modes to the high frequency modes that are strongly scattered by the point defects. Callaway's expression for the lattice thermal conductivity is

$$k_g = \frac{\kappa}{2\pi^2 v} \left(\frac{\kappa T}{\hbar}\right)^3 \left(I_a + \frac{I_b^{\ 2}}{I_c}\right) \tag{2}$$

where κ and $\hbar$ are the Boltzmann and reduced Planck constants, v is the speed of sound and

$$I_a = \int_0^{\theta/T} \frac{\tau_c\ x^4\ e^x}{(e^x-1)^2}\ dx \tag{3}$$

$$I_b = \int_0^{\theta/T} \frac{\tau_c}{\tau_N}\ \frac{x^4\ e^x}{(e^x-1)^2}\ dx \tag{4}$$

$$I_c = \int_0^{\theta/T} \frac{1}{\tau_N}\left(1 - \frac{\tau_c}{\tau_N}\right)\frac{x^4\ e^x}{(e^x-1)^2}\ dx \tag{5}$$

Here τ_c is the combined relaxation time obtained as the reciprocal of the sum of the reciprocal relaxation times of the various scattering processes, τ_N is the relaxation time for N-processes, $x = \hbar\omega/\kappa T$, and ω is the frequency. The term $I_b^{\ 2}/I_c$ in Eq. (2) accounts for the difference between τ_c and the effective total relaxation time due to the three-phonon N-processes.

If the reciprocal relaxation times for point defect scattering, U-processes, and N-processes are written respectively as $A\omega^4$, $BT\omega^2$, and $\alpha BT\omega^2$, where A, B, and α are constants, it follows from Eqs. (2) – (5) that the ratio of the lattice thermal conductivity of the disordered alloy at temperature T to that of the "virtual" crystal at some fixed temperature T' is given by

$$\frac{k_g(T)}{k_u(T')} = \frac{I_2(\theta/T) + I_4^2(\theta/T)/[I_6(\theta/T) + I_8(\theta/T)/x_o^2]}{J_2(\theta/T') + J_4^2(\theta/T')/J_6(\theta/T')} \tag{6}$$

where

$$I_n(\theta/T) = \int_0^{\theta/T} \frac{x^n e^x \, dx}{(e^x-1)^2 \left[1 + \dfrac{x^2}{x_o^2 \,(1+\alpha)}\right]} \tag{7}$$

$$J_n(\theta/T') = \int_0^{\theta/T'} \frac{x^n e^x \, dx}{(e^x-1)^2} \tag{8}$$

θ is the Debye temperature, and x_o is the reduced frequency at which the relaxation times for U-processes and point-defect scattering are equal.

The constant in the relaxation time for point-defect scattering is given by

$$A = \frac{a^3 \, \epsilon}{4\pi \, v^3} \tag{9}$$

where a^3 is the volume per atom, v is the speed of sound, and ϵ is the perturbation parameter given by

$$\epsilon = y_L \left[\left(\frac{M_L - M}{M}\right) + \gamma \left(\frac{V_L - V}{V}\right)\right]^2 + y_H \left[\left(\frac{M_H - M}{M_H}\right) \right.$$

$$\left. + \gamma \left(\frac{V_H - V}{V}\right)\right]^2 \tag{10}$$

Here y_L, M_L, and V_L are the atomic fraction, mass, and volume of the lighter element, y_H, M_H, and V_H are the corresponding values for the heavier element, and M, V, and γ are the average values of the atomic mass, atomic volume, and the Grüneisen parameter. M and γ are obtained by linear interpolation and V is estimated from Vegard's law

$$V^{1/3} = y_L \, V_L^{1/3} + y_H \, V_H^{1/3} \tag{11}$$

The mass defect terms are based on the results of Klemens [6] and Tavernier [7] who respectively considered the case of a light atom in a heavy matrix and a heavy atom in a light matrix. The distortion term and the form of ϵ are based on the results of Ackerman and Klemens [8].

The values of θ for the upper limits of the integrals in Eq. (6) were estimated from the value of $k_u(T')$ by inverting the modified Liebfried-Schlömann equation, i.e. from

$$\theta = 260 \left[\frac{(\gamma + 0.5)^2 \, k_u(T') \, T'}{MV^{1/3}} \right]^{1/3} \tag{12}$$

where the coefficient has been changed to obtain agreement with the values of the lattice thermal conductivity derived from measurements on very dilute alloys.

In using Eq. (6) to calculate the lattice component the greatest uncertainty was associated with the lattice thermal conductivity of the virtual crystal, $k_u(T')$. Here again it was necessary to rely upon experimental results at one temperature T'. The procedure followed was to interpolate between the values for the solvent and solute elements. Values of the lattice component of very dilute alloys of Cu and Ag derived from experimental data are available. Values for Ni and Pd were first estimated using the modified [9] Liebfried-Schlömann equation multiplied by a correction factor. The calculated values of k_g were then compared with values derived from experimental data, and the values for the elements were then adjusted to obtain agreement over the range of compositions.

III. LATTICE THERMAL CONDUCTIVITY

The Cu-Ni system has been studied extensively, there being 153 sets of experimental data available. While only 49 of these are for Ni-rich alloys and 23 of those 49 are single data points, it is thought that there was sufficient data to test adequately the reliability of the calculated values of the lattice thermal conductivity on both sides of this system. Also, while few measurements have been made on Ag-Pd alloys above about 100 K, measurements have been made from liquid helium temperatures to about 100 K for 12 compositions spaced more or less evenly from 2 to 95 percent Pd and all of these measurements appear to be reliable. For this system it should be noted that, due to the very strong phonon-electron interaction in Pd, the maxima of the lattice component for alloys containing more than 70 percent Pd occur above 100 K so that the only basis for the conclusion that the calculated values of the lattice thermal conductivity of these Pd-rich alloys are reliable is that they are consistent with the trend of the values at temperatures below that of the maxima.

Figures 1,2 and 3,4 show the lattice thermal conductivities of Cu-Ni and Ag-Pd alloy systems, respectively. The maximum of the

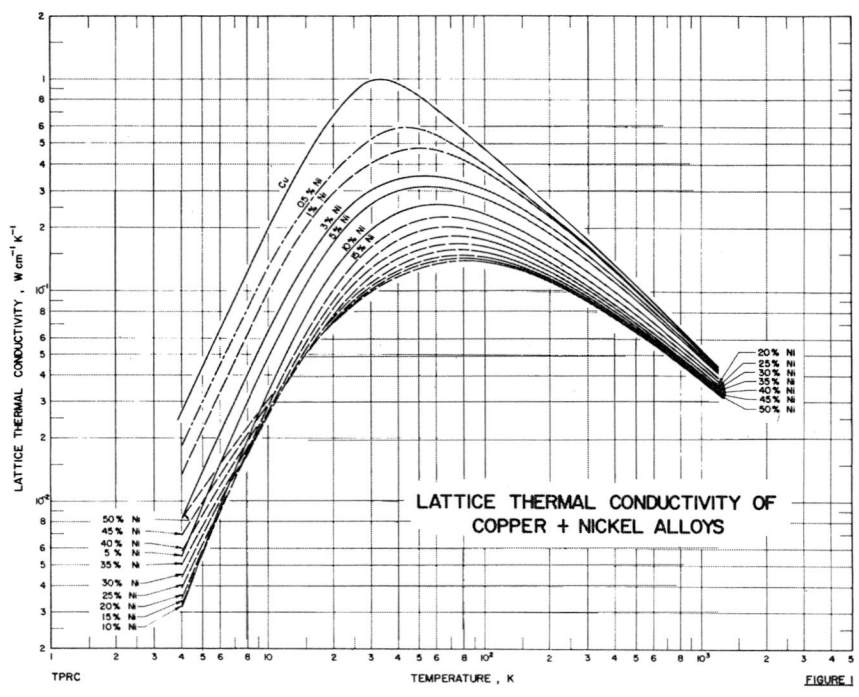

LATTICE THERMAL CONDUCTIVITY OF
COPPER + NICKEL ALLOYS

TPRC   TEMPERATURE , K   FIGURE I

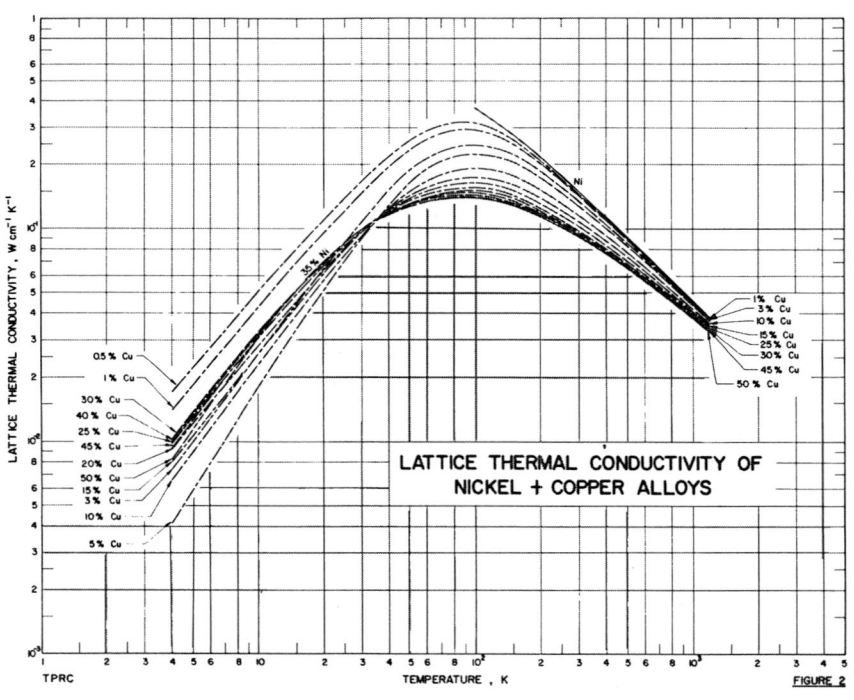

LATTICE THERMAL CONDUCTIVITY OF
NICKEL + COPPER ALLOYS

TPRC   TEMPERATURE , K   FIGURE 2

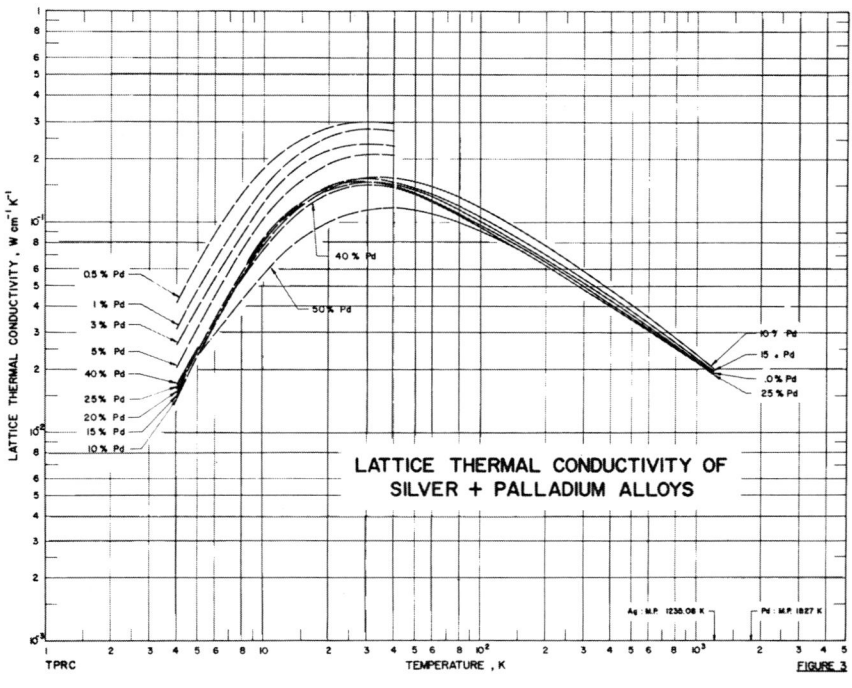

LATTICE THERMAL CONDUCTIVITY OF
SILVER + PALLADIUM ALLOYS

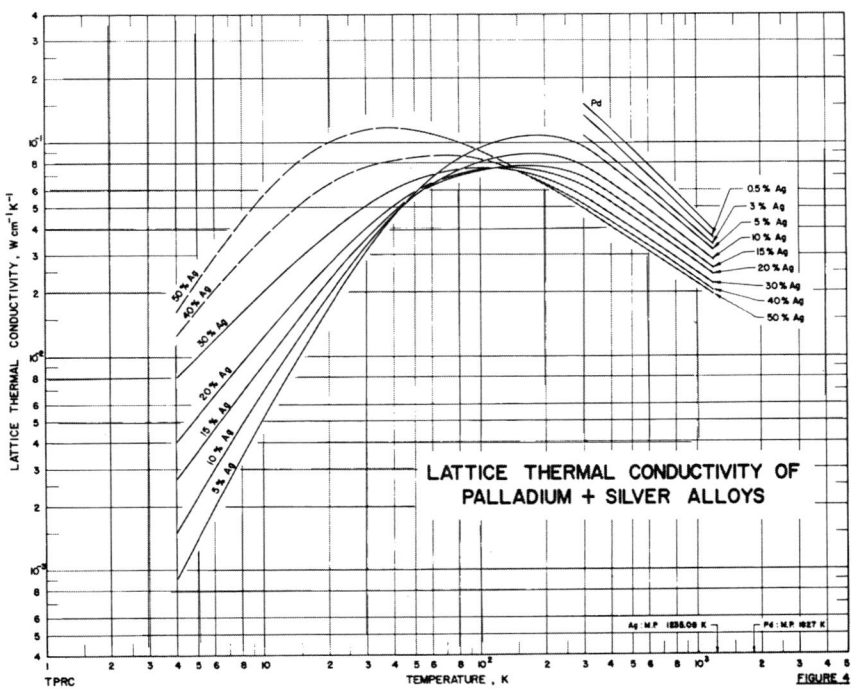

LATTICE THERMAL CONDUCTIVITY OF
PALLADIUM + SILVER ALLOYS

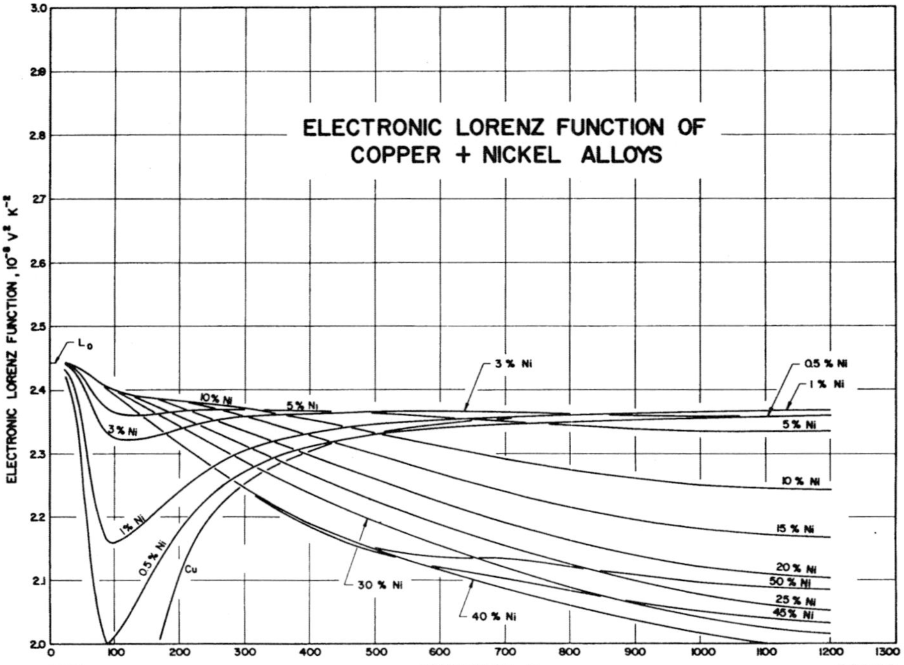

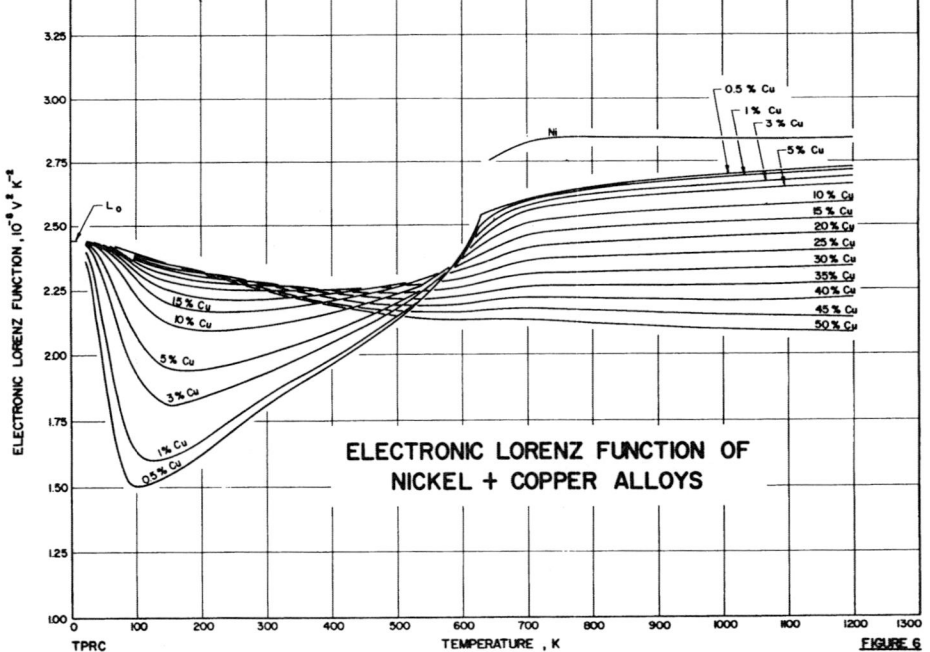

ELECTRONIC LORENZ FUNCTION OF COPPER + NICKEL ALLOYS
ELECTRONIC LORENZ FUNCTION, 10^{-8} V^2 K^{-2}
TEMPERATURE, K
L_o
3 % Ni
0.5 % Ni
1 % Ni
10 % Ni
5 % Ni
1% Ni
Cu
15 % Ni
20 % Ni
50 % Ni
25 % Ni
45 % Ni
30 % Ni
40 % Ni

TPRC FIGURE 5

ELECTRONIC LORENZ FUNCTION OF NICKEL + COPPER ALLOYS
ELECTRONIC LORENZ FUNCTION, 10^{-8} V^2 K^{-2}
TEMPERATURE, K
L_o
Ni
0.5 % Cu
1 % Cu
3 % Cu
5 % Cu
10 % Cu
15 % Cu
20 % Cu
25 % Cu
30 % Cu
35% Cu
40 % Cu
45 % Cu
50 % Cu
1% Cu
0.5% Cu

TPRC FIGURE 6

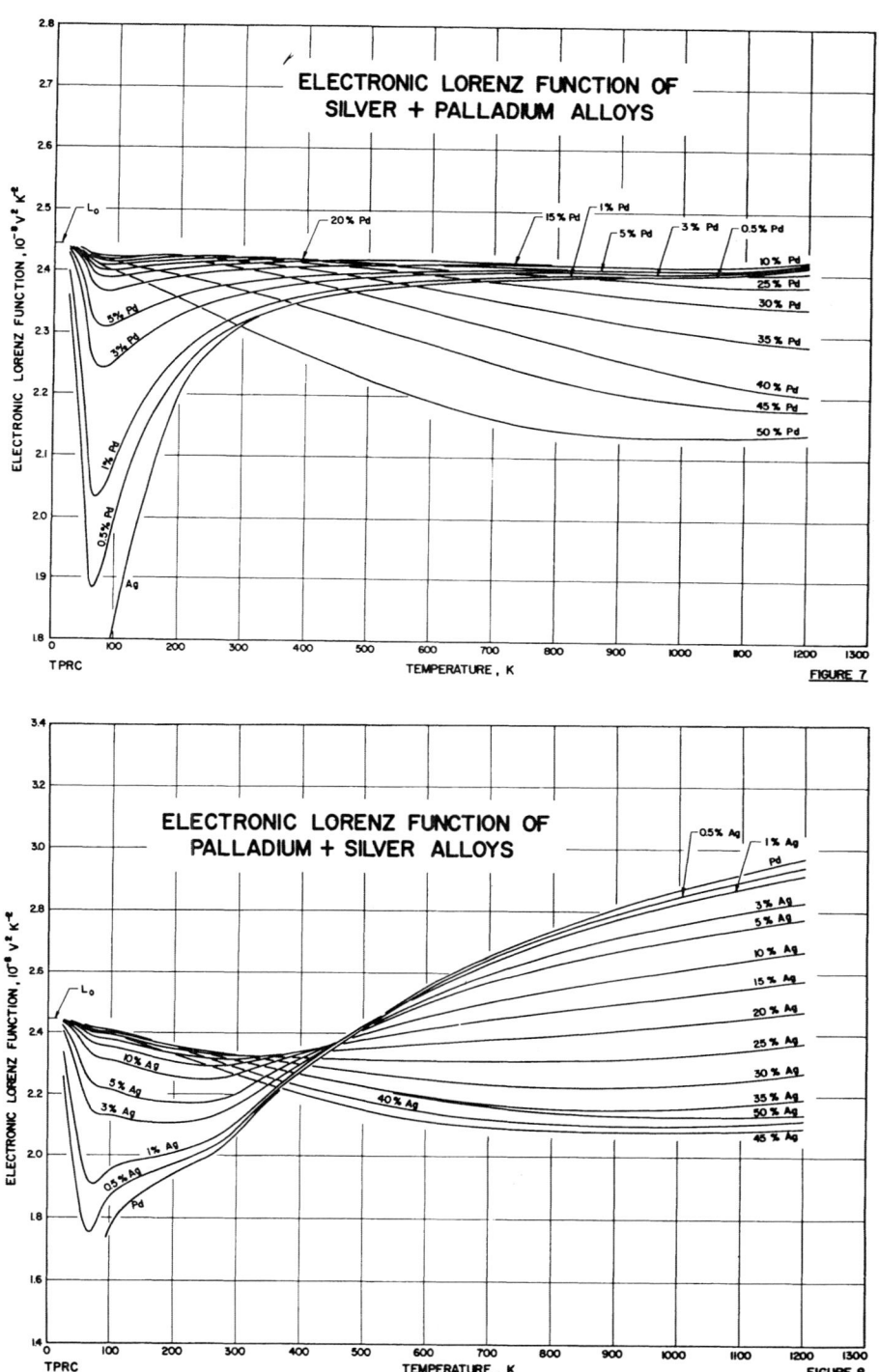

FIGURE 7

FIGURE 8

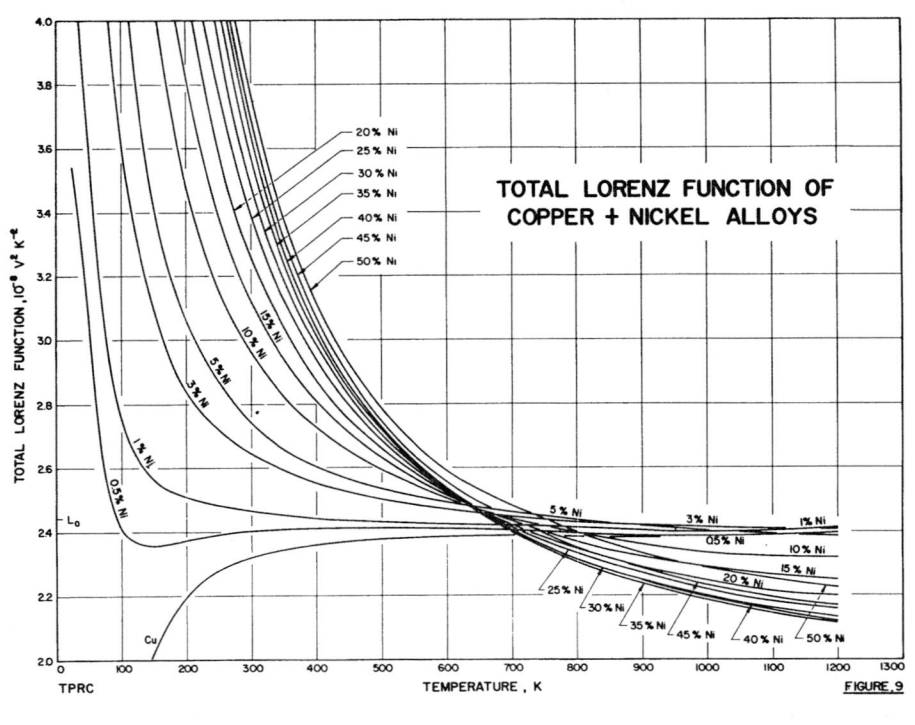

TOTAL LORENZ FUNCTION OF
COPPER + NICKEL ALLOYS

FIGURE 9

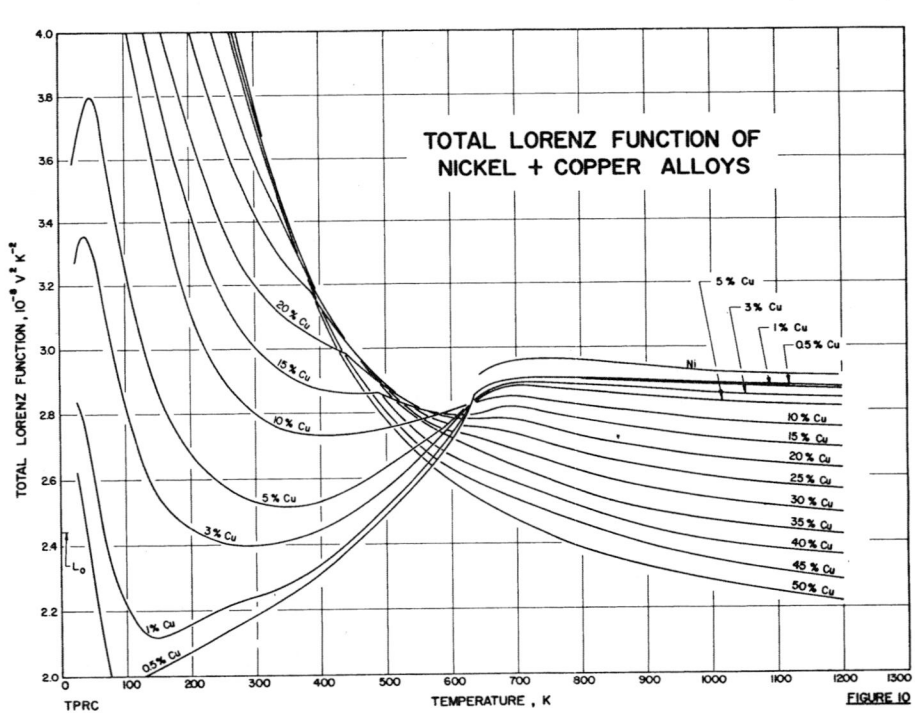

TOTAL LORENZ FUNCTION OF
NICKEL + COPPER ALLOYS

FIGURE 10

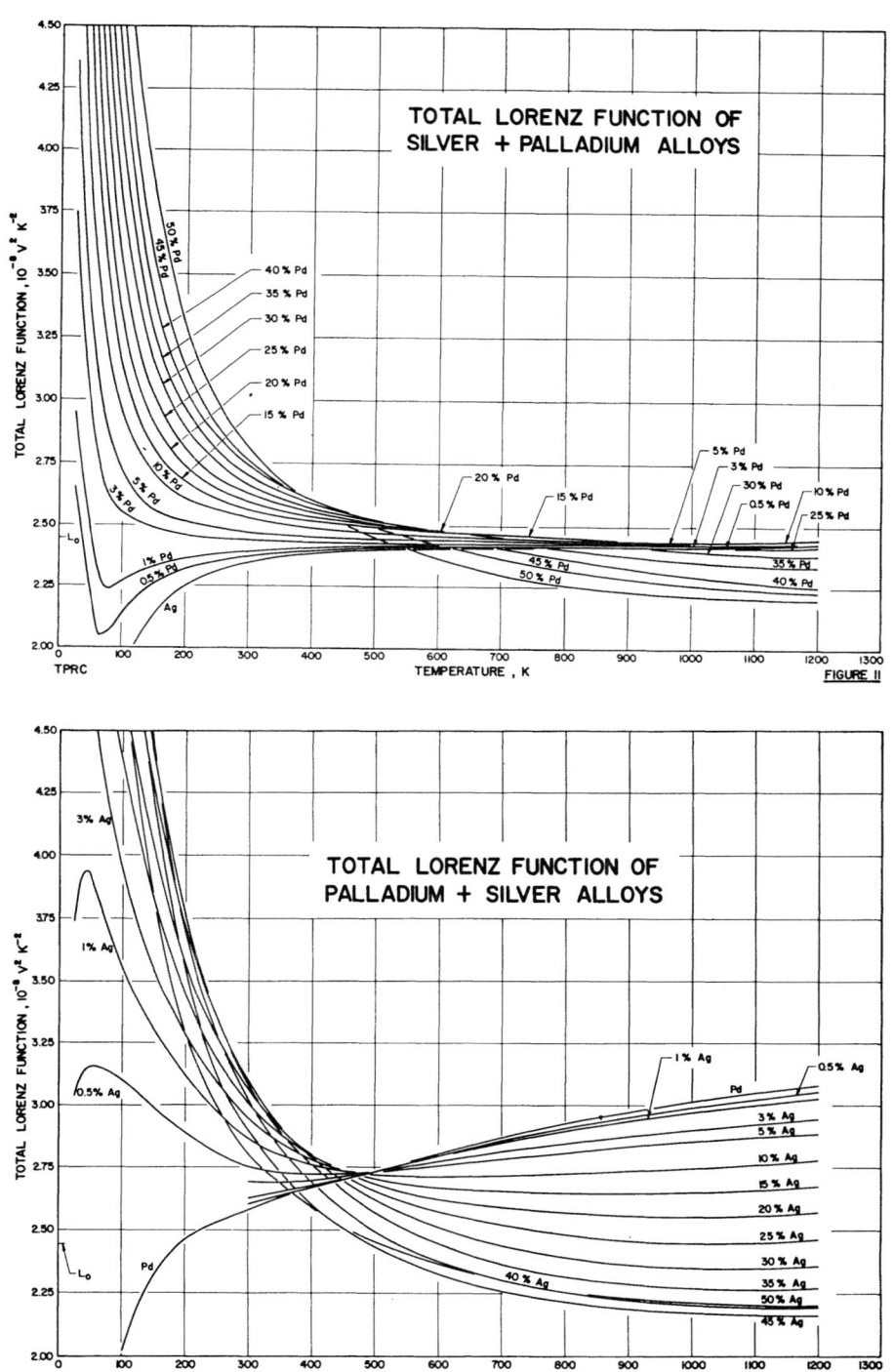

TOTAL LORENZ FUNCTION OF
SILVER + PALLADIUM ALLOYS

FIGURE 11

TOTAL LORENZ FUNCTION OF
PALLADIUM + SILVER ALLOYS

FIGURE 12

lattice conductivity of the Pd-5% Ag alloy shown in Fig. 4 is seen to occur at a much higher temperature than that of the Ag-5% Pd alloy shown in Fig. 3 but the magnitudes of the maxima differ by only a factor of 2; from this it would seem that the phonon-electron interaction is much stronger but U-processes are weaker in Pd-rich alloys. On the other hand, the maximum of the lattice conductivity of the Ni-5% Cu alloy shown in Fig. 2 occurs at a somewhat higher temperature than that of the Cu-5% Ni alloy shown in Fig. 1 and are comparable in magnitude.

IV. LORENZ FUNCTION

The electronic Lorenz function (or ratio) is defined by the equation

$$L_e = \frac{k_e \, \rho}{T} \tag{13}$$

whereas the total Lorenz function (or ratio) is defined by

$$L = \frac{k \, \rho}{T} \tag{14}$$

Figures 5, 6 and 7, 8 show the electronic Lorenz functions (or ratios) of Cu-Ni and Ag-Pd alloy systems, respectively. It is seen that at high temperatures the Lorenz functions of some of the Ni-rich and Pd-rich alloys are greater than L_0, the classical value, but for the Cu-rich and Ag-rich alloys the Lorenz functions are below the classical value. Actually, these Lorenz function values should be compared not with L_0 but with L_0-S^2 where S is the thermoelectric power; in some of these alloys S^2 can be 15 percent or more of L_0 at high temperatures. The remaining discrepancies can be accounted for by electron-electron scattering, which always reduces the Lorenz function, and band structure effects which can either reduce or, as is apparently the case for some of the Pd-rich and Ni-rich alloys, increase it.

Figures 9, 10 and 11, 12 show the total Lorenz functions (or ratios) of Cu-Ni and Ag-Pd alloy systems, respectively. Comparison of these figures with the corresponding figures for the electronic Lorenz functions indicates that for these alloys the lattice component is a large part of the total thermal conductivity at low temperatures and increases in importance with alloy concentration, and that the lattice component is insignificant above about 600 K.

V. REFERENCES

1. Mott, N.F., Proc. Phys. Soc. (London), 47, 571 (1935).

2. Mott, N. F. and Jones, H., The Theory of the Properties of Metals and Alloys, Clarendon Press, Oxford (1936); Dover Publications, New York, 196-8 (1958).

3. Klemens, P. G., Phys. Rev., 119, 506 (1960).

4. Klemens, P. G., White, G. K., and Tainsh, R. J., Phil. Mag., 7, 1323 (1962).

5. Callaway, J., Phys. Rev., 113, 1046 (1959).

6. Klemens, P. G., Proc. Phys. Soc. (London), A68, 1113 (1955).

7. Tavernier, J., Comptes Rendus, 245, 1705 (1957).

8. Ackerman, M. W. and Klemens, P. G., J. Appl. Phys., 42, 968 (1971).

9. Steigmeir, E. F. and Kudman, I., Phys. Rev., 141, 767 (1966).

ACKNOWLEDGMENT

This work was supported by the Office of Standard Reference Data, National Bureau of Standards, U. S. Department of Commerce, Washington, D. C. under Contract No. 3-35707.

NICROSIL II AND NISIL THERMOCOUPLE ALLOYS: PHYSICAL PROPERTIES

AND BEHAVIOR DURING THERMAL CYCLING TO 1200 K [*]

J. P. Moore, R. S. Graves, M. B. Herskovitz, K. R. Carr
and R. A. Vandermeer
Oak Ridge National Laboratory, Metals and Ceramics
Division and Instrumentation and Controls Division
P. O. Box X, Oak Ridge, Tennessee 37830

INTRODUCTION

Nickel base alloys of many compositions have been used extensively for thermocouple elements for many years. One of the most
prevalent is designated "type K" by the Instrument Society of
America.[1] The positive thermoelement of this pair is Chromel,
which has a nominal composition of

Ni + 9.3% Cr + 0.5% Si + minor additions,

where all percentages are with respect to weight. The negative
thermoelement is Alumel which has the nominal composition

Ni + 3% Mn + 2% Al + 1% Si + 0.5% Co + (minor additions).

The two main effects that reduce the accuracy of this thermocouple
are preferential loss of the alloying components in an oxidizing
atmosphere and short-range order in the Chromel during thermal
cycling.[2,3] Two Ni base alloys were proposed for a thermocouple
pair which would minimize both effects. These alloys are

Ni + 14.2 (± 0.1) Cr + 1.4 (± 0.05) Si   (Nicrosil II)
and
Ni + 4.4 (± 0.2) Si + 0.1 (± 0.05) Mg + 0.1 (± 0.03) Fe   (Nisil)

where all compositions are in weight percent.[4,5]

The physical properties of a thermocouple assembly are dependent on the properties of the individual thermoelements. Therefore,

[*]Research sponsored by the Energy Research and Development Administration under contract with the Union Carbide Corporation.

259

this paper presents results of the thermal conductivity, λ, electrical resistivity, ρ, Seebeck coefficient, S, and mean coefficient of thermal expansion, $\bar{\alpha}$, for two alloys near the optimum compositions of Nicrosil II and Nisil. In addition, a breakage problem with MgO insulated, 304-sheathed, and grounded Nicrosil II versus Nisil thermocouples is described. All of this is described in much greater detail elsewhere.[6]

SPECIMEN DESCRIPTION

Measurements of ρ, λ, S and $\bar{\alpha}$ required specimens of larger diameter than usually found in thermocouple wire. The only available materials of sufficient size for property measurements were out of specifications[2] by small amounts. Therefore, the material we shall refer to as "Nisil (type)" was Ni + 4.3 Si + 0.1 Mg, which, with the exception of an absence of 0.1% Fe, agreed with the nominal composition stated previously. The alloy we refer to as "Nicrosil II (type)" was Ni + 15.5 Cr + 1.5 Si, which was within 0.05% of the Si specifications for Nicrosil II but had a higher Cr content. All properties measured should be approximately the same as the properties of optimum composition alloys with the possible exception of the Seebeck coefficients.

RESULTS

All properties are identified with (+) and (−) representing Nicrosil II and Nisil, respectively. Smoothed values of λ, ρ, and S are given in Table 1.

Thermal Expansion

All $\bar{\alpha}$ data were within ± 3% of

$$\bar{\alpha}\ (+) = (9.44 + 0.00633T + \frac{815}{T}) \times 10^{-6}\ (K^{-1})\ 400 < T < 1100\ K$$

and

$$\bar{\alpha}\ (-) = (12.2 + 0.0034T) \times 10^{-6}\ (K^{-1})\ 400 < T < 1100\ K. \tag{1}$$

The $\bar{\alpha}$ (+) results were within experimental uncertainty of other results[7] on a Ni base alloy with 19.3% Cr, 0.64% Si, and 0.31% C. Both $\bar{\alpha}$ (+) and $\bar{\alpha}$ (−) were about 15% below $\bar{\alpha}$ of 304 stainless steel.[8]

Absolute Seebeck Coefficient (Uncertainty = ± 0.5 μvK^{-1})

Values obtained for S(+) and S(−) are shown in Fig. 1. The S(+) increased smoothly with increasing temperature over the measurement range from 80 to 400 K. The S(−) results decreased smoothly from about −6μvK^{-1} at 80 K to −15.8μvK^{-1} near 230 K, increased to a broad maximum near 290 K, and then decreased smoothly to the

Table 1
Smoothed λ, ρ, and S Results for Alloys Near the Compositions of
Nicrosil II and Nisil with ρ and λ Corrected for $\bar{\alpha}$

	Nicrosil II (type)			Nisil (type)		
T(K)	S(+)	ρ(+)	λ(+)	S(−)	ρ(−)	λ(−)
80	+3.81	95.70	0.1006	− 6.29	22.98	0.1418
120	5.26	96.09	0.1029	− 9.39	24.65	0.1672
160	6.57	96.50	0.1068	−11.90	26.54	0.1880
200	7.71	96.92	0.1113	−14.09	28.69	0.2062
240	8.71	97.35	0.1176	−15.48	31.04	0.2223
280	9.54	97.78	0.1257	−15.12	33.50	0.2362
300	9.92	98.00	0.1304	−15.26	34.75	0.2477
320	10.24	98.22	0.1354	−15.63	35.59	0.2562
360	10.81	98.66	0.1459	−17.01	37.01	0.2742
400	11.26	99.11	0.1566	−19.08	38.28	0.2922

upper end of the measurement range. The smooth curve drawn through
the S(−) results represents data taken with a gradient near 2 K cm^{-1}.

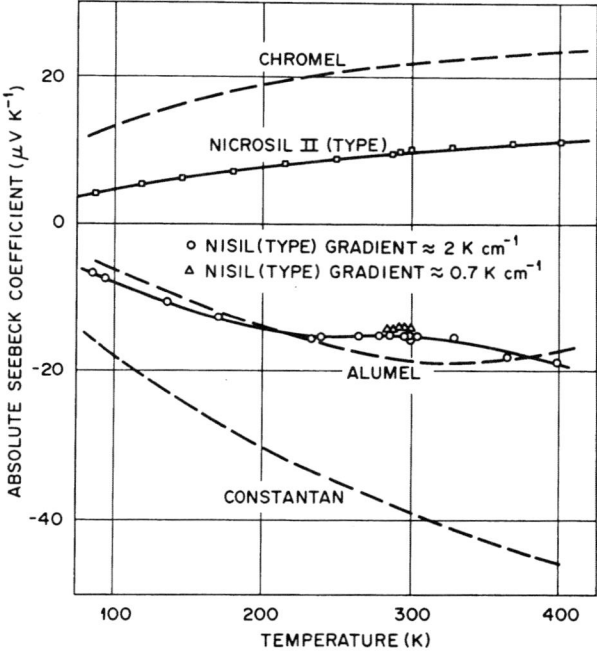

Fig. 1. Experimental values of S(+) and S(−) with S values for
Chromel, alumel, and constantan shown for comparison.

Electrical Resistivity (Uncertainty = ± 0.4%)

Experimental values of $\rho(+)$ and $\rho(-)$ were corrected for thermal expansion and are shown in Fig. 2 for a comparison to each other and to the electrical resistivity of pure Ni. Since $\rho(+)$, $\rho(-)$, and $\rho(Ni)$ differ appreciably, the scale of Fig. 2 is different for each material. The electrical resistivity of the Nicrosil II increased smoothly with increasing temperatures from 80 to 400 K and all data were within ± 0.05% of

$$\rho(+) = 94.79 + 0.010732 \ T. \tag{2}$$

The electrical resistivity of the Nisil was a factor of 3 or 4 below that of Nicrosil II but the slope $d\rho(-)/dT$ was larger than $d\rho(+)/dT$. There was a distinct slope change in $\rho(-)$ near 300 K where the broad maximum was observed in $S(-)$.

Thermal Conductivity (Uncertainty = ± 1.2%)

Figure 3 shows that $\lambda(+)$ was low but relatively uniform from 80 to 400 K whereas $\lambda(-)$ was much higher and had a distinct break near 300 K where a local maximum was observed in $S(-)$ and a sharp slope change occurred in $\rho(-)$. The $\lambda(-)$ shows some hysteresis near 300 K, but Fig. 3 is too insensitive to show this.[6]

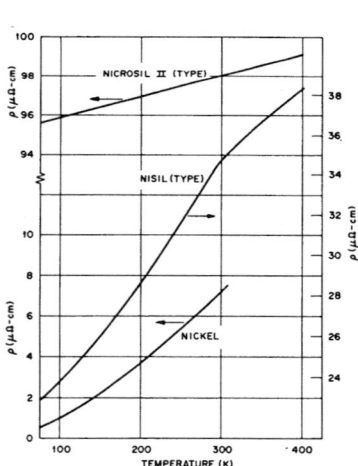
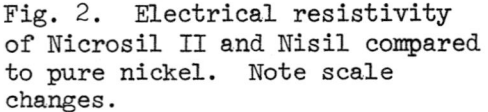

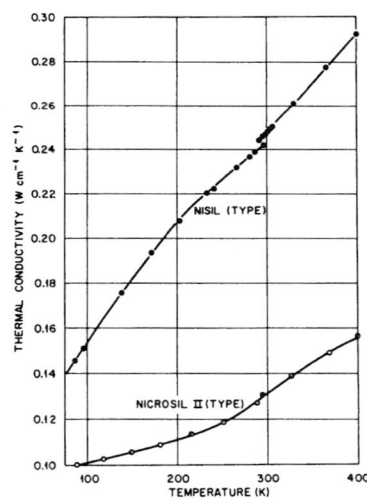

Fig. 2. Electrical resistivity of Nicrosil II and Nisil compared to pure nickel. Note scale changes.

Fig. 3. Thermal conductivity of Nicrosil II and Nisil from 80 to 400 K.

DISCUSSION OF RESULTS

Absolute Seebeck Coefficient

If the Nicrosil II and Nisil type alloys were used as thermo-couple elements referenced at 273.15 K, the output of the thermo-couple at the boiling point of water would be

$$\int_{273.15}^{373.15} \{S(+) - S(-)\} \, dT = 2521 \ \mu v$$

which is below the 2790 μv average obtained by Burley with thermo-couples near optimum composition. This lower value would be expected since the Nicrosil II type material had a higher Cr content (15.5%) than does true Nicrosil II (14.2%) and Wang et al.[9] have shown that the S of Ni + 14.2% Cr is higher than the S of Ni + 15.5% Cr.

Figure 1 includes some curves representing the S of chromel, constantan, and alumel and it can be seen that the thermal emf of a Nicrosil II versus Nisil thermocouple would be lower than thermo-couples using combinations of these other alloys. The lower sensi-tivity is not important, however, since more accurate equipment for measuring the thermal emf have become available in the last few years.

The local minimum and maximum in S(−) at 230 and 290 K, respec-tively, were caused by the paramagnetic-ferromagnetic transition which occurred near 290 K. Hansen indicates that the Curie temper-ature,[10] T_c, of Ni is lowered to about 320 K by the 4.3 wt.% addi-tion of Si and that the small amount of Mg present in the Nisil lowered T_c still further by a few degrees. The smoothness of S(+) reflects the absence of a transition over the temperature range from 80 to 400 K. The large chromium content of the alloy lowers T_c below 80 K so that the Nicrosil II alloy is paramagnetic over the entire measurement range.[10]

Electrical Resistivity

The electrical resistivity of a ferromagnetic, antiferromagnetic, or paramagnetic metal is due to scattering of conduction electrons by several mechanisms. The localized spin model would give

$$\rho(T) = \rho_i + \rho_p(T) + \rho_s(T)$$

where ρ_i, $\rho_p(T)$, and $\rho_s(T)$ are due to electron scattering by impuri-ties, lattice vibrations, and disordered spins, respectively. Nicro-sil II is paramagnetic and, qualitatively, the Nicrosil II electrical resistivity can be interpreted as that of a paramagnetic metal where

ρ_s is approximately constant and where there is a large ρ_i due mostly to the high Cr content. On the localized spin model, ρ_s and ρ_i of Nicrosil II should be constant over the measurement range from 80 to 400 K since there are no transformations over that range.

The Nisil electrical resistivity is similar to that of nickel (see Fig. 2) except that T_c has been lowered from about 630 K for Ni to 290 K for the Nisil by the Si addition. From 80 to about 290 K, $\rho(-)$ increases rapidly with increasing temperature as the aligned electron spins are thermally disordered. Above T_c, the slope of $\rho(-)$ is lower and nearly constant at about 0.035 $\mu\Omega cmK^{-1}$ which is close to the slope of 0.036 for paramagnetic Ni.[11]

Thermal Conductivity

The total λ of a metal is normally expressed as the sum of an electronic term, λ_e, and a lattice term λ_ℓ, where λ_e is related to ρ through a Lorenz function, $L(T)$. Although there are large uncertainties, there are 2 limiting cases that can be examined. If λ_ℓ is assumed to be zero, overall (net?) values for $L(T)$ of 4 x Lo and 1.5 x Lo are calculated from experimental data on ρ and λ. These values are too large to be tenable. Since most electron scattering is elastic for the mechanisms here, a second case would be to assume that $L(T) = Lo$ and calculate λ_ℓ. This approach yields λ_ℓ values close to those obtained by others on ordered and disordered Ni_3Fe.[12]

BREAKAGE OF SHEATHED-GROUNDED NICROSIL II VERSUS NISIL THERMOCOUPLES DURING THERMAL CYCLING TO 1200 K

Introduction

One of the most important requirements of a thermocouple is mechanical integrity. Even if the thermocouple elements are homogeneous, retain calibration on thermal cycling and are resistant to deleterious environments, inability of the thermocouple to survive thermal cycling without breaking cancels all assets. Therefore, the number of thermal cycles required to break or open either thermoelement of 304 SS sheathed, MgO insulated, grounded hot junction Nicrosil II versus Nisil thermocouple assemblies was measured. The hot junctions of the thermocouple assemblies were inserted into a furnace at 1200 K and then withdrawn and cooled with a blast of cold air which caused maximum heating and cooling rates of 100 K sec^{-1} and 60 K sec^{-1}, respectively.[13] The cycling period for this operation was 60 seconds.

Thermocouple wire used in this test had ρ values of 95.9 for Nicrosil II and 34.5 for the Nisil as compared to the 97.7 and 34.6 (all in $\mu\Omega cm$) of materials mentioned previously. Photomicrographs

of the thermocouple wire after swaging into a 0.03174 cm outside
diameter thermocouple assembly are shown in Figs. 4a and 4b. Grain
sizes in the Nicrosil II were about 20 to 60 microns whereas grains
in the Nisil were as large as 200 microns. These patterns were typical of the entire swaged thermocouple length and were not limited
to the region possibly effected by hot junction fabrication.

Results

During thermal cycling, most Nicrosil II versus Nisil thermocouples broke and most of these breaks were in the Nisil thermoelement. These breaks occurred as low as 287 cycles whereas Chromel-
Alumel thermocouples survive 2×10^4 cycles. Figure 4c shows a typical break in a Nisil thermoelement. This break appears to be caused
by mechanical stresses on a material with low ductility from the
large grain sizes. The stresses were caused by the difference in $\bar{\alpha}$
of the stainless steel sheath and the thermocouple wires. Low density of the MgO insulation may also have been a factor in the wire
breakages.[14]

This problem can probably be overcome for Nicrosil II versus
Nisil since it was overcome for Chromel-Alumel. Two possibilities
for doing this are use of sheath material with an $\bar{\alpha}$ near $\bar{\alpha}(+)$ and
$\bar{\alpha}(-)$ and/or improvement of the Nisil wire strength and ductility by
inhibiting grain growth.

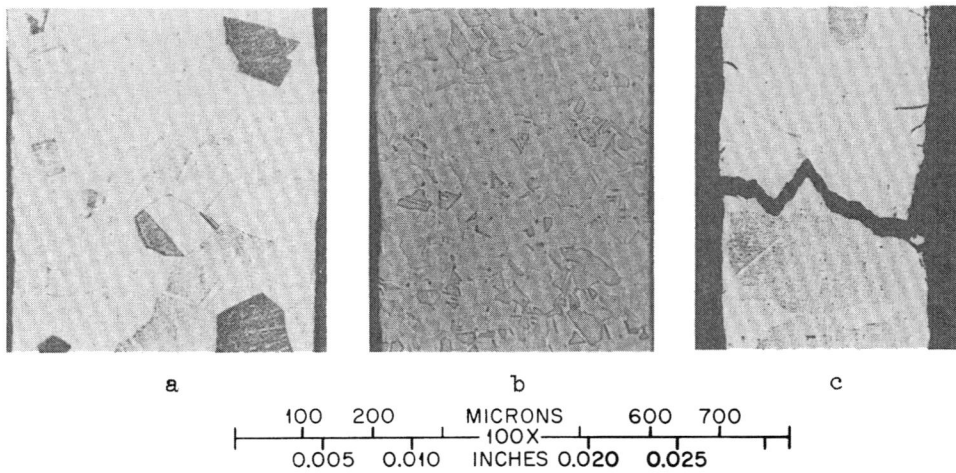

4. Photomicrographs of Nicrosil II and Nisil thermocouple wires:
a. Nisil wire inside the sheath in the pre-test condition;
b. Nicrosil wire inside the sheath in the pre-test condition;
c. Nisil wire from a thermocouple which failed after 2091 cycles.

ACKNOWLEDGMENTS

The authors wish to acknowledge T. G. Kollie for manuscript reviews and Carol Carter for secretarial assistance.

REFERENCES

1. USA Standards Institute, USA Standard C96.1 - 1964 (1964).
2. N. A. Burley, "Nicrosil and Nisil: Highly Stable Nickel-Base Alloys for Thermocouples," Temperature, its Measurement and Control in Science and Industry, Vol. 4, ISA, 1972.
3. T. G. Kollie, K. R. Carr, J. L. Horton, M. B. Herskovitz, and C. A. Mossman, ORNL-TM-4862, March, 1975.
4. C. D. Starr and Teh Po Wang, U. S. Patent No. 3,776,781 (December, 1973).
5. N. A. Burley, Project Manager; Joint NBS/DSL Project on Nicrosil/ Nisil; Cryogenics Division; Institute for Basic Standards; private communication to M. B. Herskovitz of ORNL on April 11, 1974.
6. J. P. Moore, R. S. Graves, M. B. Herskovitz, K. R. Carr, and R. A. Vandermeer, ORNL-TM-4954, 1975.
7. Thermophysical Properties of High Temperature Solid Materials, Vol. 2, Y. S. Touloukian (Ed.), (The MacMillan Company, N. Y. 1967) 1168.
8. Ibid., Vol. 3, Ferrous Alloys, 211.
9. T. P. Wang, C. D. Starr and N. Brown, Acta Mettallurgica, 14, 649, 1966.
10. Constitution of Binary Alloys, M. Hansen (Ed.), McGraw-Hill Book Company, Inc., New York, 1958, 541.
11. Thermophysical Properties of High Temperature Solid Materials, Vol. 1, Y. S. Touloukian (Ed.), (The MacMillan Company, N. Y. 1967) 696.
12. J. P. Moore, T. G. Kollie, R. S. Graves and D. L. McElroy, J. Appl. Phys., 42(8), 3114 (1971).
13. M. B. Herskovitz, ORNL-TM-3907, September 1972.
14. E. L. Babbe, A. I., NAA-SR-10511, April 1965.

THERMAL CONDUCTIVITY VARIATIONS IN STEEL AND THE EROSION BY HOT
GASES

A. R. Imam and R. W. Haskell

Watervliet Arsenal

Watervliet, New York 12189

Abstract - Variations in composition and heat treatment for an SAE
4330 type steel are considered relative to their effect on thermal
conductivity. Also considered is how these variations effect the
erosion resistance of the steel subject to a sonic velocity pulse.
For example, it is noted that an increase in the silicon content
of this type of steel leads to a decrease in its thermal conductivity and resistance to erosion.

The case of variations in the amount of temper has been found to
be anomalous. The thermal conductivity is hardly changed by such
variations yet significant changes in the resistance of the steels
are produced in this manner. These changes are explained through
modification of the kinetics of austinization.

A one dimensional finite element heat transfer model is used to
correlate changes in thermal conductivity and other thermal parameters with changes in surface temperature and heat input caused by
the thermal pulse. To supplement these calculations calorimeter
measurements were made to determine the actual heat inputs due to
the thermal pulse. The model shows how decreases in thermal conductivity for these steels results in higher surface temperatures
and hence lower resistance to erosion.

I INTRODUCTION

When materials - metallic and non-metallic - are exposed to high
temperature, high velocity gaseous impulses such as those experienced
in rockets and fire arms, a loss of material is observed. In

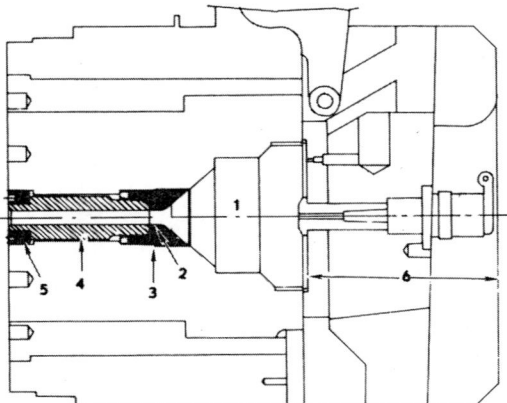

1.   Pressure Vessel (Chamber)
2.   Nozzle
3.   Nozzle Holder
4.   Vent
5.   Nut to hold vent and nozzle in position
6.   Breech plate and firing mechanism

Figure 1 - Erosion Test Fixture

general this loss is a function of several variables such as duration
of impulse, temperature, velocity, oxidative character and the density
of the gases. On the material end the mechanical, the chemical and
the thermal properties of the material exposed, determine the extent
of loss. Thus, different materials under such exposures and environments erode differently. The mechansim of this erosion, though not
clear, nevertheless can be classified in three main categories, namely,
loss due to chemical reaction between gases and the material producing
unstable or low boiling products; or the abrasive action of high velocity hot combustion products which is mostly gaseous but is mixed with
unburnt solid particles of the initial propellant. Another important
mechanism for erosion is the rapid heating of the material to its
fusion temperature, and thereafter its removal by high velocity gases.
It is this mechanism that was found to be the main cause of erosion
of SAE 4330 type steel; and it is in this mechanism where thermal
conductivity of material exposed appears to be the all important
property. The importance of thermal conductivity in steels is further
enhanced by the fact that small variations in the compositions of
steel give rise to large variations in its thermal conductivity which
in turn gives rise to still larger differences in the erosion rates
of these steels. In the following sections, a brief outline of the
erosion measurement procedure will be presented followed by the

TABLE I

Erosion Rate Versus Chemistry of 4330 Type Steel

Chemistry #	Loss g/rd	Si	Mn	Mo
1	0.29	0.08	0.32	0.23
2	0.34	0.50	0.30	0.25
3	0.14	0.06	1.03	1.00
4	0.42	0.30	1.00	0.92
5	0.33	0.05	1.08	0.20
6	0.36	0.55	1.12	0.22
7	0.11	0.08	0.28	0.90
8	0.47	0.49	0.28	0.96

results of erosion variations in different steels. These differences will then be shown to be related to the differences in the observed thermal conductivity; heat input calculations and their results will be presented in support of the above argument. Later in this paper the anamolous behavior of hardened steel will be presented whereas lowering of erosion rate can not be explained on the basis of thermal conductivity.

II PROCEDURE

a. Erosion Test Fixture

To measure the erodability of steel, a massive steel fixture was designed, a schematic of this is shown in Figure I. In this diagram the space 1 is the pressure vessel, where 450 gms of nitrocellulose is ignited. The combustion products build up about 20-21 ksi in a short interval of about 5 milliseconds. These gases then pass first through the nozzle. The purpose of the nozzle is to build up the pressure rapidly and consistantly cycle after cycle. To achieve this, the nozzle is made out of a refractory material. The gases then, at a sonic velocity, at the flame temperature, pass through the test vent. The hollow cylinder vent is about 7 inches long and has a maximum outer diameter of about 2.75 inches and an inside uniform diameter of 0.735 inches. In general, two vents are machined from any material under test. Each vent is then subjected to ten thermal impulses. After five impulses, the vent is cleaned and is checked for the weight loss as well as for the change in the internal diameter. For a given test material, in this way, an average weight loss per round or impulse is measured, averaged over twenty rounds.

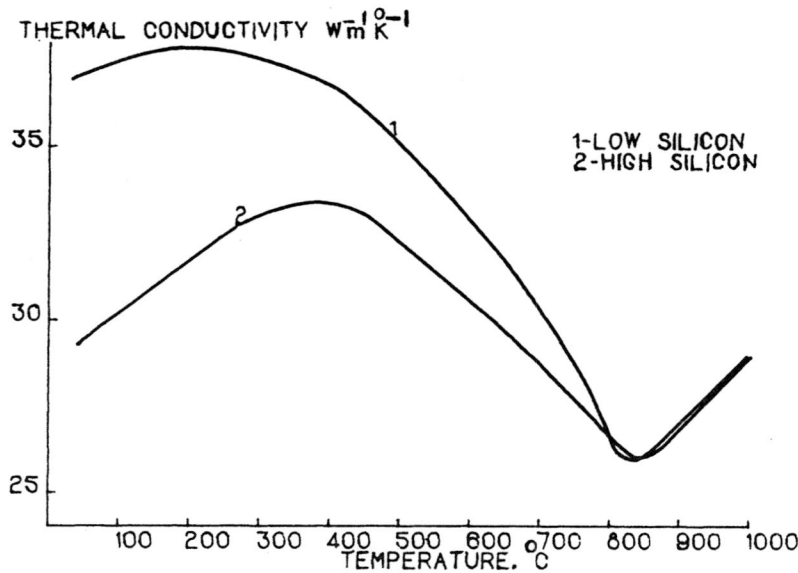

Figure 2 - Thermal Conductivities of Two 4330 Type Steels

b.   Chemistry Variations and Erosion

Using this method, eight chemistries, representing a variation of SAE 4330 steel were subjected to thermal cycles. The erosion rate observed in each chemistry are presented in Table 1. The second column shows the average loss in weight per round in a vent. The last three columns show weight percent of silicon, manganese and molybdemun. The other alloying elements are nickel 2.6 percent, vanadium about 0.1 percent, sulphur and phosphorous less than 0.01 percent and carbon 0.3 percent and rest iron: the concentration of these elements are the same in all chemistries. All the chemistries were heat treated to a martensitic structure with a hardness of Rc 34-35 - this corresponds to about a tensile strength of 130-140 ksi.

The results of this testing shows that the effect of silicon is most pronounced as is obvious from the results of chemistry number seven and eight in Table 1. An increase of 0.4 percent in silicon enhances the erosion rate by a factor of four. Such a drastic influence due to such a small change in silicon content can not be explained by considering the chemical reaction between gases and steel. It will be shown, in what follows, that such differences in the erosion rates can only be understood on the basis of heat transfer from the gases to steel, which is essentially a function of thermal conductivity of steel, provided that the other parameters that of gas and hardness of steel be the same. Further, such small differences

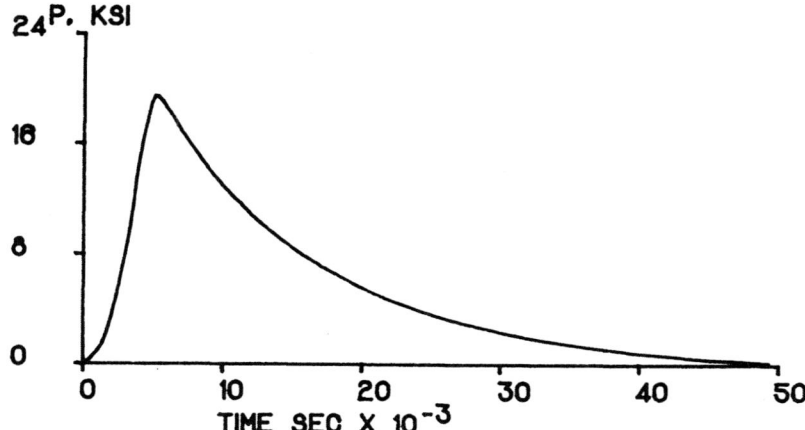

Figure 3 - Pressure-Time Trace for One Cycle

in the composition of steel give rise to rather large differences
in its thermal conductivity.

    c.   Chemistry Variations and Thermal Conductivity

    The influence of alloying content of steel on its thermal conduc-
tivity is known to exist for quite some time. An earlier equation
due to Masumoto[1], that satisfactorily provides the thermal conductivity of some steels as a function of certain alloying material is:

$$\frac{1}{K} = 5.74 + 2.43 \ (\%C) + 5.08 \ (\% \ Si) + 2.46 \ (\% \ Mn)$$

K is the thermal conductivity of steel, as is obvious from this
equation,the effect of silicon is twice as large as that of carbon
and manganese. Further, the increase in the silicon content causes
a decrease in the thermal conductivity of steel. Later works have
further confirmed this behavior. A recent paper by Speich[2] et al on
tempering of steel supports these conclusions. The effect of silicon
on thermal conductivity of steels that we tested is shown in Figure 2.
The curve 1 is the plot of thermal conductivity of a low silicon -
(.1%) - steel against temperature and the curve 2 is the graph for
a high silicon - (0.5%) steel. The maximum difference between the
thermal conductivity of the two steels is about 25% of the lower values.
The two curves come together at the transformation temperature of about
800°C, which implies that in the austenitic state, there is hardly
any influence of minor changes in chemistry on the thermal conductivity of steel. Such differences in the thermal properties are
sufficient to induce a difference of about 50°C in the maximum inside
surface temperature of the vent in a single cycle. As this takes

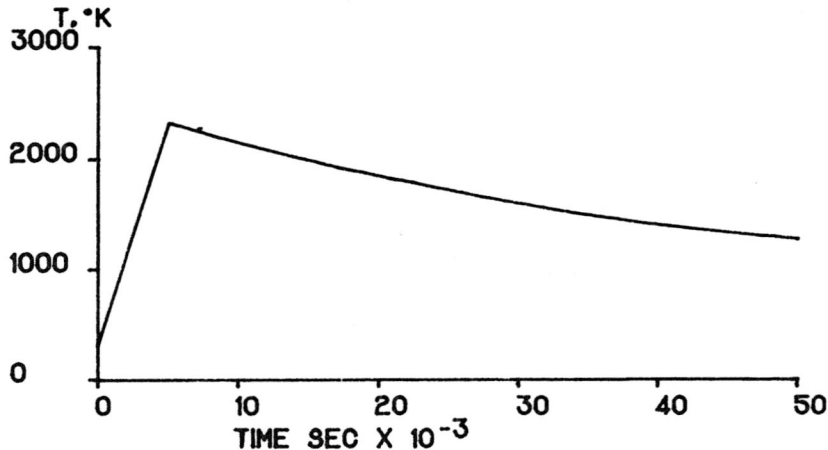

Figure 4 - Temperature-Time for a Cycle

place very close to the melting point of the steel, subsequently the differences in erosion rate can be highly exaggerated.

III HEAT INPUT CALCULATIONS

To calculate the heating during a single cycle a pressure-time trace was recorded: as stated earlier, 450 gms of nitrocellulose is ignited to produce a single thermal impulse. This cycle lasts about 25-30 milliseconds. A typical pressure-time trace for the ignition chamber is shown in Figure 3. The peak pressure consistantly stays between 20 to 21 ksi. From this observed pressure-time curve, using a modified Van der Waal's equation viz., P(V-nb) = nRT also known as Abel's equation, temperature-time curve for the chamber gases is calculated. The change considered in the process is adiabatic. The calculated temperature-time trace is shown in Figure 4.

The validity of the assumptions involved in these calculations was checked by calculating the flux of gases during a cycle using Bernoulli's equation. The flux-time curves is shown in Figure 5. The area under the curve represents the total amount of gases through the vent. This amount is in agreement with 450 gms, the initial amount of nitro-cellulose used in producing the cycle. This information of gas temperature, pressure and density is used in the following difference[3] formulation of heat transfer equation:

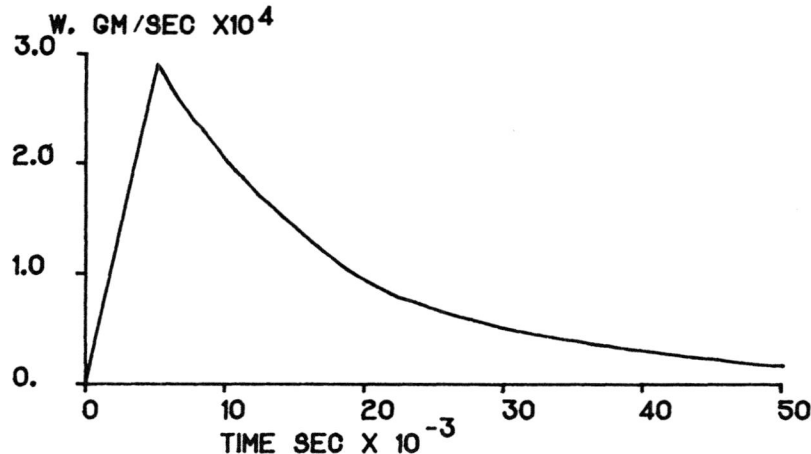

Figure 5 - Flux-Time for a Cycle

$$T_i^{n+1} = 1/2 \; (T_{i+1}^n + T_{i-1}^n)$$

$$T_o^{n+1} = (\Delta B T_\infty^n + T_1^n) \; (1 + \Delta B)$$

Biot Modulus, $\Delta B = (h \Delta x) / k$

$$\Delta x = (2 \alpha \tau)^{1/2}$$
$$k = \rho c \alpha$$

n = time interval
i = space interval
i = 0, surface
k = thermal conductivity
α = thermal diffusivity
ρ = density
h = heat transfer coefficient
τ = size of time interval
Δx = size of space interval

Note[4]: The time interval (τ) fixes the space interval (Δx) for a given diffusivity. The time interval in this work was taken to be 8.1×10^{-6} sec., further details are available in reference 4.

IV RESULTS AND DISCUSSIONS

a. Surface Temperature

The results of these calculations, using an average value of thermal conductivity for low and high silicon steel, is shown in Figure 6. The abscissa is the distance along the radius on the in-side diameter of the vent. The zero on this scale is the inside sur-face and positive direction is inside the wall, thus the abscissa in-dicates the distance in the wall from the inside surface of the vent. The graph, therefore, is a temperature profile of the wall of the vent. The maximum bore surface temperature for high silicon is about 50°C higher than that of the low silicon material. The melting

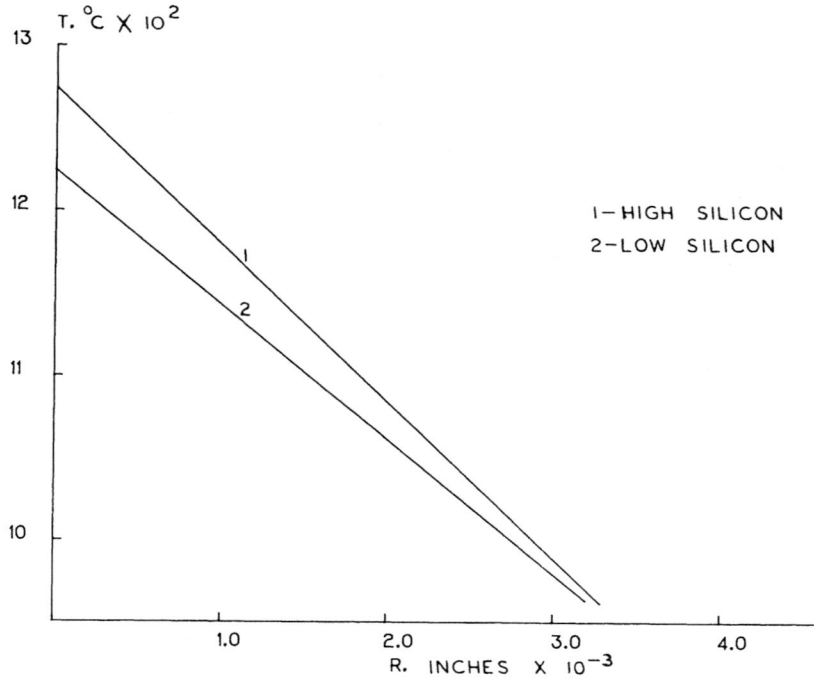

Figure 6 - Temperature Profile of the Vent Wall for a Single Cycle

point of these steels is not too far from 1300°C and thus under the
high pressures the difference in the material loss in the two cases
is very significant. The curves ultimately cross each other at
about 3.5×10^{-3} inches from the surface indicating a higher temperature inside the wall for low silicon material. The reason for
this is the larger heat input for the low silicon or higher thermal
conductivity material. The total heat input for the higher thermal
conductivity material is about 68 calories/cm^2, where as for the
lower thermal conductivity material it is about ten percent less.
This behavior is consistent with the measured heat input in different
thermal conductivity material vents.

b.   Effects of Location Microstructure

Another indication of a higher in-wall temperature for higher
thermal conductivity material is provided by the metallurgical examination of these thermally cycled surfaces. The microstructure
of steel before exposure to an impulse is tempered-martensite. The
heating of the surface and the inside material transforms the martensite into austenite wherever the temperature has reached above

1. Heat affected zone
2. Main Matrix - Martensite (unaffected by heat)

Figure 7 - Typical Microstructure of Steel After Exposure to a
Thermal Cycle   Magnification 250X

the 800°C mark - the transformation temperature. At the end of the
cycle, these heated layers of material are quenched rapidly by conducting the heat away to the bulk of the material. The austenite
transforms back into untempered martensite, thus giving rise to a
structure distinguishable from the main martrix or the structure of
the bulk material. Such a structure is shown in Figure 7. It was
observed that the depth of the heat affected zone agrees well with
those calculated values of 6 to 7 x 10^{-3} inches. The different depths
of heat affected zone can thus be attributed to differences in the
thermal conductivity of material if all other parameters such as
mechanical properties and the thermal impulse are same.

c. Variations of Mechanical Properties

We now turn to a situation where the above analysis did not provide the answer. To study the effect of mechanical properties of
steel, one specific chemistry was selected and ten vents were fabricated in such a way that each set of two had a different temper
giving rise to a different hardness, viz., Rc 30, 35, 40, 45 and 50.
These were then exposed to the thermal impulse described above. The
erosion rate in each case was observed for one single cycle only.
The reason for using a single cycle, obviously, is the transformation

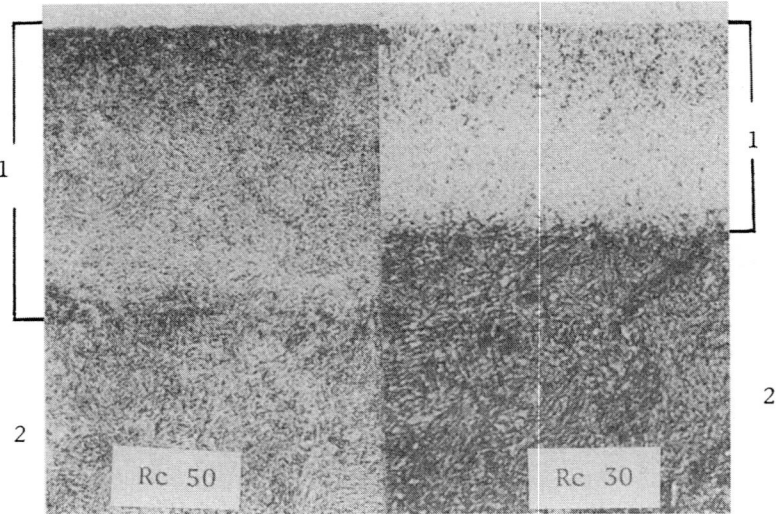

1. Heat affected zone
2. Main Matrix - Martensite (unaffected by heat)

Figure 8 - Heat Affected Zone.   Comparison for Hard and Soft Steels
                                        Magnification 250X

of the vent bore surface to untempered martensite in all cases in one
cycle. The measurement of the erosion rates thus observed shows that
the soft material erodes at a rate about four times that of the
hardest material. This indicates a higher thermal conductivity for
the harder material. This is further indicated by the comparison of
the depth of the heat affected zone shown in Figure 8. The harder
material has about 1.5 times bigger heat affected zone than the softer
material. The apparent difference in the appearance of two cases is
the contrast between the heat affected zones and the main matrix. In
case of the harder material, the main matrix structure is only slightly
tempered martensite, whereas in case of the softer material the bulk
is rather over-tempered. The two indications, that is, the low erosion
rate and bigger heat affected zone leads one to believe that the harder

material has a higher thermal conductivity. Actual measurements between 40°C to 100°C show a thermal conductivity of 39-40 $W_m-1_k^{o-1}$ for material at Rc 30 and 38.5 to 39.4 for material at Rc 50. This is the afore stated anamoly in the behavior of steels towards the high thermal impulse. The case of large heat affected zone can, perhaps, be explained by the lowering of transformation temperature in case of harder material, as carbon is not fully precipitated in the harder material as against the complete precipitation of carbon in the soft material. The higher dissolved carbon content of steel lowers the transformation temperature and thus giving rise to deeper heat affected zone in the vent wall. But the fact hard steel erodes less than the softer, can not be explained on this basis.

V CONCLUSIONS

1. Low alloy steels, in general, on exposure to hot gaseous impulses suffer loss of metal fairly rapidly, depending on pressure, initial density and maximum temperature of gases. The essential mechanism of this loss occurs through rapid heating of the metal close to or upto its melting point and its subsequent removal by high velocity fluid. The one material property that is most influential in this process is the thermal conductivity of steels.

2. Thermal conductivity of such steels, apparently, is very sensitive to the small elemental variations of alloying components, particularly, silicon and carbon.

3. It is concluded that these variables in thermal conductivity of steels gives rise to very large variations in the erosion rates of steels, when the mechanical properties are kept constant.

4. The variation of mechanical property also influences the erosion rate, for first few cycles, where harder material erodes much less than the softer material. This difference in erosion rates can not be explained on the basis of thermal conductivity.

VI ACKNOWLEDGEMENTS

The authors wish to thank and acknowledge Dynatech Corporation R/D for the measurement of thermal conductivities of two steels presented in Figure 2.

The help and support provided by the Wright Malta Corporation for testing of erosion is also appreciated.

REFERENCES

1. Austin, J.B., "The Flow of Heat in Metals", published by American Society for Metals, Cleveland, Ohio, 1942, p. 32 and 70.

2. Speich, G.R., and Leslie, W.C., "Tempering of Steel", Metallurgical Transactions, Vol 3, May 1972, p. 1043-1054.

3. Arpaci, Vedat S., "Conduction Heat Transfer", Addison-Wesley Publishing Company, 1966.

4. Haskell, R. W., and Imam, A.R., "Erosion and Heat Transfer Characteristics of Tungsten Coated Steel", Proceedings of Fifth International Conference on Chemical Vapor Deposition, 1975, (In press).

THERMAL CONDUCTIVITY OF SILICONE OILS OF THE POLYMETHYLPHENYL

SILOXANE TYPE

D T Jamieson and J B Irving

National Engineering Laboratory

East Kilbride, Scotland

Experimentally determined thermal conductivity values are presented for five polymethylphenyl siloxanes. An empirical procedure
is described for calculating the thermal conductivity of an oil
from its density and viscosity at 25°C to within the experimental
accuracy of ±3 per cent.

INTRODUCTION

Silicone oils, especially those of the polymethylphenyl
siloxane type, have many desirable characteristics including high
thermal stability, which make them particularly suitable as high-
temperature heat-transfer liquids. Thermal conductivity is a
dominant physical property required for almost every type of
design involving heat transfer and yet there exist only a few
experimental conductivity measurements for these oils.

Silicone oils have the general formula

$$R\!-\!\underset{\underset{R}{|}}{\overset{\overset{R}{|}}{Si}}\!-\!O\!\left[\!\underset{\underset{R}{|}}{\overset{\overset{R}{|}}{Si}}\!-\!O\!\right]_n\!\underset{\underset{R}{|}}{\overset{\overset{R}{|}}{Si}}\!-\!R$$

The chain length, n, of the polymer is rarely given and the size of
molecule is assessed from the kinematic viscosity of the oil. In
general, the higher the viscosity, the larger the value of n. Oils

T A B L E   1

EXPERIMENTAL VALUES OF THERMAL CONDUCTIVITY

Silicone oil	Density at 25°C g cm⁻³	Viscosity at 25°C ν cSt	Thermal conductivity mW m⁻¹K⁻¹				% Phenyl	
			0°C	50°C	100°C	150°C	Supplied by manufacturer	From Fig. 1
MS 510	0.985	50	148.2	139.9	-	130.2	9.5	12.5
MS 510	0.993	500	156.9	147.3	145.2	-	9.5	10.0
MS 510	0.999	12 500	158.5	149.5	144.1	-	9.5	10.0
MS 550	1.07	125	143.3	136.0	-	126.8	48	50
MS 710	1.11	500	140.7	137.1	-	132.0	62	61.5
PFMS-4	1.090	658.7 @ 20°C	(λ = 150.7 - 0.123t)*				-	55
PFMS-2/5L	1.008	16.93 @ 20°C	(λ = 133.1 - 0.137t)*				-	27

*Data from Rastorguev(3)

in which R = CH$_3$ are known as polydimethyl siloxanes (PMS) but polymethylphenyl siloxanes (PMPS) are less clearly defined. Most manufacturers describe their products as polydimethyl siloxanes with some of the methyl groups replaced with phenyl groups (C$_6$H$_5$). For all types of silicone oil, the kinematic viscosity and density at 25°C are usually given, but for the polymethylphenyl oils the percentages of phenyl is rarely reported.

In this paper a method is presented for predicting the percentage of phenyl in an oil from its viscosity and density. New experimental values of thermal conductivity for five polymethylphenyl siloxanes are presented, and these results are used to provide an empirical method for predicting the thermal conductivity of any polymethylphenyl siloxane oil over a range of temperature.

EXPERIMENTAL RESULTS

Thermal conductivities were measured with a hot-wire cell, using a 1000 Hz steady-state method. The cell is calibrated against water and argon, and the results were verified with several other liquids of known conductivity as described elsewhere [1]. The experimental technique involves passing known currents through the platinum filament of the cell and measuring the corresponding temperatures. The repeatability of measurement is ±0.5 per cent and the overall accuracy is better than ±3 per cent over the entire temperature range.

The conductivities of three MS 510 oils, one MS 550 oil, and one MS 710 oil (all supplied by Midland Silicones Ltd) were measured at temperatures from 0°C to 150°C. The results are presented in Table 1 along with manufacturer's published data for viscosity and density at 25°C. The number of phenyl groups, expressed as a weight percentage, is also shown in Table 1. This information was provided by the manufacturer but is not normally available. Since conductivity varies with composition it is important that a method is also available for characterizing the samples from readily available data. The only data that are consistently supplied by all manufacturers are kinematic viscosity and density at room temperature.

CHARACTERIZATION OF POLYMETHYLPHENYL SILOXANE OILS

In Fig. 1 the logarithm of kinematic viscosity is plotted as a function of density, both at 25°C, producing a family of curves for different values of percentage phenyl. The first curve is for % phenyl = 0 which, of course, corresponds to pure polydimethyl siloxanes, and this smooth curve is constructed from a large number of data points. The remaining curves were drawn parallel to this

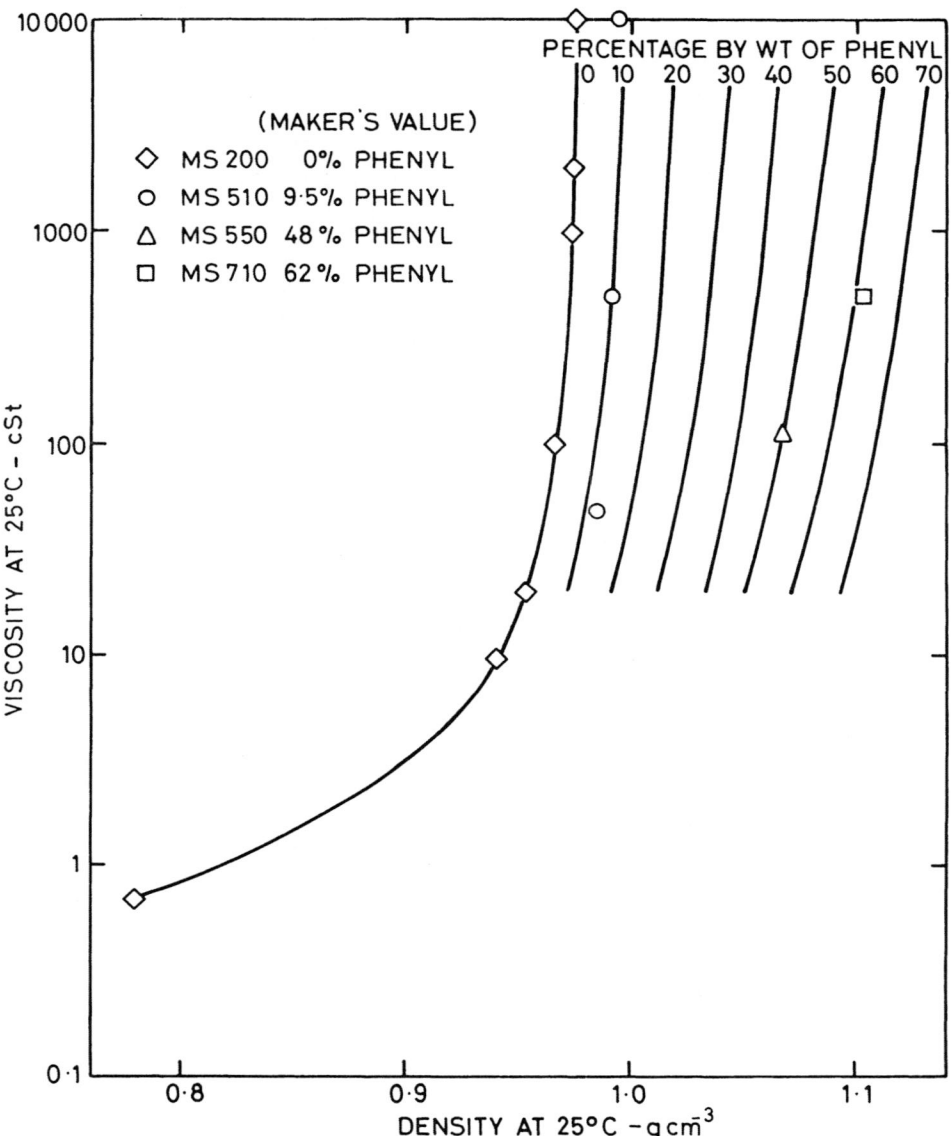

FIG. 1   KINEMATIC VISCOSITY AGAINST DENSITY FOR PMPS OILS

first well-established one, displacement being determined by the
limited data which are available for polymethylphenyl siloxanes.
These curves represent the data reasonably well, and this is shown
in Table 1 where the values for percentage phenyl from Fig. 1 are
compared with the manufacturer's values.

Thus the percentage of phenyl in any polymethylphenyl siloxane
can be found from Fig. 1 from a knowledge of its viscosity and
density.

PREDICTION OF THERMAL CONDUCTIVITY

In a previous paper [2] we presented a method for predicting the
thermal conductivity of polydimethyl siloxanes from a knowledge of
their viscosities. With polymethylphenyl siloxanes, however, such
a straightforward correlation is not possible because of the variation of these properties with phenyl groups. The percentage of
phenyl is rarely reported, and since this affects conductivity, the
first step is to characterize a polymethylphenyl siloxane by determining the proportion of phenyl groups it contains.

It will be shown that a simple graphical correlation exists
between a polymethylphenyl siloxane and the polydimethyl siloxane
of the same viscosity, and so by using the existing prediction
method for the latter, the thermal conductivity of the former is
achieved with a knowledge of percentage phenyl.

Polydimethyl Siloxane

The thermal conductivity of PMS is represented by the following equation [2]

$$\lambda_t = \lambda_{50} + \alpha(t - 50). \tag{1}$$

This shows that conductivity is a linear function of temperature,
with λ_{50} and α related to kinematic viscosity as follows

$$\lambda_{50} = 157.6\left[1 - \frac{4.52}{(\nu_{25} + 11.45)}\right] \tag{1a}$$

$$\alpha = -0.52 + 0.0025\lambda_{50}. \tag{1b}$$

This equation fits all the experimental data with an average error
of ±1.5 per cent and a maximum error of 4 per cent.

Polymethylphenyl Siloxane

Oils of the PMPS type are also linearly decreasing functions
of temperature. Graphical correlations between percentage phenyl
and reduced values of both λ_{50} and α at <u>constant viscosity</u> are
shown in Fig. 2. The reduced quantities are

$$\left[\frac{\lambda_{50}(\text{PMPS})}{\lambda_{50}(\text{PMS})}\right] \quad \text{and} \quad \left[\frac{\alpha(\text{PMPS})}{\alpha(\text{PMS})}\right].$$

In constructing Fig. 2 the weight percentage of phenyl was taken
from Fig. 1, and the λ_{50}(PMS) and α(PMS) values were calculated
from equations (1a) and (1b).

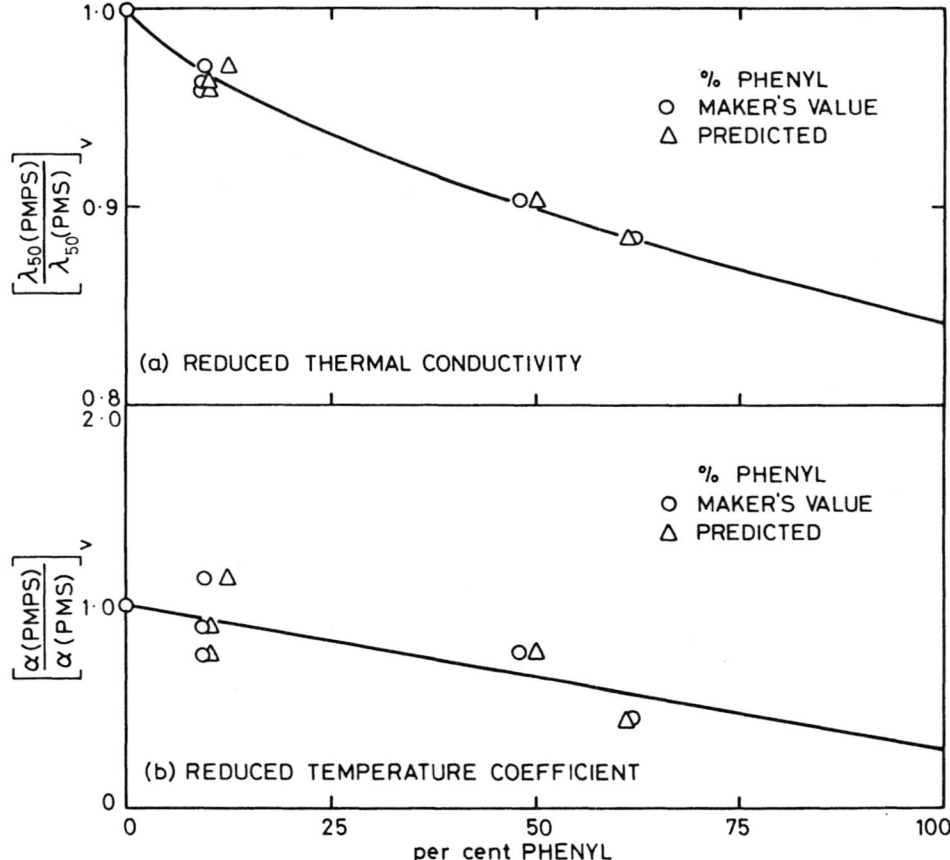

FIG. 2   REDUCED THERMAL CONDUCTIVITY
AND TEMPERATURE COEFFICIENT FOR PMPS OILS

T A B L E   2

COMPARISON OF MEASURED AND PREDICTED THERMAL CONDUCTIVITY VALUES

Thermal conductivity in mW m^{-1}K^{-1}

Silicone oil	Viscosity at 25°C cSt	0°C			50°C			100°C			150°C		
		meas	pred	diff %	meas	pred	diff %	meas	pred	diff %	meas	pred	diff %
MS 510	50	148.2	147.5	-0.5	139.9	140.0	+0.1	-	-	-	130.2	125.0	-4.0
MS 510	500	156.9	156.7	-0.1	147.3	150.6	+2.2	145.2	144.5	-0.5	-	-	-
MS 510	12 500	158.5	158.0	-0.3	149.5	151.8	+1.5	144.1	145.7	+1.1	-	-	-
MS 550	125	143.3	141.0	-1.6	136.0	136.5	+0.4	-	-	-	126.8	127.5	+0.6
MS 710	500	140.7	142.0	+0.9	137.1	138.0	+0.6	-	-	-	132.0	130.0	-1.5
PFMS-4	658.7 @ 20°C	-	-	-	144.6	139.0	-3.9	138.4	135.0	-2.5	132.3	131.0	-1.0
PFMS-2/5L	16.93 @ 20°C	-	-	-	126.3	123.0	-2.6	119.4	115.0	-3.7	112.6	108.0	-4.0

PFMS measurements reported by Rastorguev(3)

diff % = (pred-meas) x 100/meas

In summary, the method for predicting the thermal conductivity of polymethylphenyl siloxanes is to use viscosity and density values to find the percentage phenyl from Fig. 1, and then to use this value in Fig. 2 to find the reduced λ_{50} and α quantities. It then remains only to calculate the values for the equivalent PMS oil using equations (1a) and (1b) before finally substituting the PMPS values of λ_{50} and α into equation (1).

ACCURACY OF PREDICTION METHOD

The thermal conductivities predicted by the proposed method are offered for comparison with the experimental values in Table 2, and the percentage differences are also tabulated. For the MS oils the predictions are nearly all within experimental error. The r.m.s. error is 1.4 per cent, with a maximum error of 4 per cent.

CONCLUSIONS

1   The percentage of phenyl in a polymethylphenyl siloxane can be found from kinematic viscosity and density data (at 25°C) with enough accuracy to be used in the prediction of thermal conductivity.

2   The thermal conductivity of a polymethylphenyl siloxane can be predicted over a wide temperature range to within ±3 per cent.

ACKNOWLEDGEMENTS

This paper is presented by permission of the Director, National Engineering Laboratory, Department of Industry. It is British Crown copyright reserved.

REFERENCES

[1]   IRVING, J. B., JAMIESON, D. T. and PAGET, D. S. The thermal conductivity of air at atmospheric pressure. Trans. Instn chem. Engrs, 51(1), 10-13, (1973).

[2]   JAMIESON, D. T. and IRVING, J. B. Thermal conductivity of silicone oils of the polydimethyl siloxane type. Tech. Report No IP 74-012. Institute of Petroleum, London, (1974).

[3]   RASTORGUEV, Yu. L. and NEMZER, V. G. Density and thermal conductivity of silicon organic liquids (in Russian). Teplo. Svoistva. Zhidk; Mater. Vses. Teplofiz. Konf. Svoistva Vesch. Vys. Temp. 3rd, pp 155-158, (1970).

THE THERMAL CONDUCTIVITY OF REFRIGERANT 13 ($CClF_3$)

J.E.S. Venart, N.Mani*, and R.V. Paul**

Department of Mechanical Engineering

The University of New Brunswick
Fredericton, N.B. Canada

ABSTRACT:

An absolute transient line source technique was
utilized to obtain the thermal conductivity of R-13 over
the temperature range 100 to 425K at pressures from 0.17
to 21MN m^{-2} including the critical region. The accuracy
of the data was established by measurements for CO_2 and
is thought to better than ±2% outside the critical
region. Analysis of the critical isochore data yielded
power law exponents of Λ = 0.0018 ±0.0002 and ϕ = -0.57
±0.08. The saturation liquid values, which were significantly below those obtained by steady-state apparatus,
supports the conclusion that a substantial fluid radiation conductivity exists for this fluid.

INTRODUCTION:

Most refrigerants are fluorocarbon compounds and
find extensive application in the heat transfer field.
There is, however, a wide disparity in the existing
thermal conductivity values of these compounds (1) and
the establishment of reliable data covering a large
range of temperatures and pressures is essential. A
thorough investigation of the influence of anomalous
effects in the normal temperature of usage of fluids is
also important.

Precise determination of thermal conductivity for
the commonly used refrigerants is very difficult because
of their low thermal conductance and heat capacity.
Thermal conductivity of most refrigerants lies in the

* Scientific Software, Calgary, Alberta, Canada
** Cohos, Dellesalles and Evamy, Calgary, Alberta, Canada

range of 6 to 7 mW/M-K in the vapor state (one bar) and
100 to 130 mW/M-K in the liquid state (1, 2, 3). Hence
the precision of measurement for most techniques is very
low especially in the vapor state. For transient determinations, the small heat capacity of refrigerants in the
vapor state gives a low value of heat capacity ratio, ω,
and a very high correction factor to the transient line
source technique as will be discussed later.

As part of a continuing programme on the refrigerants (4, 5) Refrigerant 13 was chosen for this investigation. This fluid has the following physical properties
(6, 7):

Freezing point	92 K
Critical temperature	302 K
Critical pressure	39.2 atm
Critical density	0.578 gm/cc
Decomposition temperature	about 430 K

METHOD:

Theory:

A transient hotwire cell consists of an electrically
heated wire, suspended vertically in the fluid medium of
which the thermal conductivity is to be measured. The
fluid is contained in a cylindrical enclosure maintained
at a constant temperature. The transient behaviour of
this hotwire is idealized by the solution of the transient
one-dimensional conduction analysis.

Mani (8) gives a detailed account of the mathematical analysis, the various corrections to the ideal
model and of the experimental apparatus used in this
investigation: simply stated the equation expressing
the conservation of energy for the wire is:

$$\rho \ C_p \ \frac{\partial T}{\partial t} = - q \qquad\qquad 1$$

where   q   is the heat flow, expressed as

$$q = - \lambda \ \frac{dT}{dr} \qquad\qquad 2$$

with boundary conditions:

$$T(r, 0) = T_\infty , \ \text{a constant,} \qquad\qquad 3a$$

$$\lim_{r \to \infty} T(r, t) = T_\infty, \qquad \text{and} \qquad\qquad 3b$$

$$\lim_{r, a \to 0} \ r(\frac{\partial T}{\partial r}) = \frac{Q_l}{2\pi\lambda} , \ \text{a constant for } t \geq 0 \qquad 4$$

where Q_l is the heat generated per unit length of the
source.

The solution to equation 1 is

$$T(r, t) - T_\infty = - \frac{Q_l}{4\pi\lambda} \ E_i(- r^2/4\alpha t) \qquad\qquad 5$$

For small values of the argument, x, the exponential integral may be approximated by:

$$- E_i(- r^2/4\alpha t) \simeq - \ln(r^2/4\alpha t) - \gamma = \ln(4\alpha t/Cr^2) \; ,$$

with less than 0.2 percent error for
$$(r^2/4\alpha t) < 0.008$$
or
$$(\alpha t/r^2) > 30$$

Equation 5, therefore, reduces to:

$$T(r, t) - T_\infty = \frac{Q_l}{4\pi\lambda} \; \ln \frac{4\alpha t}{Cr^2} + \ldots\ldots \qquad\qquad 6$$

The temperature rise at the surface (r=a) of the heat source in the time interval from t_1 to t_2 is then:

$$T(a, t_2) - T(a, t_1) = \Delta T_w = \frac{Q_l}{4\pi\lambda} \ln(t_2/t_1) \qquad\qquad 7$$

The above equation does not depend on the thermal diffusivity of the fluid, and the thermal conductivity can be calculated directly.

The actual transient hotwire, utilized in an experimental thermal conductivity cell, does not however, conform exactly to the idealities of the above model and corrections must be included to account for departures from the model; i.e., the wire has finite length, diameter, conductivity and specific heat, the fluid may additionally participate in the heat exchange by radiation, convection and accommodation.

The solution of this physically more realistic model of the system is very involved. Mani (8) discusses the influence and the extent of each of the deviations of the physical model from the ideal one.

Experimental Application:

The temperature rise of the wire is measured indirectly as a function of time by monitoring the change in resistance of the wire. The hotwire thus acts, not only as a heat source, but also as a thermometer.

For a small temperature rise, ΔT_w, in a time interval, Δt, the change in resistance of the wire can be expressed as:

$$\Delta R = \Delta T_w \lim_{\Delta T_w \to 0} \frac{\Delta R}{\Delta T_w} = \Delta T_w \left(\frac{dR}{dT}\right) \qquad\qquad 8$$

If the time interval is sufficiently small such that ΔT_w is small, dR/dT can be assumed to be a constant during Δt_w. From Ohm's law

$$\Delta R = \Delta V/I - V\Delta I/I^2 \qquad\qquad 9$$

Under the condition of constant current ($\Delta I = 0$) the temperature rise can be expressed as:

$$\Delta T_w = \Delta V / (dR/dT) \ I \qquad\qquad 10$$

Thus it is important in the experiment to keep the current constant and to record the transient voltage, V_t, across the wire between the potential leads accurately.

The heat flux dissipated from the wire is given by:
$$Q_L = VI/L \qquad\qquad 11$$

The transient voltage difference ΔV in a time interval from t_1^* to t_2^* is utilized to determine the corresponding temperature rise ΔT_w. The thermal conductivity value based on equation 7 is then

$$\lambda = \frac{VI^2(1 + V_{tm}/V)(dR/dT)}{4\pi L \quad \Delta V} \ln\frac{t_2^*}{t_1^*} \left(1 + \frac{CP(t_2^*) - CP(t_1^*)}{\ln(t_2^*/t_1^*)}\right)$$

$$12$$

where V_{tm} is the mean transient voltage at a suitable mean time and $CP(t)$ the necessary specific heat connection.

A software system of the computer facility, utilized for rapid measurement and processing the data, is programmed to calculate the value of thermal conductivity λ from equation 12.

Experimental Apparatus:

Thermal conductivity measurements were made with an apparatus developed in the department (8) based on the above principles. **Accurate** and rapid measurements of the absolute thermal conductivity of liquids and dense gases in the temperature range from 75 K to 700 K and at pressures up to 700 bars are possible. The system consists of:

1. a hotwire thermal conductivity cell enclosed in a pressure bomb,
2. a high pressure system for maintaining the specified pressure during each test,
3. temperature controlled baths for the maintenance of a constant temperature,
4. an electrical system consisting of a fast response transistor switch to initiate the transient, electrical circuits and instrumentation for calibrating the cell resistance and for measuring the temperature current instability, steady and transient voltages and,
5. a computerized data acquisition system for rapid measurement of the experimental parameters during transient and their subsequent processing.

The design and functional aspects have been discussed by Mani (8).

The dimensions of the platinum wire used are:
diameter = 0.00254 cm
length   = 7.7576 cm (between potential leads).
The wire is enclosed in a high pressure bomb made
of stainless steel designed to withstand pressures up to
700 bars at temperatures from 80 to 500K. The bomb is
kept in a thermostat for precise temperature control and
accurate temperature measurement. The thermostat consists of two double walled glass dewars, one inside the
other.

Three different liquids were used as bath liquids
for the different temperature ranges.

In the range 100 K to 280 K, the inner dewar was
filled with liquid pentane and the outer one with liquid
nitrogen. The space between the two walls of the inner
dewar can be maintained at pressures down to 10^{-5} mm Hg,
to act as a variable thermal shunt between the liquid
nitrogen and the pentane.

For temperatures from 280 K to 350 K the inner dewar
was removed, and the outer dewar was filled with distilled water.

Duroline-40 was used instead of distilled water
for temperatures above 350 K and up to 425 K.

An immersion heater is provided for heating the
bath and an agitator for maintaining uniform temperature.
Temperatures were measured by a Rosemount platinum resistance thermometer (Model 104MB, specification 12 ACCC).
A Rosemount Model 414L Linear Bridge was used to convert
the voltage signal from platinum resistance thermometer
such that the bridge output signal in millivolts equals
the temperature of the sensor in degrees centigrade,
accurate to ±0.1 percent of the reading. The digital
voltmeter under computer control was then used for
direct readout of temperature by recording the bridge
output signal.

The high pressure system for pressurising the test
fluid consists of a Ruska Pump and a Budenberg dual
pressure dead weight tester for monitoring and balancing
the gas pressure with a Ruska differential pressure null
detector. The balance point could be ascertained with
an accuracy of 0.01 psi. A thermopump consisting of a
5000 psi explosion proof sampling cylinder was additionally used for pressurising the cylinder. The sampling cylinder with the Refrigerant in it was heated or
cooled externally so as to increase or decrease system
pressure.

The electrical system consists of:
1. a D.C. circuit for stability and low noise level,
together with a constant current source capable of delivering 25 to 50 milli amps;

2. an integrating digital voltmeter with a resolution
of ±1 μV coupled to a 1 MHz clock. Both are controlled
by computer for rapid measurement of the transient
voltages;
3. a fast response transistor switch under computer
control to initiate the transient and enable precise
zero time setting, and
4. facilities for calibrating the various instruments
and other components used.

Besides the DVM and the timer the system further
consists of a computerized measurement system such to
execute the following tasks under software control:
1. initiates the experiment;
2. take multiple readings with DVM;
3. measures the time lag between initiation of the
experiment and first reading of the DVM;
4. measures the time period between readings of the DVM;
5. processes the data to obtain statistical averages and
standard deviations yielding the thermal conductivity of
the test fluid.

EXPERIMENTAL RESULTS:

The freezing point of Refrigerant 13 is -181°C
(92 K). The lowest temperature at which measurements
were taken was chosen as -175°C (98 K) which ensures
that the refrigerant is still in the liquid state. The
maximum temperature is limited by the decomposition of
the refrigerant. This temperature is specified as
higher than 150°C for continuous exposure in the presence
of copper or steel (6). 150°C (423 K) was thus chosen
as the highest test temperature.

Between the limits of 98 K and 423 K, 12 other
temperature settings were chosen.

The critical temperature of Refrigerant 13 is 302 K.
Temperatures 301 K and 308 K were chosen close to the
critical temperature, the former in the two-phase region
and the latter in the vapor phase.

The pressure range chosen was limited by the experimental setup. The minimum pressure was taken as 25 psig
as pressures lower than this were very unsteady. The
maximum pressure chosen was 2000 psig.

THERMAL CONDUCTIVITY RESULTS

Table 1* shows the experimental results of thermal
conductivity for Refrigerant 13 at different temperatures.
These values are represented graphically in Figure 1 as a
function of pressure with temperature as an independent

* Complete sets of tabular data are available, at a
nominal charge from the Depository of Unpublished Data,
National Science Library, National Research Council of
Canada, Ottawa, Canada, K1A 0S2.

parameter. The isotherms are extrapolated to the corresponding saturation pressures to obtain the saturation
curve.

Figure 2, a representation of thermal conductivity
against density at various temperatures is also derived
from Figure 1. Figure 3 shows the representation on a
larger scale in the subcritical region.

A qualitative examination of these figures indicates
that:
1. Outside the critical region thermal conductivity
increases with temperature and density. Isotherms at
423 K and 373 K are almost linear with density and
pressure.
2. The effect of pressure on thermal conductivity in
the liquid region is very small at low temperature.
3. In the vicinity of the critical point, thermal conductivity exhibits an increase as the temperature and
density approach their critical values. This anomalous
behaviour can be seen to be more pronounced in Figure 4
where isotherms show sharp maxima at the critical
density. The isotherm at 302 K (critical temperature)
shows the maximum peak in thermal conductivity at
critical density. Anomalous behaviour of a similar
nature has been observed for carbon dioxide, argon,
methane and oxygen (9,10,11,12).

Accuracy and Dependability of the Results:

There are no data available on the thermal conductivity of Refrigerant 13, covering the range of temperatures and pressures reported here. In order to establish the accuracy and reliability independently, measurements on carbon dioxide were obtained utilizing the
same experimental setup.

Thermal conductivity measurements of carbon dioxide
by Sengers (9) obtained with a parallel plate apparatus
are widely accepted as accurate. Values for carbon
dioxide were measured with the transient hotwire apparatus used for the present work on Refrigerant 13, at
two temperatures: 25°C and 40°C. The results were
compared with the values obtained by Sengers for these
two isotherms. A maximum deviation of less than 2.5%
occurs in the region of critical density. The deviation
away from the critical region is very small: 0.15 to
1.00 percent.

Comparison with Available Data:

Only a few thermal conductivity measurements on
Refrigerant 13 are reported in the literature. Thermal
conductivity values obtained by Djalaian (13), Tsvetkov
(14), Grassman et al (15), Tauscher (16) and Liley
(17) have been compared with the present measurements.
All are measurements except those of Liley (17) who

reports values for both saturated liquid and vapour
Refrigerant 13. Deviations of up to +20 and -30 percent
occur in comparing the present measurements to these
authors; further discussion of this fact is not possible
here.

Thermal Conductivity in the Critical Region:

Thermal conductivity, as in the case of thermodynamic properties like specific heat, compressibility
factor, etc., increases suddenly as the critical state
is approached. This anomalous increase in thermal conductivity has been observed for a number of fluids. The
present investigation establishes this fact for Refrigerant 13 also; the measurements show an increase in thermal
conductivity value for the critical isotherm at the
critical density. The spikes in the thermal conductivity
values can be noticed even at temperatures as high as
348 K; about 45 K above the critical temperature. The
spikes converge as the critical isotherm is approached.

Critical Region Analysis:

In recent years considerable work is being done in
describing the anomalies in transport and thermodynamic
properties by simple power law relationships (18, 19, 20,
21, 22, 23   ). Widom (24) was the first to propose
that all critical anomalies can be properly incorporated if the equation of state has a certain homogeneity
(or scaling) property asymptotically near the critical
point.

In this formulation, an anomalous transport or thermodynamic property $X(\rho, T)$ that diverges along the critical isochore with an exponent ψ.

$$X(\rho_c, T) = X_0 |\varepsilon|^{-\psi}, \text{ where } \varepsilon = \frac{T - T_c}{T_c} \qquad 13$$

can be written in the form

$$\frac{A_x^*(\rho) \ x \ (\rho, T)}{X_0 |\Delta \rho^*|^{-\psi/B}} = f_x(x) \qquad 14$$

where $f_x(x)$ is again a function of the scaling parameter
X. The factor $A_x^*(\rho)$ accounts for any asymmetry of X
around the critical density ρ_c.

The experimental thermal conductivity λ can be
divided into an anomalous part $\Delta\lambda$ and an ideal part λ_{id}

$$\lambda = \delta\lambda + \lambda_{id} \qquad 15$$

The ideal or background thermal conductivity is
estimated empirically by extrapolating data away from
the critical point into the critical region. For this

purpose it is convenient to consider a so-called excess
thermal conductivity,

$$\bar{\lambda} = \lambda(\rho, T) - \lambda(0, T) \qquad\qquad 16$$

which measures the excess of the actual thermal con-
ductivity $\lambda(\rho, T)$ at density ρ over the value $\lambda(0, T)$
in the low density limit $\rho \to 0$ at the same temperature.
Once $\bar{\lambda}$ is established as a function of ρ from data outside the critical region, the ideal thermal conductivity
in the critical region can be calculated as

$$\lambda_{id} = \bar{\lambda}(\rho) + \lambda(0,T) \qquad\qquad 17$$

To investigate whether $\Delta\lambda$ satisfies a scaling law
relation of the form (14), the temperature dependence
of $\Delta\lambda$ along the critical isochore is correlated to a
power law of the form:

$$\Delta\lambda(\rho_c, T) = \Lambda \, \varepsilon^{-\psi} \qquad\qquad 18$$

Following the above the thermal conductivity values
for 373 K and 423 K were plotted as a function of density;
a single curve could be used to represent $\bar{\lambda}$ at both
temperatures and the ideal thermal conductivity was calculated at temperatures of 302, 308, 313, 323 and 348 K,
using this data and the experimental dilute gas values.
The values of $\Delta\lambda$ on the critical isochore were then
fitted to the power law Equation 18 with the result

$\Lambda = 0.0018 \pm 0.0002$

$\psi = -0.57 \pm 0.08$

The value of the critical exponent ψ obtained agrees
well with the predicted value of -0.59 ± 0.10 (25).

CONCLUSIONS:

Thermal conductivity measurements on Refrigerant 13
using a transient line source apparatus were obtained
in the temperature range 100 K to 425 K at pressures up
to 150 atmospheres.

Refrigerant 13, like other pure fluids shows sharp
increases in thermal conductivity in the critical region.
The peak value occurs at the critical density. This
anomaly in the critical region can be represented by a
scaling law of the form:

$$\Delta\lambda = \Lambda \, \varepsilon^{-\psi}$$

with the coefficients

$\Lambda = 0.0018$

$\psi = -0.57$

ACKNOWLEDGEMENTS:

This work was performed with the support of Atomic
Energy of Canada (CRNL and WNRE) and the National Re-
search Council of Canada while the authors were at the
University of Calgary. Mrs. C. Pitt typed this manuscript.

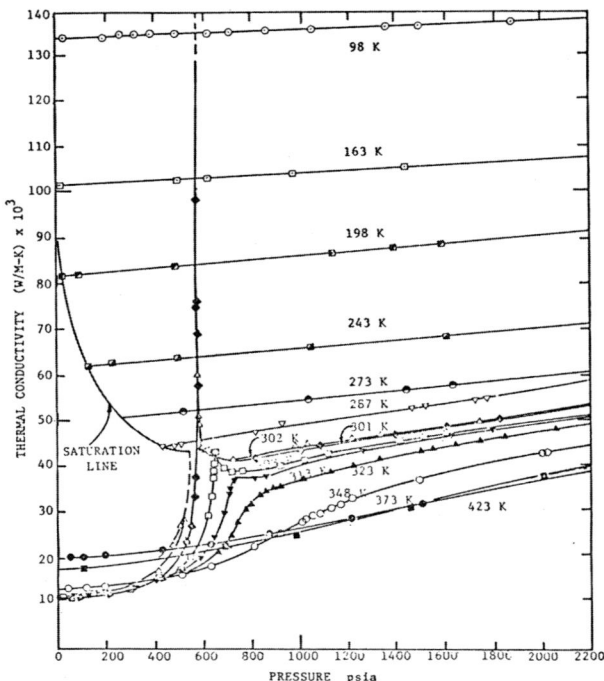

FIGURE 1: λ R 13; isotherms

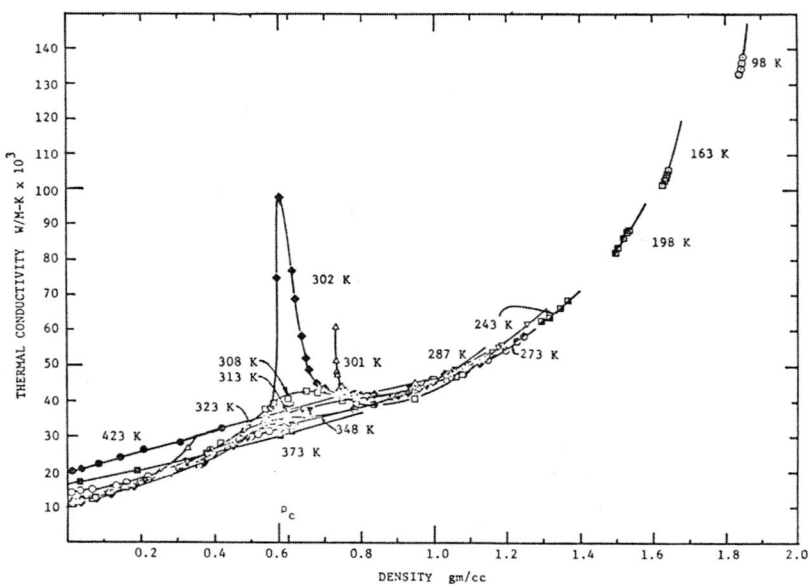

FIGURE 2: λ R 13; isotherms vs. ρ

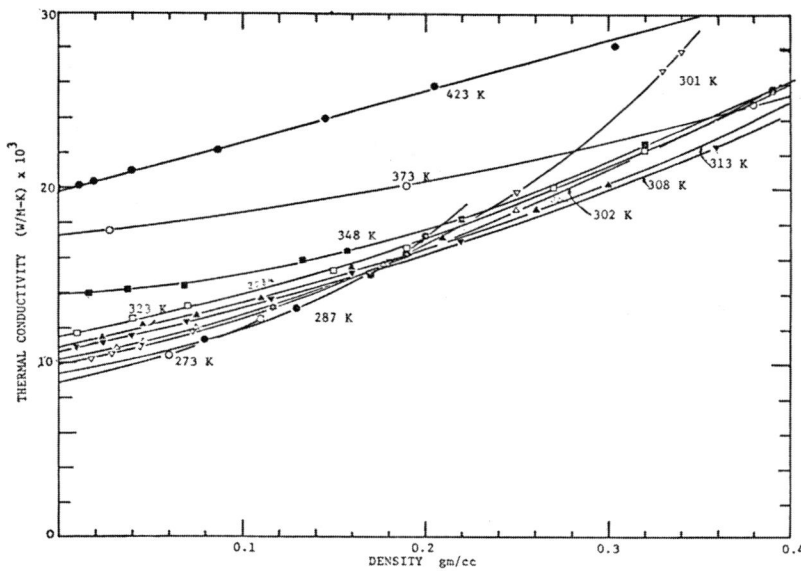

FIGURE 3: λ R 13; low pressure

REFERENCES:
1. A. Lamb, "Thermal Conductivity of Liquid Freons - A
    Critical Review", Unpublished Project Report, Univ.
    Calgary, Dept Mech Engg, Apr 1972.
2. R.C. Downing, "Estimation of the Liquid Thermal
    Conductivity of Fluorinated Refrigerants", ASHRAE J,
    Apr 1965.
3. R.C. Downing, "Transport Properties of 'FREON'
    Fluorocarbons", DuPont Freon Tech Bull C-30, 1965.
4. N. Mani and J.E.S. Venart, "Thermal Conductivity
    Measurements on Liquid and Dense Gaseous Methane
    (120 to 400K, 25 to 700 bar), Advances in Cryogenic
    Engineering 18 (1973) p. 280-288.
5. J.E.S. Venart and N. Mani, "Thermal Conductivity of
    Refrigerant-12 (300-600K, 0.2-20MN/m²)" (Trans CSME,
    to be published).
6. DuPont 'FREON' Product Information B-2, "Freon
    Fluorocarbons. Properties and Applications", 1969.
7. Albright and J.J. Martin, "Thermodynamic Properties
    of Chlorotrifluoromethane", Ind Engg Chem, 44 1952.
8. N. Mani, "Precise Determination of the Thermal
    Conductivity of Fluids using Absolute Transient
    Hotwire Technique", PhD Thesis, Univ Calgary, 1971.
9. A. Michels, J.V. Sengers and Van der Gulik, "The
    Thermal Conductivity of Carbon Dioxide in the Critical
    Region. Part II", Physica, 28 ;062.
10. V.P. Sokolova and I.F. Golubev, "Thermal Conductivity
    of Methane at Different Temperatures and Pressures",
    Thermal Engg, 14 (4) 1967.
11. K. Kellner, "The Critical Point Exponent of the

Thermal Conductivity of Fluids", British J Appl Physics, Series 2, 2 1969.

12. L.A. Weber, "Thermodynamic and Related Properties of Oxygen from the Triple Point to 300 K at Pressures to 3330 Atmospheres", Natnl Bureau Standards Report 9710, June 1968.

13. W.H. Djalalian, "Measurements of Thermal Conductivities of Liquid Refrigerants at Low Temperatures", Inst Int du Froid Annexe au Bul Commissions 2, Turin, Italy, Sept. 1964.

14. O.B. Tesetkov, "Use of a Regular Regime for Studying Thermal Conductivity of Liquid Freons", Inzh Fiz Zh Akad Nauk, Belorussk, SSR, 9 (1) 1965.

15. P. Grassman, W. Tauscher and A. Chiquillo, "Measurement of Thermal Conductivity of Refrigerants and Salt Solutions", Proc ASME Symp Thermphysical Properties, 1968.

16. W. Tauscher, "Messung der Warmleit fahigkeit Flussiger Kaltemittel mit einem Instationaren Hitz Drahtverfahren", PhD Thesis, Diss Nr 4144, Eidgenossisischen Techn Hochschule, Zurich, 1968.

17. P.E. Liley, "Thermophysical Properties of Refrigerants", Table 14, Chapter 14, ASHRAE Handbook of Fundamentals, 1972.

18. D.P. Needham and H. Ziebland, "The Thermal Conductivity of Liquid and Gaseous Ammonia and its Anomalous Behaviour in the Vicinity of the Critical Point", Int J Heat Mass Trans, 8 1965.

19. K. Kellner, "The Critical Point Exponent of the Thermal Conductivity of Fluids", British J Appl Physics, Series 2, 2 1969.

20. J.V. Sengers and P.H. Keyes, "Scaling of the Thermal Conductivity Near the Gas Liquid Critical Point", Physical Review Letters, 26 (2) Jan 1971.

21. R.C. Hendricks and A. Baron, "Prediction of the Thermal Conductivity Anomaly of Simple Substances in the Critical Region", ASME Publ No. 7-HT-28, Aug 1971.

22. M.L.R. Murthy and H.A. Simon, "Measurement of Thermal Conductivity Anomaly in the Critical Region of Carbon Dioxide", Proc Fifth Symp on Thermophysical Properties, ASME, New York, 1970.

23. M.L.R. Murthy and H.A. Simon, "Measurements of Thermal Conductivity Divergence in the Supercritical Region of CO_2", Physical Review A, 2 (4) Oct 1970.

24. J.M.H. Levelt Sengers, "Scaling Predictions for Thermodynamic Anomalies Near the Gas-Liquid Critical Point", Ind Engg Chem, 9 (3) 1970.

25. B. Chu and J.S. Lin, "Small Angle Scattering of X-Rays from Carbon Dioxide in the Vicinity of its Critical Point", J Chem Physics, 53 Dec 1970.

THERMAL CONDUCTIVITY OF POLYSILOXANE IN INTERMEDIATE

TEMPERATURE RANGE

Akira Sugawara and Ichiro Takahashi

Department of Mechanical Engineering
Faculty of Engineering, Yamagata University
Yonezawa, Yamagata, Japan

ABSTRACT - The thermal conductivity of polysiloxane, which is base material of silicone rubber, was measured with an accuracy of ±1.0 % over the temperature range from 230 to 410 °K.

For measuring thermal conductivity, the newly developed dual combined method was adopted.   The standard material used in this experiment was non-crystal, pure fused, quartz with the highest possible purity 99.9999 % .

As a new result of the experiment, it became evident that the thermal conductivity of polysiloxane decreases linearly from 0.257 to 0.167 Watt/m.°K with increasing temperature.

The polysiloxane, which is base material of silicone rubber, has been widely used in the various fields of science and engineering, because its physical properties, such as thermophysical, mechanical and electrical, are very stable in the intermediate temperature range.   Among these properties, thermophysical, in particular thermal conductivity, has recently become important with increasing demand for heat transfer calculations.

However, the few systematic studies on the thermal conductivity of polysiloxane have been done in the past, and so reliable and available data on the conductivity are surprisingly meager [1],[2] .

Therefore, in the present investigation, the exact values for the thermal conductivity of polysiloxane were measured precisely in the temperature range from 230 to 410 °K.   Measurement was carried out by the dual combined method (steady state and axial heat flow

method) which has a very high accuracy $\pm$ 1.0 %.   The principle of
this method has been explained by Sugawara [3],[4].   The apparatus
used in this experiment, which was newly developed, has the following significant advantage ;   the guard heater was designed and
fabricated so that the axial temperature distribution of the guard
heater may automatically match those of the standard material and
the test specimen with the precision of 1/20 °C.   Thus, the very
high accuracy of this experiment could be obtained.

The shapes of both the standard material and the test specimen are circular plate with 100.0 mm in diameter and 10.00 mm
thick.   The specimen is sandwiched between two standard materials
in close contact and a steady heat flow through the plates establishes
a steady linear flow through the plate centers.   The temperature
in the specimen and the standard materials were measured by potentiometer in conjunction with the copper-constantan thermocouples
having 0.06 mm in diameter.

The standard material used for the conductivity measurement
was a flat plate of non-crystal, pure fused quartz, whose conductivity and purity were given by Sugawara [5],[6].

The test specimen was supplied by Shin-etsu Chemical Co. Ltd.,
Japan.   The silica content of this sample was 38.8 weight per
cent, which was measured by the method of gravimetric analysis.
The value obtained by this analysis is little higher than the
theoretical value (37.9 weight per cent) calculated from the
structural formula of polysiloxane.   The structural formula of
polysiloxane is shown in figure 1.   In this formula, n is degree
of polymerization.   Specific weight is 1.188 at room temperature
20°C.

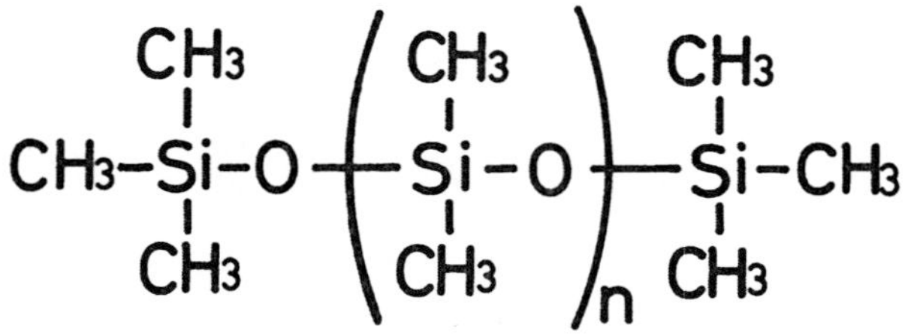

Fig.1   Structural formula of polysiloxane

The result of the measurement, the variation of thermal conductivity with temperature, is shown in figure 2.   As illustrated

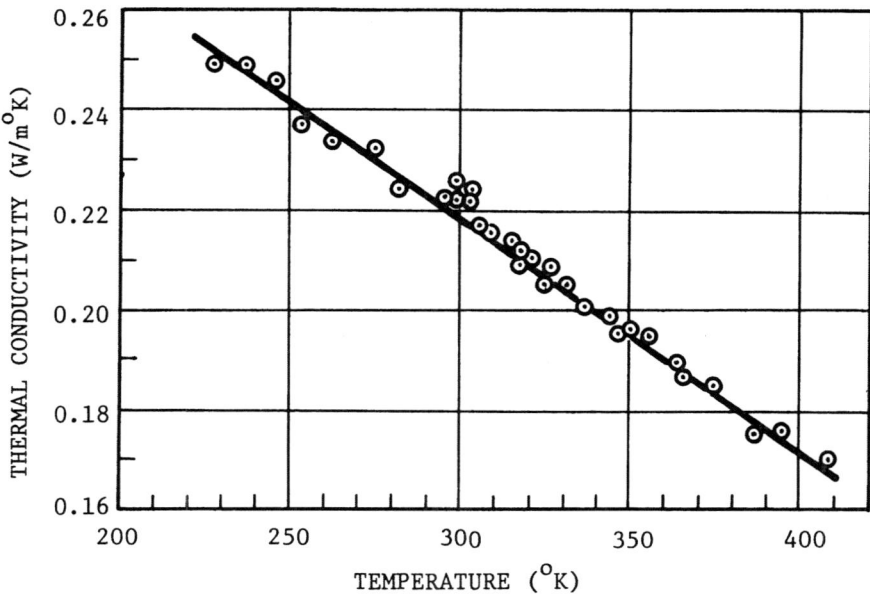

Fig.2   Variation in thermal conductivity of
        polysiloxane as a function of temperature

in this figure, the measured points do not scattered greatly, and
the conductivity values decrease linearly and gradually with increasing temperature.   Therefore, it is easy to determine the
most probable value of thermal conductivity as is shown by the
straight line in figure 2.

As a new result, the experimental formula for the thermal
conductivity of polysiloxane was derived from the figure as
follows ;

$$\lambda = 0.372 - 0.00050\,\theta$$

where, λ = thermal conductivity in Watt/m.^{o}K , and
        θ = temperature in ^{o}K, and θ is applicable from 230 to
            410 ^{o}K.

The effect of radiation on thermal conductivity can not be
observed in this experiment.

At the present investigation, it is difficult to explain theoretically the mechanism of heat transport in polysiloxane, because
the degree of influencing factors, such as the length of chain,
degree of polymerization n and a small amount of chemical impurity,
on the heat transport (or thermal conductivity) can not be estimated exactly.

References

1) K.Eierman, Kunststoffe, 51, 512 (1961)
2) K.Eierman and K.H.Hellewege, J.Polymer Sci., 57, 99 (1962)
3) A.Sugawara, J.Appl.Phys., 36, 2375 (1965)
4) A,Sugawara, J.Appl.Phys., 39, 5994 (1968)
5) A.Sugawara, Science of Machine (Japan), 21, 512 (1969)
6) A.Sugawara, Physica 41, 515 (1969)

THERMAL CONDUCTIVITY OF PARTICLE SUSPENSIONS

Avtar Singh Ahuja

Georgia Institute of Technology

Box 37654, Atlanta, Georgia 30332

This paper concerns the thermal conductivity measurements of stationary and flowing suspensions of 50 µ and 100 µ polystyrene spheres in aqueous glycerine and aqueous sodium chloride. The thermal conductivity of neutrally and nonneutrally buoyant stationary suspensions was determined by the transient technique (accuracy 3-4%). The thermal conductivity of neutrally buoyant suspensions satisfies the Maxwell formula. However, the thermal conductivity of nonneutrally buoyant suspensions (density difference of about 1.5%) is measured to be higher (at high particle concentrations) than predicted by the Maxwell formula. This is attributed a priori to the wake created by the gravity settling of particle assemblages.

Employing the Graetz solution corresponding to uniform wall temperature (accuracy 10%), the effective thermal conductivity of suspensions in laminar motion is seen to be a function of the shear rate (unlike single-phase fluids) and several other factors, and is as much as three times the thermal conductivity of stationary suspensions. Based on the shear induced particle rotations and the entrained fluid, a model for the augmentation in heat transport has been proposed and the experimental data has been analyzed.

In this paper, the salient features of the thermal conductivity measurements in stationary and flowing suspensions are described. To the author's knowledge, there is no theoretical work available on the thermal conductivity of suspensions in laminar motion. However, there is a good deal of work available on the thermal conductivity of stationary suspensions. Before describing

the thermal conductivity measurements, some classical formulas
are briefly reviewed.

Maxwell[1] derived a formula for the electric conductivity of
a composite system, monodispersing spheres of conductivity k_p
in a liquid of conductivity k_f, as

$$(k_s - k_f)/(k_s + 2k_f) = \phi \ (k_p - k_f)/(k_p + 2k_f), \qquad (1)$$

where k_s is the electric (or thermal) conductivity of the suspension, and ϕ the volume fraction of solid spheres. This formula
applies to dilute suspensions. Since electric conduction and
thermal conduction are analogous in differential equations and
boundary conditions, Eq. (1) also gives the thermal conductivity
of heterogeneous media.

Bruggeman[2] extended Maxwell's formula to concentrated suspensions (of spheres) and obtained the relation as

$$(k_s/k_f)^{1/3}(1 - \phi) = (k_p - k_s)/(k_p - k_f). \qquad (2)$$

I. CONDUCTIVITY MEASUREMENTS IN STATIONARY SUSPENSIONS

The thermal conductivity values of suspensions have been
determined by the steady state method (Ref. 3) by using silica
gel to prevent particle settling. The transient technique described by Grassman et al. (Refs. 4, 5) for single-phase nonconducting fluids has been modified by making measurements in electrically conducting fluids such as blood (Ref. 6). This study gives
one more application of the transient or unsteady state method for
measuring the thermal conductivity of particle suspensions which
may be neutrally or nonneutrally buoyant, electrically conducting
or nonconducting.

To the author's knowledge, thermal conductivity data for nonneutrally buoyant stationary suspensions are not available in the
literature. In a nonneutrally buoyant suspension, settling particles cause the flow of the surrounding fluid, thereby enhancing
the heat transfer and the effective thermal conductivity. This
simple intuition is borne out by the measurements reported herein.

The theory of the transient technique is well known[7]. This
technique is simpler, quicker, and more accurate compared to the
steady state method.

The measurements of thermal conductivity in suspensions prepared in aqueous sodium chloride (which are electrically conducting) were made by a platinum-film type sensor (Ref. 6), coated

with quartz for electrically insulating the film from the suspension. The measurements by means of a sensor of diameter 152 μ and length 0.4 cm, are accurate to within 4% (Ref. 6).

The measurements of thermal conductivity in suspensions prepared in aqueous glycerine (which are electrically nonconducting) were made by a chemically pure solid platinum wire of diameter 25 μ and length 5.6 cm and are accurate to 1% (Ref. 8).

The depth of penetration in particle suspensions for the platinum-film-type sensor is about 750 μ (Ref. 6) and for the platinum wire is 2700 μ (ref. 8). The temperature rise for the platinum-film-type sensor is 4°C and for the platinum wire is 1.2°C (Ref. 9).

The measurements were made by making use of the relative method which consists of one sensor immersed in the particle suspension and another similar sensor immersed in water whose thermal conductivity is known. Both the sensors were mounted vertically.

The polystyrene spheres of two diameters, 50 μ and 100 μ, were used. The suspending liquids used were (i) 5.2 wt % aqueous sodium chloride; (ii) 7.4 wt % aqueous sodium chloride; and (iii) 20 wt % aqueous glycerine.

The particle suspensions prepared in 7.4 wt % aqueous sodium chloride and in 20 wt % aqueous glycerine are very nearly neutrally buoyant. The measurements of heat conductivity in them (Ref. 8) (which are omitted in this report) satisfy Maxwell's (or Bruggeman's) formula. The particle suspensions prepared in 5.2 wt % aqueous sodium chloride are nonneutrally buoyant by about 1.5% (particles are denser). The values of thermal conductivity, as shown in Fig. 1, are higher than those for the corresponding neutrally buoyant suspensions by about 7% at 40% concentration. This effect seems to be due to (i) a very large number of solid particles, and (ii) their gravity sedimentation.

The augmentation in thermal conductivity in a sedimenting suspension is a function of the settling velocity, v, particle diameter 2a, kinematic viscosity ν_f and the thermal diffusivity α_f of the suspending phase, and the particle volume fraction ϕ. Mathematically, the augmentation in thermal conductivity can be expressed in dimensionless form as:

$$k_{nn}/k_n - 1 = F(\phi, 2va/\nu_f, 2va/\alpha_f),\tag{3}$$

where k_{nn} and k_n, respectively, are the thermal conductivities of the nonneutrally and neutrally buoyant suspensions, F is some function, $2va/\nu_f$ is the Reynolds number, and $2va/\alpha_f$ is the Peclet number.

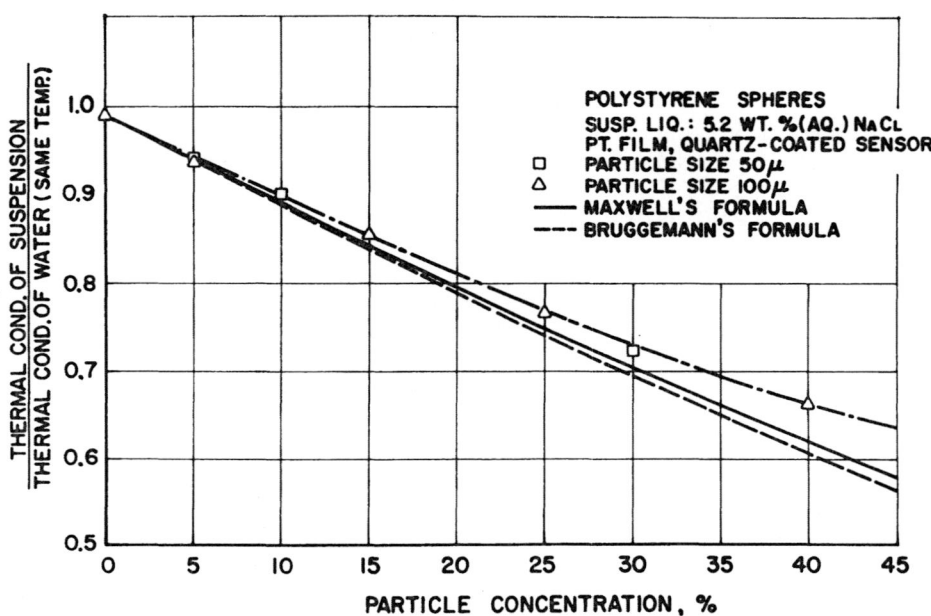

FIG. 1. Thermal conductivity measurements of nonneutrally buoyant and electrically conducting suspensions.

The augmentation effect observed in nonneutrally buoyant stationary suspension (Fig. 1) is not large enough for determining the nature of the function F. Heat and mass transfer are analogous under certain assumptions. Since the gas diffusion coefficient D_f (which replaces α_f in Eq. (3)), for example, for oxygen, is 2 orders of magnitude less than the thermal diffusivity α, the augmentation effect in diffusion coefficient of the suspension will be much more than in the thermal conductivity of the suspension and the nature of the function F can be determined.

II.   CONDUCTIVITY MEASUREMENTS IN FLOWING SUSPENSIONS

It is well known that the thermal conductivity of single-phase fluids (such as water or air) in laminar flow is independent of the state of the motion. However, the thermal conductivity of turbulently flowing single-phase fluids is dependent on the

average shear rate, in accordance with the Prandtl Mixing Length
Hypothesis (Ref. 10).

When a suspension is in laminar motion in a tube, the particles, under the influence of shear field, rotate and thereby
entrain the surrounding liquid by friction. If the shear field
is strong enough and the particles are big enough, the particles
act as a centrifugal fan, in that the fluid particles are thrown
outward by the centrifugal force from and around the area near the
equator of the suspended particle and return to an area around
the poles to preserve the law of material continuity. This churning superimposed on the rotational motion of the surrounding fluid
- the three-dimensional boundary layer - will render the thermal
conductivity of the suspension to be dependent on several parameters of the suspension flow. They are: angular velocity (ω)
and radius (a) of the particle, kinematic viscosity (ν_f) and
thermal diffusivity (α_f) of the surrounding fluid, tube radius
(R) and length (L), and the volume concentration (ϕ) of the particles. Mathematically, the augmentation in thermal conductivity
can be expressed in dimensionless form as

$$k_{mov}/k_{stat} - 1 = F(\phi, \omega_a^2/\nu_f, \omega a^2/\alpha_f, R/a, L/a). \qquad (4)$$

where k_{mov} and k_{stat}, respectively, are the thermal conductivities
of the moving and stationary suspensions, F is some function, the
group $\omega a^2/\nu_f$ is the Reynolds number and the group $\omega a^2/\alpha_f$ is the
Peclet number. Equation (4) has also been arrived at from a
dimensional analysis (Ref. 11) by starting from Navier-Stokes
and heat convection equations.

Pressure drop measurements[12] (accuracy 5%) were made in suspensions in laminar motion in a tube to determine the validity
of Poiseuille's law. The friction factor for suspensions of concentrations up to 11% was found to be corresponding to Poiseuille
flow.

The thermal conductivity measurements (accuracy 10%) were
made in an apparatus (Ref. 12) calibrated with tap water, by
employing Graetz solution corresponding to uniform wall temperature. Polystyrene particles of two diameters (50 μ and 100 μ),
two tubes (diameters = 1 mm and 2 mm, and lengths = 40 cm and
55 cm), and two suspending liquids (5.2 wt % aqueous sodium
chloride and 20 wt % aqueous glycerine) were used.

The results are shown in Fig. 2 for a particle concentration
of 8.8%; other parameters are indicated in the legend. Keeping
Eq. (4) in view, the data in Fig. 2 collapses on one curve in
Fig. 3 on the indicated abscissa. The doublet collision frequency
ratio is simply $(a_p/a_{100})^3$, where a_p is the radius of the particle

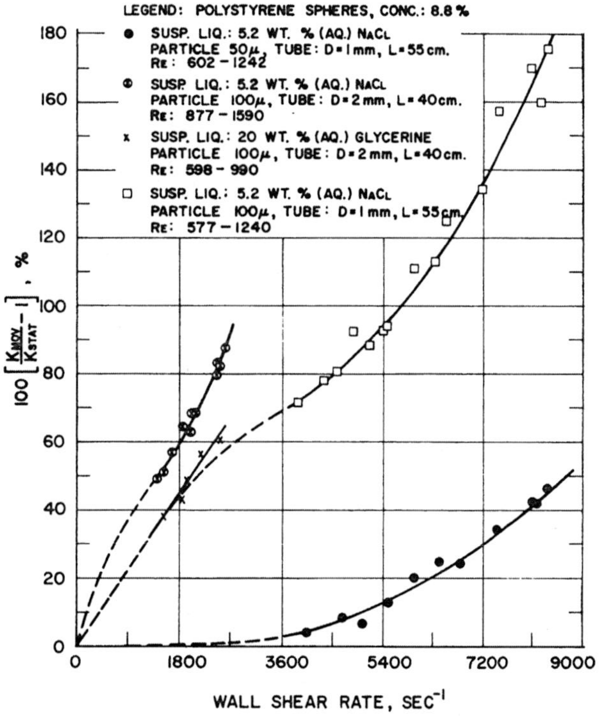

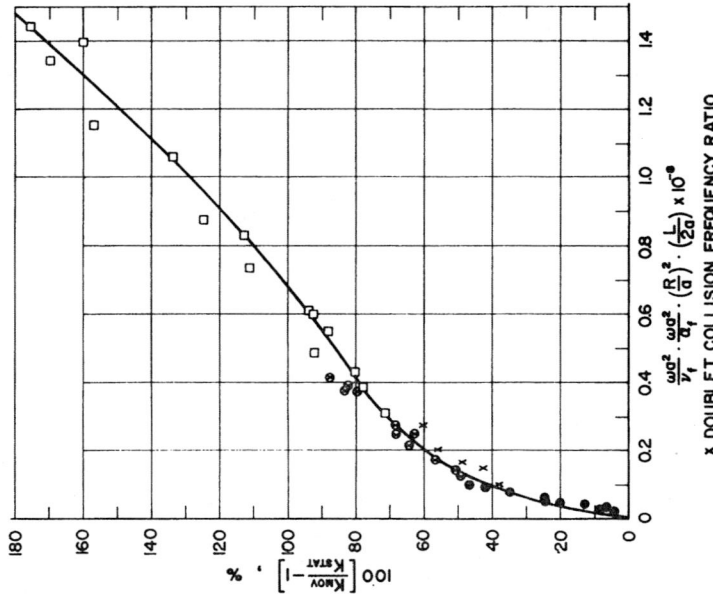

FIG. 3. Data of Fig. 2 is replotted on the indicated abscissa. For legend, see Fig. 2.

FIG. 2. Augmentation in the thermal conductivity increases with the increasing shear rate. For a given shear rate, the percent augmentation is higher, the larger the particle size and the tube size.

and a_{100} is the radius of the 100 μ particle. For example, in
Fig. 2, a_p for a 50 μ particle is 25 μ and a_{100} for the 100 μ
particle is 50 μ. A detailed discussion of the connection between
the doublet collision frequency ratio and the augmentation in
thermal conductivity of flowing suspensions is given elsewhere
(Ref. 13). Data for other concentrations (Ref. 12, 13) exhibit
the same trend as in Figs. 1 and 2.

From Eq. (4) it is seen that the augmentation in the thermal
conductivity of flowing suspension, among other parameters, depends
upon the angular velocity of the particles. The angular velocity
of the particle flowing in a tube varies linearly with the radial
distance from the tube axis (with the exception of a slight non-
linearity near the wall) - it is zero at the axis and maximum at
the tube wall. In view of the augmentation model of the churning
of the surrounding fluid, the greater the radial distance of the
particles from the tube axis, the greater the angular velocity
and, from Eq. (4), the greater the augmentation in the thermal
conductivity of the flowing suspension.

III. CONCLUDING REMARKS

There is a great amount of work available in the literature
on the augmentation or enhancement of heat transfer by a variety
of methods. Reference 14 lists about 500 publications and Ref.
15 about 100 publications, The method described herein is a novel
method. It has been shown that the motion of the particles and
the entrained fluid (gravity settling of the particles in nonneutrally buoyant suspension and shear induced particle rotations
in the tube flow of suspensions) with respect to the fluid at
infinity modifies the heat transfer in the flowing suspensions.

The enhanced transport of heat in the laminar flow of particle
suspensions is significant in industry in the effective utilization
of heat energy, that is, in the greater heat carrying capacity of
heat enchangers for the same friction factor. Since, under certain
realizable assumptions, heat transport and mass transport are
analogous phenomena, the results of this paper can be applied to
the flowing blood for explaining the enhanced transport of platelets,
thereby contributing to the understanding of the formation of blood
clots and thrombi.

[1] J. C. Maxwell, A Treatise on Electricity and Magnetism (Dover, New York, 1954), Vol. 1, p. 440.

[2] D. A. G. Bruggeman, Ann. Phys. (Leipz.) 5, 636 (1935).

[3] F. A. Johnson, U. K. Atomic Energy Authority Research Group, Harwell, Berkshire, Report No. AERE R/R 2578, 1958 (unpublished).

[4] P. Grassman, W. Straumann, F. Widmer, and W. Jobst, in Second Symposium on Thermo-Physical Properties (Princeton University, 1962), p. 447.

[5] P. Grassman and W. Straumann, Int. J. Heat Mass Transfer 1, 50 (1960).

[6] A. S. Ahuja, J. Appl. Physiol. 37, 765 (1974).

[7] H. S. Carslaw and J. C. Jaeger, Conduction of Heat in Solids (Oxford, London, 1956).

[8] A. S. Ahuja, J. Appl. Phys. 46, 747 (1975).

[9] A. Singh, Ph.D. Thesis (University of Minnesota, Minneapolis, 1968). (This work is that of the author, who, after completing it, changed his name.)

[10] H. Schlichting, Boundary Layer Theory (McGraw-Hill, New York, 1960).

[11] A. S. Ahuja, J. Appl. Phys. (in press).

[12] A. S. Ahuja, J. Appl. Phys. 46, 3408 (1975).

[13] A. S. Ahuja, J. Appl. Phys. 46, 3417 (1975).

[14] Augmentation of Convective Heat and Mass Transfer, edited by A. E. Bergles and R. L. Webb (ASME, New York, 1970).

[15] A. E. Bergles, Appl. Mech. Rev. 26, 675 (1973).

ANOMALOUS THERMAL BEHAVIOR OF AN AMORPHOUS ORGANIC SEDIMENT

Richard McGaw

U.S. Army Cold Regions Research and Engineering Lab.

Hanover, New Hampshire 03755

Using transient heating in the cylindrical probe configuration, the thermal conductivity of four organic lake sediments was measured. Each specimen contained over 90% water as a fraction of total weight. Organic matter in three of the specimens was fibrous and not greatly decomposed; at a temperature of 5C, thermal conductivity of these ranged from 2 - 5% higher than that of pure water. A fourth specimen was gelatinous and gave conductivity values from 1% to 20% less than that of water, with an apparent dependence on heat input. The insitu environment of this specimen was known to be different from that of the others.

Test data for the four specimens are reported, and possible structural models for the manner of heat conduction in the anomalous material are considered. It is tentatively concluded that gelatinous cellulose matter is an aggregation of non-conducting amorphous particles in a conducting medium (water).

MAGNETIC FIELD EFFECT ON THE THERMAL CONDUCTIVITY OF HIGH PURITY COPPER AND A NiCrFe ALLOY

L. L. Sparks

Cryogenics Division, National Bureau of Standards

Boulder, Colorado 80302

A program is underway to study the effect of a magnetic field on the thermal conductivity of technically important materials such as Inconels, stainless steels, and superconductor stabilizing materials such as copper and aluminum. The apparatus used to measure the thermal conductivity in magnetic fields up to 80 kOe and in the temperature range 4 to 20 K is described. Results obtained for the NiCrFe alloy (Inconel 718) and a high purity copper specimen will be presented. These results indicate that for the alloy the field effect is small, as expected, since lattice conduction is dominant. The effect of an 80 kOe is to reduce the conductivity by 8% at 5 K and 3% at 19 K. The effect of an 80 kOe field is to reduce the conductivity of the high purity copper specimen by 77% at 5 K and 73% at 20 K.

Chapter IV

Gases and Fluids

THERMAL CONDUCTIVITY OF THE NOBLE GASES AND OF THEIR MIXTURES

J. Kestin

Brown University

Providence, RI

The lecture begins with a discussion of the presently available theoretical basis for the calculation of the thermal conductivity of monatomic gases and mixtures of them containing an arbitrary number of components in arbitrary proportions. This makes full use of the Chapman-Enskog solution of the Boltzmann equation and its extension to mixtures due to Muckenfuss and Curtiss. This combined with the extended law of corresponding states due to Kestin, Ro and Wakeham provides a *complete solution* for low densities, except for some second-order effects in mixtures which remain to be worked out.

Extensive comparisons with recommended correlations and direct measurements contained in the open literature reveal in them errors from 5% to 55% as long as steady-state instruments of various types are considered. J. W. Haarman's transient, hot-wire instrument and our new instrument reduce this band to as little as 0.9% and 0.7% respectively.

The view is expressed that steady-state instruments have failed to eliminate, or to correct for, natural convection in all cases, and thermal diffusion in mixtures.

INTRODUCTION

From the point of view of practical applications, it is necessary to be able to calculate ("predict") the thermal conductivity, λ, of pure and arbitrarily complex mixtures of gases, because the volume of experimental work required to determine the function

$$\lambda = \lambda(T,\rho,x_1,\ldots,x_{n-1}) \qquad\qquad (1)$$

for a mixture of n components with molar fractions $x_1,\ldots x_n$ would exceed all reasonable bounds. The function itself depends on the molecular structure of the components, i.e. on the intermolecular potential and a number of molecular constants for each gas.

In the interest of economy of space we shall restrict ourselves to the discussion of the five monatomic gases and of their mixtures, all at low densities, because for this case there exists an essentially complete solution of the problem.

PURE MONATOMIC GASES AT LOW DENSITY

In the limit $\rho \to 0$ all equilibrium and transport properties of the five monatomic gases He, Ne, Ar, Kr, Xe, and of the 26 different multicomponent systems which can be formed with them are completely determined by the single-particle distribution function, which is obtained by solving the Boltzmann equation [1-3], terminated at the second term in the Chapman-Enskog expansion. Each property is expressed in terms of collision integrals which are functionals of the pair-potential

$$\phi(r) = \epsilon \; f(r/\sigma) \; . \qquad\qquad (2)$$

The scaling factors σ and ϵ as well as the form of the interaction function, f, are, in principle, different for each gas. The application of theory assures thermodynamic consistency between all properties.

The theory, though complete, is limited as far as practical applications are concerned for the following reasons: (a) In spite of many efforts, neither the interaction parameters σ and ϵ, nor the interaction function, f, are really known for any molecule. (b) In spite of the fact that many heuristic forms for f have been proposed, none has proved adequate to reproduce the most accurate data available. (c) In spite of statements to the contrary, attempts to determine the pair-potential from measured values of macroscopic properties according to the scheme

property → functional → potential

is mathematically non-unique.

As a result of an extensive analysis of a large body of accurate data, *but exclusive of any data on thermal conductivity,* J. Kestin, S. T. Ro and W. A. Wakeham [4,5] assumed that the func-

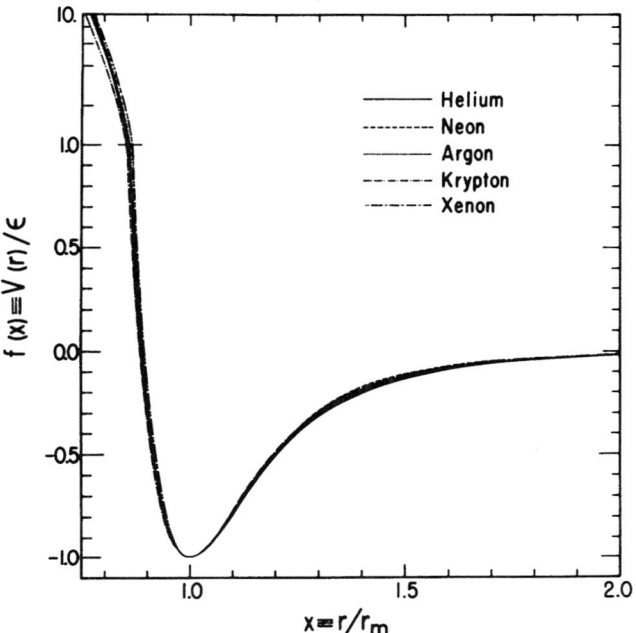

Figure 1 Scaled potentials from molecular beams [6].

tion f is the same for all monatomic gases. This has led them to
the formulation of an extended law of corresponding states and to
empirical forms of the most important collision integrals for the
monatomic gases. The aptness of such an assumption has been subsequently confirmed by the molecular-beam experiments of
J. M. Farrar, T. P. Schafer and Y. T. Lee [6] displayed in Fig. 1.

According to this modified theory, the properties of all five
monatomic gases are calculated with the aid of the standard Chapman-
Enskog formulae together with a number of empirical functionals
given in detail in ref. [5]. The properties of each individual gas
are characterized by the values of the constants displayed along
the diagonals in Table I.

The functional

$$\bar{\Omega}_{22}(T^*) = \Omega^{(2,2)^*}(T^*)/f_{\lambda}(T^*) \tag{3}$$

TABLE I.   Scaling Parameters

(a) Optimum values of the scaling factor σ, Å

	He	Ne	Ar	Kr	Xe
He	2.556	2.644	2.904	3.230	3.595
Ne		2.707	3.091	3.161	3.310
Ar			3.291	3.401	3.455
Kr				3.518	3.595
Xe					3.848

(b) Optimum values of the scaling factor ϵ/k, K.

	He	Ne	Ar	Kr	Xe
He	11.2_9	21.1_2	55.2_4	32.1_3	22.8_7
Ne		45.5_8	63.7_0	87.8_1	90.9_8
Ar			153.6_1	182.1_8	234.0_3
Kr				211.3_8	277.1_2
Xe					285.2_7

for thermal conductivity is

$$\bar{\Omega}_{22}(T^*) = \exp\left\{0.45667 - 0.53955\ (\ln\ T^*) + 0.18265\ (\ln\ T^*)^2 + \right.$$

$$\left. - 0.03629\ (\ln\ T^*)^3 + 0.00241\ (\ln\ T^*)^4\right\} /$$

$$\left\{1 + 0.0042\ [1 - \exp\ 0.33\ (1-T^*)]\right\} ; \qquad (4)$$

it differs from the functional

$$\Omega_{22}(T^*) = \Omega^{(2,2)*}(T^*)/f_\eta(T^*)$$

for viscosity by a factor

$$F(T^*) = 1 + 0.0042 \left[1 - \exp 0.33 (1-T^*) \right] = f_\lambda(T^*)/f_\eta(T^*), \quad (5)$$

which varies very little with temperature and which has been shown to be practically universal and independent of the pair potential $\phi(r)$.

Equation (4) together with the standard equation

$$\lambda(T) = \frac{75}{64} \left(\frac{R^3}{\pi N^4} \right)^{1/2} \frac{(T/M)^{1/2}}{\sigma^2 \Omega^{(2,2)^*}(T^*)/f_\lambda(T^*)} \quad (6)$$

and the characteristic factors from Table I allows us to calculate the thermal conductivity of the noble gases.

Since the preceding data have been obtained without reference to any measurements of thermal conductivity, the scheme of calculation can be looked upon as a method of "predicting" thermal conductivity in a manner which assures thermodynamic consistency with the measurements of a number of other properties. In particular, the calculation embodies in it the well-known Eucken relation between viscosity and thermal conductivity, according to which the group

$$Eu = \frac{M \lambda(T)}{2.5 \, C_v \, \eta(T) \, F(T^*)} = 1 \quad \text{(exactly)} \quad (7)$$

is a constant which has the theoretical value of unity.

The accuracy of this extended law of corresponding states has been tested extensively [4,5,7,8]; it has been found that it represents many properties (still excluding thermal conductivity) with an accuracy comparable to that of the best measurements in each case. From this we conclude that *calculated* values of the thermal conductivity of the noble gases must be assigned the highest level of confidence.

COMPARISON WITH MEASUREMENTS FOR PURE GASES

We now propose to conduct a systematic comparison between measurement and the preceding calculation based on the extended law of corresponding states. In order to reduce all comparisons to a common basis, it is convenient to calculate the group Eu rather than the conductivity itself. It is emphasized that such a comparison is based on much more than just measurements of viscosity, because the molecular constants σ and ϵ in Table I have been obtained by an optimization procedure which has involved a larger number of properties.

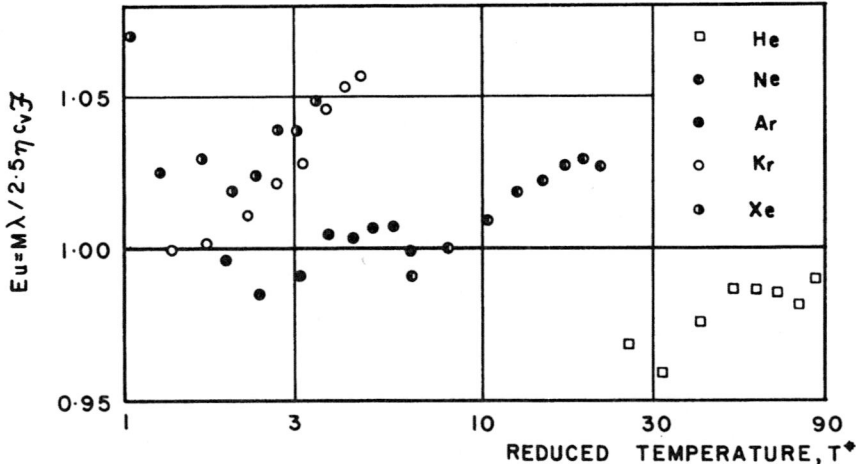

Figure 2   Eucken factor for pure gases [8].

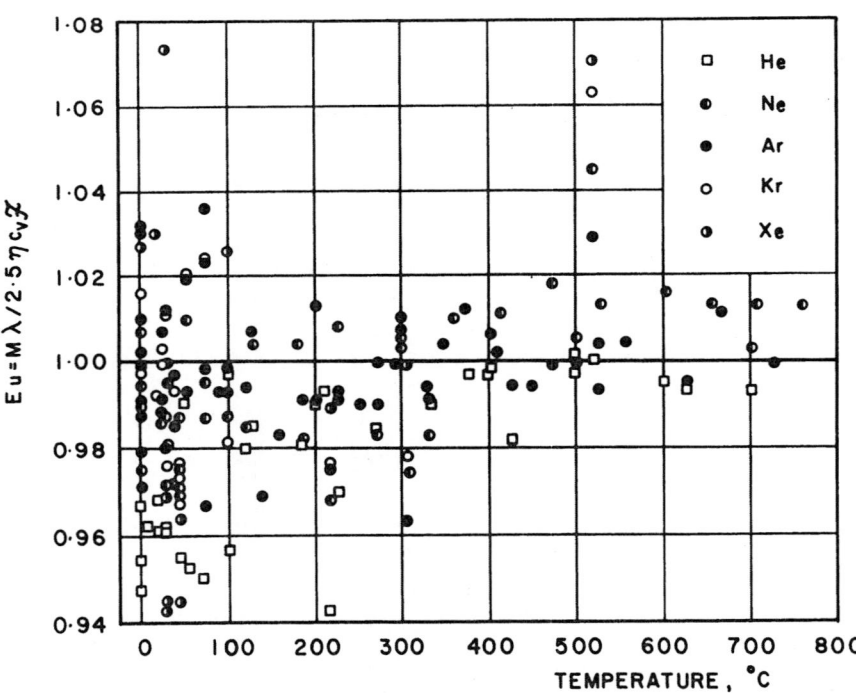

Figure 3   Eucken factor for pure gases from various sources.

The diagrams in Figs. 2 and 3 contain comparisons with selected data found in the open literature. Figure 2 shows a comparison for the data on thermal conductivity recommended by N. B. Vargaftik *et al.* [5]. It is seen that these correlations contain inaccuracies ranging from +8% to -5%, covering a total span of 13%. The diagram in Fig. 3 represents the data measured by various authors. Depending on the instruments used, the bands of uncertainties range from 14% for the older data to 5% for the data of B. Vodar and his team. It is useful to note that in almost all cases the measurements on helium show the largest deviations, testifying to the fact, known to experimenters, that measurements on this gas "are difficult" to perform. The data of J. W. Haarman [9,10] show an improvement by an order of magnitude in the reduction of the band of uncertainty. It is remarkable that even the data on helium are contained in the 0.9% band.

It is difficult to escape the conclusion that most measurements of thermal conductivity contained in the open literature suffer from unaccounted for systematic errors which exceed those indicated in the respective publications. We express the opinion that the most important effect is that due to the natural convection which it is almost impossible to avoid in steady-state instruments and which is very difficult to correct for accurately. This supposition is reinforced by the fact that the measurements among themselves exhibit discrepancies of the same order as the full uncertainty band.

GENERAL REMARKS CONCERNING MIXTURES

The discussion of the thermal conductivity of mixtures (as opposed, for example, to that of viscosity) is complicated by the fact that the flux of heat is coupled to the flux of mass. This is reflected on two levels. Theoretically, a distinction must be made between two conductivities, λ_0 and λ_∞. The former characterizes the relation between the temperature gradient and heat flux in the absence of a concentration gradient. The thermal conductivity λ_∞ characterizes the same relation in the "steady state" which results when the mass fluxes vanish thus imposing concentration gradients which are uniquely related to the temperature gradient. The difference between the two may reach as much as 1-2%. Experimentally, it is necessary to include the preceding coupling in the theory of the instruments which must now be based on the simultaneous solution of one heat and n-1 mass transfer equations. Normally, however, experimental results on mixtures are calculated on the basis of the same equations as those for pure gases, and corrections due to local changes in composition are neglected.

Our ensuing discussion is somewhat preliminary in that the required corrections and distinctions have not been introduced.

We shall assume that reported measurements, even in pulsed instruments, represent the "steady-state" conductivity, λ_∞, because the relaxation times for heat and mass transfer in gaseous mixtures are of the same order of magnitude.

The theory of the thermal conductivity of mixtures was given by C. Muckenfuss and C. F. Curtiss [11] who corrected certain shortcomings contained in ref. [1]. The theory leads to the conclusion that the thermal conductivity, λ_∞, of a mixture is given by the ratio of two determinants

$$\lambda_\infty \equiv \lambda_{mix} = 4 \begin{vmatrix} L_{11}, & \cdots, & L_{1\nu}, & x_1 \\ \cdot & & & \\ \cdot & & & \\ \cdot & & & \\ L_{\nu 1}, & \cdots, & L_{\nu\nu}, & x_\nu \\ x_1, & \cdots, & x_\nu, & 0 \end{vmatrix} \Bigg/ \begin{vmatrix} L_{11}, & \cdots, & L_{1\nu} \\ \cdot & & \cdot \\ \cdot & & \cdot \\ \cdot & & \cdot \\ L_{\nu 1}, & \cdots, & L_{\nu\nu} \end{vmatrix} \quad (8)$$

in which the elements L_{ij} are explicit functions of the mole fraction $x_1, \ldots, x_n$, the molecular weights $M_1, \ldots, M_n$ and of a number of functionals of the interaction potential between pairs (both like and unlike) of molecules.

Equation (8) gives the first approximation in terms of the standard expansion in Sonine polynomials. Even though higher approximations can be obtained, we shall base our discussion on eq. (7) in view of the fact that our discussion is still somewhat preliminary.

BINARY MIXTURES

A proper interpretation of the theory shows, contrary statements notwithstanding, that the validity of eq. (7) does not extend to mixtures of monatomic gases. The thermal conductivity of a binary mixture is given by the equation

$$\lambda_{mix} = \frac{1 + Z_\lambda}{X_\lambda + Y_\lambda} \; , \quad (9)$$

where X_λ, Y_λ, Z_λ are explicit functions of the mole fractions x_1, x_2, $(x_1 + x_2 = 1)$, of the molecular weights M_1, M_2, and of functionals of *three* interaction potentials, two for each of the interaction of like molecules, ϕ_{11}, ϕ_{22}, and one for the interaction of unlike molecules, ϕ_{12}. All potentials are of the form of eq. (2), but, in principle, the factors σ_{11}, σ_{22}, σ_{12}, ε_{11}, ε_{22}, ε_{12} as well as the interaction functions, f, are different for each type of binary interaction.

Again, in spite of considerable past efforts, it has not proved possible to indicate, theoretically or empirically, a reliable method for the determination of the potential ϕ_{12} when the pure potentials are known. Nevertheless, it has proved possible [4,5] to continue to apply the law of corresponding states, postulating similarity between ϕ_{11}, ϕ_{22}, and ϕ_{12}. The resulting values of σ_{12} and ε_{12} are given in the off-diagonal positions of Table I; these refer to the ten binary mixtures which can be formed with the five noble gases.

The potential ϕ_{12} defines the interaction thermal conductivity

$$\lambda_{12} = \frac{75}{64} \left(\frac{R^3}{\pi N^4}\right)^{1/2} \frac{[T(M_1+M_2)/2M_1M_2]^{1/2}}{\sigma_{12}^2 \, \Omega^{(2,2)*}(T*)} \quad , \tag{10}$$

which is, except for the reduced molecular mass $2M_1M_2/(M_1+M_2)$, identical with the pure-component eq. (6). Since an analogous relation applies to the interaction viscosity, η_{12}, it is again possible to form an Eucken-type consistency relation. However, this is one that relates λ_{12} to η_{12} and not λ_{mix} to η_{mix}. This relation has the form

$$Eu = \frac{[2M_1M_2 / (M_1+M_2)] \, \lambda_{12}(T)}{(15/4)R \, \eta_{12}(T)} = 1 \quad . \tag{11}$$

It is clear at this stage that the use of the data from Table I together with the appropriate theoretical formulae permits us to "predict" the thermal conductivity of binary mixtures in a thermodynamically consistent way without reference to the measurement of thermal conductivity of pure components or of mixtures.

COMPARISON WITH MEASUREMENTS FOR BINARY MIXTURES

The diagram in Fig. 4 contains a plot of the factor Eu from eq. (11) for selected measurements on three binary mixtures. It is noted, not unexpectedly, that the departures have become larger than those for the pure components. The total spread is of the order of 14%. A certain deterioration must be expected owing to the fact that λ_{12} is obtained from λ_{mix} by a lengthy calculation susceptible to truncation errors. It can also be verified that the imposition of small percentage errors on λ_{mix} induces double that percentage error in λ_{12}. Similarly, the mixed parameters σ_{ij}, ε_{ij} are somewhat less reliably known than σ_{ii}, ε_{ii}.

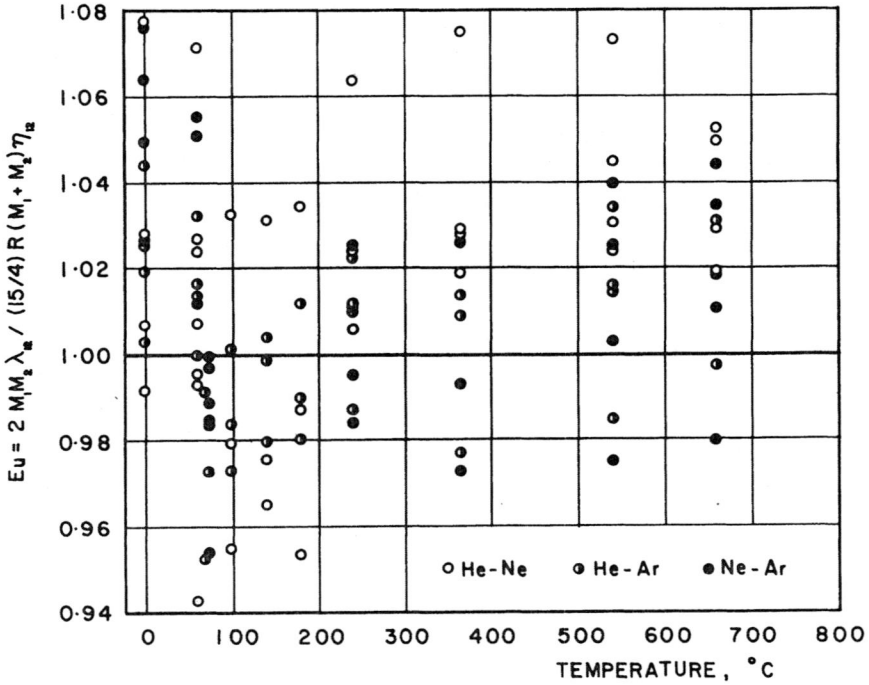

Figure 4   Eucken factor for three binary mixtures.

MULTICOMPONENT MIXTURES

Owing to the fact that in the limit $\rho \to 0$ only *binary* collisions need to be taken into account, the formula for λ_{mix} contains
a number of functionals of the potentials ϕ_{ij} for unlike interactions. The hypothesis of similarity continues to be applicable, and
no new quantities appear. In particular, the factors σ_{ij} and ε_{ij}
from Table I retain their validity, since every

$$\lambda_{ij} = \frac{75}{64} \left(\frac{R^3}{\pi N^4}\right)^{1/2} \frac{[T(M_i + M_j)/2M_i M_j]^{1/2}}{\sigma_{ij}^2 \, \Omega^{(2,2)*}(T^*)} \tag{12}$$

is the same as the corresponding λ_{12} in eq. (10).

The total number of different systems which can be produced
with the noble gases is 31 (5-pure, 10-binary, 10-ternary,
5-quaternary, 1-quinquepartite). It is, therefore, not surprising
that measurements on many of them are still unavailable, particularly if we further ponder the very large number of compositions
needed to cover a mixture of 3, 4 or 5 components. For this reason,

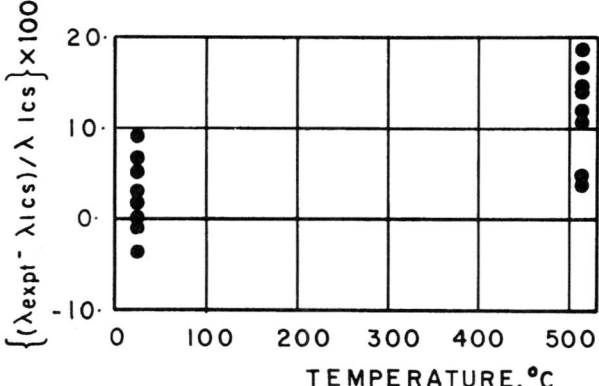

Figure 5   Deviation plot for ternary mixture [12].

we confine our comparison to the ternary mixture measured by
H. von Ubisch [12].   Instead of attempting to calculate the three
interaction quantities λ_{ij} and to represent them in the form of
three Eucken-type relations, it is now preferable to construct the
direct deviation plot of Fig. 5.   The diagram confirms and rein-
forces the previous conclusions.

We restate our belief that calculated values of the thermal
conductivity of arbitrary multicomponent mixtures of the noble gases
are superior to those obtained by direct experiment.

THE NEW INSTRUMENT

We now operate a new instrument which measures the thermal
conductivity by the transient hot-wire method.   The instrument
posesses several features which are identical with those pioneered
by J. W. Haarman [9,10].   However, as a later version, it incorporates several refinements.   The instrument has been described in
detail by J. J. de Groot, J. Kestin and H. Sookiazian [13], and a
complete theory of its operation has been worked out by J. J. Healy,
J. J. de Groot and J. Kestin [14].   The most important characteristics of this instrument are twofold: (a) It almost completely
eliminates the onset of natural convection because it completes a
measurement in a time which is very short compared to the characteristic starting times for this undesirable process.

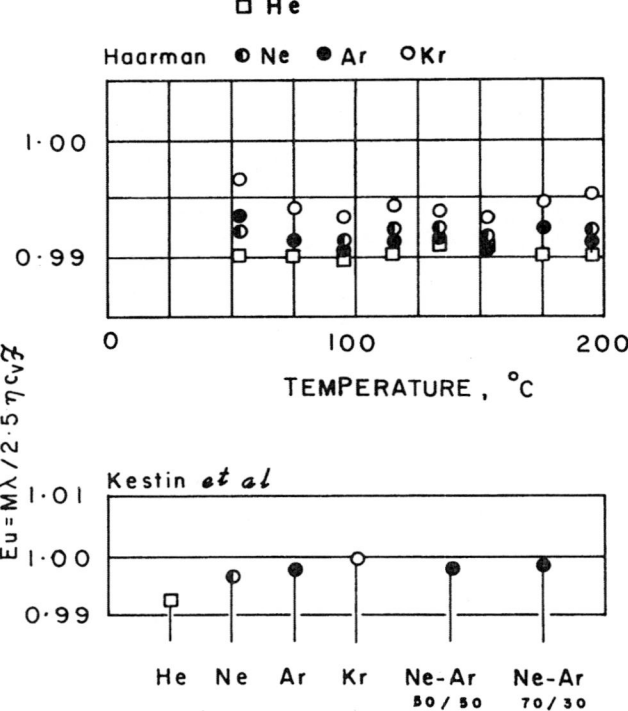

Figure 6   Results from two transient instruments [9,13].

(b) Owing to an optimal choice of dimensions and characteristics, its operation is remarkably close to the simplest mathematical model for it, making it superfluous to apply a large number of corrections whose magnitude has been reduced to insignificant proportions.

Measurements performed earlier by Haarman and recently by us fully confirm our conclusions. On the one hand, if we assume that the instrument is capable of producing measurements which are accurate to 0.2%, we obtain a good confirmation of the law of corresponding states, as seen from Fig. 6. Alternatively, if we accept the view that calculations lead us to very accurate values of the thermal conductivity of monatomic gases and their mixtures, the deviation plots of Fig. 6 give us the assurance that transient hot-wire instruments in general, and our instrument in particular, are capable of high precision and yield very reliable, standard values of the thermal conductivity of gases (polyatomic as well as monatomic).

ACKNOWLEDGMENTS

The instrument was designed and built with funds provided by the NSF under Grants GK-2133X, GK-37070, and GK-38891 to Brown University.

The author wishes to thank the Organizers of the Fourteenth International Conference on Thermal Conductivity for their kind invitation to prepare and present this summary lecture.

Thanks are due to the author's collaborators, Doctors Healy, de Groot, Sokolov and Wakeham and to Messrs. Anderson and Paul whose efforts brought this project to fruition.

Last but not least, the author wishes to thank Messrs. Khalifa and Sookiazian who helped him materially in the preparation of this lecture.

REFERENCES

[1] J. O. Hirschfelder, C. F. Curtiss and R. B. Bird, "Molecular Theory of Gases and Liquids", (John Wiley & Sons, New York 1954).

[2] S. Chapman and T. G. Cowling, "The Mathematical Theory of Non-Uniform Gases", (Cambridge University Press, 1952).

[3] J. H. Ferziger and H. G. Kaper, "Mathematical Theory of Transport Processes in Gases", North Holland/American Elsevier (1972).

[4] J. Kestin, S. T. Ro and W. A. Wakeham, Physica 58, 165 (1972).

[5] J. Kestin and E. A. Mason, article in "Transport Phenomena", AIP Conf. Proc. No. 11, ed. J. Kestin, p. 137 (1973).

[6] J. M. Farrar, T. P. Schafer and Y. T. Lee, article in "Transport Phenomena", AIP Conf. Proc. No. 11, ed. J. Kestin, p. 279 (1973).

[7] J. Kestin, S. T. Ro and W. A. Wakeham, J. Chem. Phys. 56, 4119 (1972).

[8] J. Kestin, J. Chem. Phys. 60, 3728 (1974).

[9] J. W. Haarman, Physica 52, 605 (1971).

[10] J. W. Haarman, article in "Transport Phenomena", AIP Conf. Proc. No. 11, ed. J. Kestin, p. 193 (1973).

[11] C. Muckenfuss and C. F. Curtiss, J. Chem. Phys. 29, 1273 (1958).

[12] H. von Ubisch, Ark. Fys. 16, 93 (1959).

[13] J. J. de Groot, J. Kestin and H. Sookiazian, Physica 75, 454 (1974).

[14] J. J. Healy, J. J. de Groot and J. Kestin, (to be published).

A THEORY FOR THE COMPOSITION DEPENDENCE OF THE THERMAL CONDUCTIVITY OF DENSE BINARY MIXTURES OF MONATOMIC GASES

R. DiPippo

Southeastern Massachusetts University
North Dartmouth, Massachusetts 02747

J.R. Dorfman

University of Maryland
College Park, Maryland 02740

J. Kestin, H.E. Khalifa, and E.A. Mason

Brown University
Providence, Rhode Island 02912

A method, based on the modified Enskog theory (MET), as extended to binary mixtures by H.H. Thorne, is presented which allows the computation of the thermal conductivity of binary mixtures of monatomic gases up to densities of about 200 amagat with an uncertainty no greater than 2 percent when compared with the best available measurements.

The method makes use of measurements of the thermal conductivity for the pure components up to the same density as for the mixtures together with the zero-density thermal conductivity at one composition for a given binary system. Needed information on the second virial coefficient, its temperature derivative, and the associated interaction quantities are obtained from a universal correlation developed by Kestin et al, which is valid for all monatomic gases.

The technique has been applied to three binary systems: helium-neon, helium-argon, and neon-argon. The calculated values

329

of thermal conductivity compare very favorably with recent experimental results from the Laboratoire des Hautes Pressions (LHP).

1.   BASIC EQUATIONS

For ease of reference, we first present a summary of the fundamental equations for the thermal conductivity of dense monatomic gases and their mixtures.

1.1 Thermal conductivity of pure monatomic gases at high density

High density thermal conductivity of simple gases may be found from eq.(1) which follows from Enskog[1]:

$$\lambda_i = (1 + \frac{3}{4} \alpha_{ii} n \chi_i)^2 \frac{\lambda_i^o}{\chi_i} + \frac{2}{\pi} \alpha_{ii}^2 n^2 \chi_i \lambda_i^o , \tag{1}$$

where n = molar gas density,

   λ_i^o = zero-density thermal conductivity,

   χ_i = an average radial distribution function, described in Section 16.21 of Chapman and Cowling[1],

   α_{ii} = a factor related to the second virial coefficient which expresses the shortening of the mean free path owing to the finite size of the molecules.

Repeated indices used in eq.(1), and other equations later, do not imply summation. Since one must have a reliable means of obtaining both α_{ii} and χ_i prior to application of eq.(1), most earlier theories have dealt with schemes for computing these factors[1,2,3].

1.2 Thermal conductivity of binary mixtures of monatomic gases at high density

Enskog's theory has been extended to binary mixtures by Thorne and is reported in Chapman and Cowling[1]. The equation which we give here is written in compact working form:

$$\lambda_{mix} = \frac{2\lambda_{12}^o A_{12}^*}{5\chi_{12}|G|} \{\underline{\xi}^T G \underline{\xi}\} + \kappa_{12}. \tag{2}$$

The term $|G|$ is the determinant of a symmetrical 2 x 2 matrix, $G \equiv g_{ij}$, i, j = 1, 2, in which

$$g_{ii} = \frac{x_i}{x_j} \left[\frac{6}{4} r_j^2 + r_i^2 (\frac{5}{4} - \frac{3}{5} B_{12}^*) + \frac{4}{5} r_i r_j A_{12}^* \right] + \frac{2}{5} A_{12}^* \frac{\lambda_{12}^o \chi_{jj}}{\lambda_j^o \chi_{12}} , \tag{3a}$$

$$i, j = 1, 2 \text{ and } i \neq j$$

and $g_{12} = g_{21} = -r_1 r_2 (\frac{11}{4} - \frac{3}{5} B_{12}^* - \frac{4}{5} A_{12}^*)$,   (3b)

where $r_i = M_i / (M_1 + M_2)$.

The term $\underline{\xi}$ represents a vector of two elements, $\underline{\xi} = \xi_i$, i = 1, 2, in which

$\xi_i = 1 + \frac{3}{4} \alpha_{ii} n x_i \chi_{ii} + 3 r_i r_j \alpha_{12} n x_j \chi_{12}$, i, j = 1,2 and i $\neq$ j. (4)

The term κ_{12} is the bulk conductivity contribution and is given by

$$\kappa_{12} = \frac{2n^2}{\pi} \left[\lambda_1^o \alpha_{11}^2 x_1^2 \chi_{11} + 8 r_1 r_2 x_1 x_2 \lambda_{12}^o \alpha_{12}^2 \chi_{12} + \lambda_2^o \alpha_{22}^2 x_1^2 \chi_{22} \right].$$ (5)

The terms A_{12}^* and B_{12}^* are functions of the collision integrals:

$A_{12}^* = \Omega_{12}^{(2,2)*} / \Omega_{12}^{(1,1)*}$; $B_{12}^* = (5\Omega_{12}^{(1,2)*} - 4\Omega_{12}^{(1,3)*}) / \Omega_{12}^{(1,1)*}$. (6a,b)

According to the modified Enskog theory (MET)[2,3], the factor α_{ij} may be obtained from the second virial coefficient, $B_{ij}(T)$:

$\alpha_{ij} = \frac{4}{5} (B_{ij} + T \frac{dB_{ij}}{dT})$.   (7)

While these basic equations are fresh at hand, it is useful to distinguish the terms which are obtained from zero-density (or low density) experimental data or correlations from the terms which are associated with high density phenomena. In the first category are λ_1^o, λ_2^o and λ_{12}^o, the pure gas and interaction thermal conductivities at zero density, along with the functions A_{12}^*, B_{12}^* and α_{ij}, all of which are functions of the reduced temperature $T_{ij}^* = kT/\varepsilon_{ij}$. Besides the density n, χ_{ii} implicitly reflects a part of the density dependence of λ. Moreover, the factor α appears only in the expression for the high-density thermal conductivity. We shall devote the next section of this paper to the determination of these essential elements of the theory.

2.   METHODOLOGY

In general terms, we propose to use eq.(1) together with experimental values for λ_i^o and λ_i plus values for α taken from a universal correlation for pure monatomic gases[4], and to solve the equation for the average radial distribution function χ_i as a function of density n at any particular temperature. In this way we shall guarantee perfect agreement between our theory and experimental results for the pure gases. The essence of the technique thus turns on the combination rules that we employ to account for the mixture properties.

2.1 Determination of λ_{12}^{o}, A_{12}^{*} and B_{12}^{*}

The zero-density terms are readily obtained from well-known expressions. The interaction thermal conductivity λ_{12}^{o} emerges from a solution of the zero-density mixture equation for thermal conductivity:

$$\lambda_{mix}^{o} = \frac{1 + Z_{\lambda}}{X_{\lambda} + Y_{\lambda}} \quad . \tag{8}$$

The functions, X_{λ}, Y_{λ} and Z_{λ}, each of which depends upon λ_{12}^{o} among other parameters, may be found on p. 535 of Ref. 3.

Since the interaction term is calculated for each composition for a given binary system at a given temperature, it is possible to check the constancy of λ_{12}^{o} with x. Although one would not anticipate, theoretically, any dependency of λ_{12}^{o} on x (we consider only the first-order Chapman-Enskog expressions), experimental uncertainties may cause some random variations in λ_{12}^{o}. To retain consistency when applying our theory (and comparing it to experimental results), one ought to use that λ_{12}^{o} which corresponds to the particular composition of the mixture under study.

Both A_{12}^{*} and B_{12}^{*} may be taken from the extended law of corresponding states of Kestin, et al[4],[5] which provides simple expressions for the appropriate collision integrals. The interaction force parameters ε_{12}/k needed to produce the reduced temperature T_{12}^{*} may also be taken from the compilation in Ref. 4.

It should be noted that the interaction thermal conductivity λ_{12}^{o} also could be calculated from the extended law of corresponding states, if one were willing to ignore the random scatter in the experimentally-determined λ_{12}^{o} with composition. This approach was tested and found to be slightly inferior to the method recommended here, the latter being more consistent with the experimental results used in the comparison.

2.2 Determination of α_{ij} and χ_{i}

The factor α_{ii} is calculated straightforwardly from eq.(7) with the aid of the extended law of corresponding states[4],[5]. Numerical values for the average radial distribution function χ_{i} are found for each pure component gas from experimental data on $\lambda_{i}(n,T)$, $\lambda_{i}^{o}(n = 0,T)$, n, and α_{ii}. With this input information, eq.(1) may be solved for χ_{i}:

$$\chi_{i} = \frac{-8}{19} \frac{(3/2)\alpha_{ii}n\lambda_{i}^{o} - \lambda_{i}}{\alpha_{ii}^{2}n^{2}\lambda_{i}^{o}} - \frac{16}{19} \left[\frac{(3\alpha_{ii}n\lambda_{i}^{o}/2 - \lambda_{i})^{2}}{4\alpha_{ii}^{4}n^{4}(\lambda_{i}^{o})^{2}} - \frac{19}{16\alpha_{ii}^{2}n^{2}} \right]^{1/2} \tag{9}$$

The structure of eq.(9) indicates that χ_i will become complex at and beyond some high density unless α_{ii} is kept below a certain value. However, for the gases and density range covered in this paper, no such imaginary values were encountered, and the χ_i functions were well-behaved, taking on values lying between 1.00 and 1.09.

2.3 Combination rules for χ_{ii}

The following combination rules for the χ_{ii} functions needed to compute λ_{mix} from eq.(2) are founded upon Thorne's relations as quoted on p. 292 of Chapman and Cowling[1]. The equations provided by Thorne were expressed in terms of the rigid sphere diameters σ_1, σ_2 and σ_{12}, i.e.

$$\chi_{ii} = 1 + \frac{5}{12} \pi x_i n \sigma_i^3 + \frac{\pi}{12} x_j n (\sigma_i^3 + 16\sigma_{12}^3 - 12\sigma_{12}^2 \sigma_i) + \ldots \; , \tag{10a}$$

$$i, j = 1, 2 \text{ and } i \neq j,$$

and $\quad \chi_{12} = 1 + \frac{\pi}{12} x_1 n \sigma_1^3 (8 - 3\sigma_1/\sigma_{12}) + \frac{\pi}{12} x_2 n \sigma_2^3 (8 - 3\sigma_2/\sigma_{12}) + \ldots \; ,$

$$\tag{10b}$$

with $\sigma_{12} = (\sigma_1 + \sigma_2)/2$. For the pure components, by truncating after the linear term in n, eq.(10a) can be solved for σ_i^3 to yield

$$\sigma_i^3 = \frac{12}{5\pi} \frac{(\chi_i - 1)}{n} . \tag{11}$$

Substituting for σ_1, σ_2 and σ_{12} in eqs.(10a,b) one obtains:

$$\chi_{ii} = x_i \chi_i + x_j \left[1 + \frac{1}{5}(\chi_i - 1) + \frac{2}{5} \{(\chi_i - 1)^{1/3} + (\chi_j - 1)^{1/3}\}^3 - \frac{3}{5}(\chi_i - 1)^{1/3} \{(\chi_i - 1)^{1/3} + (\chi_j - 1)^{1/3}\}^2 \right] , \tag{12a}$$

$$i, j = 1, 2 \text{ and } i \neq j,$$

and $\chi_{12} = x_1 \chi_1 + x_2 \chi_2 + \frac{3}{5} \{x_1 (\chi_1 - 1) - x_2 (\chi_2 - 1)\} \times$

$$\left[\{(\chi_2 - 1)^{1/3} - (\chi_1 - 1)^{1/3}\} / \{(\chi_2 - 1)^{1/3} + (\chi_1 - 1)^{1/3}\} \right] . \tag{12b}$$

The function χ is very closely related to the equation of state. In the MET this relation takes the form

$$\frac{5}{4} \alpha\chi = (B + T \frac{dB}{dT}) + (C + T \frac{dC}{dT})n + \ldots \quad . \tag{13}$$

Thus a combination rule for χ which is correct to first order in density is, in effect, based on an equation of state that extends only to the third virial coefficient C.

3.   ASSESSMENT OF THE METHOD AND RESULTS

As mentioned earlier, experimental values for thermal conductivity were used to obtain values for χ_i. The data of Garrabos[6] at a temperature of 30°C were selected for this purpose. The χ_i values so obtained were then correlated by a means of a simple, second-order polynomial in density:

$$\chi(n; \ 30^\circ C) = a_o + a_1 n + a_2 n^2. \tag{14}$$

Calculations were performed for three pure gases: helium, neon and argon. The values obtained from the correlation depart from the experimental points by no more than 0.1% in the worst case. The values taken for the scaling parameters σ_{ij} and ε_{ij}/k may be found in Ref. 4.

The theory was applied to three binary systems: helium-neon, neon-argon and helium-argon. Figure 1 shows a sample result on the neon-argon system at 30°C in terms of the excess thermal conductivity λ_{xs} as a function of the mole fraction of neon, where

$$\lambda_{xs} = \lambda_{mix} - x_1 \lambda_1 - x_2 \lambda_2. \tag{15}$$

The zero-density interaction thermal conductivity λ_{12}^o is listed in Table 1, along with the appropriate value for α_{ij}. The agreement between the theory and the experimental data is seen to be very good even at the highest pressure.

Summarizing the results of the calculations, we can say that for the helium-neon system, the calculated λ_{mix} agrees with the experimental findings of Garrabos to within 0.3% over the full span of mole fraction and at pressures up to 200 bar (about 37% of the critical density); for the neon-argon system to within 0.5% of the experimental values at densities up to 42% of the critical value; for the helium-argon system to about 2% for densities up to about 50% of the critical. Generally, the agreement is excellent at low to moderate densities with the maximum disagreement between the theory and the measurements appearing at the highest densities as one would expect.

ACKNOWLEDGEMENT

The work described in this paper was performed with funds provided by the National Science Foundation under Grant GK 38891 to Brown University.

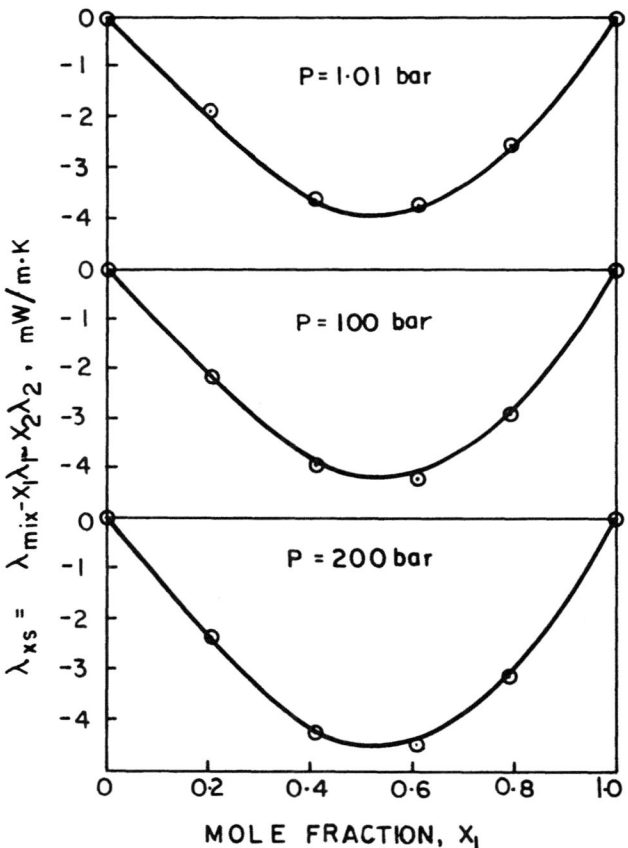

Fig. 1. Excess Thermal Conductivity, λ_{xs}, for Neon-Argon at $30°$C
——— Present Theory, o Garrabos' Data[6]

TABLE 1 Zero-density interaction thermal conductivity λ_{12}^{o} for neon-argon at 30°C. (α_{ij} = 22.126 cm^3/mol)				
x_{NE}	0.205	0.410	0.610	0.790
λ_{12}^{o} x 10^3	33.3	31.9	32.2	32.9
Units: $\{\lambda_{12}^{o}\}$ = mW·m^{-1}·K^{-1}				

REFERENCES

1) S. Chapman and T.G. Cowling, The Mathematical Theory of Non-Uniform Gases, Cambridge University Press, 1964.

2) H.J.M. Hanley, R.D. McCarthy and E.D.G. Cohen, Physica 60, 322 (1972).

3) J.O. Hirschfelder, C.F. Curtiss and R.B. Bird, Molecular Theory of Gases and Liquids, Wiley, 1954.

4) J. Kestin and E.A. Mason, An article in "Transport Phenomena", AIP Conference Proceedings, No. 11; J. Kestin, ed., 1973.

5) J. Kestin, S.T. Ro and W.A. Wakeham, Physica 58, 165 (1972).

6) Y. Garrabos, Doctoral Thesis, Laboratoire des Hautes Pressions, Paris, France, 1972.

THERMAL CONDUCTIVITIES OF GASEOUS MIXTURES OF HELIUM WITH

THE HYDROGEN ISOTOPES AT 77.6 and 283.2 K

Anathony A. Clifford, Lynne Colling,
Eric Dickinson, and Peter Gray
The University of Leeds
Department of Physical Chemistry
Leeds LS2 9JT, U.K.

The thermal conductivities of gaseous mixtures of helium with
hydrogen or deuterium have been measured at 77.6 and 283.2 K. The
results are shown to be accurately predictable on the basis of the
Hirschfelder-Eucken equation using a potential function for the
interaction between a helium atom and a hydrogen molecule recently
calculated quantum-mechanically by Tsapline and Kutzelnigg.

Since binary mixtures of helium with the hydrogen isotopes
form one of the simplest sets of systems containing polyatomic
molecules, it is logical that the thermal conductivities of these
mixtures should be fully understood at a molecular level before we
try to interpret the behaviour of more complex systems. A complete
analysis of the mixtures has been hindered both by the presence of
confusing experimental data in the literature and by the absence
of a sufficiently accurate intermolecular potential function for
the helium-hydrogen pair interaction.

With varying results and conclusions, several research groups
have made thermal conductivity measurements on these systems.
Minima have been reported in the plots of thermal conductivity
versus composition (up to 7 per cent below the value for pure
helium), and to explain the larger negative deviations it has been
necessary to invoke anomalous cross-relaxation effects. Recently,
however, some extensive and precise measurements by Cauwenbergh
and van Dael (1) have located a minimum for helium + hydrogen which
is only 1 per cent below the pure helium value and which is
explainable on the basis of the Hirschfelder-Eucken equation (2)
subject to adjusting the helium-hydrogen potential within certain
reasonable bounds.

The Hirschfelder-Eucken equation should work well for mixtures of helium with hydrogen or deuterium as the rate of rotational-translational relaxation is known from sound absorption measurements (3) to be slow.* Hence, with reliable intermolecular potential functions it should be possible to predict the thermal conductivities of these mixtures with good precision. An accurate quantum-mechanical calculation of the helium-hydrogen interaction energy has been performed recently by Tsapline and Kutzelnigg (6) for three different orientations of the hydrogen molecule. The energies of the linear and C_{2v} configurations are expressed as analytic functions of the distance between the helium nucleus and the hydrogen molecule mid-point, and the energies calculated for one intermediate angular arrangement show that the potential is adequately represented by a two-term Legendre expansion.

In this work we report measurements of the thermal conductivity of helium + hydrogen and helium + deuterium at 77.6 and 283.2 K, and compare them with the predictions of the Hirschfelder-Eucken equation using the spherically symmetric part of the Tsapline and Kutzelnigg potential to represent the helium-hydrogen interaction.

Experimental

The apparatus is of a comparative type, and is calibrated by converting reliable absolute values of the noble gas viscosities (7) into thermal conductivities through the standard theoretical relations (8). The design is such that the output depends as far as possible on the thermal conductivity - and no other property - of the gas.

The method, which is described in detail elsewhere, consists in allowing a controlled amount of heat to flow into a gas cell within which the temperature gradient is measured by two thermistors (9). The cell is shown diagrammatically in Fig. 1; it is kept in a constant temperature environment, and contains a platinum heater wire A (shown end-on)and two thermistors B and C. The control and measurement unit is shown schematically in Fig. 2. An electronic clock switches a controlled current repetitively to the heater A for equal "on" and "off" time intervals of 1-3 minutes. The thermistors B and C form two arms of a stable, thermostatted bridge circuit. The clock instructs two "sample-and-hold" circuits to record the output of the bridge at the end of each "heater-on" and"heater-off" period. The amplified difference of these two values forms the final output of the apparatus.

---

* And, although it has been suggested that rotational resonant exchange may be of importance (4), we believe its effect to be insignificant (5)

to gas supply

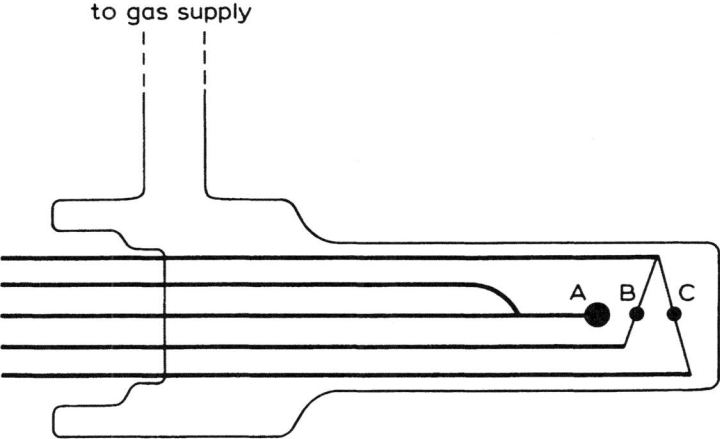

*Fig. 1   The thermal conductivity cell: A, platinum-wire heater;
B and C, thermistors.*

The precision of the output varies with its numerical value.
By taking a number of readings, averaged data are obtained which
are precise to 0.1 per cent at 283.2 K and 0.5 per cent at 77.6 K.
After calibration, the probable resulting errors in the thermal
conductivities are less than 0.5 per cent at 283.2 K and 1.5 per cent
at 77.6 K. The results, in terms of the excess thermal conductivity,
are shown later in Figs. 3-6, compared with theoretical predictions.

Discussion of Results

The detailed results will be published later (10);   here we
compare them qualitatively with those of other workers.   It is
convenient to define an excess thermal conductivity λ^E, where

$$\lambda^E = \lambda_{mix} - x_1\lambda_1 - x_2\lambda_2,$$

λ_{mix} is the thermal conductivity of the mixture, and x_i and λ_i are
the mole fraction and thermal conductivity of pure component i.

Since the quantity λ^E changes only slowly with increasing
temperature, we can use it to compare our results with the data of
other workers at slightly different temperatures.   In terms of
λ^E our results for helium + hydrogen and helium + deuterium at
283.2 K are clearly in close agreement (±0.1 per cent in λ_{mix}) with
those of Cauwenbergh and van Dael (1).   Further, those for helium +
hydrogen at 283.2 K are also in good agreement (±0.2 per cent in

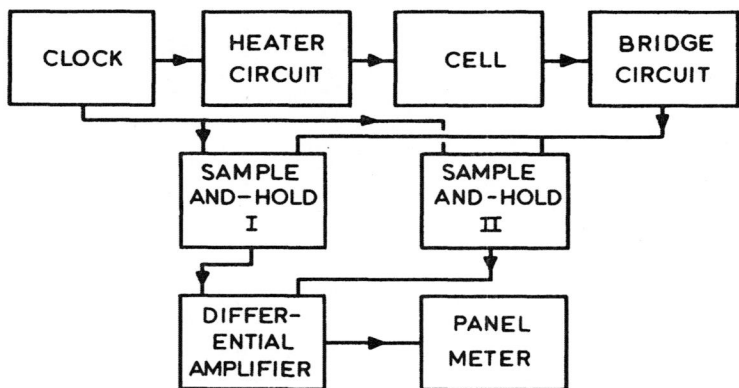

Fig. 2 The thermal conductivity control and measurement unit.

λ_{mix}) with those of Cotton (11), and in fair agreement with the
graphical representations of Hansen *et al.* (12) and Minter (13).
Agreement with other published data at 283.2 K is poor, however,
and our results at 77.6 K differ significantly from those of
Mukhopadhyay and Barua (14).

In making predictions from theory, we first assume that the
intermolecular potentials for the various hydrogen isotopes, and
their *ortho* and *para* modifications, are the same. Since the rate
of rotational-translational relaxation is low, λ_{mix} can be
calculated from the Hirschfelder-Eucken equation:

$$\lambda_{mix} = \lambda_{trans,mix} + (\lambda_2 - \lambda_{trans,2})\ 1 + (x_1 D_{22}/x_2 D_{12})^{-1},$$

where $\lambda_{trans,mix}$ is the translational thermal conductivity of the
mixture (15), D_{22} is the self-diffusion coefficient for hydrogen,
D_{12} is the mutual diffusion coefficient, and $\lambda_{trans,2}$ is the
translational thermal conductivity of pure hydrogen. The
quantities D_{22}, D_{12}, $\lambda_{trans,mix}$ and $\lambda_{trans,2}$ are calculated from
intermolecular potential functions. λ_2 is set equal to its
experimental value.

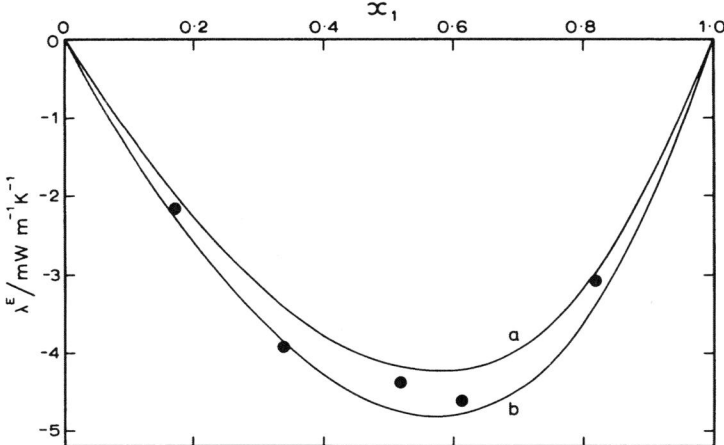

Fig. 3   *Excess thermal conductivity λ^E of helium (1) + hydrogen (2) at 77.6 K as a function of mole fraction x_1:   a, theoretical potential (ref. 6); b, molecular beam potential (ref. 19).*

Lennard-Jones (12:6) intermolecular potentials were used for the helium-helium and hydrogen-hydrogen interactions in conjunction with parameters which accurately reproduce the experimental viscosities at the two experimental temperatures. Collision integrals for the Tsapline and Kutzelnigg potential at 283.2 K were calculated classically from the programme of O'Hara and Smith (16); the quantum corrections of de Boer and Bird (17) (less than 0.5 per cent) were applied. The collision integrals at 77.6 K were calculated from the phase-shifts using the programme of Lantzsch and Walaschewski (18).

Comparisons of experiment and prediction are shown in Fig. 3 - 6. Agreement is within 0.3 per cent at 283.2 K and 0.7 per cent at 77.6 K. Also illustrated are predictions from a helium-hydrogen potential derived from molecular beam measurements (19) and the Lennard-Jones (12:6) potential with Lorentz-Berthelot combining rules. As well as reproducing accurately our thermal conductivity data, the Tsapline and Kutzelnigg potential also predicts the experimental viscosities of Kestin and Yata (20) for helium + hydrogen mixtures at 303.2 K to within 0.3 per cent. We conclude therefore that the thermal conductivities of mixtures of helium with the hydrogen isotopes can be completely understood, within experimental error, on the basis of the Hirschfelder-Eucken

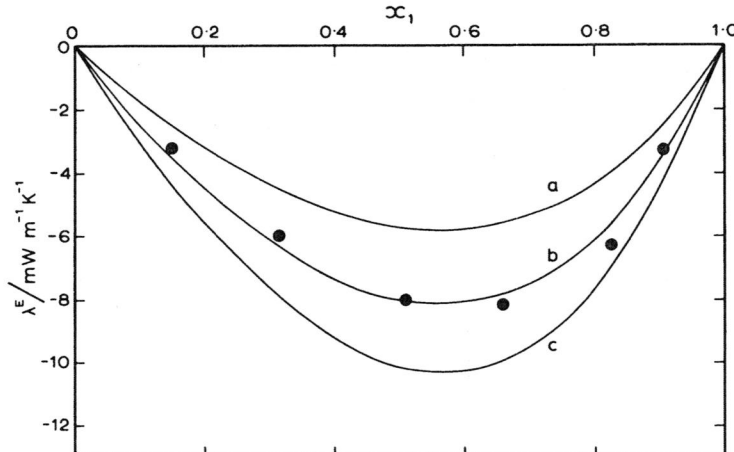

*Fig. 4   Excess thermal conductivity λ^E of helium (1) + hydrogen
(2) at 283.2 K as a function of mole fraction x_1:   a,
Lennard-Jones (12:6) Lorentz-Berthelot potential;   b,
theoretical potential (ref. 6);   c, molecular beam
potential (ref. 19).*

equation in conjunction with a realistic helium-hydrogen pair
potential.   That is, in these mixtures, translational and internal
energy are transported independently.   The translational energy
of hydrogen is carried as if by a monatomic gas;   and the internal
energy by diffusion.

1.   H. Cauwenbergh and W. van Dael, Physica, 1971, 40, 347.
2.   J.O. Hirschfelder, Sixth International Combustion Symposium
     (Rheinhold, New York, 1957), p. 351.
3.   R.M. Jonkman, G.J.Prangsma, I. Ertas, H.F.P. Knaap and
     J.J.M. Beenakker, Physica, 1968, 38, 441.
4.   C. Nyeland, E.A. Mason and L. Monchick, J. Chem. Phys., 1972,
     56, 6180.
5.   A.A. Clifford and E. Dickinson, J. Chem. Phys., submitted for
     publication.
6.   B. Tsapline and W. Kutzelnigg, Chem. Phys. Lett., 1973, 23,
     173.
7.   J.T.R. Watson, Viscosity of Gases in Metric Units (H.M.S.O.,
     Edinburgh, 1972).
8.   S. Chapman and T.G. Cowling, The Mathematical Theory of
     Non-Uniform Gases, 3rd ed. (Cambridge Univ. Press, 1970).

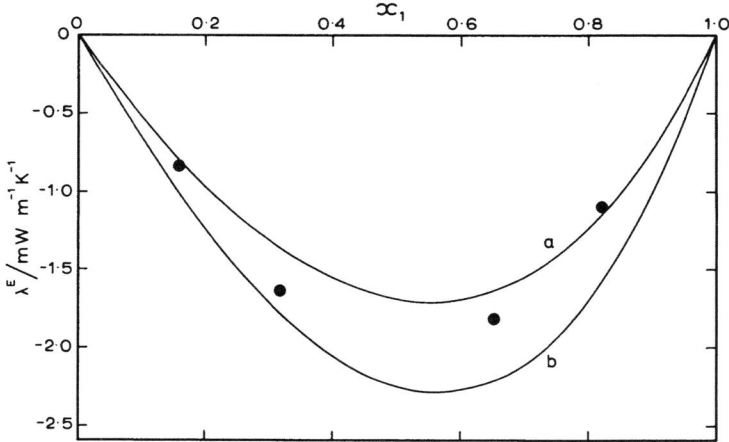

Fig. 5 Excess thermal conductivity λ^E of helium (1) + deuterium (2) at 77.6 K as a function of mole fraction x_1: a, theoretical potential (ref. 6); b, molecular beam potential (ref. 19).

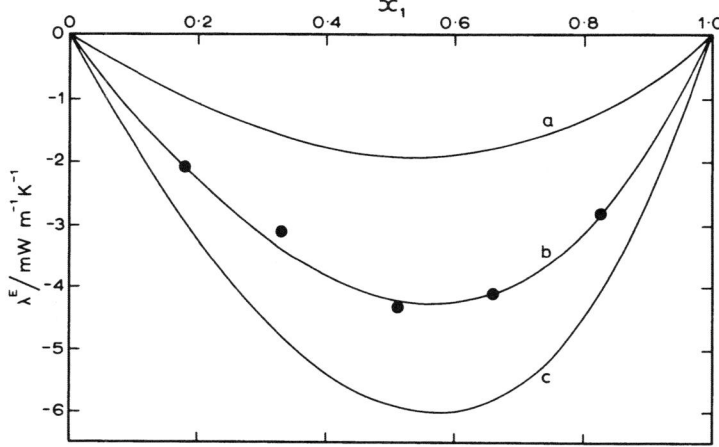

Fig. 6 Excess thermal conductivity λ^E of helium (1) + deuterium (2) at 283.2 K as a function of mole fraction x_1: a, Lennard-Jones (12:6) Lorentz-Berthelot potential; b, theoretical potential (ref. 6); c, molecular beam potential (ref. 19).

9. A.A. Clifford, L. Colling, P. Gray, D.A. Tong and G.S. Tough, J. Phys. E., 1974, 7, 283.
10. A.A. Clifford, L. Colling, E. Dickinson and P. Gray, J.C.S. Faraday I, in press.
11. J.E. Cotton, Ph.D. Dissertation (Univ. of Oregon, 1962).
12. R.S. Hansen, R.R. Frost and J.A. Murphy, J. Phys. Chem., 1964, 68, 2028.
13. C.C. Minter, J. Phys. Chem., 1968, 72, 1924.
14. P. Mukhopadhyay and A.K. Barua, Brit. J. Appl. Phys., 1967, 18, 635.
15. J.O. Hirschfelder, C.F. Curtiss and R.B. Bird, Molecular Theory of Gases and Liquids (Wiley, New York, 1954).
16. H. O'Hara and F.J. Smith, Comp. Phys. Comm., 1971, 2, 47.
17. J. de Boer and R.B. Bird, Physica, 1954, 20, 185.
18. B. Lantzsch and K. Walaschewski, unpublished work.
19. R. Gengenbach and C. Hahn, Chem. Phys. Lett., 1972, 15, 604.
20. J. Kestin and J. Yata, J. Chem. Phys., 1968, 49, 4780.

THE THERMAL CONDUCTIVITY OF SIMPLE

DENSE FLUID MIXTURES

W. McElhannon and E. McLaughlin

Department of Chemical Engineering

Louisiana State University, Baton Rouge, La. 70803

ABSTRACT

Many of the advances in recent years in the theory of the equilibrium properties of dense fluids have stemmed from initial work on systems of hard spheres, which provide the first approximation to the results for real systems. In the case of transport properties, examination of the theory for dense hard sphere systems has likewise led to very useful results which have only become possible because of the advances in the equilibrium theory.

The theory of the thermal conductivity of dense hard sphere systems was worked out by Thorne, who extended the Enskog theory to dense hard sphere mixtures. However, these equations, which are complex, could not until recently be evaluated or examined for the functional form of the concentration dependence due to a lack of knowledge of the contact pair distribution function. The equilibrium theory advances now make this evaluation possible. This work is discussed, with particular attention paid to the form of the mixing law which turns out to have the composition dependence given in terms of the volume fraction rather than the mole fraction. This result is used to show that for most size and mass ratios, the excess thermal conductivity is negative.

INTRODUCTION

Precise knowledge of the "law" governing the composition dependence of the thermal conductivity of a liquid or dense fluid mixture in terms of the values of the pure components is not at present

available. A number of expressions have been proposed which have
met with varying degrees of success, and in general, these have been
based on a composition dependence given in terms of either the mole
fraction or weight fraction of the components. For the case of a
binary mixture, the most obvious expression is

$$\lambda_{mix} = x_1^2\lambda_1 + 2x_1x_2\lambda_{12} + x_2^2\lambda_2 \tag{1}$$

where λ_1 and λ_2 are the thermal conductivities of pure species 1 and
2, x_1 and x_2 the respective mole fractions, and λ_{12} is a "cross-
coefficient" which has to be calculated but which is constant for a
particular mixture at a particular temperature and pressure. Equation (1) is analogous to the expression for the second virial coefficient of a gas mixture B_{mix} in terms of the second virial coefficients of the pure components, B_{11} and B_{22}:

$$B_{mix} = x_1^2 B_{11} + 2x_1x_2 B_{12} + x_2^2 B_{22} \tag{2}$$

where B_{12} is the cross-term coefficient arising from the hetero-
molecular interaction and for which the method of calculation is
known.

   Generalization of (1) to a multicomponent mixture of n species
can of course be readily made and introduces additional cross-term
coefficients

$$\lambda_{mix} = \sum_{i=1}^{n} \sum_{j=1}^{n} x_i x_j \lambda_{ij} \tag{3}$$

Even for simple binary mixtures, a constant value of λ_{12} is generally
not obtained when experimental results are analyzed in terms of equation (1) as shown below[1] for the system CCl_4/C_6H_6. Here the coefficient progressively increases as the composition of the component
with the higher conductivity increases. The question then arises
as to whether or not any of the alternative empirical laws are any
more adequate and what the actual form of the "laws" might be.

Table 1

The Cross-term Coefficient λ_{12} for
the Mixture CCl_4-C_6H_6 at $20^\circ C$

x_1 (mole fraction of C_6H_6)	.8551	.5716	.3742	.1359
$\lambda_{12}\times10^2$ ($Jm^{-1}s^{-1}K^{-1}$)	.811	.951	.956	.997

Moelwyn-Hughes proposed a mixing law for transport properties which for thermal conductivity of simple mixtures was given by Venart[2] and can be written

$$\lambda_{mix} = (x_1\lambda_1 + x_2\lambda_2)(1 - x_1x_2 \frac{w}{kT}) \qquad (4)$$

with w the "interchange energy" for a strictly regular solution and k is Boltzmann's constant. This association with thermodynamic properties prompts one to reexamine the mixing laws by looking at the excess transport properties and observing whether they follow the behavior of any of the excess thermodynamic properties.

The Excess Thermal Conductivity

The excess thermal conductivity λ_{xs} can be defined by

$$\lambda_{xs} = \lambda_{mix} - x_1\lambda_1 - x_2\lambda_2 \qquad (5)$$

and of all the mixtures examined the majority of the results indicate that λ_{xs} is negative. For example, of 59 binary mixtures of organic liquids examined by Jamieson and Hastings[3], only nine mixtures indicated the possibility of a positive excess thermal conductivity. If λ_{mix} were represented by equation (4), then λ_{xs} would be

$$\lambda_{xs} = (x_1\lambda_1 + x_2\lambda_2)(-x_1x_2 \frac{w}{kT}) \qquad (6)$$

which is positive if the interchange energy is negative. In general, however, among systems which show negative excess thermal conductivities are systems which exhibit both positive and negative deviations from Raoult's Law and hence both positive and negative w's. Therefore, the excess thermal conductivity does not correlate with the excess thermodynamic properties. The reason why it is generally negative has, therefore, to be sought from other reasoning. This problem will now be examined on the basis of hard sphere theory where only size and mass effects are important.

THE THORNE-ENSKOG EQUATIONS

Pure Fluids

Thermal conductivity theory was significantly improved by Enskog when, for dense fluids, he recognized the separate effects of kinetic and collisional energy transfer[4]. In a denser fluid, the collisional frequency increases, with a consequently greater collisional effect.

Analytically, the results of dense fluid theory can be expressed for the separate effects:

$$\frac{\lambda_{kinetic} Nv}{\lambda_o b_o} = (\frac{1}{y} + \frac{3}{5}) \tag{7}$$

$$\frac{\lambda_{coll} Nv}{\lambda_o b_o} = \frac{3}{5} (1 + \frac{3}{5} y) + \frac{32}{25} \frac{y}{\pi} \tag{8}$$

where $y = p/nkT - 1$, λ_o is the thermal conductivity of the corresponding dilute gas at the same temperature T, and b_o is the hard sphere second virial coefficient.

Mixtures

H. H. Thorne extended Enskog's results for a pure fluid to that for a binary mixture[5]. Thorne's resultant equation, not given here, can be reduced to the following form:

$$\lambda/\lambda_1 = f_\lambda [x_1, r, R, (v_1^*/v), P/T] \tag{9}$$

where $r = \sigma_{22}/\sigma_{11}$ and $R = m_2/m_1$ for σ_{ii} and m_i the molecular diameter and mass respectively for species i. The thermal conductivity of species i is λ_i, and the ratio of volume occupied by the molecules of species i to that of the system is (v_i^*/v).

If n_i is the ith species' number density, and

$$n = n_1 + n_2; \quad \sigma_{12} = (\sigma_{11} + \sigma_{22})/2 \tag{10}$$

then pressure is related to the number densities:

$$P/kT = \sum_i n_i + \frac{2}{3} \pi \sum_{ij} n_i n_j \sigma_{ij}^3 g_{ij}(\sigma_{ij}) \tag{11}$$

where $g_{ij}(\sigma_{ij})$ are contact pair distribution functions.

Evaluation of equation (9) requires the equation of state of a mixed dense hard sphere fluid - an unknown relationship. However, Lebowitz[6] has obtained an expression in the Percus-Yevick[7] approximation which gives

$$g_{12} = (\sigma_{22} g_{11} + \sigma_{11} g_{22})/2\sigma_{12} \tag{12}$$

where

$$g_{ii} = \left\{ (1 + \tfrac{1}{2}\xi) + \tfrac{3}{2}\tfrac{\pi}{6} n_j \sigma_{jj}{}^2 (\sigma_{ii} - \sigma_{jj}) \right\} (1 - \xi)^{-2} \qquad (13)$$

(for $i \neq j$)

$$\xi = b_{11}{}^* + b_{22}{}^* = v_1{}^* (x_1 + x_2 r^3)/v \qquad (14)$$

since $b_{ii}{}^* = \pi n_i \sigma_{ii}{}^3/6 = n_i v_i{}^*$.

The inexact nature of the Percus-Yevick approximation is illustrated by the discrepancy between the pressure derived from the compressability equation p^c and that from the virial equation p^v.[8]

$$p^c v_1{}^* = \frac{\xi(1+\xi+\xi^2)}{(1-\xi)^3 (x_1+x_2 r^3)}$$

$$- \frac{3 x_1 x_2 (1-r^2)\xi^2 [(1+r)+r\xi\{(x_1+x_2 r^2)/(x_1+x_2 r^3)\}]}{(x_1+x_2 r^3)^2 \ (1-\xi)^3} \qquad (15)$$

whereas

$$\frac{p^v v_1{}^*}{kT} = \frac{p^c v_1{}^*}{kT} - \frac{3\xi^4}{(1-\xi)^3} \left\{ \frac{x_1 + x_2 r^2}{x_1 + x_2 r^3} \right\}^2 . \qquad (16)$$

ξ, the ratio of the volume of the molecules to the volume of the container, can be found for fixed $p^v v_1{}^*/kT$ (or $p^c v_1{}^*/kT$), composition, and diameter ratio by solving equations (15) and (16) algebraically. For fixed $p^v v_1{}^*/kT$, r, and x_1, ξ and consequently $v_1{}^*/v$ can be obtained from equations (15) and (16). Therefore, equation (9) can be solved[9].

In the particular case of a dense fluid, the collisional frequency is high, and the effect due to the collisional contribution is the major one - the kinetic contribution being low.

Examination of equation (8) shows the first term arises from the distortion of the velocity distribution function from the locally Maxwellian form due to the temperature perturbation on the fluid[10]; while the second term is the contribution of the local

Maxwellian distribution itself. A similar analysis of the full form of equation (9) isolates the contribution of the local maxwellian distribution to a single term. Since Dahler also showed that the second term of equation (8) is equivalent to the Longuet-Higgins, et al. expression[11] for the thermal conductivity of a pure dense fluid, it follows that kinetic and distortion contributions in their calculations are neglected. For the case of mixtures, Kandiyoti and McLaughlin showed that the expression of Longuet-Higgins et al. for a hard sphere binary mixture of equal sized particles:

$$
\lambda_{mix} = \sigma k n y (kT/\pi)^{1/2} [x_1^2 m_1^{-1/2}
$$
$$
+ 4 x_1 x_2 \{ 2 m_1 m_2 / (m_1 + m_2)^3 \}^{1/2} + x_2 m_2^{-1/2}] \tag{17}
$$

can be generalized from Thorne's equation to a system of disparate molecules:

$$
\frac{\lambda_{mix}}{\lambda_1} = \frac{v_1^{o\,2}}{v} \frac{1}{g_1^o} [x_1^2 g_{11} + \{ \frac{R}{8(1+R)^3} \}^{1/2} x_1 x_2 g_{12} (1+r)^4
$$
$$
+ R^{-1/2} x_2^2 g_{22} r^4 \tag{18}
$$

where g_1^o is the contact pair distribution at σ_1 for pure species 1.

Using equation (8), an expression for thermal conductivity of pure species i can be derived:

$$
\lambda_i = \frac{4}{9} (\pi k T m_i)^{1/2} (n_i^o)^2 g_i^o (\sigma_{ii}) \sigma_{ii}^4 \frac{3k}{2m_i} \tag{19}
$$

Equations (18) and (19) result in the following expression for thermal conductivity of the mixture:

$$
\lambda_{mix} = G_1 \phi_1^2 \lambda_1 + 2 \phi_1 \phi_2 \lambda_{12} + G_2 \phi_2^2 \lambda_2 \tag{20}
$$

where

$$
G_i = g_{ii}/g_i^o , \quad \phi_i = x_i v_i^o / v
$$

and

$$
\lambda_{12} = \frac{4}{9} (2\pi k T \frac{m_1 m_2}{m_1 + m_2})^{1/2} n_1^o n_2^o g_{12} (\sigma_{12}) \sigma_{12}^4 \frac{3k}{m_1 + m_2} \tag{21}
$$

for

$$\sigma_{12} = \frac{1}{2}(\sigma_{11} + \sigma_{22}). \tag{22}$$

It is worth noting that the volume fraction ϕ arising naturally in the equations is not the typical volume fraction defined in terms of the partial molar volume. As a consequence, the sum of the volume fractions does not equal unity but a function of r.

EVALUATION

We now have an expression for the thermal conductivity of a dense fluid mixture of hard spheres and a means of determining whether the excess thermal conductivity is ever positive.

Using the convenient, rather than the traditional, composition variable, we define excess thermal conductivity:

$$\lambda_{xs} = \lambda_{mix} - \phi_1\lambda_1 - \phi_2\lambda_2 \tag{23}$$

Case 1

Equal Size Molecules (r = 1)

Since r is unity, all the distribution functions are equal, the sum of the volume fractions is unity, and λ_{xs} has the same sign as the following quantity:

$$\lambda_{xsr} = 2^{5/2} \frac{R}{(R + 1)^{3/2}} - R^{1/2} - 1 \tag{24}$$

Therefore, λ_{xsr} is a "reduced" excess thermal conductivity. Since every physical mass ratio is covered for $0 < R \leq 1$, λ_{xsr} was evaluated for each R at increments of 0.05. It was always negative.

Case 2

No Variables Restricted to One Value

The resultant excess thermal conductivity when "reduced" and nondimensionalized by division of common factors yields the expression

$$\lambda_{xsr} = \phi_1(G_1\phi_1-1)(n_1^{\circ}/n_2^{\circ})g_1^{\circ}R^{1/2} + \phi_2(G_2\phi_2-1)(n_2^{\circ}/n_1^{\circ})g_2^{\circ}r^4$$

$$+ 2^{5/2}\phi_1\phi_2g_{12}(\frac{r+1}{2})^4 \frac{R}{(R+1)^{3/2}} \tag{25}$$

Note that for equal size molecules, this case checks with the previous case (except for a multiple).

Every conceivable physical situation of hard sphere molecules is covered by allowing the following range of variables: r from zero to one, R from zero to infinity, x_1 from zero to one, and $p^V v_1^*/kT$ from zero to infinity. In actuality, molecules in a binary mixture seldom vary in size or mass by more than a ratio of five-to-one. Macromolecules present special problems and alternate possibilities for solution that are not covered in this paper. For the solution of equation (25), r was varied from zero to one, R from one-fifth to five, x_1 from zero to one, and $p^V v_1^*/kT$ from one to five. Under all computer calculated situations for these variable ranges, the excess thermal conductivity proved to be negative within computer accuracy.

CONCLUSION

Hard sphere theory applied to the collisional contribution to thermal conductivity indicates no basis for ever expecting a positive excess thermal conductivity in binary systems.

REFERENCES

[1] McLaughlin, E., Chem. Rev., 1964, <u>64</u> (4), 389.

[2] Venart, J.E.S., "Proc. 4th Symposium on Thermophysical Properties" Ed. J. R. Moszynski, A.S.M.E. 1968, 292.

[3] Jamieson, D. T., and Hastings, E. H., "Proc. 8th International Conference on Thermal Conductivity", Ed. C. Y. Ho and R. E. Taylor. Plenum Press, 1969, 631.

[4] Chapman, S. and Cowling, T. G., <u>Mathematical Theory of Non-Uniform Gases</u>, Cambridge University Press, 1953.

[5] Thorne, H. H., Quoted in [4]

[6] Lebowitz, J. L., <u>Phys. Rev. A</u>, 1964, <u>133</u>, 895.

[7] Percus, J. K. and Yevick, G. J., <u>Phys. Rev.</u>, 1958, <u>110</u>, 1.

[8] Rowlinson, J. S., <u>Rep. Prog. Physics</u>, 1965, <u>28</u>, 169.

[9] Kandiyoti, R., and McLaughlin, E., 1969, <u>Molec. Phys.</u>, <u>17</u>, 643.

[10] Dahler, J. S., 1957, <u>J. Chem. Phys.</u>, <u>27</u>, 1428.

[11] Longuet-Higgins, H. C., Pople, J. A., and Valleau, J. P., 1950, <u>Transport Processes in Statistical Mechanics</u> (London: Interscience Publishers), 73.

MOLECULAR ORIENTATION EFFECTS ON THERMAL CONDUCTIVITY

AND THERMAL DIFFUSION

Louis Biolsi* and E. A. Mason**

*Chemistry Department, Univ. of Missouri-Rolla
 Rolla, Missouri 65401
**Brown University
 Providence Rhode Island 02912

The effects of molecular angular momentum (spin polarization) on diffusion and thermal diffusion in a multicomponent polyatomic gas mixture are considered. Formal results for the diffusion and thermal diffusion coefficients are obtained. An expression for the thermal conductivity coefficient, λ_∞, when diffusion is independent of spin polarization but thermal diffusion depends on spin polarization is also obtained. The Wang Chang-Uhlenbeck approach to the kinetic theory of gases with internal states is used in the work presented here.

The kinetic theory of a single component polyatomic gas was developed by Wang Chang and Uhlenbeck [1] and extended to gas mixtures by Monchick, Yun, and Mason [2]. In this approach the translational motion of the molecules is treated classically and the internal degrees of freedom are treated quantum mechanically. In 1961 Kagan and Afanas'ev [3] showed that, if the molecules possess rotational degrees of freedom, there are two independent vector quantities associated with the transport properties; the linear velocity vector and the angular momentum vector (or operator). Contributions to the expressions for the transport properties due to the dependence on the angular momentum vector are called spin polarization effects. Such effects can be interpreted as arising from the polarization of molecular angular momentum caused by the partial alignment of the angular momentum vectors of the rotating molecules due to gradients in the gas [4]. Classical model calculations [4,5] indicate that this effect is important for transport properties that depend sensitively on inelastic collisions such as thermal conductivity and thermal diffusion.

A formal quantum mechanical approach to spin polarization effects on the thermal conductivity of a single component gas was

developed by McCourt and Snider [6] and extended to multicomponent
gas mixtures of uniform composition by Biolsi and Mason [7]. The
work reported here involves the development of the formal theory of
spin polarization effects on diffusion, thermal diffusion, and thermal conductivity in a multicomponent gas mixture in which the composition does not remain uniform (due to thermal diffusion). The
approach of Wang Chang and Uhlenbeck [1] is used to obtain these
results.

The heat flux vector, $\vec{q}$, can be written as

$$\vec{q} = \vec{q}_{tr} + \vec{q}_{int} \tag{1}$$

$$= kT\sum_{q}(\frac{5}{2} + \bar{\varepsilon}_q)n_q<\vec{V}_q> - \lambda_o\frac{\partial}{\partial\vec{r}}T +$$

$$nkT\sum_{qi}\int \vec{V}_q f^o_{qi}[(W_q^2 - \frac{5}{2}) + (\varepsilon_{qi}-\bar{\varepsilon}_q)]\sum_{q'}(\vec{C}_q^{q'}\cdot\vec{d}_{q'})d\vec{V}_q$$

where

$$\lambda_o = \lambda_{otr} + \lambda_{oint}. \tag{2}$$

The various symbols are defined in reference [2]. The first
term on the right in equation (1) represents heat flux due to mass
transport and the second term represents heat flux due solely to a
temperature gradient. Thus λ_o is the thermal conductivity of a gas
mixture at uniform composition. However, since thermal diffusion
occurs when a mixture is placed in a temperature gradient, there will
also be a heat flux due to diffusion. This effect is included in
the third term on the right in equation (1) [2].

An expression for λ_o is obtained by expanding the vector associated with the temperature gradient, $\vec{A}_{qi}$, in a complete set of
functions; i.e. [7]

$$\vec{A}_{qi} = A^1_{qi}\vec{W}_q + \frac{1}{4} \{[\vec{J}_q;\vec{J}_q\cdot\vec{W}_q]_+;A^2_{qi}\}_+ \tag{3}$$

where

$$A^r_{qi} = \sum_{n,p,t} a^r_{qnpt}S^{(n)}_{3/2}(W_q^2)R^{(0)}_p(\varepsilon_{qi})P^{(r-1)}_t(m^2). \tag{4}$$

The effect of spin polarization is included in the second term
on the right in equation (3). The various symbols are defined in
references [2] and [7]. Expressions for λ_{otr} and λ_{oint} are given in
terms of the expansion coefficients a_{qnpt}. These expansion coeffi-

cients are determined from the set of integral equations given by

$$R_{qnpt} = -\frac{k}{12} \sum_{q'} \sum_{n'p't'} \frac{g_{qnpt}g_{q'n'p't'}}{x_q x_{q'}} \{a^1_{q'n'p't'}\tilde{L}^{npt,n'p't'}_{qq'00}$$

$$+ a^2_{q'n'p't'}\tilde{L}^{npt,n'p't'}_{qq'01}\} \qquad (5)$$

and

$$\frac{1}{3} R_{qnpt} = -\frac{k}{12} \sum_{q'} \sum_{n'p't'} \frac{g_{qnpt}g_{q'n'p't'}}{x_q x_{q'}} \{a^1_{q'n'p't'}L^{npt,n'p't'}_{qq'01}$$

$$+ a^2_{q'n'p't'}L^{npt,n'p't'}_{qq'11}\}. \qquad (6)$$

The symbols R_{qnpt} and g_{qnpt} are defined in reference [7] and the $L^{npt,n'p't'}_{qq'ab}$ are given in the appendix of [7].

Explicit expressions for λ_{otr} and λ_{oint} are obtained by using a trial function. The smallest number of terms necessary to give all the physical effects is contained in the trial function with the terms

$$n = p = t = 0 \qquad n = t = 0; \; p = 1 \qquad p = t = 0; \; n = 1.$$

Using this trial function, the final results are [7]

and

$$[\lambda_{0int}]_1 = 4 \left| L_{qq'ab}^{npt,n'p't'} \begin{array}{c} 0 \\ \hline \begin{bmatrix} x_q \\ \hline \hline x_q \\ \hline 0 \end{bmatrix} \\ \hline \begin{bmatrix} x_q/3 \\ \hline x_q/3 \end{bmatrix} \end{array} \right. + \frac{1}{3} \left| L_{qq'ab}^{npt,n'p't'} \begin{array}{c} 0 \\ \hline \begin{bmatrix} x_q \\ \hline \hline x_q \\ \hline 0 \end{bmatrix} \\ \hline \begin{bmatrix} x_q/3 \\ \hline x_q/3 \end{bmatrix} \end{array} \right| \quad (8)$$

$$\left| 0 \; \left| \; 0 \; \left| \; x_{q'} \; \right| \; 0 \; \left| \; 0 \; \right| \; 0 \; \right| \; 0 \qquad 0 \; \left| \; 0 \; \left| \; 0 \; \right| \; 0 \; \left| \; 0 \; \right| \; x_{q'} \; \right| \; 0 \right|$$

$$\left| L_{qq'ab}^{npt,n'p't'} \right|$$

The symbol $L_{qq'ab}^{npt,n'p't}$ is defined in reference [7] and the symbol $[\;]_1$ denotes the first approximation.

Now consider diffusion. The diffusion velocity can be written as [2]

$$\langle \vec{v}_q \rangle = \frac{n^2}{\rho n_q} \sum_{q'} m_{q'} D_{qq'} \vec{d}_{q'} - \frac{1}{n_q m_q T} D_q^T \frac{\partial}{\partial \vec{r}} T \qquad (9)$$

where $D_{qq'}$ is the multicomponent thermal diffusion coefficient.

An expression for $D_{qq'}$ is obtained by expanding the vector associated with the diffusion gradients, $\vec{C}_q^q$, in a complete set of functions; i.e. [2,3,6,7]

$$\vec{C}_q^{q'} = c_q^{q'1} \vec{W}_q + \frac{1}{4} \{[\vec{J}_q ; \vec{J}_q \cdot \vec{W}_{q+}] ; c_q^{q'2}\}_+ \qquad (10)$$

where

$$\vec{C}_q^{q'r} = \sum c_{qnpt}^{q'r} S_{3/2}^{(n)} (W_q^2) R_p^{(0)} (\epsilon_{qi}) P_t^{(r-1)} (m^2). \qquad (11)$$

Using the expansion in equation (10), the multicomponent diffusion coefficient can be written as

$$D_{qq'} = x_q \frac{\rho}{m_{q'}} \sqrt{\frac{kT}{2m_q}} (c_{q000}^{q'q1} + \frac{1}{3} c_{q000}^{q'q2}) \qquad (12)$$

subject to the auxiliary condition

$$\sum_q n_q \sqrt{m_q} (c_{q000}^{q'1} + \frac{1}{3} c_{q000}^{q'2}) = 0. \qquad (13)$$

Also, using the expansion for $\vec{A}_{qi}$ given in equation (3), the multi-component thermal diffusion coefficient can be written as

$$D_q^T = n_q m_q \sqrt{\frac{kT}{2m_q}} \left(a_{q000}^1 + \frac{1}{3} a_{q000}^2 \right) \tag{14}$$

subject to the auxiliary condition

$$\sum_q n_q \sqrt{m_q} \left(a_{q000}^1 + \frac{1}{3} a_{q000}^2 \right) = 0 . \tag{15}$$

The effects of spin polarization are contained in the second term on the right in equations (12) and (14).

Explicit expressions for $D_{qq'}$ and D_q^T are obtained by finding the expansion coefficients. This requires the choice of a trial function. A trial function for diffusion is obtained by considering only the term $n = p = t = 0$ [2]. This leads to the following set of integral equations for the c's;

$$\frac{3}{2} \sqrt{\frac{2kT}{m_q}} (\delta_{qh} - \delta_{qk}) = \frac{75}{32} \sum_{q'} \frac{n_q n_{q'}}{x_q x_{q'}} \frac{k^2 T}{\sqrt{m_q m_{q'}}} \{ c_{q'000}^{h,k1} \tilde{L}_{qq'00}^{000,000} +$$

$$c_{q'000}^{h,k2} \tilde{L}_{qq'01}^{000,000} \} \tag{16}$$

and

$$\frac{1}{2} \sqrt{\frac{2kT}{m_q}} (\delta_{qh} - \delta_{qk}) = \frac{75}{32} \sum_{q'} \frac{n_q n_{q'}}{x_q x_{q'}} \frac{k^2 T}{\sqrt{m_q m_{q'}}} \{ c_{q'000}^{h,k1} L_{qq'01}^{000,000} +$$

$$c_{q'000}^{h,k2} L_{qq'11}^{000,000} \} . \tag{17}$$

The usual choice of a trial function for thermal diffusion is obtained by considering the terms [2]

$$n = p = t = 0 \qquad n = t = 0; \ p = 1 \qquad p = t = 0; \ n = 1$$

Using this trial function and the set of integral equations given by equations (5) and (6), the coefficient of thermal diffusion can be written as

$$[D_{q''}^T]_1 = -\frac{8}{5}\frac{m_{q''}}{k} x \tag{18}$$

$$L_{qq'ab}^{npt;n'p't'} \begin{bmatrix} 0 \\ x_q \\ x_q \\ 0 \\ x_q/3 \\ x_q/3 \end{bmatrix} + \frac{1}{3} L_{qq'ab}^{npt;n'p't'} \begin{bmatrix} 0 \\ x_q \\ x_q \\ 0 \\ x_q/3 \\ x_q/3 \end{bmatrix}$$

$$[x_q'\delta q'q'' \mid 0 \mid 0 \mid 0 \mid 0 \mid 0 \quad 0 \qquad 0 \mid 0 \mid 0 \mid x_q'\delta q'q'' \mid 0 \mid 0 \quad 0]$$

$$\left| L_{qq'ab}^{npt;n'p't'} \right|$$

A comparison of equations (1) and (9) shows that the heat flux and the diffusion velocity are related. This becomes clearer if the heat flux is written somewhat differently. It can be written as [2]

$$\vec{q} = kT \sum_q (\frac{5}{2} + \bar{\varepsilon}_q)n_q<\vec{v}_q> - \lambda_o \frac{\partial}{\partial \vec{r}} T - nkT \sum_q \frac{1}{n_q m_q} D_q^T d_q. \tag{19}$$

The dependence of both the diffusion velocity and the heat flux on the mass and temperature gradients is explicitly displayed by equations (9) and (19). It is useful to separate these effects.

If the diffusion coefficient is essentially independent of spin polarization, equation (9) can be written as [2]

$$\vec{d}_q = \sum_{q'} \frac{x_q x_{q'}}{[D_{qq'}]_1} (<\vec{v}_{q'}> - <\vec{v}_q>) +$$

$$\frac{\partial}{\partial \vec{r}} T \sum_{q'} \frac{x_q x_{q'}}{[D_{qq'}]_1 T} (\frac{D_{q'}^T}{n_{q'} m_{q'}} - \frac{D_q^T}{n_q m_q}). \tag{20}$$

This set of equations is called the Stefan-Maxwell equations and $[D_{qq'}]_1$ is the first approximation to the binary diffusion coefficient when the effects of spin polarization are negligible; i.e. [2]

$$[D_{qq'}]_1 = \frac{3}{16n\mu} \frac{kT}{\Omega_{qq'}^{(1,1)}} \tag{21}$$

where μ is the reduced mass and $\Omega_{qq'}^{(1,1)}$ is a collision integral [2].

The Stefan-Maxwell equations have the same form as is obtained in the absence of spin polarization. However, while diffusion is assumed to be independent of spin polarization, thermal diffusion depends on spin polarization; the dependence is given by equation (18). The assumption that spin polarization effects can be ignored in diffusion processes but not in thermal diffusion is supported by classical model calculations on the isotopes of hydrogen gas [4,5]. These model calculations indicate that spin polarization effects are negligible for diffusion but quite significant for thermal diffusion.

Using equation (20), equation (19) can be written so that the dependence on the diffusion velocity and the temperature gradients are separated; i.e.

$$\vec{q} = kT \sum_q (\frac{5}{2} + \bar{\varepsilon}_q) n_q \langle \vec{v}_q \rangle - [\lambda_\infty]_1 \frac{\partial}{\partial \vec{r}} T$$

$$+ kT \sum_q \sum_{q'} \frac{x_{q'}}{m_q} \frac{[D_q^T]_1}{[D_{qq'}]_1} (\langle \vec{v}_q \rangle - \langle \vec{v}_{q'} \rangle) \tag{22}$$

where

$$[\lambda_\infty]_1 = [\lambda_o]_1 - \frac{k}{2} \sum_q \sum_{q'} \frac{x_q x_{q'}}{[D_{qq'}]_1} n (\frac{[D_q^T]_1}{n_q m_q} - \frac{[D_{q'}^T]_1}{n_{q'} m_{q'}})^2 . \tag{23}$$

The thermal conductivity coefficient, $[\lambda_\infty]_1$, is associated with heat transfer in a steady state in the absence of chemical reactions [2]. Equations (22) and (23) have exactly the same form as they do in the absence of spin polarization [2]. However, although $[D_{qq'}]_1$ does not depend on spin polarization, $[\lambda_o]_1$ and $[D_q^T]_1$ do depend on spin polarization. Thus $[\lambda_\infty]_1$ also depends on spin polarization. The formal result for $[D_{qq'}]_1$ is given by equation (21), the formal result for $[D_q^T]_1$ (including the effects of spin polarization) is given by equation (18), and the formal result for $[\lambda_o]_1$ (including the effects of spin polarization) is given by equations (7) and (8).

In order to obtain numerical results for the transport coefficients, the $L_{qq';ab}^{npt;n'p't'}$ must be evaluated. This requires the choice of an interaction potential. However the $L_{qq';ab}^{npt;n'p't'}$ can be separated in a formal way into contributions from elastic and inelastic collision processes [2,7].

Several important assumptions are made in the Wang Chang-
Uhlenbeck approach to kinetic theory. Perhaps the most important
is the assumption of inverse collisions which is correct if the
internal states (vibrational and/or rotational) are nondegenerate
or if the cross section is degeneracy averaged (i.e. for rotational
states) [8]. Inverse collisions do not eliminate the effect of
spin polarization [4,5]. Thus, although an estimate of the magnitude of the spin polarization effect requires numerical calculations, the model calculations mentioned earlier [4,5] suggests that
the effect is important for transport properties that depend sensitively on inelastic collision processes such as thermal conductivity and thermal diffusion.

REFERENCES

1. Wang Chang, C. S. and G. E. Uhlenbeck, "Transport Phenomena in
   Polyatomic Gases", University of Michigan Engineering Research
   Rept. No. CM-681 (July 1951).

2. Monchick, L., K. S. Yun, and E. A. Mason, J. Chem. Phys. 39,
   654 (1963).

3. Kagan, Y., and A. M. Afanas'ev, JETP 14, 1096 (1962).

4. Sandler, S. I., and J. S. Dahler, J. Chem. Phys. 47, 2621 (1967).

5. Sandler, S. I., and E. A. Mason, J. Chem. Phys. 47, 4653 (1967).

6. McCourt, F. R., and R. F. Snider, J. Chem. Phys. 41, 3185 (1964).

7. Biolsi, L., and E. A. Mason, J. Chem. Phys. 63, 10 (1975).

8. Waldmann, L., Handb. Phys. 12, XXX (1958).

VISCOSITY AND HEAT CONDUCTIVITY OF FLUORINE GAS: COMPUTED VALUES FROM 100-2500 K AND 0.25-2.0 ATMOSPHERES*

Kaare I. Oerstavik and T.S. Storvick

Chemical Engineering Department
University of Missouri, Columbia, Missouri   65201

ABSTRACT

The computed values of the viscosity and heat conductivity of fluorine gas are reported in the range 100-2500K and 0.25-2.0 atmospheres pressure. The Chapman-Enskog first approximation equations for the thermal equilibrium binary mixture of fluorine atoms and molecules are used in the calculations.[1] The procedures developed by Mason and Monchick[2] to account for inelastic encounters for the fluorine molecules and Parker's[3] assignment of the change in rotational collision numbers with temperature are used. Experimental viscosity and heat conductivity data are reproduced over the temperature range where measurements are available.

Fluorine is a strong oxidizing agent that when combined with hydrocarbons provides a wide range of stable components used as refrigerants and polymers. Recently, it has been used as one component in chemical laser systems. The transport properties of fluorine gas are difficult to measure and it is the purpose here to report the calculated properties at temperatures from 100-2500K and pressures from 0.25-2.0 atmospheres. Over this range of conditions the fluorine molecules nearly all dissociate. It is this molecule-atom mixture for which the properties are reported.

The chemical equilibrium of fluorine gas was computed using the thermodynamic data given in the JANAF Tables.[4] The concentration of F^+, F^- and electrons was computed to be less than 10^{-23} (mole fraction) at 2500K and these species were neglected. The equilibrium compositions of the $F_2(g)$ and $F(g)$ over the temperature and pressure range of this study are given in Table 1.

The classical first order Chapman-Enskog solution of the Boltzmann equation for binary mixtures forms the starting point for computing transport properties.[1] The effect of the fluorine molecule inelastic collisions were included using the procedures developed by Mason and Monchick.[2] These procedures are familiar and can be found in the references cited.

It is necessary to have the parameters for the intermolecular potential function used to compute the Chapman-Enskog collision integrals to calculate the transport properties. The Lennard-Jones (12-6) potential has most often been used for these calculations but it is well known that the same energy and distance scaling parameters do not reproduce both the equilibrium and transport properties or treat these properties over a wide temperature range. Johnston[5] has developed an empirical correlation for estimating the Lennard-Jones parameters based on Badger's force constant rule. This rule states that the logarithim of the force constant is a linear function of the intermolecular separation for all "clusters" and stable molecular configurations and that the value of this constant depends only upon the row location of the atom in the periodic table. This empirical procedure gives the unknown Lennard-Jones parameters for fluorine atoms that are almost identical to the parameters for neon.

An analysis by Hanley and Klein[6] has shown that the (m,6,8) potential does accurately reproduce both equilibrium and transport properties of gases over a wide temperature range with one set of parameters. Parameters for fluorine molecules have been reported for this potential.[7] The fluorine atom potential form was assumed to be nearly the same as neon as indicated for the Lennard-Jones parameters. The distance and energy scaling parameters for the $m = 12$, $\gamma = 2$ potential were determined using neon viscosity data. It has been shown that the $m = 11$, $\gamma = 3$ potential best represents the inert gas properties,[6] but combining rules for m and γ in the (m,6,8) model have not been worked out. On this basis, the following parameters were used to compute the transport properties:

$$m = 12 \quad \gamma = 3 \quad (m,6,8) \text{ potential}$$

Interaction	$\varepsilon/k(°K)$	$\sigma(cm \times 10^{-8})$
F_2---F_2	138	3.32
F---F	82.4	2.554
F_2---F	106.6	2.936

The unlike pair energy (ε/k) and distance (σ) scaling parameters were obtained using the geometric and arithmetic combining rules respectively.

The inelastic encounters between diatomic molecules make a significant contribution to the heat conductivity of a gas. Mason and Monchick[2] have shown how the formal kinetic theory for polyatomic molecules of Wang Chang, Uhlenbeck, and De Boer[8] can be used to account for a major part of this correction. The irreducible integrals in these equations were identified as relaxation times for energy exchange between translational motion and the internal vibrational and rotational modes of motion in the molecule. In the fluorine calculations the vibrational modes were neglected because the vibrational relaxation times are longer than the rotational relaxation times. This assumption becomes weak at the highest temperature.

Parker[3] has shown that the rotational collision number increase with temperature. This model was used to account for the change in the rotational collision numbers with temperature.

The Mason-Monchick procedure also requires an estimate of the diffusion rate of internal energy through the gas. This is often assumed to be the same as the self diffusion coefficient for the gas and therefore can be simply computed using the kinetic theory equations. Sandler[9] has shown that inelastic collisions effect the ratio of the energy diffusion to the self diffusion and he has provided an estimate of this correction. This correction was also used in the calculation of the fluorine properties.

The calculated viscosity and heat conductivity for fluorine gas are given in Table 1. The viscosity measurements reported by Haynes[10] (T = 90 to 300 K) are reproduced with a maximum error of 2.5%. The viscosity measurements of Franck and Stöber (T = 90 to 471 K) are reproduced with a maximum deviation of 9%. The measurements of Kanda[12] are significantly different than the other two data sets and these measurements were not considered. The fluorine molecule potential parameters were obtained by using Haynes' measurements which accounts for the excellent fit to these data.

Franck[13] reports the only fluorine heat conductivity measurements that were found (T = 100 to 800 K). These data were reproduced to $\pm$ 5% by the computed values which is within the probable experimental error of the measurements.

TABLE 1. Calculated Viscosity and Heat Conductivity of
          Fluorine Gas.

Table Entries
$\left[\begin{array}{l} \text{Viscosity (gm/cm-sec) x } 10^{-6} \\ \text{Mole Fraction } F_2 \\ \text{Heat Conductivity (Cal/cm-sec-°K} \\ \qquad\qquad x10^{-6}) \end{array}\right.$

Temp. (°K)	Pressure (Atm.)			
	0.25	0.5	1.0	2.0
100	83.1 1.0000 20.3	83.1 1.0000 20.3	83.1 1.0000 20.3	83.1 1.0000 20.3
300	227.1 0.9999 61.5	227.1 0.9999 61.5	227.1 1.000 61.5	227.1 1.000 61.5
500	335.3 0.9951 99.1	335.2 0.9965 99.1	335.1 0.9975 99.1	335.0 0.9982 99.1
700	430.0 0.9243 130.1	428.1 0.9459 130.3	426.8 0.9614 130.6	425.8 0.9726 130.8
1000	594.5 0.5322 179.8	579.8 0.6379 173.1	568.2 0.7267 169.7	559.6 0.7975 168.6
1300	775.8 0.1707 271.6	754.5 0.2669 250.9	731.5 0.3808 231.8	709.9 0.4988 217.4
1600	918.5 0.0529 346.6	906.0 0.0964 332.0	887.4 0.1647 311.6	863.5 0.2591 288.1
1900	1035. 0.0210 399.9	1029. 0.0403 391.9	1018. 0.0749 378.4	1000. 0.1319 358.3
2200	1141. 0.0104 444.0	1137. 0.0204 439.4	1130. 0.0392 430.8	1118. 0.0730 416.3
2500	1240. 0.0060 484.1	1237. 0.0119 481.0	1233. 0.0233 475.3	1224. 0.0445 464.9

It is difficult to assign error limits to the calculated fluorine properties. The intermolecular potential between the various excited states of fluorine should be computed from spectroscopic measurements to provide more accurate fluorine properties, but this was beyond the scope of this study. Details of the procedures used in the calculations to obtain Table 1 are available.[14]

*Partial financial support for this work was received from the Engineering Experiment Station (KIO) and the National Science Foundation (TSS).

[1] S. Chapman and T.G. Cowling, The Mathematical Theory of Non-Uniform Gases, third edition (Cambridge Press, 1970); J. O. Hirschfelder, C.F. Curtiss and R.B. Bird, Molecular Theory of Gases and Liquids, (John Wiley and Sons, New York, 1954) Chapter 8.

[2] E.A. Mason and L. Monchick, J. Chem. Phys. 35, 1676 (1961); Ibid, 36, 1622 (1962).

[3] J.G. Parker, Phys. Fluids, 2, 449 (1952).

[4] D.R. Stull and H. Prophet, JANAF Thermochemical Tables, Second Edition (U.S. Dept. of Commerce, NSRDS-NBS 37, June, 1971).

[5] H.S. Johnston, J. Am. Chem. Soc. 86, 1943 (1964).

[6] H.J.M. Hanley and M. Klein, J. Phys. Chem. 76, 1743 (1972).

[7] H.J.M. Hanley and R. Prydz, J. Phys. Chem. Ref. Data 1, 1101 (1972).

[8] C.S. Wang Chang, G.E. Uhlenbeck and J. De Boer, "The Heat Conductivity and Viscosity of Polyatomic Gases", Studies in Statistical Mechanics, (North-Holland Pub. Co., Amsterdam, 1964). Part C.

[9] S.I. Sandler, Phys. Fluids 11, 2549 (1968).

[10] W.M. Haynes, Physica 76, 1 (1974).

[11] E.U. Franck and W. Stöber, Z. Naturforschung 7a, 822 (1952).

[12] E. Kanda, Bul. Chem. Soc. Japan, 12, 463 (1937).

[13] E. U. Franck, Z. Electrochem., 55, 636 (1951).

[14] K.I. Oerstavik, "Viscosity and Thermal Conductivity of Fluorine Gas to 2500 K and 2 Atm.", M.S. Thesis, University of Missouri-Columbia, August, 1975.

QUANTUM-MECHANICAL CORRECTIONS TO TRANSPORT CROSS-SECTIONS FOR HARD

SPHERES

Josef G. Solomon   Sigurd Larsen

336 Newbold Ave.   and   Department of Physics

Moorestown, N.J. 08057   Temple Univ., Phila., Pa. 19122

We present a derivation of the classical and first quantum-mechanical correction terms in a high-energy expansion for the transport cross-section $Q^{(1)}$. This complements a similar calculation of one of us (S.L.) for the viscosity cross-section $Q^{(2)}$. In both cases, the first correction is found to be proportional to $1/(k\sigma)^{4/3}$, which is a term non-analytic in h (i.e., $h^{4/3}$).

Numerical work is also presented for $Q^{(1)}$ and $Q^{(2)}$ which establishes the form of the asymptotic expansions, demonstrates the presence of such terms as $\ln(k\sigma)/(k\sigma)^2$ (involving $h^2 \ln h$) and gives excellent estimates of a number of coefficients.

INTRODUCTION

We are investigating cross-sections in order to be able to make statements about transport coefficients. For hard spheres, we are especially interested in:

1. high-temperature behavior of the Boltzmann transport coefficient (the quantum-mechanical correction);

2. how symmetry effects decay with temperature.

These two questions involve different parts of the cross-sections. If we write, for example,

$$Q_{Bose} = Q_{Boltzmann} + Q_{exchange}$$

367

where Q could be either $Q^{(1)}$ or $Q^{(2)}$, which we shall discuss, then $Q_{Boltzmann}$ and $Q_{exchange}$ are rather different in character. Typically, $Q_{exchange}$ behaves as an exponentially damped sinusoid, and $Q_{Boltzmann}$ decays monotonically to the classical value. (See Figure 1.) One can study $Q_{exchange}$ analytically, using the Watson-Sommerfeld transformation (more familiar from its use in connection with Regge poles), and also numerically.

The related behavior of the exchange phase-shift sum for the second virial implies an asymptotically exponential decay of the symmetry effects of the form $e^{-(\ )T}$. This does not appear to be true for the transport coefficients--there we have rather an exponential approach to a power law.

In this talk we concentrate on $Q_{Boltzmann}$. Some work has already been done on $Q^{(2)}$ (Boyd and Larsen, Physical Review A, 4, 1155 (1971)). What we present here is analytical work bearing on $Q^{(1)}$, and numerical work bearing on both $Q^{(1)}$ and $Q^{(2)}$

The results are as follows:

$$\frac{Q^{(2)}}{Q^{(2)}_{classical}} = 1 + \frac{10.376805}{x^{4/3}} + (-17.000...)\frac{\ln x}{x^2} + (\ )\frac{1}{x^2} \quad (1)$$

$$+ (\ )\frac{1}{x^{8/3}} + \ldots$$

$$\frac{Q^{(1)}}{Q^{(1)}_{classical}} = 1 + \frac{3.458935}{x^{4/3}} + (-2.000)\frac{\ln x}{x^2} \quad (2)$$

$$+ (\ )\frac{1}{x^2} + (\ )\frac{1}{x^{8/3}} + \ldots$$

In Figure 2, the solid curve represents the exact expression, evaluated numerically. The dashed curve represents the two-term asymptotic formula. "Hidden" in the solid curve is the curve representing the extended asymptotic formula. For x > 100, the deviation of the extended asymptotic formula from the exact expression is $\sim 10^{-7}$ if four terms are included; with five terms, it is $\sim 10^{-10}$.

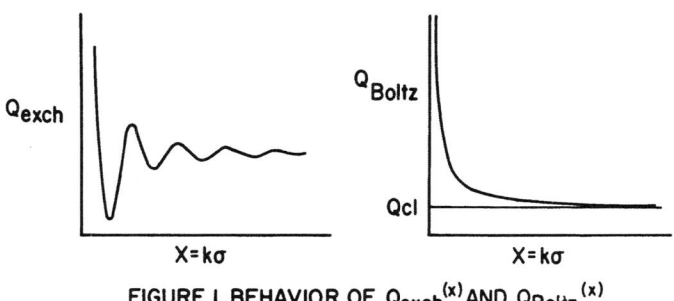

FIGURE I BEHAVIOR OF $Q_{exch}^{(x)}$ AND $Q_{Boltz}^{(x)}$

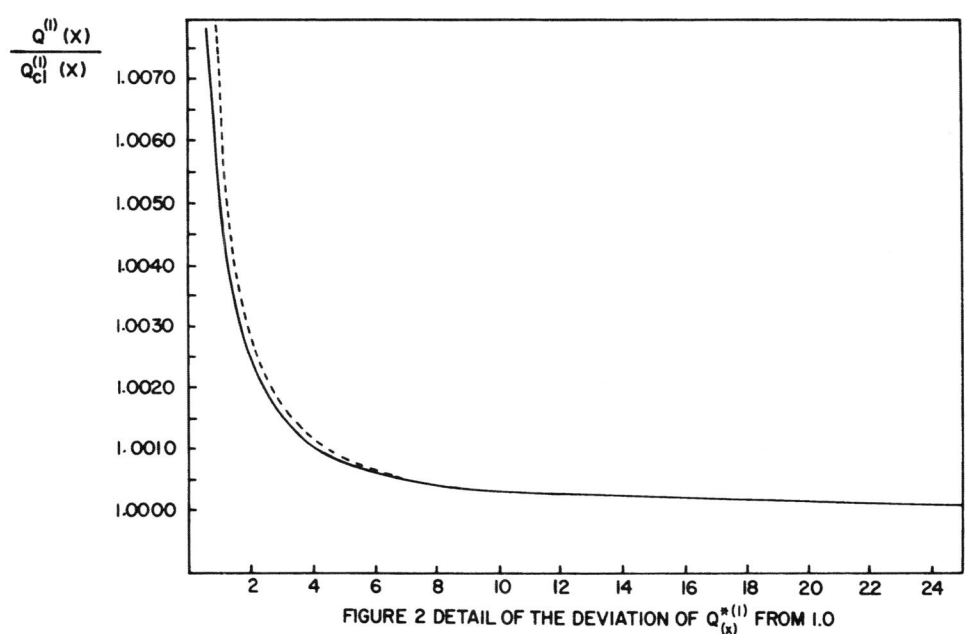

FIGURE 2 DETAIL OF THE DEVIATION OF $Q_{(x)}^{*(1)}$ FROM 1.0

ANALYTICAL WORK

The exact expression for $Q^{(1)}$ is

$$Q^{(1)} = (\frac{2\pi}{k^2}) \sum_{\ell=0}^{\infty} (\ell+1) \sin^2 (\delta_{\ell+1} - \delta_\ell) \tag{3}$$

For hard spheres, δ_ℓ is given by

$$\tan \delta_\ell = \frac{J_{\ell+\frac{1}{2}}(k\sigma)}{N_{\ell+\frac{1}{2}}(k\sigma)}$$

where k is the wave number, and σ is the hard-sphere diameter. Substituting, and dividing by the classical value of $Q^{(1)}$, we get

$$Q^{*(1)}(x) = \frac{Q^{(1)}(x)}{Q^{(1)}_{classical}(x)} = \frac{16}{\pi^2 x^4} \sum_{\ell=0}^{\infty} \frac{\ell+1}{H^{(1)}_{\ell+3/2}(x) \, H^{(2)}_{\ell+3/2}(x) \, H^{(1)}_{\ell+1/2}(x) \, H^{(2)}_{\ell+1/2}(x)} \tag{4}$$

where $x = k\sigma$, and we have made use of the identities

$$H^{(1)}_\ell(x) = J_\ell(x) + i \, N_\ell(x) \qquad H^{(2)}_\ell(x) = J_\ell(x) - i \, N_\ell(x) \tag{5}$$

This equation is still exact--but it is rather formidable. We can, however, evaluate it numerically for any value of x. By a considerable amount of tedious manipulations (see Boyd and Larsen, op cit), it can be shown that $Q^{*(1)}$ can be approximated by the following integral:

$$Q^{*(1)}(x) = \frac{16}{\pi^2 x^4} \int_0^\infty \frac{\ell \, d\ell}{\left[H^{(1)}_\ell(x) \, H^{(2)}_\ell(x) \right]^2} + 0(\frac{1}{x^2}) \tag{6}$$

We would like to use an asymptotic expansion on the Hankel functions appearing in the denominator. The problem is that the usual asymptotic expressions for Bessel functions have "islands of validity". (See Figure 3.) That is, the more accurate the asymptotic expansion is to be, the narrower is the region where the expansion is useful. The solution to this problem is the use of the uniform asymptotic expansions of Langer. The trouble with this approach is that the Hankel functions are now expressed in terms of Bessel functions of order 1/3.

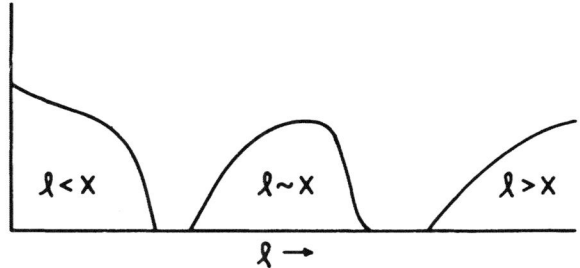

FIGURE 3 "ISLANDS OF VALIDITY"

We surmount this obstacle by employing two "tricks":

Trick #1--We add and subtract a term equivalent to the class-ical result (i.e., $Q^* = 1$), and then use the uniform asymptotic expansions.

Trick #2--The remaining contribution comes mainly from a "band" around $\ell \sim x$. We make explicit use of this.

RIGOROUS RESULT

$$Q^{*(1)}(x) = 1 + \frac{(A'+B')}{x^{4/3}} + O(\frac{1}{x^2}) \tag{7}$$

where

$$A' = \frac{16}{\pi^2} 3^{1/3} \int_0^\infty dz \; z^{1/3} \left[\frac{1}{z^2 (H_{1/3}^{(1)}(z) \; H_{1/3}^{(2)}(z))^2} - \frac{\pi^2}{4} \right] \tag{8}$$

and

$$B' = \frac{16}{\pi^2} \, 3^{1/3} \int_0^\infty dz \; z^{-5/3} \left\{ \frac{1}{\left[\frac{1}{\pi} K_{1/3}(z)\right]^2 + \left[I_{1/3}(z) + I_{-1/3}(z)\right]^2} \right\}^2 \quad (9)$$

These expressions for A' and B' are exactly 1/3 of the corresponding coefficients A and B which appear in the asymptotic expansion of $Q^{*(2)}$ (Boyd and Larsen, op cit). These authors evaluated the coefficients numerically; from their results, it follows that $A' + B' = 3.458935$, with a possible error of 1 in the last place.

NUMERICAL FITTING

To arrive at a more complete asymptotic description of the cross-sections valid at high energies, we attempted to fit different analytical forms to accurate numerical data for $Q^{*(1)}(x)$ and $Q^{*(2)}(x)$. These data were arrived at by using the exact formula for $Q^{*(1)}(x)$ given by Eq. (13), and its equivalent for $Q^{*(2)}(x)$, and computing the necessary Bessel and Neumann functions to 10 significant figures. The terms included in the proposed asymptotic formulas were suggested by our analytical investigations of some of the corrections, as well as by the nature of our error bounds.

Fitting $Q^{*(1)}(x)$

For the range $400 \le x \le 2900$, our best results were as follows:

$$Q^{*(1)}(x) = C_0 \cdot 1 + C_1 \cdot \frac{1}{x^{4/3}} + C_2 \cdot \frac{\ln x}{x^2} + C_3 \cdot \frac{1}{x^2} + \ldots \quad (10)$$

where

$$C_0 = 1.0000000000 \quad C_1 = 3.45892934 \quad C_2 = -1.99929$$

$$C_3 = -0.156649 \quad (11)$$

C_0 should of course be exactly 1.0. The predicted value for C_1 is 1/3 the value predicted by Boyd and Larsen for the corresponding coefficient in the asymptotic expansion for $Q^{*(2)}(x)$; that value was 10.376805, hence our predicted value for C_1 here is 3.458935.

The question arises: Is the logarithmic term really necessary? To find an answer to that question, we endeavored to fit $Q^{*(1)}(x)$ without the logarithmic term. The results were as follows:

$$C_0 = 1.0000000 \quad C_1 = 3.446968566 \quad C_2 = -15.02908$$

$$C_3 = 242.1953 \qquad (12)$$

Observing that omission of the logarithmic term costs us three significant figures in the 1.0 and in C_1, we conclude that it is essential to retain the logarithmic term.

The close agreement of the (original) computed coefficients with the predicted values led us to subtract 1.0 from $Q^{*(1)}$ and repeat the calculation. The best results were as follows.

$$C_1 = 3.458935 \quad C_2 = -1.999927 \quad C_3 = -.015230 \qquad (13)$$

Fitting $Q^{*(2)}(x)$

Numerical fitting for $Q^{*(2)}(x)$ was done in the same manner as for $Q^{*(1)}(x)$. The best results were as follows:

$$Q^{*(2)}(x) = C_0 \cdot 1.0 + C_1 \cdot \frac{1}{x^{4/3}} + C_2 \cdot \frac{\ln x}{x^2} + C_3 \cdot \frac{1}{x^2} + \dots$$

where   (14)

$$C_0 = 1.0000000000 \quad C_1 = 10.376755 \quad C_2 = -16.9885$$

$$C_3 = 25.056888 \qquad (15)$$

Fitting without the logarithmic term produced results even more striking than for $Q^{*(1)}(x)$:

$$C_0 = 1.000000 \quad C_1 = 10.275100 \quad C_2 = -101.313$$

$$C_3 = 2014.99391 \qquad (16)$$

Again, we conclude that it is essential to retain the logarithmic term in the expansion.

Fitting $Q*^{(2)}(x) - 1.0$ produced the following results.

$$C_1 = 10.376784 \quad C_2 = -16.99272 \quad C_3 = 25.0868964 \qquad (17)$$

Fitting to $Q*^{(2)}(x) - 1.0$ in the range $400 \leq x \leq 1900$ produced better results:

$$C_1 = 10.3768056 \quad C_2 = -16.997969 \quad C_3 = 25.128 \qquad (18)$$

(It should be noted that Boyd and Larsen, in their preliminary fitting, indicated values for C_2 and C_3 of -17.000 and 25.149, respectively.)

CONCLUSIONS

We have demonstrated the existence of non-analyticities that appear in the transport coefficients but not in the virial expansion. It is not yet clear where these terms arise in the physics of the situation, but the excellence of the numerical agreement between the asymptotic expansion and the exact expression leaves little doubt that the non-analytic terms are really there.

The direction that further work should take is clearly indicated. Numerically, we should extend the accuracy to which we have computed $Q*^{(1)}$ and $Q*^{(2)}$, try to find other terms in the asymptotic expansions, and in general try to refine the accuracy of the present coefficients. Analytically, the goal would be to derive the coefficients of the logarithmic term in the two expansions. (From our numerical work, the coefficients certainly appear to be -2 and -17, respectively.)

DETERMINATION OF THERMAL CONDUCTIVITY OF GASES FROM HEAT

TRANSFER MEASUREMENTS IN THE TEMPERATURE-JUMP REGIME*

S. C. Saxena and B. J. Jody

Department of Energy Engineering
University of Illinois
Chicago, Illinois 60680

ABSTRACT

The thermal conductivity values are computed for helium and neon in the temperature range 500-2500K from heat transfer data taken in the temperature-jump regime employing a hot-wire type column instrument. These apparent pressure dependent conductivity values are corrected for the temperature-jump effect according to kinetic theory in conjunction with experimentally determined thermal accommodation coefficients. These corrected conductivity values are pressure independent and are in good agreement with the values calculated from heat transfer data in the continuum regime. This is regarded as sufficient evidence validating the kinetic theory model for heat transfer in the temperature-jump region and for the experimentally determined thermal accommodation coefficients.

*This work is supported by the National Science Foundation under Grant No. GK-12519.

INTRODUCTION

We have employed a hot-wire conductivity column to determine the thermal conductivities of gases over a wide temperature range (50°-2450°C) and at pressures up to one atmosphere. In particular, the thermal conductivities of argon [1], helium [2] and nitrogen [3] are determined from the knowledge of the electrical power required to heat the axial tungsten wire to different temperatures in vacuum and in the presence of the test gas at such pressures where both the convection and thermal accommodation effects are small. The initial design of the column and the related instrumentation is described by Chen and Saxena [3] and the later improvements are given by Jody and Saxena [2]; and Jody, Jain and Saxena [4]. The same experimental facility is successfully exploited to determine the thermal accommodation coefficients for metal-gas system viz., tungsten-nitrogen [5] and tungsten-neon [6], from heat transfer rate data pertaining to the temperature-jump regime. The instrument is also employed to obtain the electrical resistivity of metal wires by Jain and Saxena [7], hemispherical total emittance of the metal surfaces [7,8], and the tensile breaking stress of tungsten wires [4].

Here, we employ the heat transfer rate data referring to helium and neon gases in the temperature-jump region to compute the apparent conductivity values. Specifically, in this calculation the knowledge of thermal flux as a function of the temperature of the tungsten wire surface is needed for a constant pressure of the gas. These apparent conductivity values are corrected for the temperature-jump effect at the wire surface through a correction factor derived from kinetic theory involving thermal accommodation coefficient. The latter are derived from the experimental data on the pressure dependence of the thermal power dissipated from the wire to the gas for a fixed temperature difference between the hot-wire and the cold-wall. The corrected conductivity values are in good agreement with the experimentally determined values obtained from the heat transfer rate data taken in the continuum region. This is regarded as a good evidence validating the mean-free-path kinetic theory in its present form to explain the phenomenon of temperature-jump at the wire surface and the thermal accommodation coefficient values derived from experimental data. Some comments are presented concerning the surface conditions of tungsten under such conditions as employed in the present work.

THEORY

If Q_H^c is the thermal energy propagated through the gas by pure molecular conduction in the conductivity column instrument per unit length of the wire then the conductivity of the gas, k, is given by the following relation [3]:

$$k(T_H) = \frac{\ln(b/a)}{2\pi} \frac{dQ_H^c}{dT_H} (1 + K) \qquad (1)$$

Here T_H is the temperature of the hot-wire of radius $\underline{a}$ and b is the inner radius of the glass column. K represents the total correction to the conductivity value arising from the small temperature drop across the column glass wall, the change in the wire diameter as it elongates, and the fluctuations along the cold wall temperature. The total magnitude of all the three corrections is usually about one percent. For W-He system [4] its magnitude is less than 0.8 percent, while for W-Ne system it never exceeds over 0.9 percent.

If heat transfer measurements are taken for pressures smaller than those where continuum heat flow conditions exist in the column, the $k(T_H)$ values as obtained from the above relation will be smaller than the true conductivity values. Under such heat flow conditions there exist a temperature discontinuity at the wire surface which results in the pressure dependence of these apparent thermal conductivity values, Present [9], Kennard [10]. The temperature discontinuity at the wire surface is expressed by the following relation:

$$T_H - T_e = -g(\partial T/\partial r)_{r=a} \qquad (2)$$

Here, T_e is the linearly extrapolated gas temperature on the wire surface at a temperature T_H, g the temperature-jump distance, T the temperature and r the radial coordinate. The conductivity values from the heat transfer data $Q_H^{'}$ taken in the temperature-jump region may be computed from the following relation:

$$k(T_H) = \frac{\ln(b/a)}{2\pi} \frac{dQ_H^{'}}{dT_H} (1 + K + k_1) \qquad (3)$$

Here k_1 is the correction factor which accounts for the temperature-jump at the wire surface and for a monatomic gas is given by the following relation [10,11]

$$k_1 = 1.875 \frac{2 - \alpha}{\alpha} \frac{\lambda}{a \ln(b/a)} \qquad (4)$$

α is the thermal accommodation coefficient and is defined as the ratio of the actual mean-energy change of molecules colliding with a wall to the mean-energy change if the molecules came into equilibrium with the wall, Present [9]. λ is the mean-free-path and is computed from the following relation:

$$\lambda = \frac{32}{5\pi} \frac{\mu}{\rho} \left[\frac{\pi M}{8RT}\right]^{1/2} \tag{5}$$

Here μ is the viscosity of the gas of molecular weight M, ρ is the gas density and R the universal gas constant. For a metal-monatomic gas system, α is defined in terms of the incident and reflected gas stream temperatures T_i and T_r respectively, so that

$$\alpha = (T_i - T_r)/(T_i - T_H) \tag{6}$$

Harris [12] developed a mean-free-path kinetic theory for the temperature-jump at the wire and derived under well defined approximations that for a monatomic gas:

$$T_H - T_e = (Q_H/2\pi aP) (\pi MT_e/2R)^{1/2} (2 - \alpha)/2\alpha \tag{7}$$

Here P is the pressure of the gas.

The gas pressure in the temperature-jump region is in the range where Fourier gas thermal conductivity is independent of gas pressure and consequently if Q_H is held constant, the temperature distribution in the bulk of the gas and hence T_e will be independent of gas pressure. Accordingly, a plot of T_H versus P^{-1} for a constant value of Q_H will be linear if the above relation is at least quantitatively adequate to describe the heat transfer process. Jody and Saxena [2], and Jody, Jain and Saxena [4] found that this indeed is the case for W-Ne and W-He systems for the gas pressure ranges of 1.0 to 5.0 and 1.6 to 8.9 cm of mercury respectively.

The governing heat flow equation in the temperature-jump region is

$$\frac{Q_H}{2\pi} \int_b^a \frac{dr}{r} = - \int_{T_w}^{T_e} k(T)dt \tag{8}$$

For helium and neon gases $k(T)$ is accurately represented by a quadratic relation in temperature [2,4] so that

$$k(T) = A + BT + CT^2 \tag{9}$$

Integration of equation (8) in conjunction with relations of equations (7) and (9) leads to the following result:

$$\frac{2\pi a}{Q_H} \left[1 + \frac{C(T_H - T_w)^2}{12k(\bar{T})}\right] = \frac{a \ln(b/a)}{k(\bar{T}) (T_H - T_w)}$$

$$+ \quad \frac{2 - \alpha}{2\alpha} \quad \frac{k(T_H)}{k(\bar{T})} \left(\frac{\pi M T_e}{2R} \right)^{1/2} \quad \frac{1}{(T_H - T_w)} \quad \frac{1}{P} \tag{10}$$

where

$$\bar{T} = 1/2(T_H + T_w) \tag{11}$$

Equation (10) predicts that a plot of Q_H^{-1} versus P^{-1} for a constant value of (T_H-T_w) will be a straight line whose intercept and slope may be interpreted to determine $k(\bar{T})$ and α respectively. Such a plot hereafter will be referred to as a "reciprocal plot". Jody and Saxena [2] and Jody et al. [4] verified the relation of equation (10) for W-Ne and W-He systems and reported $k(T)$ and $\alpha(T)$ values. This method of α determination is referred to as the constant temperature difference method and these data are employed in the calculations described below.

RESULTS

Computed values of $k(T_H)$ from equation (3) and experimental $Q_H'(T_H)$ data taken in the temperature-jump region are reported in Table 1 for helium and neon. In each case the conductivity values are calculated at three pressure levels in the range of validity of the reciprocal plots. Also indicated in this table within parenthesis are the deviations of these computed values from k values obtained from equation (1) and Q_H^c data taken at higher pressures where continuum conditions exist and the temperature-jump correction is negligibly small. In these calculations the temperature-jump theory is subjected to a sensitive test because the magnitude of the correction factor, k_1, ranges from 2.1 to 5.7 for helium and 0.15 to 2.9 for neon.

In Tables 2 and 3 are reported the various sets of calculated conductivity values obtained according to different procedures from heat transfer data taken in the continuum and temperature-jump regimes. In column 2 of these tables are the values obtained from Q_H^c data at high pressures [4,2] according to equation (1) at arbitrarily chosen temperatures of column 1. The temperature-jump data in conjunction with equation (10) leads to conductivity values which are listed in the third column of these tables. The reciprocal plots also enable to determine $Q_H^c(T_H)$ values corresponding to $P = \infty$ and these in conjunction with equation (1) lead to the conductivity values given in column 4 of these tables. Finally, in column 5 are listed a representative set of values from Table 1 which are based on equation (3) and heat transfer data at a pressure when there is appreciable temperature-jump effect present to influence heat transfer rates. In each case we find that all different sets of conductivity values agree within a maximum deviation of about five percent.

Table I: Thermal Conductivity, in mW/cm-K, of helium and neon from data

taken in the temperature-jump region

Temp, K	He			Ne		
	P=4.30 cm Hg	P=6.42 cm Hg	P=8.92 cm Hg	P=1.0 cm Hg	P=2.0 cm Hg	P=5.0 cm Hg
500	2.05(5.1)	2.16(0.0)	2.09(3.2)	0.666(4.6)	0.665(4.7)	0.665(4.7)
700	2.71(2.2)	2.79(0.7)	2.78(0.4)	.852(2.4)	.835(4.4)	.835(4.4)
900	3.38(0.9)	3.38(0.9)	3.30(1.5)	1.02 (1.0)	.996(3.3)	.996(3.3)
1100	4.01(3.1)	3.89(0.0)	3.89(0.0)	1.16 (0.9)	1.15 (1.7)	1.15 (1.7)
1300	4.58(4.1)	4.48(1.8)	4.48(1.8)	1.29 (0.8)	1.29 (0.8)	1.29 (0.8)
1500	5.10(4.5)	4.88(0.0)	4.93(1.0)	1.41 (0.7)	1.42 (0.0)	1.42 (0.0)
1700	5.53(4.3)	5.31(0.2)	5.32(0.4)	1.52 (1.3)	1.54 (0.0)	1.54 (0.0)
1900	5.88(3.2)	5.65(0.9)	5.70(0.0)	1.61 (1.8)	1.66 (1.2)	1.66 (1.2)
2100	6.15(2.3)	5.93(1.3)	5.97(0.7)	1.71 (2.3)	1.77 (1.1)	1.77 (1.1)
2300	6.37(0.6)	6.17(2.5)	6.21(1.9)	1.81 (2.2)	1.88 (1.6)	1.88 (1.6)
mean deviation	3.0	0.8	1.1	1.8	1.9	1.9

Table 2: Thermal conductivity of helium, in mW/cm-K,

by different methods

Temp, K	Continuum, Eq. (1)	Temperature-jump, reciprocal plot and Eq. (10)	Intercept of and Eq. (1)	Eq. (3) and P = 6.4 cm Hg
500	2.16	2.17	2.17	2.16
700	2.77	2.77	2.83	2.79
900	3.35	3.33	3.41	3.38
1100	3.89	3.86	3.96	3.89
1300	4.40	4.40	4.46	4.48
1500	4.88	--	4.93	4.88
1900	5.70	--	5.86	5.65
2300	6.33	--	6.44	6.17

Table 3. Thermal conductivity of neon, in mW/cm-K,

by different methods

Temp., K	Continuum, Eq. (1)	Temperature-jump, of reciprocal plot and Eq. (10)	Intercept and Eq. (1)	Eq. (3) and P=2.0 cm Hg
	x			
500	0.698	0.692	0.702	0.665
700	0.873	0.871	0.878	0.835
900	1.03	1.03	1.03	0.996
1100	1.17	1.18	1.16	1.15
1300	1.30	1.32	1.29	1.29
1500	1.42	1.44	1.42	1.42
1700	1.54	--	1.51	1.54
1900	1.64	--	1.61	1.66
2100	1.75	--	1.70	1.77
2300	1.85	--	1.79	1.88

These calculations thus support the conjectured heat transfer model for the temperature-jump region [12,13]. Also to be emphasized is the result that α values determined from the temperature-jump heat transfer data for gas covered tungsten surface are considerably larger, about an order of magnitude, from those determined from the low pressure region heat transfer data on a bare surface. However, it is clear that these high values are the ones which are capable of representing the heat transfer rates for the metal-gas interface created at such pressures and operating conditions. The tungsten wire in the conductivity column is annealed in air at a pressure of 4×10^{-5} mm of mercury and at a temperature of 2450K. The air components interact with the tungsten surface as it is heated to 2450K and subsequently cooled to room temperature for heat transfer measurements. The available literature on the adsorption and desorption kinetics of oxygen [14-23] and nitrogen [24-26] suggests that the tungsten surface is saturated with these gases at room temperature and the extent of coverage is dependent on temperature. Efforts are being made to quantify these ideas. The presence of noble gases is not likely to interfere with the behavior of the adsobed layer as the adsorption of noble gas atoms on metal surfaces is limited to the very rough regions around the (100) pole at temperatures below 300K [27,14].

ACKNOWLEDGEMENTS

Computing services used in this research were provided by the Computer Center of the University of Illinois at Chicago Circle. Their assistance is gratefully acknowledged.

REFERENCES

[1] Chen, S.H.P. and Saxena, S.C., Mol. Phys. 29, 455-466, 1975.

[2] Jody, B.J. and Saxena, S.C., Phys. Fluids, Vol. 18, 20-27, 1975.

[3] Chen, S.H.P. and Saxena, S.C., High Temp. Sci., Vol. 5, 206-233, 1973.

[4] Jody, B.J., Jain, P.C. and Saxena, S.C., to be published.

[5] Chen, S.H.P. and Saxena, S.C., Int. J. Heat Mass Transfer 17, 185-196, 1974.

[6] Jody, B.J. and Saxena, S.C., 5th International Heat Transfer Conference, Cu4.6, 264-268, 1974.

[7] Jain, P.C. and Saxena, S.C., J. Phys. E: Scientific Instruments, Vol. 7, 1023-1026, 1974.

[8]   Chen, S.H.P. and Saxena, S.C., Ind. Eng. Chem. Fundam. 12, 220-224, 1973.

[9]   Present, R.D., Kinetic Theory of Gases, pp. 188-193, McGraw-Hill Book Co., Inc., 1958.

[10] Kennard, E.H., ibid., pp. 311-327, McGraw-Hill Book Co., New York, 1938.

[11] Faubert, F.M. and Springer, G.S., J. Chem. Phys. 58, 4080-83, 1973.

[12] Harris, R.E., J. Chem. Phys., Vol. 46, 3217-3220, 1967.

[13] Wachman, H.Y., J. Chem. Phys., 42, 1850-1851, 1965.

[14] Geus, J.W., in Physical and Chemical Aspects of Adsorbents and Catalysts, Edited by B.G. Linsen, 650pp, Academic Press, New York, 1970.

[15] Ptushinskii, Yu.G. and Chuikov, B.A., Surf. Sci. 6, 42-56, 1967.

[16] Becker, J. and Hartman, C., J. Phys. Chem. 57, 153-159, 1953.

[17] Roberts, J.K., Proc. Roy. Soc. A152, 464-477, 1935.

[18] McCarroll, B., J. Chem. Phys. 46, 863-869, 1967.

[19] Mazumdar, A.K. and Wassmuth, H.W., Surf. Sci. 30, 617-631, 1972.

[20] Schlier, R., J. Appl. Phys. 29, 1162-1167, 1958.

[21] Ehrlich, G., J. Chem. Phys. 34, 39-46, 1961.

[22] Singleton, J., J. Chem. Phys. 47, 73-82, 1967.

[23] Morrison, J.L. and Roberts, J.K., Proc. Roy. Soc. A173, 1-12, 1939.

[24] Ehrlich, G., J. Phys. Chem. 60, 1388-1400, 1956.

[25] Ehrlich, G., J. Chem. Phys. 34, 29-38, 1961.

[26] Hickmott, T. and Ehrlich, G., J. Phys. Chem. Solids 5, 47-77, 1958.

[27] Ehrlich, G. and Hudda, F., J. Chem. Phys. 30, 493-512, 1959.

ON ESTIMATING THERMAL CONDUCTIVITY COEFFICIENTS

IN THE CRITICAL REGION OF GASES

H. J. M. Hanley*

J. V. Sengers**

J. F. Ely†

The thermal conductivity of a fluid exhibits a pronounced anomalous increase in a large range of densities and temperatures around the gas-liquid critical point. In this paper we discuss an attempt to estimate the thermal conductivity in the critical region of fluids from a knowledge of the equilibrium properties and the regular behavior of the transport properties outside the critical region.

* Cryogenics Division, National Bureau of Standards,
  Boulder, Colorado 80302

** Institute for Molecular Physics, University of Maryland,
   College Park, Maryland 20742 and
   Laboratorium voor Technische Natuurkunde, Technische Hogeschool,
   Delft, Nederland

† Department of Chemical Engineering, Rice University,
  Houston, Texas 77001

I. INTRODUCTION

The critical point is a center of anomalous behavior of many
thermophysical properties: the compressibility (K_T), specific
heat (c_p), and the thermal conductivity (λ), for example, all diverge as the critical point is approached. In this work, we will
discuss briefly the behavior of the latter property -- the thermal
conductivity coefficient -- and propose an expression to estimate
its value for a particular fluid.

Table I gives a summary of the experimental data supporting
the observation that the thermal conductivity does display an
anomalous enhancement in the critical region.[1],[2] Evidence is available for helium, argon, xenon, hydrogen, carbon dioxide, ammonia,
methane, nitrogen, sulfur hexafluoride, and steam.

A discussion on experimental techniques is given in reference
2 and will not be repeated here. However one should note that in
most conventional methods a stationary temperature difference is
established across a layer of the fluid confined either between
two concentric cylinders or between two parallel plates.[2] Recently,
attempts have been made to investigate the effect with a transient
hot wire method.[3] The most serious problem encountered in all these
experiments is the high probability of convection when the critical
point is approached. As a result, experimental thermal conductivity
data in the vicinity of the critical point must be evaluated with
great care.

However, the thermal diffusivity, $\lambda/\rho c_p$, where ρ is the density
and c_p the specific heat at constant pressure per unit mass, can be
determined independently by measuring the width of the Rayleigh
line in the spectrum of scattered light.[4] This method is restricted
to a range around the critical point where the critical opalescence
is sufficiently large, but it has the advantage that the system
under study remains in thermodynamic equilibrium so that convection is avoided. Since c_p diverges strongly, the thermal diffusivity in fact vanishes at the critical point, but the light scattering
experiments have clearly established that $\lambda/\rho c_p$ approaches zero
slower than $1/c_p$ (i.e., with a smaller power of the temperature
difference with the critical temperature). The light scattering
experiments have thus convincingly confirmed that the thermal
conductivity diverges as the critical point is approached.

The most detailed and reliable experimental information is
that for carbon dioxide where the results of three different investigations, Michels and Sengers, Le Neindre et al. and Swinney
and Cummins are in mutual satisfactory agreement.

The general behavior of the thermal conductivity coefficient
is summarized by the schematic figure 1 in which the excess thermal
conductivity, $\Delta\lambda$, is plotted versus density. One usually defines
the excess in terms of density, ρ, and temperature, T, by

$$\Delta\lambda(\rho,T) = \lambda(\rho,T) - \lambda_0(T) \tag{1}$$

where $\lambda_0(T) = \lambda(o,T)$ is the thermal conductivity in the low density limit at the same temperature. Outside the critical region $\Delta\lambda$ is found to be a very weak function of temperature (helium and hydrogen excepted). However, inside the critical region, $\Delta\lambda$ becomes strongly dependent on temperature as indicated by the isotherms at T_1 and T_2. In order to account for this effect, we redefine $\Delta\lambda$ by

$$\Delta\lambda(\rho,T) = \lambda_{id}(\rho,T) + \Delta\lambda_c(\rho,T) - \lambda_0(T) \tag{2}$$

where $\lambda_{id}(\rho,T)$ is an ideal or background thermal conductivity (extrapolated from data outside the critical region and weakly dependent on temperature) and where $\Delta\lambda_c(\rho,T)$ is a *critical excess* thermal conductivity.

The critical excess thermal conductivity is treated most conveniently as a function of reduced temperature and reduced density relative to the temperature T_c and density ρ_c of the critical point

$$\Delta\tilde{T} = (T-T_c)/T_c \quad , \quad \Delta\tilde{\rho} = (\rho-\rho_c)/\rho_c \tag{3}$$

At the critical density the anamolous behavior extends over a temperature range $|\Delta\tilde{T}| \approx 1/3$ and at the critical temperature it extends over a density range $|\Delta\tilde{\rho}| \approx 2/3$.

II. AN EQUATION FOR THE CRITICAL EXCESS THERMAL CONDUCTIVITY

The compressibility, K_T, of a fluid diverges as the critical point is approached and a large compressibility means that only a small amount of energy is required to induce a change in density. Hence the critical region is accompanied by large spontaneous fluctuations in the density. The spatial extent of these fluctuations is determined by the range, ξ, of the pair correlation function, $G(r)$, which measures the probability of finding a pair of molecules separated by a distance r. In the critical region this correlation length becomes large and it becomes infinite at the critical point. Loosely speaking we may imagine that the critical region is associated with conglomerates of molecules, referred to as clusters, whose mean radius is given by the correlation length ξ. Since the correlation length becomes much larger than the range of the intermolecular forces in the critical region it is expected that the detailed shape of the molecules will not affect the mathematical character of the critical anomalies.[5] This leads to the hypothesis of universality which states that the nature of the critical anomalies, when reduced appropriately, will be the same for all gases.[6]

TABLE I

Survey of experimental information

Investigators*	Method	Substance
Guildner[a]	vertical cylinder	carbon dioxide
Michels, Sengers[b]	parallel plate	carbon dioxide
Ziebland, Burton[c,d]	vertical cylinder	nitrogen
Ikenberry, Rice[d,e]	vertical cylinder	argon, methane
Needham, Ziebland[f]	vertical cylinder	ammonia
Golubev, Sokolova[g]	vertical cylinder	ammonia, methane
Lis, Kellard[h]	vertical cylinder	sulfurhexafluoride
Bailey, Kellner[i]	vertical cylinder horizontal cylinder	argon
Kerrisk, Keller[j]	parallel plate	helium-3
Roder, Diller[k]	parallel plate	hydrogen
Murthy, Simon[l]	parallel plate	carbon dioxide
Le Neindre et al.[m]	vertical cylinder	carbon dioxide, steam
Mani, Venart[n]	transient hot wire	methane
Sirota et al.[o,p]	parallel plate	steam
Van Oosten et al.[q]	parallel plate	xenon
Swinney et al.[r,s]	light scattering	carbon dioxide, xenon, SF_6
Maccabee, White[t]	light scattering	carbon dioxide
Benedek et al.[s,u]	light scattering	xenon, SF_6
Ohbayashi, Ikushima[v]	light scattering	helium-3

* References in Appendix I.

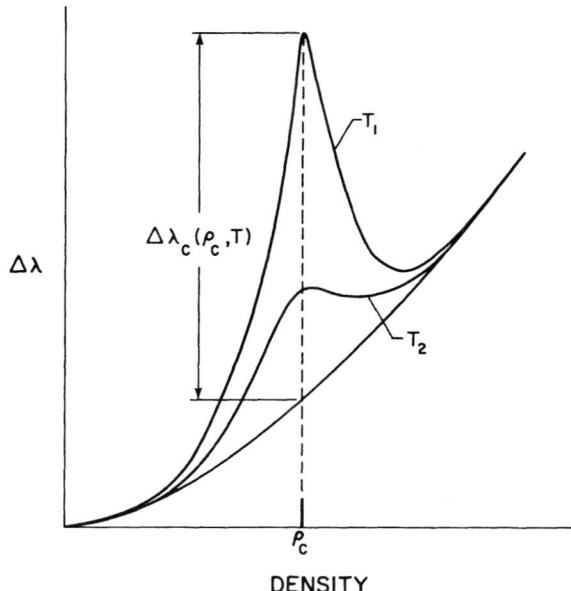

Figure 1. Schematic representation of the excess thermal conductivity in the critical region as a function of
density at several representative temperatures.
$T_c \lesssim T_1 \lesssim T_2$.

In order to discuss the anomalous behavior of the transport
properties we need to examine the dynamics of the critical fluctuations. For this purpose we decompose the density fluctuations
into pressure and entropy fluctuations. The former correspond to
the thermally excited sound waves which propagate with the sound
velocity and are not considered here. The latter, however, are
of particular importance for our purpose: we know from hydro-
dynamics[4] that the entropy fluctuations decay according to a
diffusion equation whose diffusion coefficient is equal to the
thermal diffusivity $\lambda/\rho c_p$. Hence, the anomalous part of the thermal diffusivity, i.e., $\Delta\lambda_c/\rho c_p$, is identified with the diffusion
coefficient of the critical fluctuations in the entropy.

The thermal diffusivity goes to zero at the critical point.
Thus in the critical region the entropy fluctuations decay slowly [usually referred to as the phenomenon of critical slowing
down of the fluctuations]. Based on rather general theoretical
considerations it can be argued[7] that the thermal diffusivity
will go to zero as the inverse correlation length ξ^{-1},

$$\frac{\Delta\lambda_c}{\rho c_p} \propto \frac{1}{\xi} \tag{4}$$

and the fact that $\Delta\lambda_c/\rho c_p$ indeed vanishes asymptotically as ξ^{-1} is now well established experimentally.[8] This result can be understood on an intuitive basis by invoking Einstein's relation

$$\frac{\Delta\lambda_c}{\rho c_p} = \frac{kT}{\zeta} \tag{5}$$

which relates the thermal diffusivity, as any diffusion coefficient, to a corresponding friction coefficient ζ. The asymptotic behavior as given by (4) follows then by noting that the friction coefficient ζ of the clusters will be proportional to their size ξ.

In order to calculate the thermal diffusivity, and hence the thermal conductivity, we need an estimate for the proportionality constant in (4). Such an estimate can be made by identifying ζ with the friction coefficient $6\pi\eta\xi$ of clusters with radius ξ in a medium of viscosity η. This argument was first put forward by Mistura et al. in discussing the critical slowing down of the concentration fluctuations near the critical mixing point of a binary mixture[9] and it leads to

$$\Delta\lambda_c = \frac{kT}{6\pi\eta\xi}\ \rho c_p. \tag{6}$$

Equation (6) can only be valid approximately. It was derived by Kadanoff and Swift[7] and by Kawasaki[10] who obtained (6) as the leading contribution to the first term in a perturbation procedure usually referred to as mode coupling theory. But it is known that the higher order terms can also lead to contributions that vary as ξ^{-1}. In the derivation of (6) it is also assumed that the shear viscosity η does not exhibit any anomalous behavior near the critical point. This assumption is not strictly justified,[1] but the anomalous behavior of η turns our to be sufficiently small so that it can be neglected for most practical purposes.

Sengers[1] has pointed out that the range of applicability of (6) can be extended if c_p in the equation is replaced by the difference $c_p - c_v$, so that

$$\Delta\lambda_c = \frac{kT}{6\pi\eta\xi}\ \rho(c_p - c_v). \tag{7}$$

Since c_v is only weakly divergent, this equation will lead to the same asymptotic behavior as (5) when the critical point is approached. Since

$$c_p - c_v = \frac{T}{\rho}\left(\frac{\partial P}{\partial T}\right)_\rho^2 K_T \tag{8}$$

we thus obtain

$$\Delta\lambda_c = \frac{kT^2}{6\pi\eta\xi}\left(\frac{\partial P}{\partial T}\right)_\rho^2 K_T \tag{9}$$

This modified formula was shown[1] to yield a reasonable representation of $\Delta\lambda_c$ in a range of densities and temperatures corresponding to $|\Delta\tilde{\rho}| \leq 0.25$ and $|\Delta\tilde{T}| \leq 0.03$. However, as mentioned in the introduction, the anomalous behavior of the thermal conductivity extends to densities and temperatures much farther away from the critical point. In an attempt to cover the entire range of the anomaly, we generalize (9) to[11,12]

$$\Delta\lambda_c = \frac{kT^2}{6\pi\eta\xi}\left(\frac{\partial P}{\partial T}\right)_\rho^2 K_T \, F(\Delta\tilde{T},\Delta\tilde{\rho}). \tag{10}$$

Since the empirical expression, F, should approach unity at the critical point and should vanish far away from the critical point we introduced the form

$$F(\Delta\tilde{T},\Delta\tilde{\rho}) = \exp\left[-\left(A(\Delta\tilde{T})^2 + B(\Delta\tilde{\rho})^4\right)\right] \tag{11}$$

where A^{-1} and B^{-1} are measures of the range of the critical anomaly in temperature and density, respectively. Arguing on the basis of the universal nature of the anomaly, we surmise that this range will be approximately independent of the nature of the fluid[2]. Based on an analysis of the data for carbon dioxide and (to a lesser extent) of those for methane and argon, the values A = 18.66; B = 4.25 were proposed[11].

III. COMPRESSIBILITY

The proposed equation (10) relates the critical excess thermal conductivity, $\Delta\lambda_c$, to the shear viscosity η, the correlation length ξ and the compressibility K_T: all properties which, of course, have to be estimated. The viscosity can be determined from existing empirical and semi-empirical correlation techniques[12]. A method for estimating the correlation length will be discussed in section IV. In this section, we comment on the equation of state of fluids near the critical point and, in particular, on the evaluation of K_T.

In our previous work[1,2] we have stressed how central the equation of state is to a correlation and prediction of transport properties. For example we have implied here that the thermal conductivity should be studied in terms of temperature and density, rather than in terms of the more practical variables, temperature and pressure. The equations of state mostly used in our work are modifications of the BWR(Benedict-Webb-Rubin) developed by McCarty and co-workers.[12] However, it is now well established that the thermodynamic behavior of fluids near the critical point has an essential non-analytic character; that is, the Helmholtz free energy and its derivatives in the vicinity of the critical point cannot be expanded in a Taylor series in terms of $\Delta\tilde{T}$ and $\Delta\tilde{\rho}$. Unfortunately, most empirical equations of state in the engineering literature (such as the BWR) are classical in the sense that they are analytic at the critical point. Hence, they cannot accommodate appropriately the anomalous behavior of the various thermodynamic derivatives.[13] In particular, classical equations cannot be used to describe the large compressibilities in the critical region unless one introduces an excessive number of adjustable parameters.

The current thermodynamic description[6,13,14] of a system near the critical point is based on the observation that the asymptotic behavior of various thermodynamic properties can be described in terms of power laws when the critical point is approached along specific paths in the $\Delta\tilde{T}$-$\Delta\tilde{\rho}$ plane. For example, the density along the gas and liquid branch of the coexistence curve varies asymptotically as $|\Delta\tilde{\rho}| \propto |\Delta\tilde{T}|^{\beta}$; the chemical potential $\mu(\rho,T)$ along the critical isotherm varies as $\mu(\rho,T) - \mu(\rho_c,T) \propto |\Delta\tilde{\rho}|^{\delta}$; the compressibility K_T and specific heat c_p at constant pressure vary along the critical isochore as $|\Delta\tilde{T}|^{-\gamma}$; the specific heat c_p at constant volume and the second derivative $(\partial^2 P/\partial T^2)_e$ vary along the critical isochore at $|\Delta\tilde{T}|^{-\alpha}$. The critical exponents $\alpha, \beta, \gamma, \delta$ are interrelated by the equalities[15]

$$\gamma = \beta(\delta-1)$$
$$2-\alpha = \beta(\delta+1) \ . \tag{12}$$

It is found that the compressibility and the specific heat c_p exhibit a strong anomaly since the corresponding critical exponent γ is of the order of 1.2 and thus very different from zero. The specific heat c_v, and the derivative $(\partial P^2/\partial T^2)_\rho$, in contrast, show weak anomalies with a correspondingly small exponent $\alpha \approx 0.1$.

The equation of state in the critical region assumes its simplest form when we consider the chemical potential $\mu(\rho,T)$ which is the intensive thermodynamic property conjugate to the density. In particular, near the critical point, the chemical potential can be represented by a scaled equation of state

$$[\mu(\rho,T) - \mu(\rho_c,T)] \frac{\rho_c}{P_c} = \Delta\tilde{\rho}|\Delta\tilde{\rho}|^{\delta-1}h(x) \tag{13}$$

where the scaling variable x is defined as

$$x = \frac{\Delta \tilde{T}}{|\Delta \tilde{\rho}|^{1/\beta}} \quad . \tag{14}$$

Curves of constant x in the $\Delta \tilde{T}$-$\Delta \tilde{\rho}$ plane are schematically indicated in figure 2. At the coexistence curve x assumes a constant value, $- x$; $x = 0$ represents the critical isotherm; and $x = \infty$ represents the critical isochore.

Since $\rho^2 K_T = (\partial \rho / \partial \mu)_T$, the scaled equation for the compressibility reads

$$\rho^2 K_T = \frac{\rho_c}{P_c} |\Delta \tilde{\rho}|^{1-\delta} [h(x) - \frac{x}{\beta} h'(x)]^{-1} \tag{15}$$

where $h'(x) \equiv dh(x)/dx$.

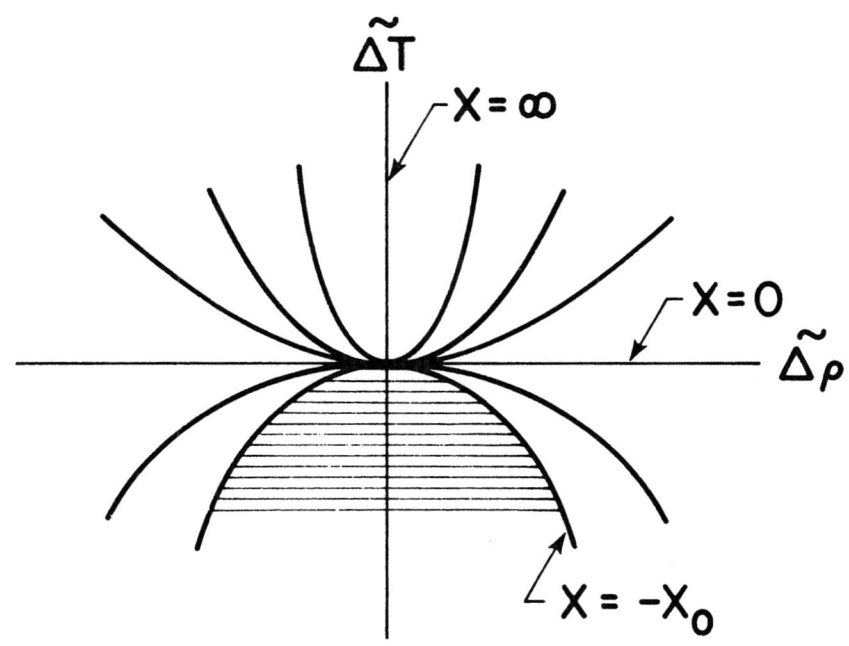

Figure 2.   Curves of the constant scaling variable
x in the $\Delta \tilde{T}$ - $\Delta \tilde{\rho}$ plane.

The function $h(x)$ is constrained by a number of conditions formulated by Griffiths[15] An explicit, though approximate, expression for the function was proposed by Vicentini-Missoni, Levelt Sengers and Green:[16]

$$h(x) = E_1 \left(\frac{x + x_o}{x_o}\right) \left[1 + E_2 \left(\frac{x + x_o}{x_o}\right)^{2\beta}\right]^{(\gamma-1)/2\beta}. \quad (16)$$

This equation contains the critical parameters ρ_c and T_c (through the definition of x), two critical exponents β and γ [$= \beta(\delta-1)$], and three constants x_0, E_1 and E_2.

The hypothesis of universality of critical behavior would imply that the critical exponents and the constant E_2 be independent of the nature of the gas. From an analysis of equation of state data Levelt Sengers et al. recently concluded[17] that

$$\begin{aligned}
\beta &= 0.355 \pm 0.04 \\
\delta &= 4.35 \pm 0.10 \\
\gamma &= \beta(\delta-1) = 1.19 \pm 0.03 \quad (17) \\
E_2 &= 0.287
\end{aligned}$$

With the values thus adopted from the critical exponents and the constant E_2, the compressibility for various fluids in the critical region is then completely determined by two parameters, namely x_0 and E_1, (in addition, of course, to the critical parameters ρ_c, T_c, P_c). Values of these parameters for a number of fluids are presented in Table II taken from reference 17.

The scaled equation (15) yields an accurate representation of the compressibility in a range of temperatures and densities around the critical point bounded by

$$|\Delta\tilde{\rho}| \leq 0.25 \qquad |\Delta\tilde{T}| \leq 0.03 . \qquad (18)$$

Inside this region a classical equation of state leads to serious errors in the compressibility and use of the scaled equation (15) is imperative in most cases; outside this region the compressibility can be calculated with reasonable accuracy from existing equations of state, such as the BWR[12]

The scaled equation (13) can be extended into an equation of state for the pressure as a function of density and temperature as we demonstrated recently for steam[18] Such an equation should in principle be used to calculate $(\partial P/\partial T)_\rho$. However, for the purpose of the present paper, we have simply estimated this derivative from existing classical equations of state. Since $(\partial P/\partial T)_\rho$ remains finite at the critical point [$(\partial^2 P/\partial T^2)_\rho$ being only weakly divergent] the resulting errors remain within reasonable bounds.

TABLE II

Parameters for the scaled equation of state (16) for a number of fluids. See reference 17.

	critical point parameters			critical region parameters	
	P_c MPa	ρ_c kg/m^3	T_c K	x_o	E_1
Ar	4.865	535.	150.725	0.183	2.34
Kr	5.4931	908.	209.286	0.183	2.34
Xe	5.8400	1110.	289.734	0.183	2.34
O_2	5.043	436.2	154.580	0.183	2.21
N_2	3.398	313.9	126.24	0.164	2.17
CH_4	4.595	162.7	190.555	0.164	2.03
CO_2	7.3753	467.8	304.127	0.141	2.06

$\beta = 0.355$ $\delta = 4.352$

$\gamma = 1.190$ $\alpha = 0.100$

$E_2 = 0.287$

IV.   CORRELATION LENGTH

We next discuss the correlation length, ξ, that appears in the basic equation (10). This quantity can be found from experiment for a given fluid since the Fourier components of the pair correlation function $G(r)$ (and hence ξ) can be determined from light or x-ray scattering data recorded as a function of the scattering angle. Unfortunately, such data are restricted to a small number of fluids at this time.[19] Moreover, even in these examples, the data are usually restricted to the critical isochore. It is, therefore, most desirable to have a procedure to calculate the correlation length for a range of densities and temperatures given other properties of the system. In turn, of course, the available data can then be used to test any estimation procedure. We note, too, that a procedure would have significance beyond the immediate objective of obtaining a value of $\Delta\lambda_c$ from (10).

The zeroth moment of the pair correlation function is related to the isothermal compressibility

$$kT\left(\frac{\partial n}{\partial P}\right)_T = 1 + n\int d\vec{r}\ G(r) \tag{19}$$

where the number density n is related to the mass density ρ by

$$n = \frac{N}{M}\rho. \tag{20}$$

(N is Avogadro's number and M the molecular weight.) Since the compressibility becomes very large in the critical region the constant 1 in (19) may be neglected so that

$$n^2 K_T = \frac{n^2}{kT}\int d\vec{r}\ G(r). \tag{21}$$

This relation indicates that K_T increases in the critical region because the correlation $G(r)$ becomes a long-range function of the intermolecular separation, r. The range of the correlation function is characterized by a correlation length ξ such that

$$\xi^2 = \frac{1}{6}\frac{\int d\vec{r}\ r^2\ G(r)}{\int d\vec{r}\ G(r)}. \tag{22}$$

From the scaled equation (15) for the compressibility in the critical region, we note that $\rho^2 K_T$, and hence $n^2 K_T$, is symmetric around the critical isochore $\rho = \rho_c$; that is, $n^2 K_T$ is an even function of $\Delta\tilde{\rho}$. From (21) and (22) we thus expect ξ also to be an even function of $\Delta\tilde{\rho}$.

In the theory of Ornstein and Zernike, the long range character of the correlation function in the critical region can be represented by[20]

$$G(r) = \frac{1}{4\pi n R^2 r}\ \exp\left[-r/\xi\right] \tag{23}$$

where R is a length parameter usually referred to as the short range
correlation length. It thus follows from (22) and (23) that the
correlation length ξ is related to the compressibility by the relation

$$\xi = R\sqrt{n\ kT\ K_T} \quad .\tag{24}$$

As mentioned above, both the correlation length ξ and the quantity
$n^2 K_T$ are functions of the density that are symmetric with respect
to the critical density. From (24) we thus expect that R will vary
in the critical region proportional to the square root of the density. This density dependence can be made more explicit by writing

$$R = R_o (n/kT)^{1/2}\tag{25}$$

and

$$\xi = R_o \sqrt{n^2 K_T}\tag{26}$$

where R_O will be treated as a density independent constant.

We know that the Ornstein-Zernike form in the correlation
function is not rigorously correct.[20] However, deviations are small
and only show up very close to the critical temperature. For the
purpose of the present paper we shall assume the Ornstein-Zernike
form[23] for all temperatures and densities. The errors thus introduced are much smaller than the accuracy with which we can estimate
the thermal conductivity.

The relations (24) or (26) enable us to calculate the correlation length from the compressibility provided that we have an
estimate for the length parameter R, or equivalently, R_O. Accordingly, we use that, in the Ornstein-Zernike theory, the short range
correlation length is related to the direct correlation function
$C(r)$ by[20]

$$R^2 = \frac{n}{6} \int d\vec{r}\ r^2\ C(r).\tag{27}$$

Allowing for nonsphericity of the molecules this equation is
generalized to[21]

$$R^2 = \frac{n}{48\pi} \int\int d\vec{r}\ d\Omega r^2 C(r,\Omega)\tag{28}$$

where Ω denotes the relative angular orientation between molecular
pairs[22] (see Appendix II). Away from the critical point the range of
the direct correlation function is related to the range of the
intermolecular forces and in the Ornstein-Zernike theory it is assumed
that the behavior of $C(r)$ [or $C(r,\Omega)$] remains the same in the critical region. Although in reality R diverges at the critical
point[20] this divergence is very weak and just barely observable.
In the spirit of the Ornstein-Zernike theory we shall assume that
R remains related to the range of the intermolecular forces.

If deviations from the Ornstein-Zernike theory are neglected, therefore, the direct correlation function is expected to vary in the critical region for large distances in the same manner as it does far away from the critical point, namely as $-\phi(r)/kT$, where $\phi(r)$ is the intermolecular potential function. For nonspherical molecules the intermolecular potential function will also depend upon the orientation Ω so that we expect more generally that

$$C(r,\Omega) = -\phi(r,\Omega)/kT \tag{29}$$

for large values of r. We note that this assumption is consistent with the mean spherical approximation for hard spheres with embedded permanent dipole moments.[21]

Ely and Hanley have shown that the properties of simple polyatomic gases at low densities can be well described in terms of an intermolecular potential of the form[23]

$$\phi(r,\Omega) = \phi_s(r) + \phi_{ns}(r,\Omega) \tag{30}$$

where $\phi_s(r)$ is a spherical core characterized by the usual parameters, ε and r_m where $\phi_s(r_m) = -\varepsilon$ [or by ε and σ where $\phi_s(\sigma) = 0$], while $\phi_{ns}(r,\Omega)$ is the nonspherically symmetric contribution to the intermolecular interaction, also characterized by ε and r_m, and by the relative molecular orientation angles.

The short range correlation length can in principle, therefore, be determined from (28) but one needs values of $C(r)$ as a function of distance, r. Since one only has at this time the expression (29) for large r, we introduced in a previous paper[11], as an ansatz, the expression for spherically symmetric molecules

$$R^2 = - \frac{n}{6kT} \int_{r_m}^{\infty} d\vec{r}\, r^2\, \phi_{att}(r) \tag{31}$$

where $\phi_{att}(r)$ is the attractive part of the potential and where the lower limit of the integration, which is in principle an adjustable parameter, was identified with the distance r_m, i.e., the transition point between the repulsive and attractive forces. In this paper we want to generalize the expression to include polyatomic, i.e., nonspherical, molecules. A straightforward first step is to use the attractive part of the general potential (30), but it is not obvious whether r_m should be used as the lower limit of integration. This limit is arbitrary for spherical molecules and even less evident for nonspherical molecules since any lower cutoff will be dependent on the relative orientation of the molecule. We, therefore, introduce, on an *ad hoc* basis, a Boltzmann factor to weight the possible orientations. Our general expression for R becomes, therefore,

$$R^2 = - \frac{n}{6kT} \frac{\int_{r_m}^{\infty} d\vec{r} \; r^2 \int \Omega \phi_{att}(r,\Omega) e^{-\Delta/kT}}{\int d\Omega e^{-\Delta/kT}} \qquad (32)$$

where

$$\Delta = - \int_{r_m}^{\infty} dr \; \phi_{att}(r,\Omega). \qquad (33)$$

We note that (32) implies that R will vary as $(n/kT)^{1/2}$ as suggested in (25). For spherical molecules it reduces, of course, to (31).

For convenience we introduce dimensionless quantities

$$T^* = kT/\epsilon \quad , \quad n^* = nr_m^3$$

$$r^* = r/r_m \quad , \quad \phi^*_{att} = \phi_{att}/\epsilon \quad , \quad \Delta^* = \Delta r_m/\epsilon \qquad (34)$$

so that (32) becomes

$$R^2 = - (\tfrac{2\pi}{3}) \frac{n^*}{T^*} \; r_m^2 \int_1^{\infty} dr^* \; r^{*4} \int d\Omega \phi^*_{att}(r^*,\Omega) e^{-\Delta^*/T^*} / \int d\Omega e^{-\Delta^*/T^*} \quad (35)$$

V. COMPARISON WITH EXPERIMENTAL CORRELATION LENGTH DATA

Hanley and coworkers have shown that second virial coefficients and the dilute gas transport properties of a variety of quadrupolar fluids can be adequately represented by a potential function $\phi(r,\Omega)$ whose attractive part is given by[23,24],

$$\phi^*_{att}(r^*,\Omega) = - (\frac{C}{r^{*6}} + \frac{\gamma}{r^{*8}}) + \frac{3}{4} \theta^{*2} (\frac{d}{r^*})^5 F(\Omega)$$

$$- \frac{9\bar{\alpha}^* \theta^{*2}}{8} (\frac{d}{r^*})^8 [(1-\kappa)G(\Omega) + 3\kappa H(\Omega)] \qquad (36)$$

where

$$C = \{m - \gamma(m-8)\}/(m-6), \quad d = \sigma/r_m \qquad (37)$$

and where $F(\Omega)$, $G(\Omega)$ and $H(\Omega)$ are functions of the orientation angle Ω presented in Appendix II. The meaning of the various terms is as follows. The first term represents the attractive part of the m-6-8 potential function introduced by Klein and Hanley[24] for $\phi_s(r)$. This potential function contains two parameters, m and γ, in addition to ϵ and r_m (or σ). The second term represents the effect of quadrupole-quadrupole interactions as a function of the reduced quadrupole moment; $\theta *^2 = \theta^2/\epsilon\sigma^5$. The third term represents a contribution from quadrupole-induced dipole interactions including anisotropy effects (measured by κ) in the molecular polarizability; $\bar{\alpha}*$ is the reduced mean polarizability, α/σ^3. [The quantity $d = \sigma/r$ appears in the nonspherical part of the potential function since θ and α are conventionally reduced in terms of σ but it is often more convenient to work in terms of r_m.] Values of the potential parameters for a number of gases, are presented in Table III†.

Substitution of the potential function (36) into the proposed expression (32) for R yields

$$R^2 = \frac{2\pi}{3} \ \frac{n*}{T*} \ r_m^2 \ [\{C + \frac{\gamma}{3} + V\}\{1 - \frac{C}{5T*} - \frac{\gamma}{7T*}\} - \frac{3}{7} \frac{V}{T*} \ (C + \frac{\gamma}{3})]/$$

$$\{1 - \frac{C}{5T*} - \frac{\gamma}{7T*} - \frac{3}{7} \ \frac{V}{T*}\} \tag{38}$$

where

$$V = \bar{\alpha}* \ \theta *^2 d^8 .$$

For spherical molecules this expression reduces to the one used in the earlier publication[11]

$$R^2 = \frac{2\pi}{3} \ \frac{n*}{T*} \ r_m^2 \ (C + \gamma/3). \tag{39}$$

In fact, the corrections due to nonsphericity are small and (39) is a good approximation for simple polyatomic molecules.

Because of the ad hoc nature of the equation (32) and, hence, (38), its use can only be justified *a posteriori* if it turns out to reproduce the experimental data for a variety of fluids. In order to conduct such a test we need not only experimental correlation length data, preferably at a number of different densities, but also reliable values for the potential function parameters.

Schmidt and coworkers have reported experimental data for the short range correlation length of argon and nitrogen as a function of density[25,26]. Chu and Lin have measured the correlation length of carbon dioxide at three different densities as a function of temperature from which we can deduce values for R.

---

† γ and α of (36) should not be confused with the critical exponents introduced in Section III.

TABLE III

Parameters for the m-6-8 potential. See references 23 and 24.

Gas	m	γ	$10^{10}\sigma$ m	$10^{10}r_m^\dagger$ m	ε/k K	$10^{24}\alpha$ cm^3	$10^{26}\theta$ esu	κ
Ar	11	3	3.292	3.669	153.0	-	-	-
Kr	11	3	3.509	3.911	216,0	-	-	-
Xe	11	3	3.841	4.281	295.0	-	-	-
O_2	10	1.0	3.45	3.904	109.5	1.568	0.40	0.234
N_2	12	2	3.54	3.933	118.0	1.737	1.4	0.134
* CH_4	11	3	3.680	4.101	168.0	-	-	-
CO_2	14	1.0	3.68	4.101	282.0	2.925	4.3	0.239

† Values of the ratio r_m/σ are given in NSRDS-NBS Monograph No. 47 (1974) Max Klein, H. J. M. Hanley, et al.

* For the purpose here, we treat methane as a spherical molecule.

    In Fig. 3 we present a comparison between these experimental
data and the values calculated from (38). The agreement is satisfactory, particularly if one realizes that the calculated values ori-
ginate from model potential parameters obtained independently from
such dilute gas properties as the viscosity coefficient, the pressure second virial coefficient and the dielectric second virial
coefficient as described elsewhere.[23,24] We also determined that the
values of R from (32) are sensitive to the model potential function
selected. Hence, one can surmise that agreement is only possible
if a valid model potential is used in conjunction with (32).

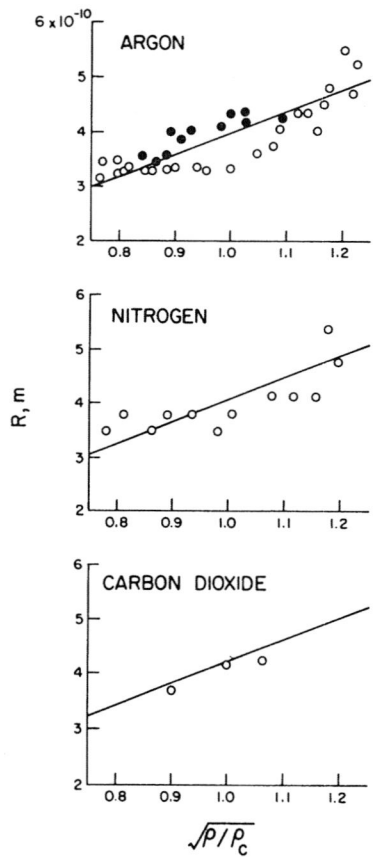

Figure 3. Calculated values of R (solid line) as a function of
         $\sqrt{\rho/\rho_c}$ compared with experimental data for argon
         open circles,[25] closed circles[26]), nitrogen[25] and
         carbon dioxide.[27]

We also note from Fig. 3 that the short range correlation length is not independent of the density as sometimes has been assumed in the literature but that it indeed varies approximately as $\sqrt{\rho}$.

VI. COMPARISON WITH EXPERIMENTAL THERMAL CONDUCTIVITY DATA

An explicit expression for the excess thermal conductivity is obtained if the correlation length (24) is substituted in (10)

$$\Delta\lambda_c(\rho,T) = (\frac{M}{\rho NkT})^{1/2} \frac{kT^2}{6\pi\eta R} (\frac{\partial P}{\partial T})_\rho^2 K_T^{1/2} \exp[-\{A\Delta\tilde{T}^2 + B(\Delta\tilde{\rho})^4\}]. \quad (40)$$

The total thermal conductivity is then obtained by combining this equation with (1) and (2). We have thus calculated the thermal conductivity of carbon dioxide, argon, nitrogen, methane and oxygen.[11,12] Details on the equation of state and on the calculation of the viscosity η and the background or ideal thermal conductivity λ_{id} are given elsewhere.[1,11,12]

In Fig. 4 we compare the experimental data[28] for CO_2 with the values thus calculated. The equation (40) appears to account for a major part of the anomalous thermal conductivity. The remaining difference of approximately 15% in $\Delta\lambda_c$ may be because the higher order terms in the mode coupling theory could modify the proportionality constant $1/6\pi$ in (6) and, hence, in (7).

Reliable data for the thermal conductivity of argon and nitrogen in the critical region are scarce and we present a comparison for these fluids in tabular form in Table IV.

In Fig. 5 we present the results for methane. We do not have thermal conductivity data for methane in the close vicinity of the critical point but the figure does illustrate the extensive character of the critical anomaly in the thermal conductivity.

In conclusion, the results indicate that the proposed equation (40) provides us with a practical procedure to obtain a realistic (to within 15% or better) estimate of the thermal conductivity of a gas in the critical region.

VII. ACKNOWLEDGMENTS

The authors are indebted to Professors P. W. Schmidt and W. A. Steele for some valuable comments. The work was supported by the Office of Standard Reference Data (U.S.) and the National Aeronautics and Space Administration, Grant NGR-21-002-344. J. V. Sengers was also supported in part by the Ir. Cornelis Gelderman fund of the Technological University at Delft. We also thank Mildred F. Birchfield for her considerable help in the preparation of this paper.

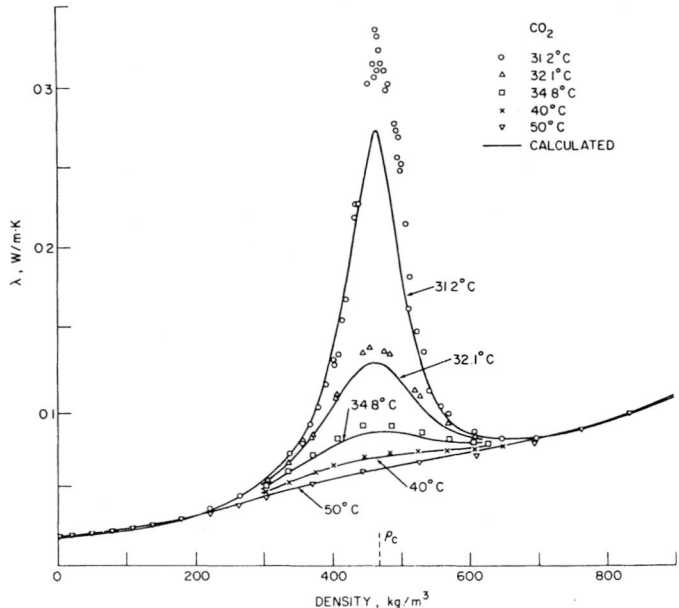

Figure 4. Calculated thermal conductivity for carbon dioxide
compared with the experimental data.[28]

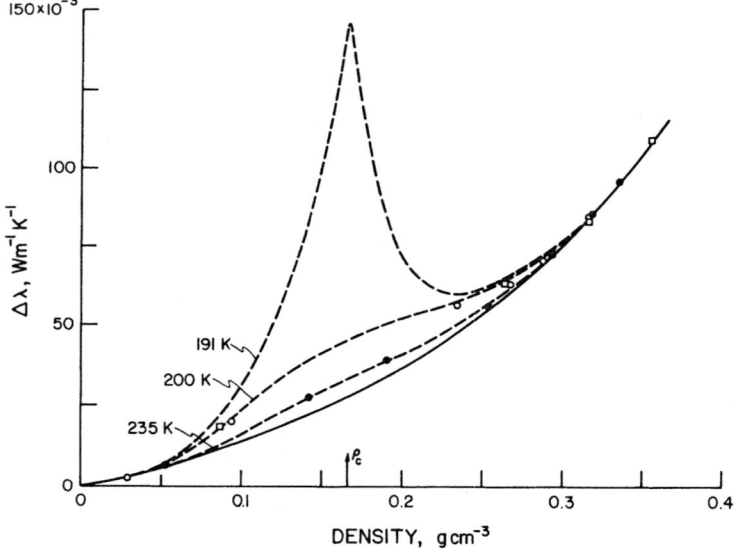

Figure 5. Calculated excess thermal conductivity for methane
compared with the experimental data. ●[29]○[29]□[30]

TABLE IV

Deviations between experimental and calculated thermal conductivity coefficients for argon and nitrogen in the critical region. Data from Ziebland and Burton, Table I.

ARGON

T K	P atm	$\Delta\tilde{T}$	$\Delta\tilde{\rho}$	percent deviation
159.2	64.2	0.06	0.010	− 2.0
164.5	72.0	0.09	0.08	−10.2
164.7	72.0	0.09	0.09	0.0
159.2	72.0	0.06	0.28	− 2.6
159.0	72.0	0.06	0.29	− 2.7
149.2	48.0	0.01	0.45	9.0
149.2	48.0	0.01	0.45	− 3.4

NITROGEN

136.9	50.3	0.09	0.05	−13.0
136.2	50.3	0.08	0.003	3.0
139.0	50.3	0.10	0.17	− 8.0
142.1	67.2	0.13	0.17	− 1.8

VIII. REFERENCES

1.   J. V. Sengers, Ber. Bunsenges. Physik. Chemie 76, 234 (1972);
     in "Transport Phenomena - 1973," J. Kestin, Ed., AIP Confer-
     ence Proceedings No. 11 (American Institute of Physics, New
     York, 1973) p. 229.

2.   J. V. Sengers in "Critical Phenomena," Proc. Intern School of
     Physics "Enrico Fermi," Course LI, M.S. Green, ed. (Academic
     Press, New York, 1971) p. 445.

3.   N. Mani and J.E.S. Venart, Advances in Cryogenic Engineering,
     Vol. 18, K.D. Timmerhaus, Ed. (Plenum Press, New York, 1973)
     p. 280.

4.   R. D. Mountain, Rev. Mod. Phys. 38, 205 (1966); D. McIntyre
     and J. V. Sengers, in "Physics of Simple Liquids,"
     H.N.V. Temperley, J.S. Rowlinson and G.S. Rushbrooke, Eds.
     (North Holland, Amsterdam, 1968) p 449; H.Z. Cummings and
     H.L. Swinney in "Progress in Optics," Vol. VIII, E. Wolf, ed.
     (North Holland, Amsterdam, 1970) p. 135.

5.   L. P. Kadanoff, Ref. 2, p. 100.

6.   J.M.H. Levelt Sengers, Physica 73, 73 (1974).

7.   L. P. Kadanoff and J. Swift, Phys. Rev. 166, 89 (1968).

8.   H. L. Swinney and D. L. Henry, Phys. Rev. A8, 2586 (1973).

9.   G. Arcovito, G. Faloci, M. Roberti and L. Mistura, Phys. Rev.
     Lett. 22, 1040 (1969).

10. K. Kawasaki, Ann. of Phys. 61, 1 (1970).

11. H.J.M. Hanley, R. D. McCarty and J. V. Sengers, NASA Contractor
     Report NASA CR-2440 (NASA, Washington, D.C., 1974).

12. H.J.M. Hanley, R. D. McCarty and W. M. Haynes, J. Phys. Chem.
     Ref. Data 3, 979 (1974).

13. J.M.H. Levelt Sengers, Ind. Eng. Chem. Fund. 9, 470 (1970);
     "Experimental Thermodynamics," Vol. II, B. Le Neindre and
     B. Vodar, Eds. (Butterworth, London, 1975) p. 657.

14. B. Widom, Physica 73, 107 (1974).

15. R. B. Griffiths, Phys. Rev. 158, 176 (1967).

16. M. Vicentini-Missoni, J.M.H. Levelt Sengers and M. S. Green, J. Res. NBS 73A, 563 (1969).

17. J.M.H. Levelt Sengers, W. L. Greer and J. V. Sengers, J. Phys. Chem. Ref. Data (in press). J.M.H. Levelt Sengers and J. V. Sengers, Phys. Rev. A (in press).

18. T. A. Murphy, J. V. Sengers and J.M.H. Levelt Sengers, Proc. 8th Int. Conf. on the Properties of Steam (in press, 1975).

19. B. Chu, Ber. Bunsenges Physik. Chemie 76, 202 (1972).

20. M. E. Fisher, J. Math. Phys. 6, 944 (1964).

21. M. S. Wertheim, J. Chem. Phys. 55, 4291 (1971).

22. J. O. Hirschfelder, C. F. Curtiss and R. B. Bird, "Molecular Theory of Gases and Liquids," (Wiley, New York, 2nd Ed, 1964) p. 30.

23. J. F. Ely and H.J.M. Hanley, Mol. Phys. 24, 684 (1972); Mol. Phys. (in press, 1975); J. F. Ely, H.J.M. Hanley and G. C. Straty, J. Chem. Phys. 59, 842 (1973); H.J.M. Hanley and J. F. Ely, J. Phys. Chem. Ref. Data 2, 735 (1973).

24. M. Klein and H.J.M. Hanley, J. Chem. Phys. 53, 4722 (1970); H.J.M. Hanley and M. Klein, J. Phys. Chem. 76, 1743 (1972).

25. P. W. Schmidt and C. W. Tompson, in "Simple Dense Fluids," H. L. Frisch and Z. W. Salsburg, Eds. (Academic Press Inc., New York, 1968) p. 31; J. E. Thomas and P. W. Schmidt, J. Chem. Phys. 39, 2506 (1963); J. Am. Chem. Soc. 86, 3554 (1964).

26. J. S. Lin and P. W. Schmidt, Phys. Rev. A10, 2290 (1974).

27. B. Chu and J. S. Lin, J. Chem. Phys. 53, 4454 (1970).

28. A. Michels, J. V. Sengers and P. S. Van der Gulik, Physica 28, 1216 (1962).

29. L. D. Ikenberry and S. A. Rice, J. Chem. Phys. 39, 1561 (1963).

30. N. Mani, Ph.D. Thesis (Univ. of Calgary, Calgary, Alberta, Canada, 1971).

APPENDIX I

References for Table I

a   L. A. Guildner, Proc. Nat. Acad, Sci., U.S.A. 44, 1149 (1958); J. Res. NBS 66A, 333, 341 (1962).

b   A. Michels, J. V. Sengers and P. S. Vander Gulik, Physica 28, 1201, 1216 (1962); A. Michels and J. V. Sengers, Physica 28, 1238 (1962).

c   H. Ziebland and J. T. A. Burton, Brit. J. Appl. Physics 9, 52 (1958).

d   J. V. Sengers, Int. J. Heat Mass Transfer 8, 1103 (1965).

e   L. D. Ikenberry and S. A. Rice, J. Chem. Phys. 39, 1561 (1963).

f   D. P. Needham and H. Ziebland, Int. J. Heat Mass Transfer 8, 1307 (1965).

g   I. F. Golubev and V. P. Sokalava, Teploenergetika 11, no. 9, 64 (1964); V. P. Sokalava and I. F. Golubev, Teploenergetika 14, no. 4, 91 (1967).

h   J. Lis and P. O. Kellard, Brit. J. Appl. Physics 16, 1099 (1965).

i   B. J. Bailey and K. Kellner, Brit, J. Appl. Physics 18, 1645 (1967); Physica 39, 1144 (1968).

j   J. F. Kerrisk and W. E. Keller, Phys. Rev. 177, 341 (1969).

k   H. M. Roder and D. E. Diller, J. Chem. Phys. 52, 5928 (1970).

l   M. L. R. Murthy and H. A. Simon, Phys. Rev. A2, 1458 (1970); Proc. 5th Symposium on Thermophysical Properties, C. F. Bonilla, ed. (ASME, New York, 1970) p. 214.

m   R. Tufeu, B. Le Neindre and P. Bury, Comptes Rendus 273B, 113 (1971); B. Le Neindre, R. Tufeu, P. Bury and J. V. Sengers, Ber. Bunsenges Physik. Chemie 77, 263 (1973).

n   N. Mani and J. E. S. Venart, Advances in Cryogenic Engineering, Vol. 18, K. D. Timmerhaus, ed. (Plenum Press, New York, 1973) p. 280. N. Mani, Ph.D. Thesis (Univ. of Calgary, Calgary, Alberta, Canada 1971).

o   A. M. Sirota, V. I. Latunin and G. M. Beljaeva, Teploenerge-tika 21, no. 8, (1974); Teploenergetika (in press).

p   A. A. Tarzimanov, Proceedings 8th International Conference on the Properties of Steam (in press).

q   J. Van Oosten, Ph.D. Thesis (Van der Waals Laboratory, Univ. of Amsterdam, 1974).

r   H. L. Swinney and D. L. Henry, Phys. Rev. A 8, 2586 (1973);
     T. K. Lim, H. L. Swinney, K. H. Langley and Th. A. Kachnowski,
     Phys. Rev. Letters 27, 1776 (1971).

s   T. K. Lim, H. L. Swinney, I. W. Smith and G. B. Benedek,
     Optics Comm. 7, 18 (1973).

t   B. S. Maccabee and D. L. Henry, Phys. Rev. Letters 27, 495
     (1971).

u   G. T. Feke, G. A. Hawkins, J. B. Lastovka and G. B. Benedek,
     Phys. Rev. Letters 27, 1780 (1971).

v   K. Ohbayashi and A. Ikushima, J. Low Temperature Physics 19,
     449 (1975).

APPENDIX II

Angular Part of the Potential

If the relative angular orientation of two rigid nonspherical molecules is as shown in the sketch[22]

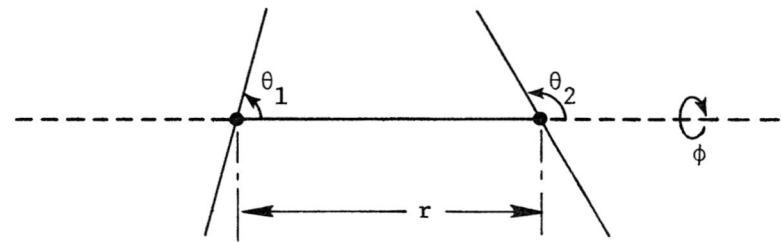

then we have for the potential of equation (36):

$$F(\Omega) = 1 - 5 \cos^2\theta_1 - 5 \cos^2\theta_2 - 15 \cos^2\theta_1 \cos^2\theta_2$$

$$+ 2 (\sin\theta_1 \sin\theta_2 \cos\phi - 4 \cos\theta_1 \cos\theta_2)^2$$

$$G(\Omega) = \sin^4\theta_1 + \sin^4\theta_2 + 4 \cos^4\theta_1 + 4 \cos^4\theta_2$$

$$H(\Omega) = [(1 - 3 \cos^2\theta_1) \cos\theta_2 + 2 \sin\theta_1 \cos\theta_2 \sin\theta_2 \cos\phi]^2$$

$$+ [(1 - 3 \cos^2\theta_2) \cos\theta_1 + 2 \sin\theta_2 \cos\theta_1 \sin\theta_1 \cos\phi]^2$$

also

$$\int_\Omega d\Omega = {}_o\!\int^\pi \sin\theta_1 d\theta_1 \; \int_o^\pi \sin\theta_2 d\theta_2 \; {}_o\!\int^\pi d\phi \quad .$$

KINETIC-THEORY RELATION OF THERMAL CONDUCTIVITY TO OTHER GAS PROPERTIES[*]

E. A. Mason

Brown University

Providence, Rhode Island

Theoretical relations among measurable quantities can be used to test the consistency of experimental data or to calculate one property from measurements of another. The kinetic theory of dilute gases relates the thermal conductivity to the viscosity and other properties. For monatomic gases the molecular weight is the only additional property needed, the relation is simple and nearly exact, and can be usefully applied.[1] The relation becomes much more complicated for polyatomic gases, and there has been a series of theoretical advances over the last thirteen years.[2-7] The crudest relation requires at least the specific heat and a self-diffusion coefficient. More accurate relations must further include two major effects - inelastic collisions and (for polar gases) the resonant exchange of rotational energy. These require further knowledge of at least a collision number for rotationally inelastic collisions and an exchange correction, which must be calculated from the molecular dipole (and perhaps quadrupole) moment and the spectroscopic rotational constants of the molecule. A discussion will be given of the bounds that can be put on the thermal conductivity in light of available knowledge of these various other properties. In some cases there are better prospects for improving our knowledge of thermal conductivity by measurement of some auxiliary property (e.g., thermal transpiration or the radiometer effect) than by additional direct measurements of conductivity.[8]

[*]Supported in part by NASA Grant NGL-40-002-059.
[1]J. Kestin, S. T. Ro, and W. Wakeham, Physica 58, 165 (1972).
[2]E. A. Mason and L. Monchick, J. Chem. Phys. 36, 1622 (1962).
[3]L. Monchick, A. N. G. Pereira, and E. A. Mason, J. Chem. Phys. 42, 3241 (1965).

[4]L. Monchick, Phys. Fluids $\underline{7}$, 882 (1964); $\underline{8}$, 1416 (1965).

[5]S. I. Sandler, Phys. Fluids $\underline{11}$, 2549 (1968).

[6]C. Nyeland, E. A. Mason, and L. Monchick, J. Chem. Phys. $\underline{56}$, 6180 (1972).

[7]W. F. Ahtye, J. Chem. Phys. $\underline{57}$, 5542 (1972).

[8]J. Thoen-Hellemans and E. A. Mason, Int. J. Engng. Sci. $\underline{11}$, 1247 (1973).

Chapter V

Experimental Methods and Numerical Analysis

ANALYSIS OF EDGE-HEAT-LOSS OF A GUARDED-HOT-PLATE APPARATUS

M. C. I. Siu

National Bureau of Standards

Washington, D.C. 20234

ABSTRACT - Analysis on the error in measured specimen thermal conductivity arising from edge-heat-loss at the periphery of any guarded-hot-plate apparatus is presented. It is shown that the error due to edge-heat-loss varies linearly with respect to the deviation of the ambient temperature from mean specimen temperature. Explicit expressions are presented for the line-heat-source guarded-hot-plate apparatus being constructed at the National Bureau of Standards.

1. INTRODUCTION

In a guarded-hot-plate apparatus, measured thermal conductivity is affected by the temperature, T_a, of the fluid (air) in the vicinity of the hot and cold plates. The amount of the error depends upon the particular guarded-hot-plate. The percentage error is known to vary with the type and thickness of edge insulation as well as the thickness of the test specimen[1,2]. In accordance with the American Society of Testing Materials Standard Method of Test for Thermal Conductivity of Materials by Means of the Guarded-Hot-Plate Apparatus, ASTM C177[3], it is recommended that the ambient temperature, T_a, surrounding the guarded-hot-plate apparatus during a test be adjusted such that $(T_m - T_a)/(T_h - T_c) \leq 0.1$, where $T_m = (T_h + T_c)/2$ and T_h and T_c are the temperatures of the hot and cold plates, respectively. The basis of this recommendation is not clearly stated and the percentage error associated with this limit is not given.

An analysis on edge-heat-loss error is presented in this paper. Application is made to the line-heat-source guarded-hot-plate apparatus presently being assembled at the National Bureau of Standards.

413

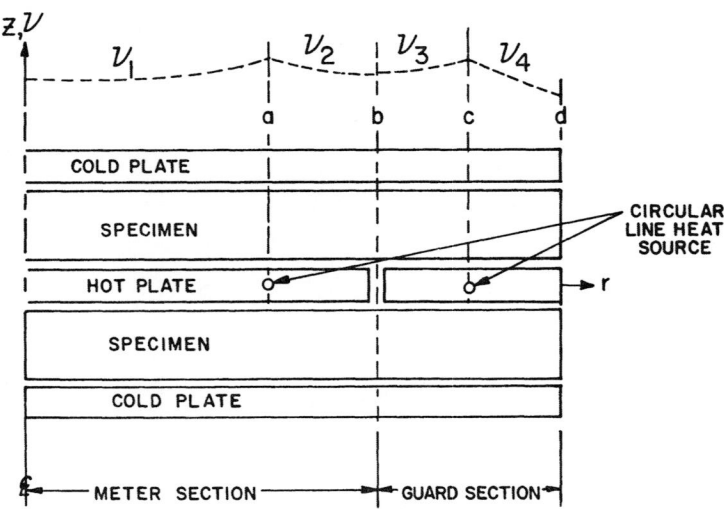

Figure 1 Schematic Diagram Showing the Critical
Geometric Parameters of the Robinson Apparatus

(NBS). The NBS apparatus has been named the Robinson apparatus after
the late Henry E. Robinson, who originated the line-heat-source concept of guarded-hot-plate design at NBS.

2.   THE ROBINSON APPARATUS

A line-heat-source guarded-hot-plate apparatus (Robinson Apparatus) utilizing the design analyses of Hahn, et al[4,5] is under construction at NBS. The apparatus uses circular hot and cold plates
and a circular line-heat-source. Figure 1 is a schematic diagram
showing the geometrical parameters which affect the temperature distributions over the surface of the hot-plate. Figure 1 shows exaggerated temperature cusps at $r = a$ and $r = c$, but Peavy's[5] analytical results show that the temperature is a well-behaved function over
the hot-plate. Robinson[5] showed that the average temperature of the
hot-plate over the meter section is, to a good approximation, equal
to the temperature at the gap $r = b$ when the line-heat-source in the
meter-section is placed at $a = b/\sqrt{2}$. More exact calculations are
given by Peavy[5] . Figure 2 shows the fraction deviation
$\delta = (\nu_g - \bar{\nu})/\bar{\nu}$ predicted by Peavy for hot-plates made from copper
and aluminum, where ν_g is the temperature at $r = b$ and $\bar{\nu}$ is the
the average temperature of the hot-plate.

Analysis[5] also indicates that the temperatures ν_2 and ν_3 can be
matched if the line-heat-source in the guard-ring is placed at

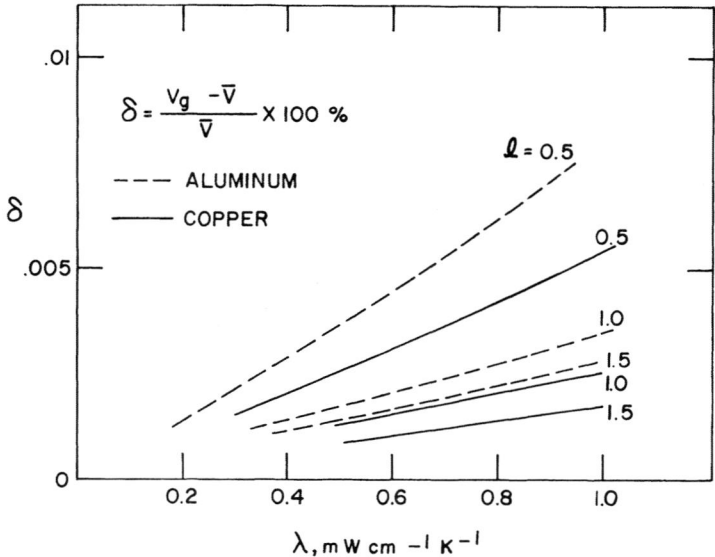

Figure 2 Predicted Fractional Percentage Difference
Between the Gap Temperature ν_g and the Average Temperature
of the Meter Section $\bar{\nu}$ for Several Specimen Thicknesses

c = 1.29 b. The hot-plate of the Robinson Apparatus satisfies the
above critical dimensions.

3. EDGE-HEAT-LOSS-ERROR ANALYSIS

For the purpose of determining the error in measured thermal
conductivity arising from edge-heat-loss at the periphery of a guarded-hot-plate apparatus, consider steady state heat flow within a test
specimen of volume V bounded by the surfaces of the hot-plate S_h, the
cold-plate S_c and edge S_e. Let λ_s denote the thermal conductivity
of the test specimen. Within the region bounded by these surfaces,
the temperature ν satisfies the equation $\text{div}(\lambda_s \nabla \nu) = 0$.
Applying divergence theorem gives

$$\int_S \hat{n} \cdot \lambda_s \nabla \nu \, d\sigma = 0$$

where $\hat{n}$ is the unit vector normal to the surface element $d\sigma$ of the
specimen and $S = S_h + S_c + S_e$ is the total surface of the specimen.
Without loss of generality, assume that the insulation is homogeneous
and λ_s is not a function of position. Let the plates be separated
into meter, m, and guard, g, sections and write $S_h = S_{hm} + S_{hg}$ and
$S_c = S_{cm'} + S_{cg'}$. Then

$$\lambda_s \int_{S_{hm}} \hat{n} \cdot \nabla v d\sigma = -\lambda_s (\int_{S_{hg}} + \int_{S_{cm}} + \int_{S_{cg}} + \int_{S_e}) \hat{n} \cdot \nabla v d\sigma \qquad (1)$$

Experimentally, thermal conductivity λ_{exp} is determined from the formula

$$q_{exp} = -A_{hm} \lambda_{exp} (T_h - T_c)/\ell \qquad (2)$$

where q_{exp} is the measured total heat flux over the meter area A_{hm} and ℓ is the thickness of the specimen. It follows from (1) and (2) that

$$\frac{\lambda_{exp}}{\lambda_s} = \frac{\ell}{A_{hm}(T_h - T_c)} (\int_{S_{hg}} + \int_{S_{cm}} + \int_{S_{cg}} + \int_{S_e}) \hat{n} \cdot \nabla v d\sigma$$

If $\nabla v = |\nabla v| \hat{n}_z$ everywhere where $\hat{n}$ is a unit vector in the z-direction and $\hat{n} \cdot \nabla v = (T_h - T_c)/\ell$, then $\lambda_{exp}/\lambda_s = 1$. In general, this is not the case. Assume that on S_e,

$$\hat{n} \cdot \nabla (v - T_a) = h(v - T_a)/\lambda_s$$

where T_a is a constant which is taken to be the ambient air temperature and h is the heat transfer coefficient. Thus

$$\frac{\lambda_{exp}}{\lambda_s} \quad \frac{\ell}{A_{hm}(T_h - T_c)} \left[I + \frac{A_e \bar{h}(\bar{v}_e - T_a)}{A_{hm}(T_h - T_c) \lambda_s} \right] \qquad (3)$$

where

$$I = (\int_{S_{hg}} + \int_{S_c}) \hat{n} \cdot \nabla v d\sigma ,$$

$$A_e \bar{h}(\bar{\nu}_e - T_a) = \int_{S_e} h(\nu_e - T_a)d\sigma \quad ,$$

A_e is the total area over S_e and $\bar{h}$ and $\bar{\nu}_e$ are the average heat transfer coefficient and average temperature, respectively, at the edge
of the specimen.

Equation (3) shows that λ_{exp}/λ_s varies linearly with respect to
the ambient temperature index $\chi = (\bar{\nu}_e - T_a)/(T_h - T_c)$.

4. EDGE-HEAT-LOSS IN THE LINE-HEAT-SOURCE APPARATUS

The above analysis is applied to a guarded-hot-plate apparatus
(Robinson Apparatus) which uses circular plates. Assume that the
temperature of the surfaces cf the hot and cold plates are maintained
at T_h and T_c, respectively. Neglecting radiation and convection
effects, a solution of $\nabla^2 \nu_A = 0$ within the insulation relative to the
ambient air temperature is[4]

$$\nu - T_a = T_h - T_a - (T_h - T_c)\frac{z}{\ell}$$

$$- \frac{2H}{\ell}\sum_{n=1}^{\infty} \frac{T_h - T_a + (-1)^{n+1}(T_c - T_a)}{\alpha_n [\alpha_n \ell I_1(\alpha_n d) + HI_0(\alpha_n d)]} I_0(\alpha_n r)\sin/(\alpha_n z)$$

where $\alpha_n = n\pi/\ell$, $H = h\ell/\lambda_s$ and $I_j(z)$ is the modified Bessel function
of the first kind of order j. It can be shown by direct integration
of this equation that the average temperature potential at the edge
$r = d$ is given by

$$\bar{\nu}_e - T_a = (T_m - T_a)(1 - \varepsilon) \tag{4}$$

i.e. $\chi = \xi (1 - \varepsilon)$, where

$$\varepsilon = \frac{4H}{\pi^2}\sum_{n=0}^{\infty} \frac{1}{(2n+1)^2} \frac{I_0(\alpha_{2n+1}d)}{[(2n+1)\pi I_1(\alpha_{2n+1}d) + HI_0(\alpha_{2n+1}d)]} \tag{5}$$

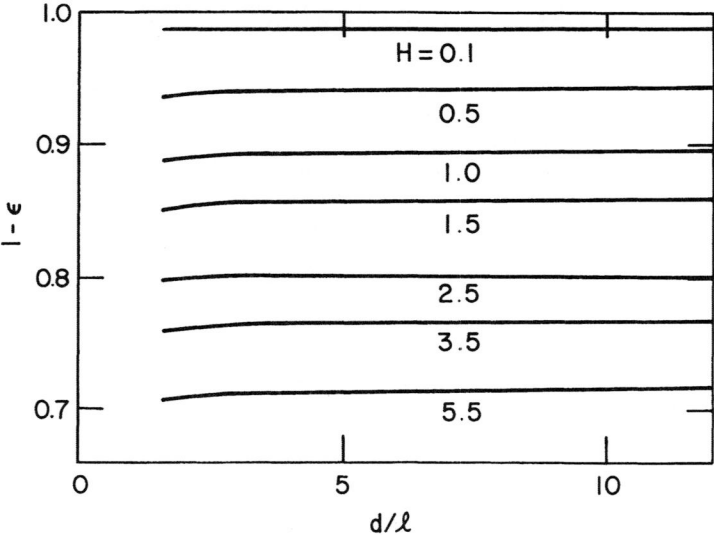

Figure 3 Correction Factor of the Mean Edge Temperature
as a Function of d/ℓ for Several Values of H

It is seen from this equation that the correction factor $(1 - \epsilon)$
depends only upon H (= $h\ell/\lambda_s$) and the ratio d/ℓ. Figure 3 shows
the variation of this correction factor with d/ℓ for several values
of H. This figure shows that for thin specimens, the correction is
rather insensitive to d/ℓ.

Direct evaluation of the integral on the left hand side of (1)
and using the relationships (2) and (3) gives

$$x = M\xi + B \qquad\qquad (6)$$

where

$$x = \frac{\lambda_{exp}}{\lambda_s} - 1$$

$$\xi = (T_m - T_a)/(T_h - T_c)$$

$$M = 8H \sum_{n=0}^{\infty} \frac{I_1(\alpha_{2n+1}b)}{\alpha_{2n+1}b[(2n+1)\pi I_1(\alpha_{2n+1}d) + HI_0(\alpha_{2n+1}d)]}$$

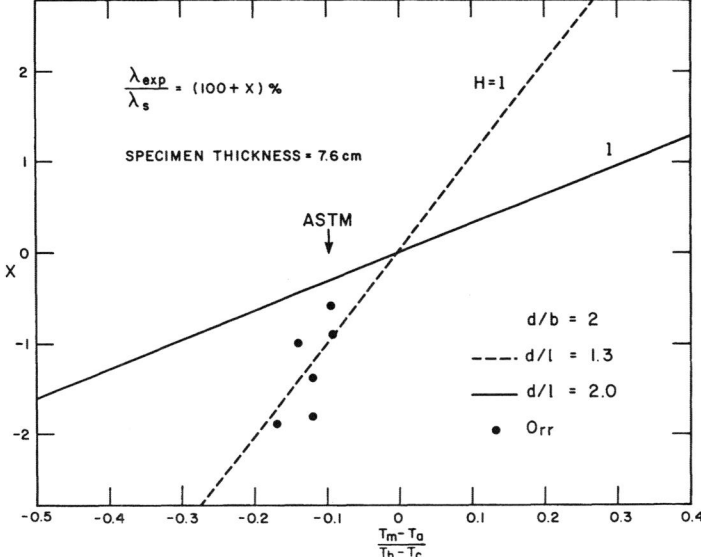

Figure 4 Calculated Fractional Error x of a 7.6 cm Thick Polyurethane Specimen Due to Edge-Heat-Loss in a Line-Heat-Source Apparatus With d/b = 2 (Robinson Apparatus) and d/b = 1.3 (Same Ratio as the Present NBS-Type Apparatus) and Values Obtained by Orr From Measurements Made on an NBS-Type Apparatus

$$B = 4H \sum_{n=1}^{\infty} \frac{I_1(\alpha_{2n}b)}{\alpha_{2n}b[2n\pi I_1(\alpha_{2n}d) + HI_0(\alpha_{2n}d)]}$$

Figure 4 shows the percentage deviation of the parameter x obtained from (6) for the Robinson Apparatus and for a line-heat-source apparatus with the same d and b values as the present NBS guarded-hot-plate apparatus. The upper limit of ξ = 0.1 as recommended by ASTM C177 is indicated by an arrow in Figure 4. Results of measurements of Orr[3] on a 7.6 cm thick polyurethane specimen with 5.7 cm thick polyurethane edge-insulation have been included in Figure 4. This figure shows that Orr's measurements corresponded to H = 1, so that h = 0.4 W m^{-2} C^{-1}. For H = 1, equations (5) and (6) give M = 0.134 and B = 0.003. Thus setting $T_a = T_m$ would result in a residual error of 0.3 percent. In the case of the Robinson Apparatus for H = 1, these values become M = 0.034 and B = 0.0002. Thus for practical purposes, at $T_a = T_m$ the error arising from edge-heat-loss is expected to be negligible in the Robinson Apparatus for the test situation under consideration.

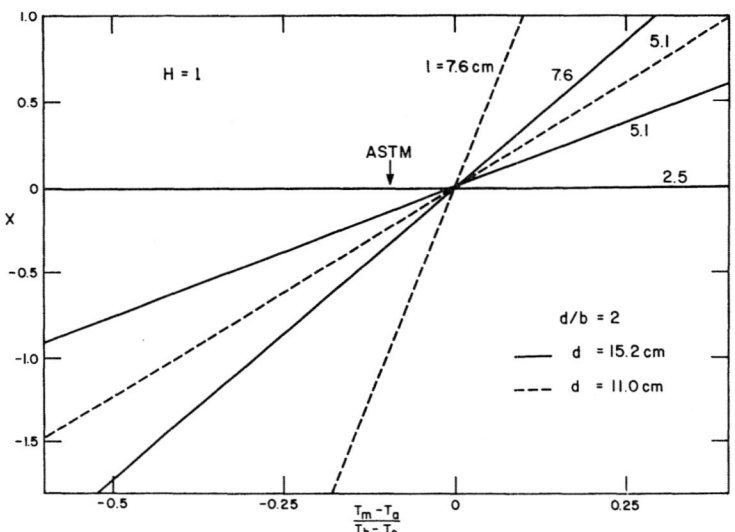

Figure 5 Calculated Percentage Error in Thermal
Conductivity Due to Edge-Heat-Loss in the Robinson
Apparatus for Several Specimen Thicknesses, ℓ

It should be noted that Orr used a square plate NBS type guarded-
hot-plate apparatus in obtaining his data. However, as shown in (3)
the linearity relation between x and χ, equivalently ξ, is independent
of the shape of the plate and differences in geometric configuration
show up as differences in $\bar{h}$. Addition of edge-insulation results in
a reduction in $\bar{h}$. Simple approximations by Hahn[4] give

$$h_d = \lambda_i/w$$

and

$$H - \lambda_i \ell/w\lambda_s$$

where h_d is the reduced heat transfer coefficient and w and λ_i are
the thickness and thermal conductivity of the edge insulation. For
Orr's test, these equations give H = 1.3 and h_d = 0.4 W m^{-2} C^{-1}.

Figure 5 gives expected errors x for various specimen thicknesses with H = 1 and for the same d/ℓ ratios shown in Figure 3.

5.   CONCLUDING REMARKS

Equation (3) shows that in a guarded-hot-plate apparatus the error x varies linearly with respect to χ and this result is independent of the geometrical configuration of the specimen. Despite the simplifying assumptions made in arriving at this result, linearity of x with χ, equivalently ξ, is found to be consistent with existing experimental data made on a square-plate guarded-hot-plate apparatus. Computations made on an ideal circular-plate guarded-hot-plate apparatus also substantiates this linear relationship. However, more experimental data is required to check on the validity of this linear relationship.

Thus, it is seen that edge-heat-loss errors can be estimated quantitatively. The upper limit $\xi = 0.1$ as recommended by ASTM C177 corresponds to an error of about one percent in the present NBS-type guarded-hot-plate apparatus. In the Robinson Apparatus, the corresponding error due to edge-heat-loss is expected to be about 0.4 percent. In a guarded-hot-plate apparatus which uses circular plates, edge-heat-loss-error depends upon d/ℓ, d/b as well as H ($= h\ell/\lambda_s$) as shown in Figures 4 and 5 and equation (6). For a given apparatus the magnitude of such errors decreases with a decrease in ℓ or H. In the Robinson Apparatus, edge-heat-loss is not expected to be an important factor for a 2.5 cm thick specimen when H = 1 and the error would be less than one percent for $\ell = 7.6$ cm even when $\xi = 0.25$, a relatively large value of ξ.

While the fractional error x varies linearly with respect to χ, independently of heat flux measurements made at the surfaces of the plates and independently of their areas, the value of λ_s depends upon S_{hm} and S_{cm}. In general, $S_{hm} \neq S_{cm}$ so that it is necessary to distinguish between thermal conductivity derived from heat flux measurements made at the cold-plate from that made at the hot-plate. This is particularly important in apparatuses other than the guarded-hot-plate apparatus which use heat flow meters to monitor heat flux.

6.   ACKNOWLEDGMENTS

The author is indebted to Dr. M. H. Hahn for stimulating discussions on the subject and Mr. B. A. Peavy for writing the computer programs from which numerical results were obtained.

7.   REFERENCES

1. Donaldson, I. G., Quart. Appl. Math. 19 (3), 205 (1961).

2. Orr, H. W., "A Study of the Effects of Edge Insulation and Ambient Temperatures on Errors in Guarded Hot Plate Measurements",

Proceedings of the Seventh Conference on Thermal Conductivity, pp. 521-526, D. R. Flynn and B. A. Peavy, Eds., NBS SP 302, 1968.

3. Method of Test for Thermal Conductivity of Materials by Means of the Guarded Hot Plate (C177-71), Annual Book of ASTM Standards, Part 18, pp. 15-28, American Society for Testing and Materials, Philadelphia, 1974.

4. Hahn, M. H., "The Line Heat Source Guarded Hot Plate for Measuring Thermal Conductivity of Building and Insulating Materials", Doctoral Dissertation, The Catholic University of America, Washington, D.C., 1971.

5. Hahn, M. H., Robinson, H. E. and Flynn, D. R., "Robinson Line-Heat-Source Guarded Hot Plate Apparatus", ASTM STP 544, pp. 167-192, American Society for Testing and Materials, Philadelphia, 1974.

DETERMINATION OF THERMOPHYSICAL PROPERTIES OF LAYERED

COMPOSITES BY FLASH METHOD

H. J. Lee and R. E. Taylor

School of Mechanical Engineering
Purdue University
West Lafayette, Indiana 47906

The thermal diffusivities of two-layer and three-layer composites were investigated by using the heat diffusion equation solved with appropriate boundary conditions and with the use of a heat pulse function closely describing the actual pulse shape. However, a Kronecker delta function for the heat pulse is used when the half-time is much larger than the heat pulse time. Furthermore, the dimensional similarity analysis among the relative heat capacity ($H_{1/2}$), the square root of the relative heat diffusion time ($\eta_{1/2}$), and the modified Fourier number ($\alpha_2 t_{1/2}/\ell_2^2$) is graphically represented. This method is in agreement with experimental results. The thermal diffusivities of tungsten carbide alloys flame spray coated on Armco iron and on 316 stainless steel, of epoxy coated on 316 stainless steel and on copper, and of brass explosively bonded to stainless steel were measured. The thermal diffusivity of distilled water was also measured by using a tantalum container. The interfacial thermal contact conductance coefficient of two-layer composites composed of 316 stainless steel and brass was found. The experimental results for these samples are given.

INTRODUCTION

The use of layered composites, especially coating materials on substrates, has rapidly increased in a number of applications such as thermal barriers, emissivity control, electrical insulation and wear, and erosion or corrosion resistance protection. Thus, layered composites have become one of the more important engineering materials in practical use. It is then an interesting and important problem to be able to predict and measure the thermophysical properties of layered composites. The flash technique for measuring thermal diffusivity of homogeneous materials can be applied to layered composites.

In Refs. 1 and 2 the heat diffusion equations for three-layer composites, two-layer composites with or without interfacial thermal contact resistance, and homogeneous materials are solved with appropriate boundary conditions and with the use of a heat pulse function closely describing the actual pulse shape; for example, the triangular heat pulse of a laser and an exponential heat pulse for a xenon-flash tube. However, a Kronecker delta function for the heat pulse is used when the half-time is much larger than the heat pulse time.

SAMPLE DESCRIPTIONS

Several different types of layered composites were used in the investigation. Composites were made using flame spraying, epoxy resin coating, explosive bonding, and soldering. Nine samples were used for measuring the diffusivity of the front layer, five samples were used to examine the three-layer case, and five samples were used for measuring the interfacial thermal contact conductance coefficient. Some of the samples were the same except that the thickness of one or the other layer was reduced by machining or the sample was reversed so that the front face became the rear face. These samples are listed and described in Table I.

Metco 31 C tungsten carbide alloys were flame spray coated on standard Armco iron for Samples 1-1 and 1-2, and on standard 316 stainless steel for Sample 1-3. The composition of Metco 31 C tungsten carbide alloy was: tungsten carbide and cobalt aggregate 35.0%, iron 2.5%, nickel 40.0%, chromium 11.0%, carbon 0.5%, silicon 2.5%, and boron 2.5% by weight.

An epoxy resin was coated onto standard 316 stainless steel for Sample 2-1 and on copper (Cabra Alloy 110, 99.9% pure) for Sample 2-2. The epoxy resin was fabricated by mixing two parts Resin 600 with one part Hardener No. 66, and five weight percent Black Color Paste 502 (Thermoset Plastic, Inc.). In addition, samples of epoxy resin only were fabricated for use in determining the properties of this material.

Samples 3-1 to 3-4 were Detaclad, which is the registered trademark for DuPont's explosion-bonded materials. A brass sheet was explosively-bonded to a stainless steel plate. Both the stainless steel and brass faces were serrated. Therefore, there was an uncertainty of 0.0012 cm in determining the effective thickness of each layer.

A piece of standard 316 stainless steel was soldered to brass to make Sample 4-1. Samples 4-1, 4-2, and 4-3 were the same except that the thickness of the 316 stainless steel layer was reduced by machining. Sample 4-4 was similar to Sample 4-1 and Sample 4-5 is Sample 4-4 reversed. The solder was Eutec Rod 157-B (Eutetic Corp.), which was 96.5% tin and 3.5% silver. The thickness of solder was less than 0.001 cm. These samples were used for the interfacial contact resistance studies.

Table I. Layered Composite Sample Descriptions

Sample No.	Layer No. 1			Layer No. 2			Layer No. 3	
	Material	Thickness (cm)	Bonding	Material	Thickness (cm)	Bonding	Material	Thickness (cm)
1-1	WC†	0.1930	Flame-sprayed	Armco Iron	0.3150	-----	-----	-----
1-2	WC	0.1306	Flame-sprayed	Armco Iron	0.3150	-----	-----	-----
1-3	WC	0.1516	Flame-sprayed	316 S.S.*	0.3150	-----	-----	-----
2-1	Epoxy	0.0447	Epoxy	316 S.S.	0.077	-----	-----	-----
2-2	Epoxy	0.0462	Epoxy	Copper	0.3302	-----	-----	-----
3-1	S.S.*	0.2802	Explosion	Brass	0.3050	-----	-----	-----
3-2	S.S.	0.1887	Explosion	Brass	0.3045	-----	-----	-----
3-3	S.S.	0.0919	Explosion	Brass	0.2998	-----	-----	-----
3-4	S.S.	0.0542	Explosion	Brass	0.2995	-----	-----	-----
4-1	316 S.S.	0.1118	Soldered	Brass	0.3157	-----	-----	-----
4-2	316 S.S.	0.0549	Soldered	Brass	0.3157	-----	-----	-----
4-3	316 S.S.	0.0261	Soldered	Brass	0.3157	-----	-----	-----
4-4	Brass	0.3312	Soldered	S.S.	0.1582	-----	-----	-----
4-5	S.S.	0.1582	Soldered	Brass	0.3312	-----	-----	-----
5-1	Solder	0.2045	Explosion	S.S.	0.0839	Explosion	Brass	0.2206
5-2	Brass	0.2206	Explosion	S.S.	0.0839	Soldered	Solder	0.2045
5-3	Solder	0.1367	Soldered	S.S.	0.0839	Explosion	Brass	0.2206
5-4	Brass	0.2206	Explosion	S.S.	0.0839	Soldered	Solder	0.1367
5-5	Tantalum	0.0025	None	Water	0.0911	None	Tantalum	0.0025

† WC is tungsten carbide alloy.

* S.S. is stainless steel.

The three-layer Sample 5-1 was made by adding a solder layer to the stainless steel side of a stainless steel brass explosion-bonded composite (Detaclad). Sample 5-3 was made from Sample 5-1 by removing some of the solder layer. Samples 5-2 and 5-4 were, respectively, Samples 5-1 and 5-3 reversed so that the front layer became the rear layer and vice-versa. For Sample 5-5, two one-mil tantalum sheets were glued to opposite sides of a plexiglass ring which was 1.27 cm O.D., 1.11 cm I.D., and 0.0635 cm thick. A small portion of the ring was cut out before the gluing operation to provide an opening for piping in the distilled water sample.

EXPERIMENTAL RESULTS

The description of the layered samples are given in Table I and the thermophysical properties of the individual layers are shown in Table II. These values were used to calculate the thermal diffusivity of the front layer (the density, specific heat, or thermal conductivity could have been chosen as the unknown), using measured thickness and half-times. The experimental results are presented in Table III. The relative volumetric heat capacity $H_{i/j}$ and the square root of relative heat diffusion time are defined as

$$H_{i/j} = H_i/H_j,$$
$$\eta_{i/j} = \eta_i/\eta_j.$$

The volumetric heat capacity H_j and the square root of heat diffusion time η_j of the jth layer are defined as

$$H_j = a\,\rho_j\,c_j\,\ell_j,$$
$$\eta_j = (\ell_j^2/\alpha_j)^{1/2},$$

where a is the cross-sectional area, ℓ is the thickness, α is the thermal diffusivity, ρ is the density, c is the specific heat at constant pressure, and subscript j denotes the jth layer. The half-time which is the time for the initiation of the pulse until the rear face temperature rise reaches one-half of its maximum, is designated by the symbol $t_{1/2}$.

Tungsten carbide alloy was flame-sprayed on a standard Armco iron sample to form Sample 1-1 and was flame-sprayed at the same time on 316 S.S. to form Sample 1-3. Sample 1-2 was formed from Sample 1-1 by reducing the thickness of the Armco iron layer. The homogeneous sample of tungsten carbide alloy (Table II) was formed from Sample 1-2 by completely removing the iron layer. Therefore, the values of the thermal diffusivity obtained on Samples 1-1, 1-2, 1-3, and the homogeneous sample should be the same. The experimental values for Samples 1-1 to 1-3 are within two percent of each other (Table III) and within five percent of those formed for the homogeneous case (Table II). In the case of the layered

Table II. Thermophysical Properties at 300 K of Materials Used
in Layered Composite Samples

Material	α (cm^2 sec^{-1})	c** (J g^{-1} K^{-1})	k[†] (W cm^{-1} K^{-1})	ρ (g cm^{-3})
Armco Iron	0.206	0.494	0.801	7.874
Brass*	0.348	0.379	1.126	8.46
Brass	0.297	0.379	0.952	8.46
Copper	1.16	0.384	3.982	8.94
Epoxy Resin	0.00128	1.46	0.00212	1.133
Solder	0.348	0.305	0.781	7.36
Stainless Steel*	0.0367	0.452	0.132	7.952
316 Stainless Steel	0.0352	0.452	0.127	7.952
Tantalum	0.247[+]	0.144[x]	0.572	16.6
WC Alloy	0.0286	0.379	0.0965	9.0
Distilled Water	0.001465***	4.31[‡]	0.00603	0.998

[†]   k was calculated from $k = \alpha \rho c$.

*   From Detaclad.

[+]   Recommended value by the TPRC Ref. 3.

[x]   From reference 4.

[‡]   Recommended value by the TPRC Ref. 5.

**   The specific heat was measured by a differential scanning calorimeter (Perkin-Elmer DSC2) unless otherwise specified.

***   Value at 298.9 K from reference 6.

samples, the values of the diffusivity tended to approach the homogeneous case as the relative diffusion time ($\eta_{1/2}$) was increased.

The thermal diffusivity values of epoxy resin (Table III) obtained from Samples 2-1 (epoxy resin-316 S.S.) and 2-2 (epoxy resin-copper) were within 7% of the thermal diffusivity of epoxy resin determined from a homogeneous sample (Table II), which was fabricated from the same batch. The differences between the epoxy resin thermal diffusivity values obtained from Samples 2-1 and 2-2 and that of the homogeneous sample was somewhat smaller for Sample 2-2 which had the larger relative diffusion time.

The values of the thermal diffusivity of the stainless steel supplied by DuPont, obtained from Samples 3-1 to 3-4, were within 3.5% of that obtained for the homogeneous sample (Table II) machined from the same block. Samples 3-2 to 3-4 were fabricated from Sample 3-1 by reducing

Table III. Thermal Diffusivity of the Front Layer, Measured Half-Time on the Rear Face, and Dimensionless Parameters of Two-Layer Composites

Sample No.	Front Layer	Rear Layer	α (cm² sec⁻¹)	α^+ (cm² sec⁻¹)	Deviation (%)	$t_{1/2}$ (sec)	$\eta_{1/2}$	$H_{1/2}$	$\dfrac{\alpha_2 t_{1/2}}{\ell_2^2}$
1-1	WC Alloy	Armco Iron	0.0294	0.0286	2.8	0.5040	1.64	0.537	1.05
1-2	WC Alloy	Armco Iron	0.0295	0.0286	3.1	0.2986	1.11	0.364	0.62
1-3	WC Alloy	316 S.S.	0.0300	0.0286	4.9	0.8941	0.53	0.457	0.32
2-1	Epoxy Resin	316 S.S.	0.00119	0.00128	-7.0	0.5927	3.04	0.268	3.52
2-2	Epoxy Resin	Copper	0.00135	0.00128	5.5	0.5960	4.21	0.0674	6.34
3-1	S.S.*	Brass*	0.0371	0.0367	1.1	0.6359	2.83	1.03	2.37
3-1	Brass*	S.S.*	0.330	0.348	-5.2	0.6450	0.353	0.97	0.30
3-2	S.S.*	Brass*	0.0361	0.0367	-1.6	0.3545	1.9	0.695	1.33
3-2	Brass*	S.S.*	0.337	0.348	-3.2	0.3517	0.524	1.44	0.36
3-3	S.S.*	Brass*	0.0359	0.0367	-2.2	0.1358	0.954	0.344	0.53
3-3	Brass*	S.S.*	0.361	0.348	3.7	0.1320	1.06	2.9	0.57
3-4	S.S.*	Brass*	0.0354	0.0367	-3.5	0.0811	0.557	0.203	0.32
3-4	Brass	S.S.*	0.350	0.348	0.6	0.0793	1.8	4.93	0.99

* From Detaclad.

+ From homogeneous layer (Table II).

† Deviation % = $\dfrac{\alpha_{\text{front layer}} - \alpha_{\text{homogeneous}}}{\alpha_{\text{homogeneous}}}$ x 100.

the thickness of both layers. Samples 3-1 to 3-4 were reversed so that the brass became the front layer. The values of thermal diffusivity of brass obtained from Samples 3-1 to 3-4 were within 6% of that measured for the homogeneous sample (Table II) machined from the same block when the samples are reversed, the mathematical solution for the half-time[1,2] remains the same. The differences in half-times obtained upon reversing Samples 3-1 through 3-4 were all within three percent, even though the magnitude of the half-time changed by a factor of eight. The thermal diffusivity values for both stainless steel and brass obtained from Samples 3-1 to 3-4 tended to approach the values obtained for the homogeneous samples as the relative diffusion time was increased.

In order to investigate the applicability of the mathematical solution for determining the interfacial thermal contact conductance coefficient[1,2], Sample 4-1 was fabricated by soldering a layer of 316 S.S. to brass. The solder layer was less than 0.001 cm. Samples 4-2 and 4-3 were obtained by machining off some of the 316 S.S. from Sample 4-1 without disturbing the soldered joint. Thus, all three samples should give the same interfacial conductance coefficient. The results, obtained using the mathematical solution and the properties for 316 S.S. and brass listed in Table II, are given in Table IV. The results for the coefficient are within an eight percent band even though the measured half-times vary by a factor of three. Sample 4-4 was fabricated in a manner similar to Sample 4-1. The contact conductance coefficient of Sample 4-4 was 4.64 W cm^{-2} K^{-1} and the measured half-time was 0.3252 seconds. When the sample was reversed (Sample 4-5), the interfacial contact conductance coefficient was 3.30 W cm^{-2} K^{-1} and the measured half-time was 0.3475 seconds. This difference in measured half-times upon sample reversal is several times larger than that observed for any of the two-layer samples without contact resistance (Table III) and is an order of magnitude greater than that observed under comparable conditions (Sample 3-2, Table III).

Table IV. Interfacial Thermal Contact Conductance Coefficient of Two-Layer Composites

Sample No.	Flashed Layer	$t_{1/2}$ (sec)	$h_{c\gamma c}$ (W cm^{-2} K^{-1})
4-1	316 Stainless Steel	0.2603	3.09
4-2	316 Stainless Steel	0.1357	3.25
4-3	316 Stainless Steel	0.0888	3.01
4-4	Brass	0.3252	4.64
4-5	316 Stainless Steel	0.3475	3.30

The observed dimensionless temperature response curve of the rear face of a two-layer composite without contact resistance follows the theoretical curve for homogeneous material (expressed as V in Refs. 1 and 2) versus dimensionless time. However, if there is significant contact resistance, the experimental curve deviates from the ideal curve.

The three-layer case was investigated using Samples 5-1 through 5-4 (Table I). One thermophysical property of any layer can be determined using the mathematical solution for three-layer composite[1,2] and the known thermophysical properties of the other two layers. This was demonstrated by calculating the diffusivity of each layer from the measured half-time and the thermophysical properties of the other two layers for Samples 5-1 through 5-4. The results are given in Table V. The maximum deviation from the results obtained for a homogeneous sample was 5.7% and the average deviation was 3%. It can be shown that the mathematical solution is symmetrical and the half-time should not be affected by reversing the sample. This is confirmed by the results of Samples 5-2 and 5-4, since the half-times associated with these samples are within 1% of those measured for Samples 5-1 and 5-3, respectively, even though the half-time was reduced 20% by reducing the thickness of the solder layer.

Table V.   Thermal Diffusivity of the Front, Center, and Rear Layer of Three-Layer Composites

Sample No.	Half-Time (sec)	Experimental α (cm²/sec)			Material	Deviation[+] %
		Front Layer	Center Layer	Rear Layer		
5-1	0.25412	0.363			Solder	4.3
5-1	0.25412		0.0371		S.S.	1.1
5-1	0.25412			0.308	Brass	3.7
5-2	0.25374	0.314			Brass	5.7
5-2	0.25374		0.0373		S.S.	1.6
5-2	0.25374			0.366	Solder	5.2
5-3	0.20699	0.332			Solder	-4.6
5-3	0.20699		0.365		S.S.	-0.5
5-3	0.20699			0.288	Brass	-3.0
5-4	0.20746	0.288			Brass	-3.0
5-4	0.20746		0.363		S.S.	-1.1
5-4	0.20746			0.329	Solder	-5.5
5-5	0.53782		0.00141		Water	-3.4*

[+] Deviation % = $\dfrac{\alpha_{experimental} - \alpha_{homogeneous}}{\alpha_{homogeneous}}$ x 100.

* Deviation from Reference 6.

DISCUSSION OF RESULTS AND CONCLUSIONS

The differential equation describing heat diffusion of a solid without internal heat generation,

$$\rho\, c\, \frac{\partial T}{\partial t} = \nabla \cdot (k\nabla T),$$

is derived in Carlsaw and Jaeger[7] where T is the temperature, t is the time, and k is the thermal conductivity. For an isotopic, homogeneous material whose thermal conductivity k can be considered as constant independent of position and temperature, the heat diffusion equation will be

$$\frac{\partial T}{\partial t} = \frac{k}{\rho c}\, \nabla^2\, T.$$

From this equation, the thermal diffusivity of the isotropic, homogeneous material can be defined as $\alpha = k/\rho c$. But, the thermal diffusivity cannot be defined in the usual way when the thermal conductivity depends upon position or temperature because of difficulties associated with moving k from behind the ∇ operator. If k depends on position only because of its temperature dependency, then $\nabla(k\nabla T)$ becomes

$$k\nabla^2\, T + \nabla k \cdot \nabla T = k\nabla^2\, T + \frac{dk}{dT}\, (\nabla T)^2$$

and one could define the diffusivity in the usual manner in terms of a variable which has the dimensions of temperature[7]. If k depends upon position, the concept of diffusivity may not be very useful. In the case of composites formed from homogeneous layers, it is customary to treat each layer separately and the concept of diffusivity of each layer is not impaired. However, the concept of an effective diffusivity of layered composites is not very meaningful in contrast to the concept of effective conductivity of these composites[1].

In the present work, the mathematical analysis of two-layer and three-layer composites without interfacial thermal contact resistance[1,2] allows the measurement of one unknown thermophysical property for the material in one of the layers if the thermophysical properties of material(s) in the other layer(s) are known. The mathematical analysis of two-layer composites with interfacial thermal contact resistance[1,2] also allows measurements of one unknown thermophysical property for the material in one of the layers if the thermophysical properties in the other layer and interfacial thermal contact resistance are known, or allows measurement of interfacial contact conductance coefficient when all thermophysical properties of each layer are known. Consistent experimental results were obtained for these three cases (Tables III, IV, and V) using the derived mathematical models[1,2].

When samples of two-layer or three-layer composites without interfacial thermal contact resistance are reversed, mathematically the dimensionless temperature rise of the rear face remains the same[1,2]. This was observed experimentally for Samples 1-1 to 3-4, and 5-1 to 5-4, since the half-times were within 2%.of each other (Tables III and V) and the observed dimensionless temperature response of the rear face for layered composites follows the theoretical curve for homogeneous material.

Mathematically reversing a sample of a two-layer composite with interfacial thermal contact resistance does not change the dimensionless temperature history of the rear face or the half-time[1,2], providing the interfacial contact conductance coefficient is the same for heat flow in the forward and reverse directions. However, the rear face temperature rise and the half-time was observed to change when Sample 4-4 was reversed (Table IV).

The dimensionless temperature rise of the rear face of Sample 4-4 was slightly higher than the theoretical value for homogeneous material at the 70% to 90% temperature rise of the rear face. The dimensionless temperature on the rear face of the reversed sample (Sample 4-5) was much lower than the theoretical value at the 70% to 90% temperature rise. The experimentally measured half-time for Sample 4-5 was 7% greater than Sample 4-4 and the interfacial contact conductance coefficient of Sample 4-5 was less (i.e., greater resistance) than Sample 4-4. Thus, the resistance to heat flow was not the same in the forward and reverse directions for this case.

In the computer procedure to determine α_1 (unknown thermal diffusivity), trial values are used along with the observed half-time. When one also considers the uncertainty in the value of the observed half-time, the optimum sample configuration is achieved when the partial derivative of V (dimensionless temperature rise of the rear face) with respect to α_1 is equal to unity at the experimentally measured half-time. This is especially important for the case of a thin layer of highly conducting material on a substrate of lower conductivity.

The pulse duration of conventional optical sources (a xenon lamp or a solid state laser, for example) is normally one or two milliseconds and, therefore, the half-time should be a minimum of 50 to 100 milliseconds for using an instantaneous heat pulse. However, the half-time of thin samples of highly conducting materials may be substantially less than this. Then a proper heat pulse function should be used to correct for the finite-pulse time effect.

When the half-time is greater than 50τ (τ = pulse time), the instantaneous heat pulse can be used with less than a one percent error. Thus, we can apply the dimensional similarity analysis for two-layer composites when the half-time is greater than 50τ using the following dimensionless parameters: $H_{1/2}$ is the relative volumetric heat capacity, $\eta_{1/2}$ is the square root of the relative heat diffusion time of the front layer to the rear one, and $\alpha_2 t_{1/2}/\ell_2^2$ is a modified Fourier number which is dimensionless time. When the half-time is shorter than 50τ, we can still apply this analysis by using a correction for the finite pulse-time effect. This analysis can be extended to high temperature diffusivity measurements by

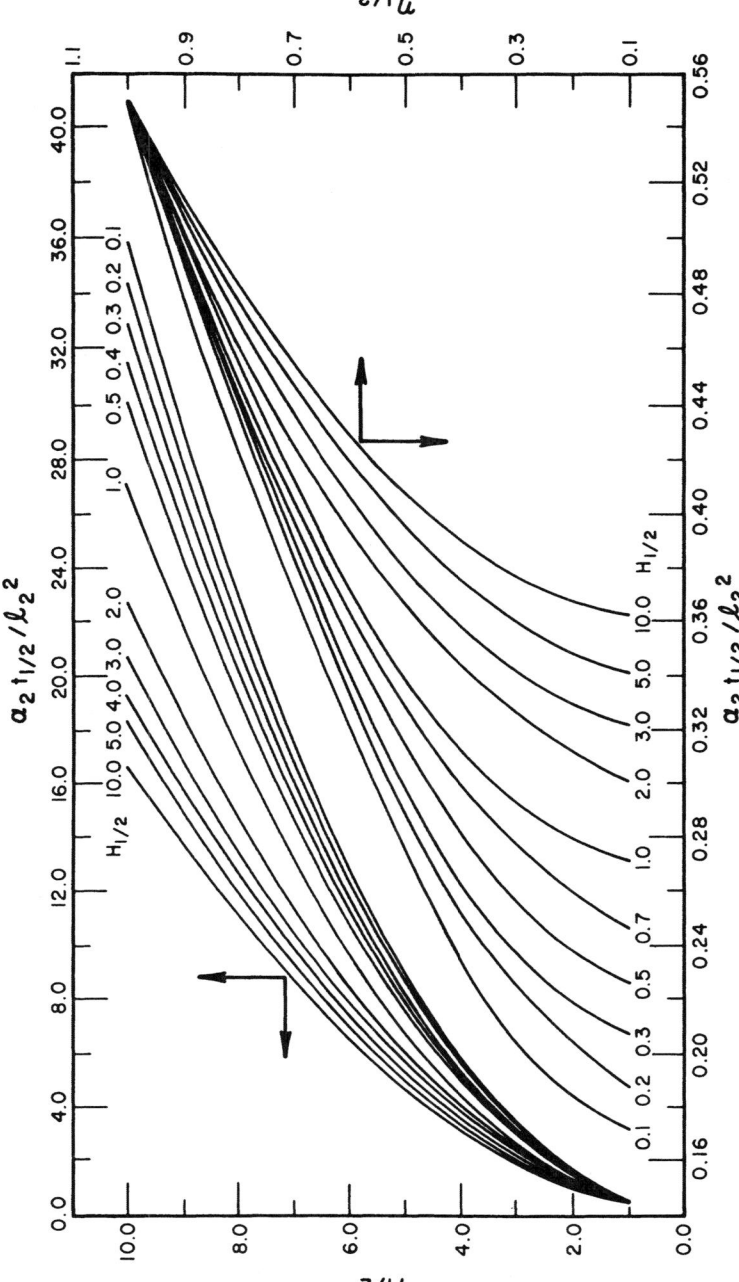

Figure 1. Square Root of Relative Heat Diffusion Time ($\eta_{1/2}$) vs Modified Fourier Number ($\alpha_2 t_{1/2}/\ell_2^2$) for a Family of Relative Volumetric Heat Capacity Values ($H_{1/2}$) in the Case of Two-Layer Composites

using a correction for the heat loss from the sample surface[1], because the temperature response on the rear face of the sample follows the theoretical curve for homogeneous material.

Figure 1 is a plot of the square root of the relative diffusion time versus the modified Fourier number at several different relative heat capacities. The half-time can be estimated for various layer thicknesses when the properties of each layer can be assumed. Furthermore, we can estimate the characteristics of the curve of V versus α_1; thus, we are able to determine the limit of desirable deviation of V from 0.5 for the computer data reduction procedure. Using Figure 1, the optimum sample geometry has been shown to be such that the relative diffusion time falls between one and three depending on the relative heat capacity. This agrees well with the experimental results (Table III).

ACKNOWLEDGMENTS

We would like to express our thanks to Mr. Weldom Vaughn for machining samples. We also wish to acknowledge the help of Messrs. Hans W. Goebel and Hans Groot in the experimental work. The work was performed as part of an NSF Grant from the Engineering Division.

REFERENCES

[1] H. J. Lee, Purdue University, Ph.D. Thesis (1975).

[2] H. J. Lee and R. E. Taylor, NTIS Rept. PB-239, 114/25L (1974).

[3] Y. S. Touloukian, R. W. Powell, C. Y. Ho, and M. C. Nicolaou, "Thermal Diffusivity," Vol. 10 of "Thermophysical Properties of Matter--The TPRC Data Series" (IFI/Plenum Data Corp., New York, 1973), p. 173.

[4] K. F. Sterret, University of Pittsburgh, Ph.D. Thesis (1957).

[5] Y. S. Touloukian and T. Makita, "Specific Heat--Nonmetallic Liquids and Gases," Vol. 6 of "Thermophysical Properties of Matter--The TPRC Data Series" (IFI/Plenum Data Corp., New York, 1970), p. 102.

[6] O. Bryngdahl, Ark. Fys. (Sweden), 21(22), 289 (1962).

[7] H. S. Carslaw and J. C. Jaeger, "Conduction of Heat in Solids" (Oxford University Press, 1959), 2nd Edition, p. 9.

EXACT SOLUTIONS FOR THERMAL CONDUCTANCES OF PLANAR AND CIRCULAR CONTACTS

T. N. Veziroglu, M. A. Huerta and S. Kakaç

University of Miami

Coral Gables, Florida 33124

This work presents a theoretical analysis of the thermal conductances of planar and circular contacts with an interstitial fluid present. By the use of orthogonal functions, exact solutions have been obtained for both cases, without making any of the simplifying and constraining assumptions used in the earlier theories. The solutions have been presented in terms of four dimensionless numbers—a conductance number, a conductivity number, a constriction number and a gap number. The analytical results have been confirmed by the existing experimental data. The solution has also been compared with the previously reported theories. It is found that the earlier theories exhibit significant deviations from the exact solutions.

INTRODUCTION

When two solid surfaces are brought into contact, they actually touch only at a limited number of spots, the aggregate area of which is usually only a small fraction of the apparent contact area. The remainder of the space between the surfaces may be filled with air or another fluid, or may be in vacuum. When heat flows from one solid to the other, flow lines converge toward the actual contact spots, since the thermal conductivities of solids are generally greater than those of fluids. This causes a thermal resistance which is usually high compared to resistances offered to heat flow away from the contact spots.

Thermal contact conductances have been the subject of many investigations, both experimental and theoretical, since such

conductances must be accurately estimated for a reliable heat
transfer analysis of a given system. For theoretical studies,
usually geometrically simple "contact elements" are considered.
The simplest of these has a two-dimensional planar configuration.
It consists of two solids of rectangular cross-sectional area
having an actual strip of contact in the middle of their bases
facing each other with an interstitial fluid between the remainder of the contact surfaces. Theoretical investigations of planar
contacts in vacuum or with interstitial insulations have been
carried out by various researchers: Kouwechoven and Sackett[1]* and
Sackett[2] evaluated the constriction resistance by assuming that
only a wedgelike part of the flow channel was effective in conducting the heat flow. Mikic and Rohsenow[3] derived expressions
for the constriction resistance, in the form of infinite series,
for two specified heat flux distributions at the actual contact.
Veziroglu and Chandra[4] obtained an exact solution for the constriction resistance in this special case of planar contacts.

The next simplest of the contact elements is the two-dimensional circular configuration. It consists of two cylindrical
solids having an actual circular contact in the middle of their
bases, and an interstitial fluid between the remainder of the
contact surfaces. Theoretical expressions for the thermal conductances of circular contacts in vacuum (i.e., contacts with
interstitial insulations) have been derived by Roess,[5] Clausing
and Chao,[6] Tovanovich and Fenech,[7] and Atkins and Fried.[8] For
the general case of contacts with interstitial fluids, Laming,[9]
Fenech and Rohsenow,[10] Shylkov and Ganin,[11] Ozisik and Hughes,[12]
and Veziroglu and Fishender[13] derived analytical expressions
for the thermal contact conductances. Theories abound because
none of them is the exact solution to the problem. They are all
attempts to obtain good approximations to the solution by making
simplifying and sometimes constraining assumptions such as
(1) taking the element radius to be infinite in calculating the
constriction resistance, (2) specifying the flux distribution over
the actual contact region, (3) assuming a uniform flux distribution over the interstitial fluid, (4) dividing the element into
two or more regions with simplified boundary conditions, etc. In
the present analysis, exact solutions are presented for the thermal
conductances of both planar and circular contacts; the results have
been confirmed by the existing experimental data in the literature.

FORMULATION OF PROBLEM

Fig. 1 shows a contact element cross-section which applies to

---

*Superscripted numbers refer to references at the end of the paper.

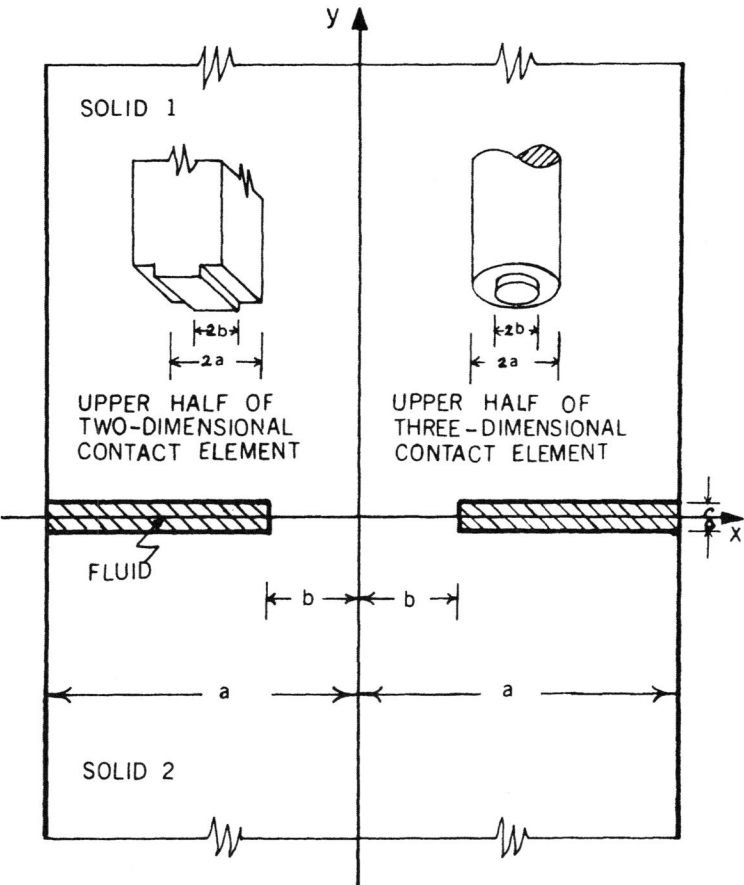

Fig. 1. Contact element cross-section.

both the planar and circular contacts. Translation of the cross-section normal to the xy-plane would generate a planar contact element, and rotation around the y-axis would generate a circular contact element, as shown by the inserts in the figure. The dimension a is the contact element half-width for planar contacts and the contact element radius for circular contacts, and b is the actual contact half-width for planar contacts and the actual contact radius for the circular contacts. The fluid layer thickness, which is small compared to the contact element dimensions, is δ. The x-axis is located so that it divides the fluid layer in the inverse ratio of the thermal conductivities of solids 1 and 2, i.e.,

$$\delta_1 = \delta k_2 / (k_1 + k_2) \tag{1}$$

and

$$\delta_2 = \delta k_1 / (k_1 + k_2) \tag{2}$$

where δ_1 and δ_2 are the fluid layer thicknesses between the x-axis and solids 1 and 2 respectively. Then the system becomes symmetrical from the heat transfer point of view with respect to a plane passing through the x-axis and normal to the y-axis. The system is also symmetrical with respect to the y-axis. Therefore, only the xoy quarter-plane need be considered. The steady state temperature (T) distribution in solid 1 satisfies Laplace's equation.

$$\frac{\partial^2 T}{\partial x^2} + \frac{p}{x} \cdot \frac{\partial T}{\partial x} + \frac{\partial^2 T}{\partial y^2} = 0 \tag{3}$$

where $p = 0$ for planar contacts, and $p = 1$ for circular contacts. The boundary conditions of the problem are:

$$\frac{\partial T(o,y)}{\partial x} = 0 \qquad (0 < y < \infty) \tag{4}$$

$$\frac{\partial T(a,y)}{\partial x} = 0 \qquad (0 < y < \infty) \tag{5}$$

$$\frac{\partial T(x,\infty)}{\partial y} = M \qquad (0 < x < a) \tag{6}$$

$$T(x,o) = 0 \qquad (0 < x < b) \tag{7}$$

and

$$k_1 \frac{\partial T(x,o)}{\partial y} = \frac{k_f}{\delta_1} T(x,o) \qquad (b < x < a) \tag{8}$$

where M_1 is the temperature slope at $y = \infty$, k_1 and k_f are the thermal conductivities of solid 1 and the interstitial fluid respectively, and the temperature along the x-axis is assumed to be zero for convenience.

SOLUTION

It can be shown that the temperature distribution which satisfies Laplace's equation (3) and the boundary conditions (4), (5), and (6) is given by

$$T(x,y) = M_1 y + G_o - \sum_{n=1}^{\infty} G_n \exp(-\lambda_n y) \, \Psi(\lambda_n x) \tag{9}$$

where the G's are constants to be determined and the λ_n's are the eigenvalues given by the positive roots of the equation,

$$\frac{\partial \Psi(\lambda_n a)}{\partial x} = 0 \tag{10}$$

The eigenfunctions of the problem are given by

$$\Psi(\lambda_n x) = \cos(\lambda_n x) \tag{11}$$

for planar contacts, and by

$$\Psi(\lambda_n x) = J_o (\lambda_n x) \tag{12}$$

for circular contacts. In equation (9) a negative sign has been used for the summation so as to make the G_n's positive. Elimination of δ_1 from the boundary condition (8), using equation (1), results in,

$$\frac{\partial T(x,o)}{\partial y} = \frac{2k_f}{\delta k_s} T(x,o) \qquad (b < x < a) \tag{13}$$

where k_s is the harmonic mean of the thermal conductivities of solids 1 and 2, and is defined as follows:

$$k_s = (2k_1 k_2)/(k_1 + k_2) \tag{14}$$

To ensure that the results will also be valid for $k_f = o$ (i.e., the contacts in vacuum), the left hand side of equation (13) can be made the same as that of equation (7) by adding $\alpha T(x,o)/a$ to both sides of equation (13). The dimensionless constant α acts as a damping factor in obtaining convergence as seen later. Division by a ensures dimensional compatability and keeps the dimensionless numbers δ/a (=B) and k_f/k_s(=K) as a group, viz., K/B. After performing the above described operation and combining the result with equation (7), one obtains,

$$T(x,o) = \begin{cases} 0 & (0 < x < b) \\[2ex] \dfrac{a}{\alpha} \dfrac{\partial T(x,o)}{\partial y} + (1 - \dfrac{2ak_f}{\alpha \delta k_s})T(x,o) & (b < x < a) \end{cases} \tag{15}$$

Substituting equation (9) in (15), multiplying both sides by $x^p dx$, integrating over the region x=o to x=b, and nondimensionalizing, one obtains,

$$g_o = \frac{p+1}{C^{p+1}} \sum_{n=1}^{\infty} \frac{g_n}{a^{p+1}} \int_0^b x^p \, \Psi_n \, dx \tag{16}$$

where $\Psi_n = \Psi(\lambda_n x)$, g the dimensionless coefficient and C the constriction number (= b/a).

To determine the dimensionless coefficients, g_n, we substitute equation (9) in (15), multiply both sides by $x^p \Psi(\lambda_m x)dx$, integrate over the region x=o to x=a, and substitute for g_o from equation [16]. The resulting expression, in dimensionless form is

$$\int_b^a x^p \Psi_m dx = g_m \alpha \int_o^a x^p \Psi_m^2 dx + \sum_{\substack{n=1 \\ n \neq m}}^{\infty} g_n \left[\left(\alpha - 2\frac{K}{B} - a\lambda_n \right) \int_b^a x^p \Psi_n \Psi_m dx \right.$$

$$\left. + \left(\alpha - 2\frac{K}{B} \right) \frac{p+1}{A^{p+1}C^{p+1}} \int_o^b x^p \Psi_m dx \int_o^b x^p \Psi_n dx \right] \quad (m=1,2,..,\infty) \tag{17}$$

Equations (17) form a set of linear equations in the unknown g_n or g_m. The approximate values of the unknowns can be calculated by considering a large number (N) of the unknowns and an equal number of equations (17). As the number N increases, the accuracy of the approximation improves. After the coefficients g_1 through g_N are calculated, the zeroth coefficient g_o can be calculated using equation (16).

The additional temperature drop, ΔT_1, produced in solid 1 as a result of the contact is given by,

$$\Delta T_1 = \lim_{y \to \infty} \left(T - y\frac{\partial T}{\partial y} \right) \tag{18}$$

Substituting equation (9) in (18), and taking the limit, one obtains,

$$\Delta T_1 = g_o a M_1 \tag{19}$$

The thermal contact conductance for the solid 1 side per unit of the apparent contact area is given by,

$$u_1 = \frac{k_1}{\Delta T_1} \frac{\partial T(x,\infty)}{\partial y} \tag{20}$$

Substituting equation (6) in (20) gives,

$$u_1 = (k_1 M_1)/\Delta T_1 \tag{21}$$

The overall thermal contact conductance per unit apparent contact area becomes,

$$u = (\frac{\Delta T_1}{k_1 M_1} + \frac{\Delta T_2}{k_2 M_2})^{-1} \tag{22}$$

since the conductance components of the solid 1 and solid 2 sides are in series. The heat flow rates in both solids are equal, i.e.,

$$k_1 M_1 = k_2 M_2 \tag{23}$$

Also the temperature drops T_1 and T_2 of both solids are inversely proportional to the thermal conductivities k_1 and k_2, or

$$k_1 \Delta T_1 = k_2 \Delta T_2 \tag{24}$$

Substituting equations (23) and (24) in (22), one obtains,

$$u = (k_s M_1)/(2\Delta T_1) \tag{25}$$

Combining equations (16), (19) and (25), the dimensionless contact conductance number, U, becomes,

$$U = \frac{a^{p+1} C^{p+1}}{2(p+1) \sum\limits_{n=1}^{\infty} g_n \int\limits_o^b x^p \psi_n dx} \tag{26}$$

Planar Contacts

Substituting equation (11) in (10), (17) and (26), and carrying out the indicated operations, the thermal conductance relationship for planar contacts in dimensionless form becomes.

$$U = \frac{C}{2 \sum\limits_{n=1}^{\infty} \frac{g_n}{n\pi} \sin(n\pi C)} \tag{27}$$

where the dimensionless constants, g_n, are evaluated from,

$$\frac{2\sin(m\pi C)}{m\pi} = g_m \left[\alpha - (\alpha - 2\frac{K}{B}) \frac{2\sin^2(m\pi C)}{m^2 \pi^2 C} + (\alpha - 2\frac{K}{B} - m\pi) \left[\frac{\sin(2m\pi C)}{2m\pi} \right. \right.$$

$$- 1+C] \Bigg] + \sum_{\substack{n=1 \\ n \neq m}}^{\infty} g_n \left[(\alpha-2\frac{K}{B} - n\pi) \; [\frac{\sin(n+m)\pi C}{(n+m)\pi} + \frac{\sin(n-m)\pi C}{(n-m)\pi}] \right.$$

$$\left. - (\alpha-2\frac{K}{B})\frac{2\sin(m\pi C)\sin(n\pi C)}{mn\pi^2 C} \right] \qquad (m=1,2,..,\infty) \qquad (28)$$

Circular Contacts

Similarly, for circular contacts the dimensionless thermal contact conductance becomes,

$$U = \frac{C}{4 \; \sum_{n=1}^{\infty} \frac{g_n}{\beta_n} \; J_1(\beta_n C)} \qquad (29)$$

where the dimensionless eigenvalues, β_n, are the positive roots of,

$$J_1(\beta_n) = 0 \qquad (30)$$

and the dimensionless constants, g_n, are evaluated from,

$$\frac{2CJ_1(\beta_m C)}{\beta_m} = g_m \left[(\alpha-2\frac{K}{B}\beta_m)[C^2 J_0^2(\beta_m C)+C^2 J_1^2(\beta_m C)-J_0^2(\beta_m)] \right.$$

$$\left. + \alpha \; J_0^2(\beta_m)-(\alpha-2\frac{K}{B})\frac{4J_1^2(\beta_m C)}{\beta_m^2} \right] - \sum_{\substack{n=1 \\ n \neq m}}^{\infty} g_n \left[(\alpha-2\frac{K}{B})\frac{4J_1(\beta_m C)J_1(\beta_n C)}{\beta_m \beta_n} \right.$$

$$\left. - (\alpha-2\frac{K}{B}\beta_n)\frac{2C}{\beta_n^2-\beta_m^2}[\beta_n J_0(\beta_m C)J_1(\beta_n C)-\beta_m J_1(\beta_m C)J_0(\beta_n C)] \right]$$

$$(m=1,2,..,\infty) \qquad (31)$$

RESULTS AND DISCUSSION

The general solutions (equations (27) and (29) for the thermal conductances of planar and circular contacts with interstitial fluids are in the form of infinite series. To obtain exact numerical results all the terms must be evaluated. This is not possible in practice. It then becomes important to know the number of series terms to be used for a given accuracy. Fortunately a closed form exact solution has been obtained for planar contacts in vacuum (i.e., K=o), with which the present theory can be compared. This special case closed form solution can be written as,

$$U_{co} = \frac{\pi}{4\ln\left[\frac{1}{\sin(\pi C/2)}\right]} \tag{32}$$

where U_{co} is the dimensionless contact conductance of the closed form solution. The percentage deviation of the series solution, defined as $100(U_o-U_{co})/U_{co}$, has been calculated for several cases by varying the constriction number C between 0.1 and 0.9 and the factor α between 100 and 900 for up to 90 terms (N) in the series. Some typical results are shown in Fig. 2. It can be seen from the figure that α acts as a damping factor in obtaining convergence and that as N increases the series solution first intersects the N-axis and then approaches the exact solution asymptotically. For C = 0.5 the relationship between α and N at the intersection points is given by,

$$\alpha = 8N - 7 \tag{33}$$

Using α given by this formula, the percentage deviation has been calculated for nine different C's and five different N's. The results are presented in Table I. As seen from the table, the percentage deviation is less than 1% even for N = 15, and reduces to 0.05% for N = 90. Such small deviations become possible by judicious use of the damping factor α.

For comparison of the solutions for the thermal conductances of circular contacts having interstitial fluids, the earlier theories given in references 9, 10, 12 and 13 have been rewritten in terms of the dimensionless numbers used and the percentage deviation of each earlier theory (U) from the present theory (U_o), defined as $100(U-U_o)/U_o$, has been calculated for several combinations of constriction number (C) and the ratio of the conductivity and gap numbers (K/B). The results are presented in Table II. In calculating the conductance number of the present theory (equation (29)), α was taken as 900 and the first 90 terms of the infinite series were used. Under these conditions, one would expect the deviation from the exact (limiting) solution to be of the order of 0.2% (see Fig. 2), assuming that a correspondence exists between the solutions of conductances for planar and circular contacts. A study of Table II shows that all the earlier theories have important deviations from the exact solution. In general, the deviations are larger for larger values of the constriction number. The theories of references 9 and 13 agree reasonably well with the exact theory for smaller values of the

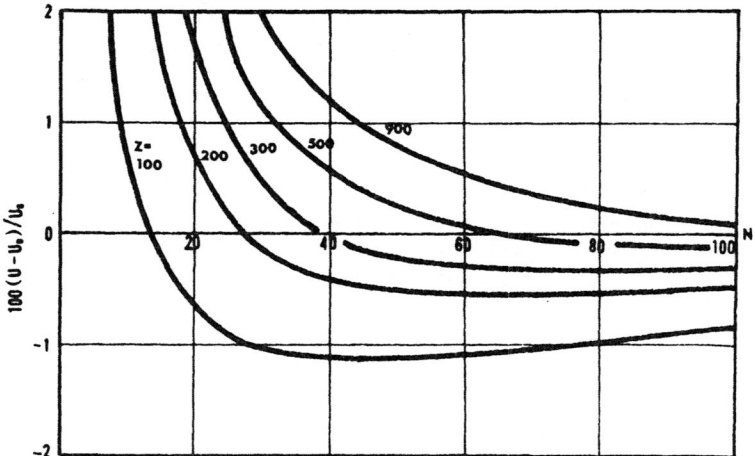

Fig. 2. Deviation of series solution of planar contact conduc-
         tance from closed form solution for K=0 and C=0.5.

Table I. Comparison of Closed Form and Series Solutions for Con-
          ductance of Planar Contacts in Vacuum

C	U_{co}	Percentage Deviation = $100(U_o - U_{co})/U_{co}$				
		N=15	N=45	N=60	N=75	N=90
0.1	0.423	0.18	0.07	0.09	0.04	0.03
0.2	0.669	0.34	0.05	0.03	0.02	0.01
0.3	0.995	0.07	0.02	0.01	0.01	0.01
0.4	1.478	0.06	0.00	0.00	0.01	0.00
0.5	2.266	0.07	0.01	-0.01	0.00	0.00
0.6	3.706	-0.04	-0.01	-0.01	-0.01	-0.00
0.7	6.805	0.23	0.03	-0.01	-0.04	-0.03
0.8	15.651	0.10	-0.00	-0.02	-0.07	-0.05
0.9	63.502	0.76	0.35	0.00	0.09	0.05

Table II. Comparison of Exact and Approximate Solutions for Conductance of Circular Contacts

C	K/B	U_o	Percentage Deviation from Exact Solution = $100(U-U_o)/U_o$			
			Rf.9	Rf.10	Rf.12	Rf.13
0.1	10^{-4}	0.07432	-1	63	-91	-8
	10^{-2}	0.0844	-1	57	-81	-6
	1	1.091	-2	14	-9	1
	10^2	101.23	-1	11	-2	0
0.3	10^{-4}	0.32583	-1	21	-82	-21
	10^{-2}	0.3371	-1	22	-80	-20
	1	1.454	-9	25	-34	-2
	10^2	110.52	-9	28	-18	0
0.5	10^{-4}	0.93749	2	15	-83	-32
	10^{-2}	0.9514	1	14	-82	-31
	1	2.330	-16	13	-61	-7
	10^2	135.14	-25	18	-44	-26
0.7	10^{-4}	3.17993	11	-48	-90	-46
	10^{-2}	3.2006	11	-48	-90	-45
	1	5.260	-14	-36	-84	-23
	10^2	201.41	-49	-13	-75	-49
0.9	10^{-4}	31.59068	-446	-82	-98	-74
	10^{-2}	31.6469	-445	-82	-98	-74
	1	37.257	-391	-77	-98	-63
	10^{-2}	570.30	-102	-47	-98	-82

constriction number. For the model used for circular contacts, reference 10 presents some experimental results for three different interstitial fluids. Fig. 3 shows a comparison of these experimental results for C = 0.25 with the theories under consideration. It can be seen that the present theory agrees well with the experiments. For the constriction number in question (i.e., C = 0.25), the theories of references 9 and 13 give reasonable agreement while the theory of reference 10 overpredicts and that of reference 12 underpredicts.

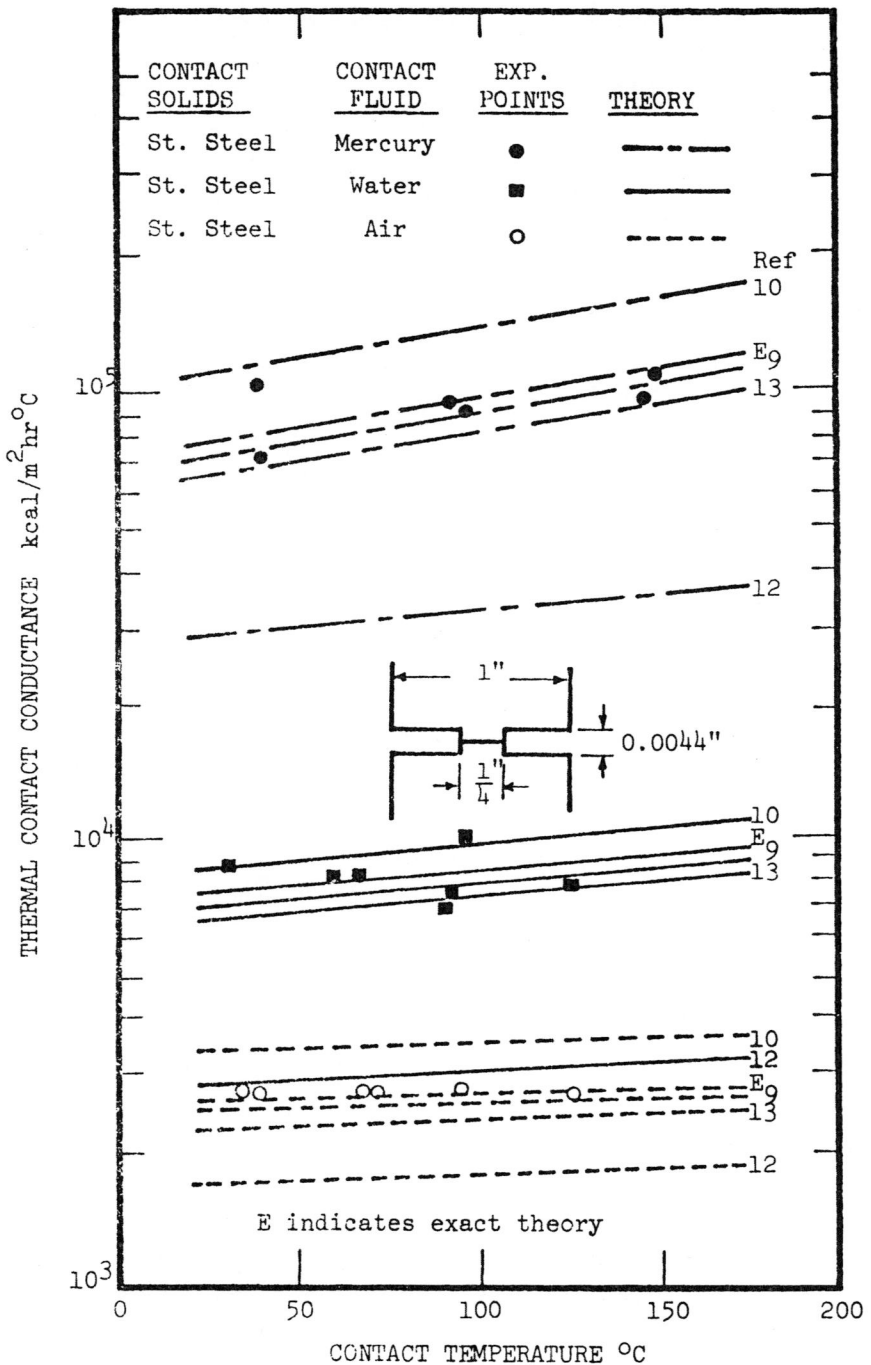

Fig. 3. Comparison of theoretical solutions with experimental
         results of Ref. 10 for C = 0.25.

CONCLUSION

Exact solutions have been obtained for the thermal conductances of planar and circular contacts with interstitial fluids. The solutions are expressed in the form of infinite series, the coefficients of which are determined by a set of linear equations. The solutions are also valid for the special cases of planar and circular contacts in vacuum.

ACKNOWLEDGMENTS

The research reported herein was sponsored by the National Aeronautics and Space Administration under Grant NGR 10-007-010 at the University of Miami, Coral Gables, Florida, and by the Scientific and Technical Research Council of Turkey under Project MAG-196 at the Middle East Technical University, Ankara, Turkey. The authors wish to express their appreciation to Dr. O. Yeşin, Dr. Y. Göğüş, Dr. H. Yüncü, K. Civci, and A. Tepebag of the Middle East Technical University, and S. Mathavan, P. Kondis, R. Chungo, and J. Myers of the University of Miami for their interest and assistance.

REFERENCES

1. Kouvenhoven, W. B., and Sackett, Jr., W. T. "Electrical Reistance Offered to Non-iniform Current Flow," Welding Research Supplement, 466s-470s (Oct. 1949).

2. Sackett, Jr., W. T. "Contact Resistance," Ph.D. Dissertation, the Johns Hopkins University (1950).

3. Mikic, B. B., and Rohsenow, W. M. Appendix D, "Thermal Contact Resistance," MIT, Heat Transfer Laboratory Report No. DSR 74542-41, Contract No. NGR 22-009-065 (Sept. 1966).

4. Veziroglu, T. N., and Chandra, S. "Thermal Conductance of Two-Dimensional Constrictions," Progress in Astronautics and Aeronautics, 21, 591-615 (1969).

5. Roess, L. C. "Theory of Spreading Conductance," Appendix A, "Thermal Resistance Measurement of Joints formed between Stationary Metal Surfaces," by Wells, N. D. and Ryder, E. A., Trans. Am. Soc. Mech. Engrs., 71 (1949).

6. Clausing, A. M., and Chao, B. T. "Thermal Contact Resistance in a Vacuum Environment," J. Heat Transfer, 27, 243-251 (May 1965).

7. Yovanovich, M. M., and Fenech, H. "Thermal Contact Conduc-
    tance of Nominally-flat, Rough Surfaces in a Vacuum Environ-
    ment," AIAA 3rd Aerospace Sciences Meeting, Paper No. 66-42,
    (Jan. 1966).

8. Atkins, H. L., and Fried, E. "Thermal Interface Conduc-
    tance in a Vacuum," 1st Annual AIAA Meeting, Paper 64-253
    (June-July 1964).

9. Laming, L. C. "Thermal Conductance of Machined Metal Con-
    tacts," Am. Soc. of Mech. Engrs. International Developments
    in Heat Transfer, Part I, 65-76 (1962).

10. Fenech, H., and Rohsenow, W. M. "Prediction of Thermal
    Conductance of Metallic Surfaces in Contact," J. Heat
    Transfer, 85C, 15-24 (1963).

11. Shylkov, Yu. P., and Ganin, Ye. A. "Thermal Resistance of
    Metallic Contacts," Intern. J. Heat Mass Transfer, 7, 921-
    929 (1964).

12. Ozisik, M. N., and Hughes, D. "Thermal Contact Conduc-
    tance of Smooth to Rough Contact Joints," American Society
    of Mechanical Engineers Paper 66-WA/HT-54 (November 1966).

13. Cetinkale (Veziroglu), T. N. and Fishenden, M. "Thermal
    Conductance of Metallic Surfaces in Contact," General Dis-
    cussion of Heat Transfer, Proc. Inst. Mech. Engrs. and
    Am. Soc. Mech. Engr., 271-275 (1951).

INVESTIGATION OF THE IMPROVEMENT OF THE THERMAL CONTACT CONDUCTANCE BETWEEN A BASEPLATE AND A HONEYCOMB PLATFORM USING VARIOUS INTERSTITIAL MATERIALS

A. Wild
HSD, Stevenage, England
and
J.-P. Bouchez
ESTEC, Noordwijk, The Netherlands

A series of tests have been performed using a representative satellite unit baseplate and aluminium honeycomb platform assembly to determine the decrease in interface temperature differentials which are achieved by introducing interstitial fillers. The compounds used, Eccotherm TC4, C6-1102 and DC 340, were all silicone greases; tests were also performed on bare junction assemblies to provide control data, and on a carbon fibre skinned honeycomb assembly, using TC4 only. The derived contact coefficient values of 0.036 and 0.027 watts/cm^2.$^\circ$C for TC4 with aluminium skinned, and carbon fibre skinned honeycomb, respectively, 0.024 watts/cm^2.$^\circ$C for C6-1102, 0.014 watts/cm^2.$^\circ$C for DC 340, and 0.0018 to 0.0027 watts/cm^2.$^\circ$C for bare junctions, were apparently constant over the test power dissipation range of 5 to 40 watts.

Due to difficulties involved in the measurement of interface temperature differentials for high contact coefficient values, a computer conductance model was constructed, and a computer matching technique developed in order to minimise test errors.

An unfilled assembly, and those containing TC4 and C6-1102, were subjected to representative vibration and thermal cycling levels. Although the bare junction contact coefficient was reduced by almost 50% by exposure to vibration, no measurable effects were noted for the filled junctions. Exposure to thermal cycling had no apparent effect on either the filled or unfilled junctions.

Additional comparative tests were performed for a number of interface materials, using small test samples. These showed that RTV-11 was superior to the silicone greases, but displayed an unacceptable dependence on interface temperature and a lack of repeatability in the joint conductance values that were obtained.

1. INTRODUCTION

It is known that the introduction of selected interstitial compounds between the mating surfaces of a joint improves its contact conductance. The influence of such interstitial materials has been studied by many investigators (1-8) and it is apparent that the test method has a significant effect on the results of contact heat transfer coefficient. The compressed cylinder column technique employed by most experimenters produces values which are at least an order of magnitude greater than those obtained by testing bolted plate samples due to the non-representative pressure distribution. It is therefore imperative that the test configuration simulates as closely as possible the joint design. An experimental program was initiated to obtain the contact conductance representative of a power dissipating component mounted on a typical honeycomb platform and subjected to an environment similar to that expected for the Orbital Test Satellite (OTS).

The contact of such a component with a platform is characterized mainly by a low interface pressure outside of the bolted area and there exists no general relationship to convert the bolt torques into an effective pressure across the interface. A literature search produced the following observations relevant to our investigation.

a)   The variation of conductance over a moderate pressure range is not great when an interface filler is used. The interface pressure is not important when using thermal greases (1, 2) but becomes important for bare metal joints (2).

b)   The use of metal shims will not result in good reproducibility for low contact pressure (3).

c)   No reasonable method of measuring the average interface pressure between a baseplate and a thin skinned honeycomb panel is available. Insertion of a measuring device modifies the joint by distorting the profile of the face skin.

d)   Torque load is usually specified for the mounting of a component and it is sensible to correlate the average conductance with this parameter. The bolt load produced for a given torque varies considerably, in particular with the insert thread variations produced by removal and retorquing of bolts. This adds another uncertainty in the bolt load (for a given torque) to the uncertainty in average interface pressure (for a given bolt load).

e)   A method to determine the pressure under the head of mounting bolts was defined by Aron and Colombo (4), but it is unlikely to correlate well with the average interface pressure because of the waviness in the baseplate and/or honeycomb panel and the insert position which may be slightly proud of or buried in the honeycomb.

f)   Flatness and roughness leads to somewhat inconclusive comments.
        They are important criteria for metal to metal contacts but
        become less important with interstitial compounds.   Roca and
        Mikic (5) state that it may not be possible to determine if
        increasing or decreasing the surface roughness indeed increases
        or decreases the total thermal resistance of a joint.

2.   TEST PROGRAM AND DESCRIPTION OF TEST SAMPLES

        As many of the factors which influence the contact conductance
are determined by spacecraft design constraints, this investigation
was primarily conducted to select a suitable space-qualified interstitial compound and study the variation of its conductance value
under representative environments.   Small sample tests were carried
out to select the suitable fillers and these were then studied in
detail using representative baseplates/honeycomb assemblies.

2.1   Small Test Samples

        The small test samples (Figure 1) were only intended to provide
a relative basis for comparison between the different candidate materials.   The small test pieces were all manufactured to the same size
(35 x 25 x 2mm) and nominal surface finish.

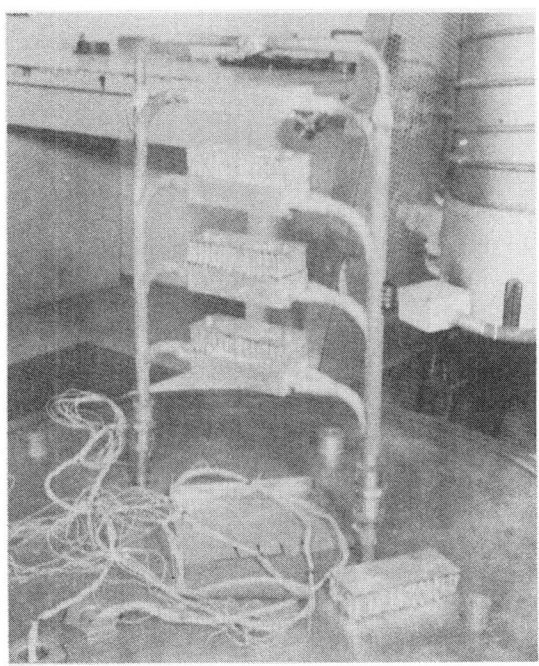

FIGURE 1 : SMALL TEST SAMPLES

Three materials were used as representative of possible component baseplates (aluminium, stainless steel and magnesium) and four samples were produced for each material to permit multiple testing. Flatness and roughness properties were measured and the aluminium and stainless test pieces were found to be smoother than the magnesium samples. The degree of variation was, however, not considered to be significant for this application. A more detailed description of the samples and of the test set-up is given by Wild (9).

A MINCO etched copper foil heater mat was bonded on each baseplate using a hardening silicone rubber compound. The power dissipation of each heater was regulated using a variable DC power supply capable of delivering over 20W. The test pieces were bolted to a small honeycomb platform with face skins either of aluminium or carbon fibre; the honeycomb itself was manufactured out of aluminium. The platforms were in turn bolted to cooling copper blocks using silicone grease as an interface filler. A cooling fluid was circulated through the blocks maintaining them at a fixed temperature.

The overall test assembly was instrumented with thermocouples coated with a silicone heat sink compound to ensure a good thermal contact. Multilayer insulation was wrapped around each test sample and the device was placed in a small vacuum chamber with a controllable shroud.

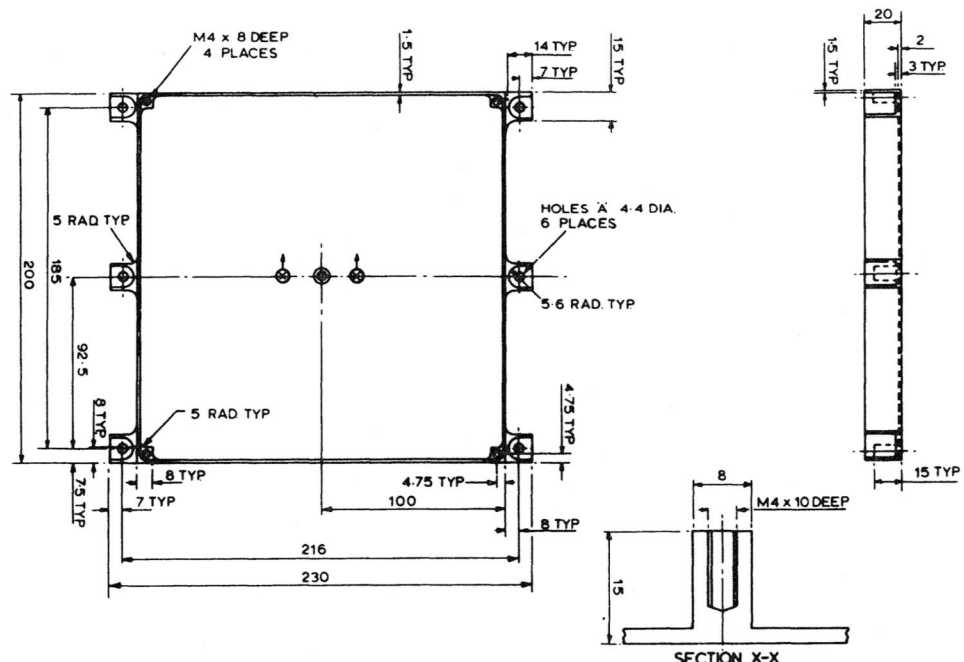

FIGURE 2 : SIMULATED BOX BASEPLATE

2.2   Simulated Baseplates

Two simulated baseplates were machined from solid aluminium and designed to be representative of an OTS unit in terms of materials, mounting foot sizes, number of feet and material thickness (Figure 2). The contact points for the mass representation were chosen to ensure a realistic transmission of the vibrational stresses to the baseplate. Required finishes of the mating faces were defined from the OTS specifications (flatness 0.05mm, roughness 3.2um) and measured for each of the baseplates (9).

A mass representation was used during the vibration testing to simulate the effects of a unit of 5kg total weight.   The mass was manufactured to give the correct centre of gravity and thermal capacity of a typical electronics unit of this base size.

The simulated baseplates were mounted on a honeycomb platform of size similar to the OTS platform (Figure 3).   Two platforms were manufactured, one with aluminium face skins and the other with face skins made out of two carbon fibre layers.

Each simulated baseplate was instrumented with four MINCO heaters which were separately controlled, allowing the possibility of non-uniform power dissipation.   Thermocouples were used for temperature readings and the overall assembly was wrapped in multilayer insulation before being placed in a Lintott thermal vacuum chamber with independently controllable shroud and cooling block supplies.

FIGURE 3 :   SIMULATED BASEPLATE TEST ASSEMBLY

3.   RESULTS AND DISCUSSION

3.1   Small Sample Test

Candidate materials were first chosen for their thermal characteristics at atmospheric pressure but they also had to satisfy stringent degassing criteria. Only six materials were selected and this number was reduced to three after small sample testing. The materials selected were:

- DC 340

- C6-1102

- Eccotherm TC 4

- RTV 11

- Choseal 1215

- Indium foil

All of the candidate materials were first tested with the aluminium baseplate and only the most promising three interstitial compounds were subjected to the same tests with stainless steel and magnesium baseplates. In all cases, a bare junction was present for comparison. The test assembly was then refurbished to incorporate the two carbon fibre face skin honeycomb test pieces which were only tested with aluminium and magnesium baseplates. Results are given in Figures 4 to 7; the indicated values are not absolute but can only be interpreted in a relative sense. The reduction of such data was then performed in a simplified manner.

3.1.1 Discussion. From the results, it is evident that RTV 11, Eccotherm TC4, DC 340 and C6-1102 improve appreciably the contact conductance.

RTV 11 appears to be the best candidate material but its contact heat transfer coefficient is a function of the power dissipation and decreases as the power increases. This was also noted in a previous experiment (10) performed with higher power densities. For this reason and the fact that it is only marginally accepted for outgassing, this compound was discarded from further testing with simulated baseplates.

Eccotherm showed more promise of repeatability with units which may have to be removed and replaced a number of times during satellite integration. It was considered the best candidate.

Indium foil did not prove to be suitable for use with honeycomb platforms.

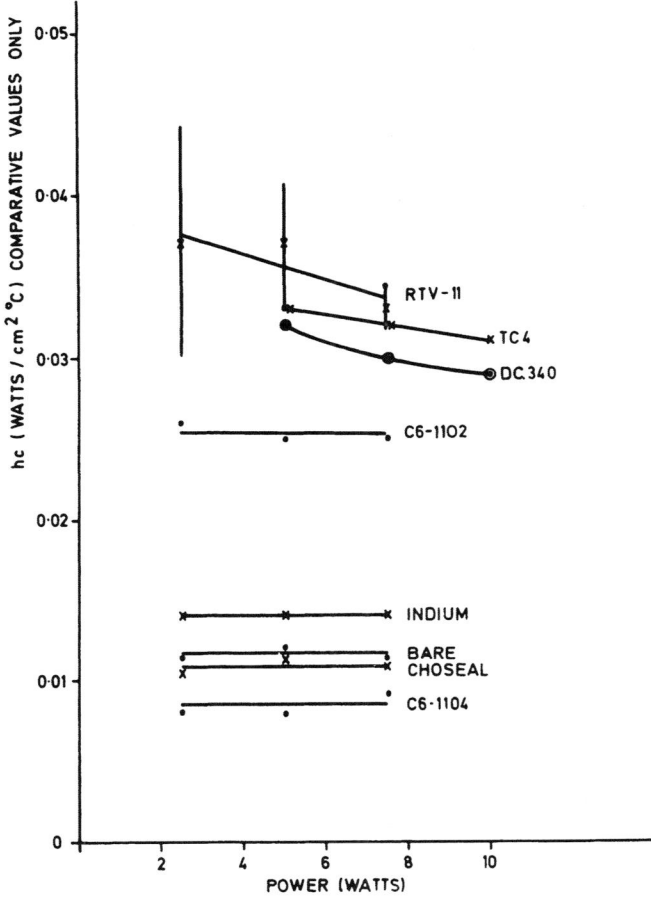

FIGURE 4 : SMALL TEST SAMPLES — ALUMINIUM BASEPLATES

3.2   Simulated Box Baseplate

3.2.1   Test Sequence.   Two aluminium baseplates were manufactured for this part of the investigation and three separate tests were
conducted.   Bolt torque was 18 lb.in for the whole sequence.

Preliminary tests were performed using two aluminium honeycomb
platforms.   DC 340 was used under one of the baseplates and the other
interface was left bare.   Uniform power levels of 5, 10, 15, 25 and
40W were applied on the filled junction while the bare junction was
subjected to power levels of 5, 10, 15, 25 and 30W.   A non-uniform
dissipation of 10W was also applied to define the variation of the
contact heat transfer coefficient under such a condition.

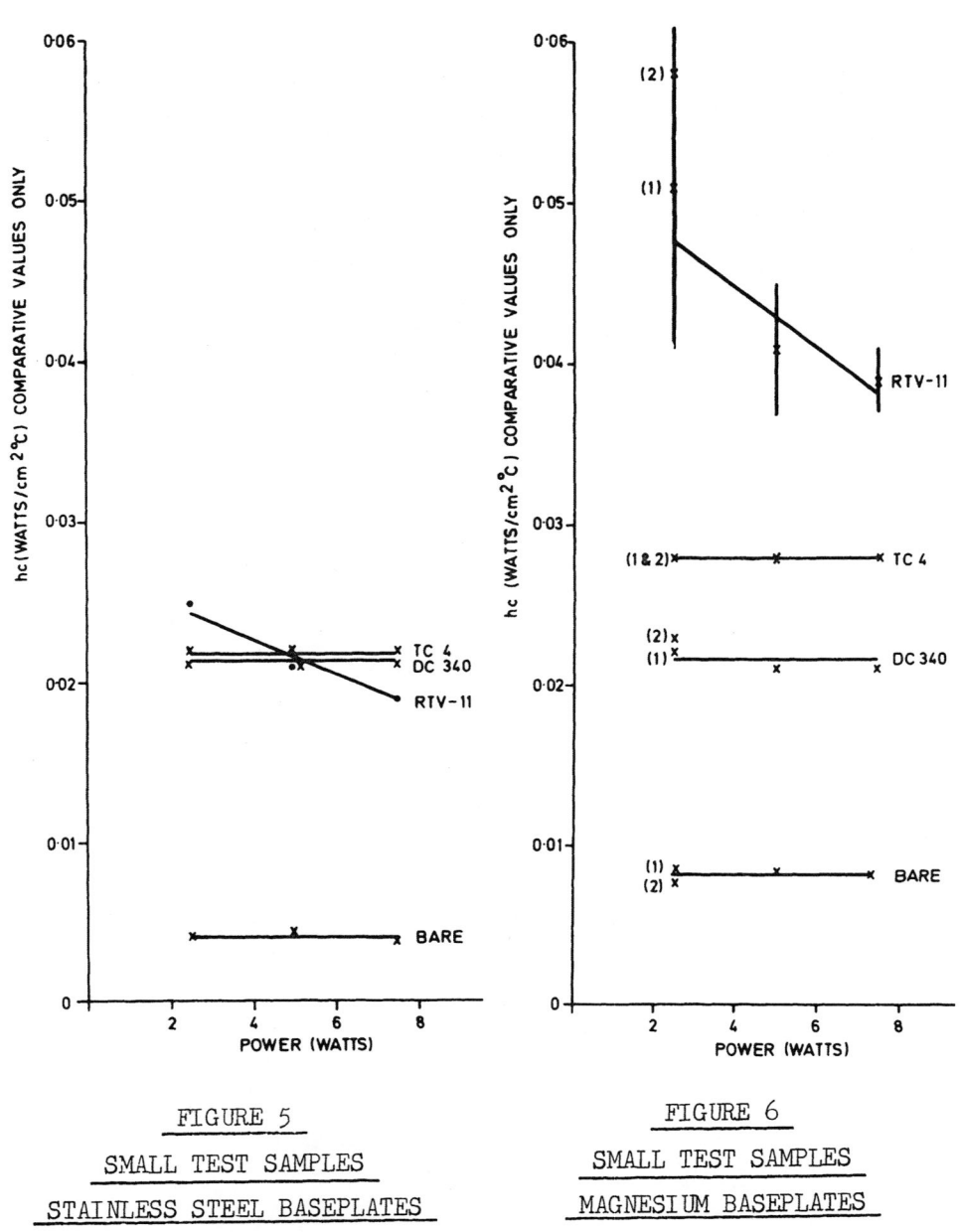

FIGURE 5

SMALL TEST SAMPLES

STAINLESS STEEL BASEPLATES

FIGURE 6

SMALL TEST SAMPLES

MAGNESIUM BASEPLATES

The assemblies were removed from the chamber and the interstitial compound was changed. The previously bare junction was disassembled and rebolted with C6-1102 (sealed with C6-1104 which is a poor thermal conductor). The joint which had previously been filled was cleaned and rebolted bare. Uniform power distributions were applied

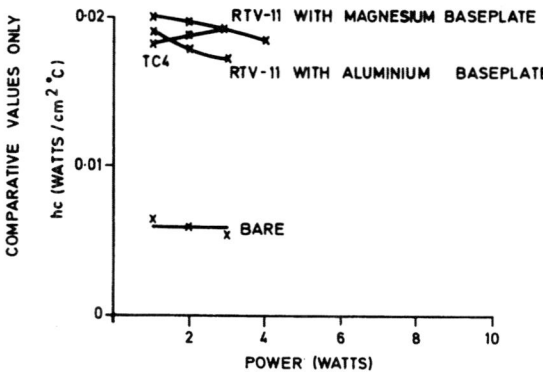

FIGURE 7 : SMALL TEST SAMPLES - CARBON FIBRE HONEYCOMB

at different levels.   The assemblies were then vibrated, subjected
to thermal cycling and then re-tested in the vacuum chamber for comparison.

     The final test configurations used TC4 for both assemblies but
with different honeycomb platforms; one with aluminium face skin and
one with carbon fibre face skin.   Uniform and non-uniform power distributions were applied.   Vibration and thermal cycling followed with
a repeat of the initial test at 10W uniformly distributed for assess-
ment of changes.

     Levels of vibration and thermal cycling tests are fully descri-
bed in (9) and were typical of qualification level of OTS.

     3.2.2 Analysis of Results.   Baseplates and heaters were ade-
quately monitored whereas it was impossible to provide sufficient
thermocouples on the honeycomb in the immediate vicinity of the baseplate to accurately determine the temperature distribution beneath
the baseplate.   The analysis was therefore performed using a small
38 node model.

     A first estimate of the interface heat transfer coefficient was
obtained by dividing the conductive power transfer by the temperature
differential, derived by subtracting the mean of the two temperature
measurements of the platform beneath the baseplate from the mean baseplate temperature (Figure 8).   An analysis was performed with the
mathematical model to obtain a complete profile of the temperature
distribution of the platform.   It had been previously noted that a

large change in the interface conductance only caused a small change
in the temperature distribution of the platform.   For an order of
magnitude increase in contact coefficient, the platform temperatures
only vary by a maximum of $1.5^{\circ}C$.   The values of the seven thermocouples located on the platform were compared with the prediction
and the four nodes directly under the baseplate were adjusted to
give 'best fit' temperatures of the upper face skin which coincided
with the general temperature profile observed during the test.   In
this way, the final heat transfer coefficient is dependent on the
mean error of seven measurements.

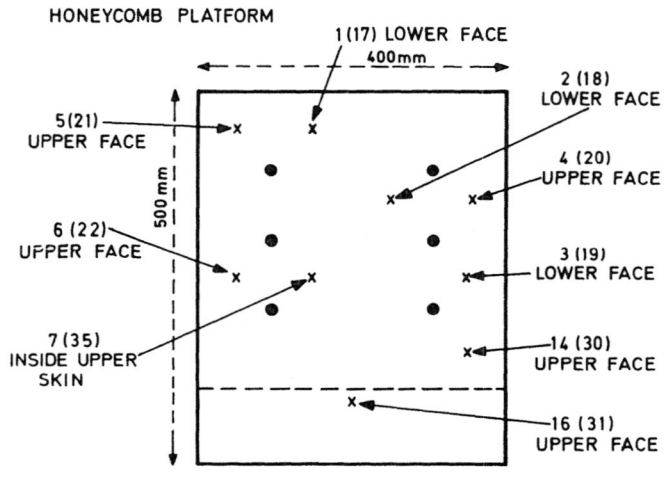

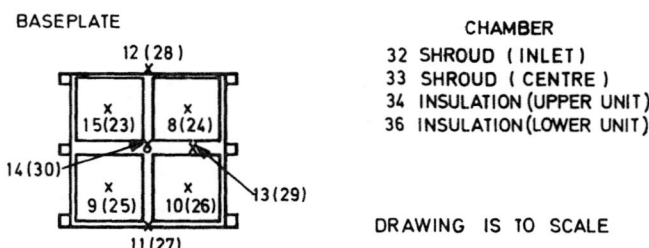

FIGURE 8 : SIMULATED BASEPLATE ASSEMBLY - THERMOCOUPLE POSITIONS

The effective contact coefficient was calculated for each of the test configurations by dividing the power dissipation by the total baseplate mating surface area (including the feet), and the derived mean temperature differential.   This yields an average effective heat transfer coefficient and it is to be noted that the local value under the feet is much higher than under the centre of the baseplate.

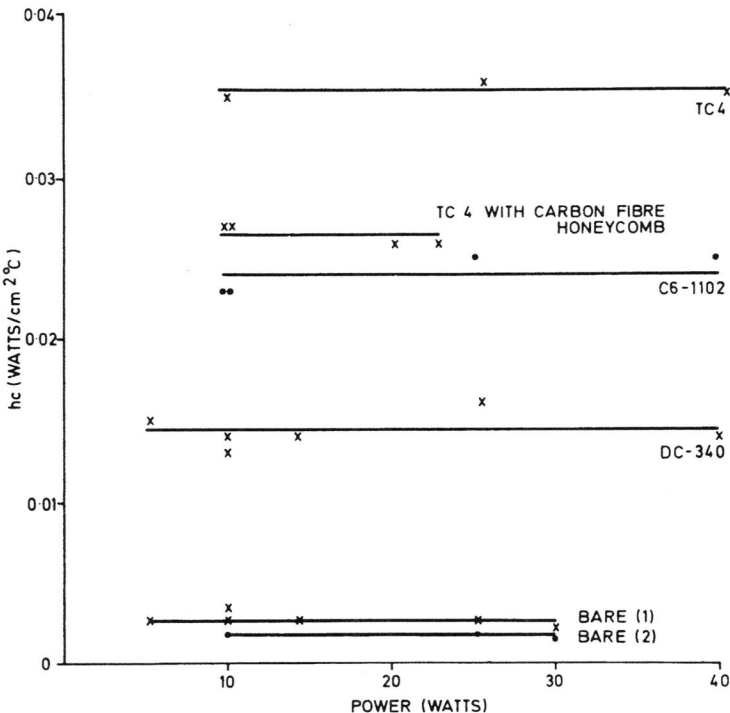

FIGURE 9

SIMULATED BASEPLATE TESTS

DERIVED CONTACT HEAT TRANSFER COEFFICIENTS

3.2.3 Comparison of Results. Results are presented in
Figure 9 and a short discussion on each filler follows.

a) Bare Junctions. Two assemblies were tested with bare junctions;
the mean derived contact coefficients are 0.0027 and 0.0018 W/
(cm^2K). This corresponds to a difference of approximately 9°C
in the mean baseplate/platform temperature differentials over
the test power distribution range and it is therefore not attributed to measurement errors. This variation seems to be due to
the non-repeatability between bare junctions caused by variation
in bolt torques, insert height and minor distorsions resulting
from dimensional variations. All these effects change appreciably the interface pressure which is the most important factor
affecting the contact coefficient of bare joints (2).   Only
one plate was subjected to vibration and it had drastic results.
The heat transfer coefficient decreased from 0.0018 to 0.0010 W/
(cm^2K). Thermal cycling had a slight recovery influence which
is considered to be within the allowable test error.

b) DC 340. Tests on this component give a mean contact heat transfer coefficient of 0.0144 W/(cm^2K) which remains constant over the
range of power dissipation studied. Vibration and thermal cycling has no appreciable effect.

c) C6-1102. For this interface filler, the mean value of the contact coefficient over the test range is 0.024 W/(cm^2K) and a
slight variation of this coefficient after vibration and thermal
cycling is within measurement errors.

d) TC4. This interface filler was used with the two different
types of honeycomb platforms with aluminium and carbon fibre
face skin. Values of 0.0355 and 0.0265 were found respectively. The test power range had to be limited for the carbon
fibre face skin due to smaller platform conductivity and the
low skin/cooler contact conductance. It is not certain if,
in this case, the resultant decrease of contact coefficient
with increasing power is related to a change in the thermal
conductivity of the carbon fibre itself, or if the variation
is to be attributed to measurement tolerances. Vibration
and thermal cycling does not appear to have any effect.

4. CONCLUSIONS

This investigation has led to the following conclusions regarding the introduction of an interstitial compound in a joint to be used under space environment.

a) Interstitial compounds increase the contact conductance by an order of magnitude when used between a component and a honeycomb platform.

b) The interface pressure is very important for a bare junction, and the values of the contact coefficient between a baseplate and a honeycomb platform is much smaller than between two solid plates.

c) Repeatability of bare junctions is poor and variations of 50% of the contact coefficient can be expected.

d) In most cases, the power level does not have any influence on the value of the contact heat transfer coefficient.

e) Vibration and thermal cycling does not deteriorate the joint if filled with an interstitial compound. This is not true however for a bare junction where 45% decrease of the coefficient was found after vibration.

f) A carbon fibre skin honeycomb has a lower, but still acceptable, value of the contact coefficient. The difference is attributed to a superior small scale flatness of the aluminium skin resulting from a lack of depressions caused by the core formation.

5. ACKNOWLEDGEMENTS

This work was performed by Hawker Siddeley Dynamics Limited under contract number 2132/73/JS from the European Space and Technology Centre located in Noordwijk, The Netherlands.

REFERENCES

1. L. S. FLETCHER,
    "A Review of Thermal Control Materials for Metallic Junctions".
    AIAA paper No. 72-284.

2. R. C. GETTY, R. E. TATRO,
    "Spacecraft Thermal Joint Conduction".
    AIAA paper No. 67-316.

3. G. R. CUNNINGTON,
    "Thermal Conductance of Filled Aluminium and Magnesium Joint
    in a Vacuum Environment".
    ASME paper No. 64-WA/HT-40.

4. W. ARON, G. COLOMBO,
    "Controlling Factors of Thermal Conductance Across Bolted
    Joints in a Vacuum Environment".
    ASME paper No. 63-WA-196.

5. R. T. ROCA, B. B. MIKIC,
    "Thermal Conductance in a Bolted Joint".
    AIAA paper No. 72-282.

6. H. L. ATKINS, E. FRIED,
    "Thermal Interface Conductance in a Vacuum".
    AIAA paper No. 64-253.

7. J. T. BEVANS, T. ISHIMOTO,
    "Prediction of Space Vehicle Thermal Characteristics".
    AFFDL-TR-65-139.

8. H. J. SAUER, C. R. REMINGTON, W. E. STEWART, J. T. LIN,
    "Thermal Contact Conductance with Several Interstitial Materials".
    N72-25927.

9. A. WILD,
    "The Investigation of Thermal Contact Conductances of Typical
    Equipment to Honeycomb Panels".
    HSD-TP-7516.

10. A. WILD,
    "Solid Radiator Plate Interface Conductance Tests".
    MOTS/053/PWN/0042/HSD.

AN OSCILLATOR TECHNIQUE TO DETERMINE THE THERMAL CONTACT CONDUC-

TANCE BETWEEN FUEL AND CLADDING IN AN OPERATING NUCLEAR FUEL ROD

H.E. Schmidt, M. Van den Berg[+], and M. Wells[++]

European Institute for Transuranium Elements

Karlsruhe, Germany

ABSTRACT

It is being proposed, that fuel-to-clad thermal contact conductances in nuclear-reactor fuel pins can be determined during
operation by measuring the phase shift of periodic temperature
variations, produced by radially inward-flowing heat waves, at the
surface and in the centre of the fuel rod. Solutions to the problem
i.e. the dependence of the phase shift on gap conductance, have
been obtained for the typical geometry of a fuel pin. The theoretical predictions were checked experimentally with a fuel-pin model
consisting of a copper-clad alumina rod. Agreement between measured
and predicted phase-shifts was excellent and proved the validity
of the concept which will be further developed for in-pile application.

1. INTRODUCTION

Considerable uncertainties in in-pile fuel temperature evaluation arise from insufficient knowledge of the thermal resistance
between fuel and cladding of a nuclear-reactor fuel rod under operating conditions.
Fuel-to-clad contact conductances are usually estimated from the
known linear power of the fuel rod and the outer cladding temperature by measuring directly central fuel temperatures or by analy-

+ Present address : AERE, Winfrith, Dorchester, Project Dragon, UK.
++ University of Surrey, Guildford, Surrey, UK.

sing post-irradiation macrographs of fuel rod sections and attributing pilot temperatures to characteristic structural features of
the fuel, like the onset of columnar grain growth in a uranium
oxide or a uranium plutonium oxide pin (see, for instance, Baily
et al., /1/). This method is not very precise, but a direct determination of the thermal gap conductance h under steady-state
conditions from measurements of heat flux and temperature drop
across the gap appears practically impossible because of the high
temperature gradient near the outer surface of the fuel rod.
Another experimental approach to the in-pile measurements of h is
based on the fact that the velocity of penetration of temperature
transients into a composite sample depends on the interfacial
thermal resistance between the sample components. In contrast to
the method based on post-irradiation inspection, cited above, this
technique would allow continuous monitoring of the evolution of h
during the course of the irradiation. Various techniques using this
principle have been proposed in which the thermal transients had
to be obtained by varying the neutron flux in the vicinity of the
test rod, Burdg et al./2/, Daenner,/3/, Perez et al./4/, but none,
to our knowledge, has ever produced satisfactory results. A transient technique has been used for out-of-pile work by Bober,/5/,
h values were determined by Robinson,/6/, by correlating reactor
noise and resulting fuel temperature fluctuations.

We have proposed a concept /7/ whereby the cylindrical fuel
rod is heated periodically by an external resistance heater attached to the cladding surface. The changes with time of the surface
and central temperatures are monitored. With the geometry given,
the phase shift between the two signals should then allow the mean
fuel conductivity and/or the gap conductance to be determined.
While a detailed description of the underlying theory is given in
/7/, the present report gives an account of the second stage of the
project, i.e. an experimental check of the validity of the mathematical model and its applicability to fuel-pin type systems.

2. RESUME OF THE MATHEMATICAL TREATMENT

The temperature distribution as a function of time, $\theta(r,t)$
in the test rod section is given by the following differential
equations:

$$\theta'' + \frac{1}{r}\theta' - \frac{1}{a_1}\dot{\theta} = \frac{-A(r)}{\lambda}, \quad r_a \leqslant r \leqslant r_b, \tag{1}$$

$$\theta'' + \frac{1}{r}\theta' - \frac{1}{a_2}\dot{\theta} = 0, \qquad r_b \leqslant r \leqslant r_c, \tag{2}$$

with the boundary conditions, in the case of sinusoidal excitation:

$$\theta_{2c} = Ve^{i\omega t} + \tau_{2c} \tag{3}$$

$$\phi_{1b} = -\lambda_1\theta'_{1b} = h(\theta_{1b}-\theta_{2b}) = -\lambda_2\theta'_{2b} = \phi_{2b} \tag{4}$$

$$\phi_{1a} = -\lambda_1\theta'_{1a} = 0. \tag{5}$$

The indices 1 and 2 refer to the fuel and to the cladding, respectively: a, b and c to the radii r_a, r_b and r_c (Fig.1); a_1 and a_2 are
the mean thermal diffusivities; and λ_1 and λ_2 are the mean thermal
conductivities of the fuel and the cladding material. A (r) is
the specific fission power distribution, ϕ designates the heat flux
in the radial direction and h is the thermal contact conductance
between the fuel and the cladding, τ_{2c}
is the mean temperature at the outer
cladding surface. With the problem thus
defined, the solution to the set of
equations (1), (2) with boundary conditions (3), (4), (5) relates the unknown
quantities a_1 and/or h to the temperature function θ (r,t), which can be
measured. For further treatment, the
temperature function θ in (1) has been
split into a time-dependent and a time-
independent part

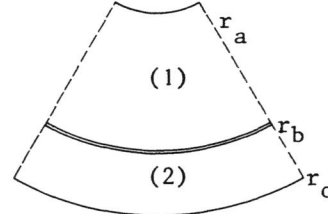

Figure 1: Cross section
of the test rod.

$$\theta(r,t) = v(r,t) + \tau(r)$$

Furthermore, for the time-dependent part, a sinusoidal modulation
was assumed

$$v(r,t) = T(r)e^{i\omega t}$$

The temperature T_{1a} at the edge of the central hole has been calculated from the heat flux $F_{1a} = -\lambda_1 T'_{1a}$ and the temperature T_{2c} by
means of a procedure known from four-terminal network theory
(M.I.T./8/). Results typical for stainless steel clad mixed ura-
nium-plutonium-oxide fuel pins are discussed in /7/.
As an example, the expected variation of phase-shift with gap
conductance in a fast-breeder fuel pin and with temperature modulation frequency is given in Fig. 2.

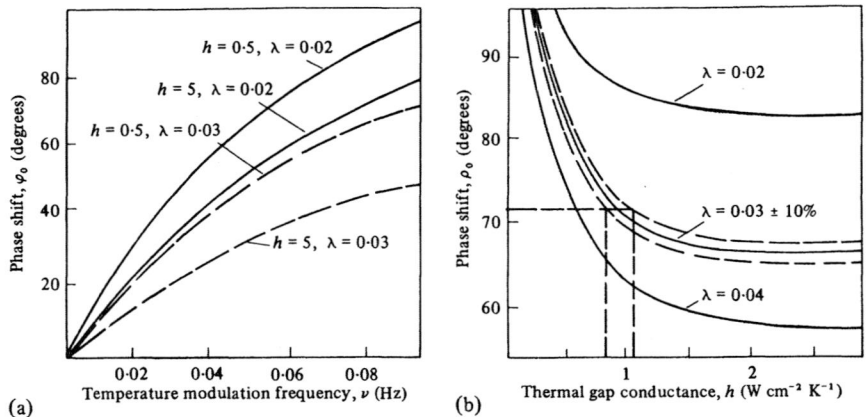

Figure 2: Phase shift φ_0 as a function of, (a) modulation frequency, (b) gap conductance, h, in W cm^{-2} K^{-1} (for ν = 0.1 Hz) for different mean fuel conductivities λ, in W cm^{-1} K^{-1}.

3. LABORATORY DEVELOPMENT WORK

Experimental Apparatus

To test the validity of the above model, an experiment was set up, in which the fuel rod was simulated by a rod of aluminium oxide clad in a copper tube equipped with an electrical heater and thermocouples to measure surface and central temperatures (Fig. 3).

A copper cladding was chosen, primarily because a copper-constantan thermocouple can be easily attached to the surface, whilst the aluminium oxide was thought representative of the ceramic nature of uranium and plutonium oxide fuel. (A uranium-plutonium oxide pin consists of a stainless steel cladding enclosing a compacted fuel of sintered uranium-plutonium oxide pellets).

The resistance heater consists of a closely wound cylindrical coil of insulated constantan wire which can be easily slid on and off the fuel pin model. A shallow groove in the cladding leading down to the thermocouple junction allows close contact between the cladding and the heater. Heater current is supplied by a Proportional Integral Differential (P.I.D.) temperature controller which is modulated by a sinusoidal wave generator.

The two thermocouples consist of 0.1 mm diameter copper and
constantan wires. The central thermocouple, (which lies opposite
the surface thermocouple), is permanently fixed and sealed, with
epoxy resin, into the aluminium oxide rod. The model lies inside
a glass bell jar. This has facilities for introducing gases into
the gap to vary its conductance.

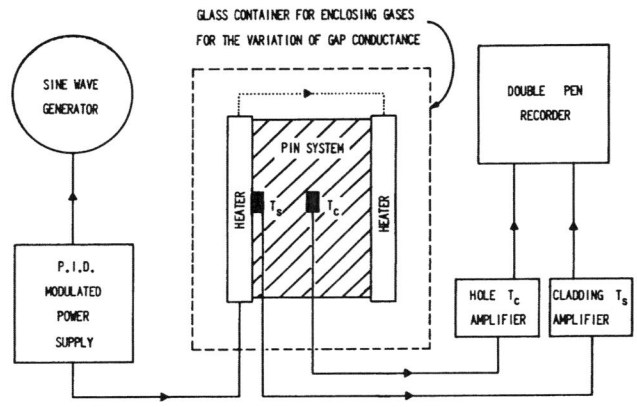

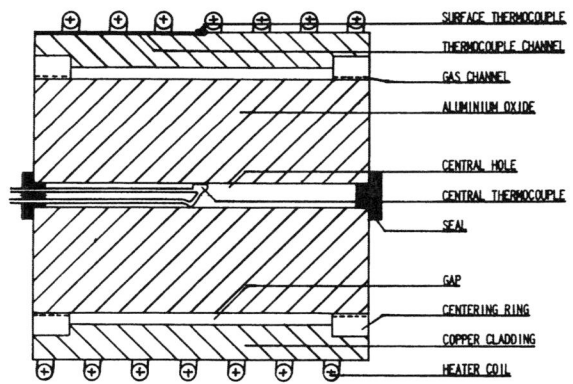

Figure 3: Experimental set-up
          Top: schematic of complete set-up
          Bottom: pin system

Procedure

The sinusoidally varying heat waves generated in the heater coil are monitored by the two thermocouples as they pass radially into the fuel pin. The signals from the thermocouples are amplified by differential amplifiers and fed into a double pen recorder (Fig.3). The phase shift observed between the two signals is calculated by measuring the distance between two corresponding points on the two pen recorder traces. (The maxima on the cooling-heating half cycles are found to be convenient points (Fig.4).)
When a sufficient number of different modulation frequencies have been taken the bell jar is repeatedly evacuated and flooded with a different gas, (previously air), so the gap conductance is changed. The results from a series of different frequency power inputs with the corresponding phase shifts are shown in Figures 5 and 6, where the computer predicted phase angles for a known gap conductance value and modulation frequency are compared with experimental data. The error associated with the experimentally determined phase shifts is about ± 6 degrees. Clearly, the experimental and computer predicted results are comparable.

Discussion

One major problem is how to measure the central hole temperature variations. The computer analysis for predicting the phase shifts relies upon the fact that the temperature variations of the hole must be monitored at r_a. The results shown in Figures 5 and 6 are obtained by having the central thermocouple gently pressing against the hole wall. Due to contact resistance between the thermocouple and the wall, the value of the phase shift is somewhat uncertain. It must be remembered that determining the temperature variations of a central hole in an active fuel pin, (which is the ultimate aim of this experiment), will require a more thorough investigation.

The scatter of experimental data, expressed in h, is at present of the order of ± 20%. Predicted and measured conductances agree about within the same tolerances.

4. CONCLUSIONS

The experiments described have confirmed the validity of the proposed model for the determination of thermal gap conductances between metal sheath and ceramic core of composite systems with cylindrical geometry.

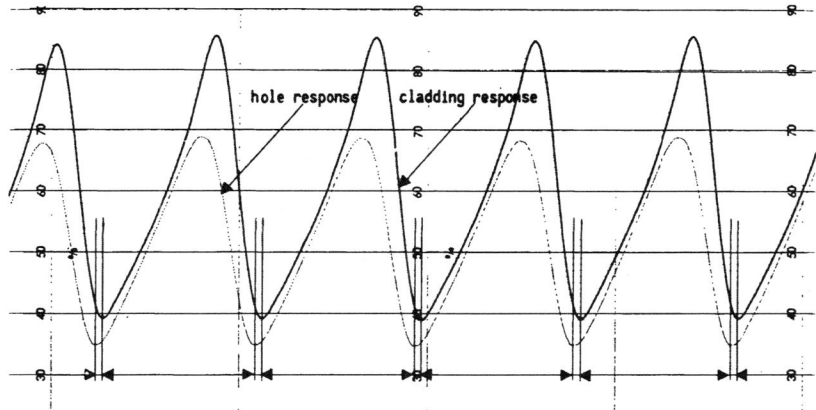

Figure 4: Typical output responses from the central
thermocouples.

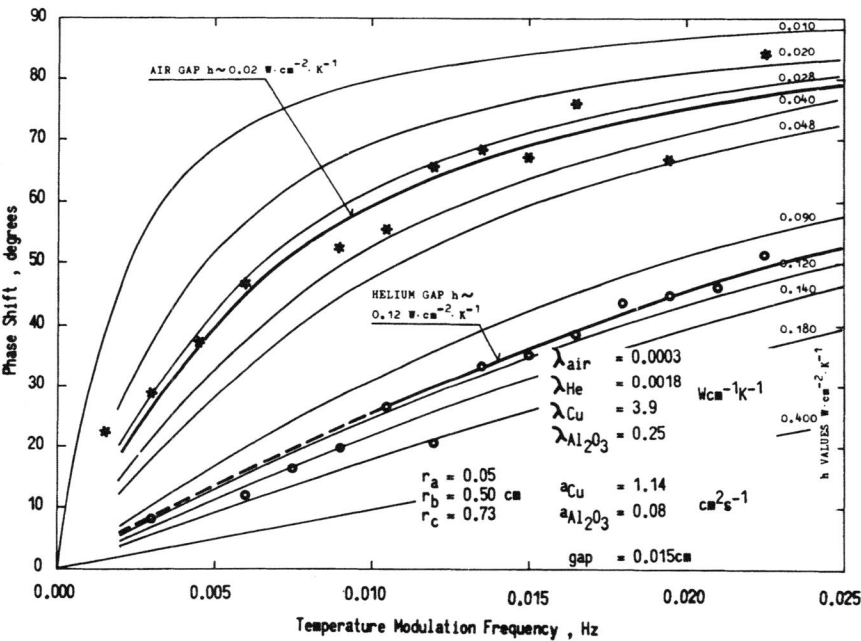

Figure 5: Phase shift as a function of modulation
frequency for a copper-alumina test rod.

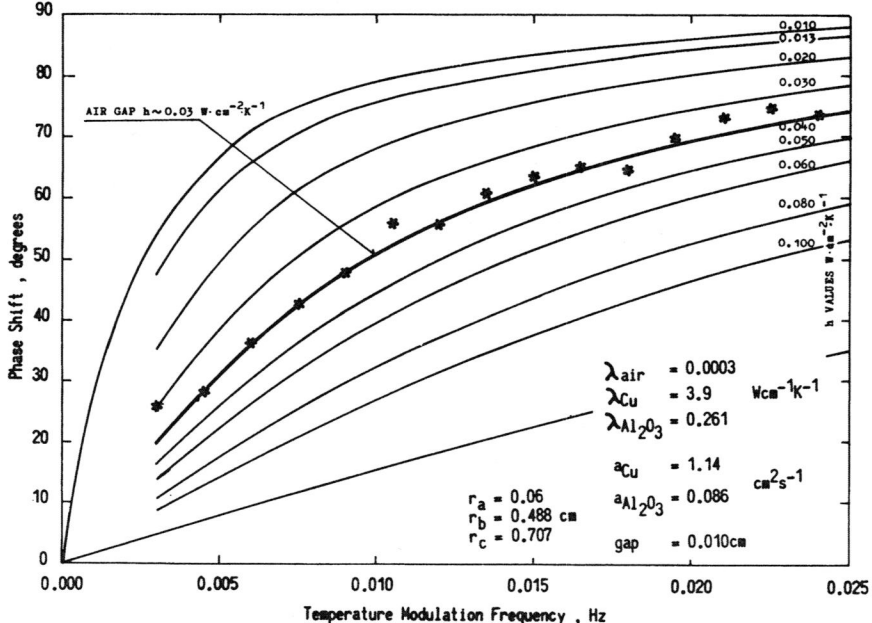

Figure 6: Phase shift as a function of modulation
frequency for a copper-alumina test rod.

The next step in making this method applicable to the in-pile
determination of fuel-clad thermal conductances in nuclear fuel
pins will be the design and test of a mock-up for an in-pile testing
device.

REFERENCES

/1/ W.E.Baily, C.N.Craig, E.L.Zebroski, Trans.Am.Nucl.Soc. 9
     (1966), 1.

/2/ E.C.Burdg, J.E.Parrette, R.L.Brockett, W.P.Chernock, CEND-
     3336-260, 1966.

/3/ W.Daenner, KfK-1125, 1970.

/4/ R.B.Perez, R.M.Carroll, O.Sisman, ORNL-4478, 1971!

/5/ M.Bober, INR-4/68-18, Thesis, University of Karlsruhe, BRD/
     West Germany, 1968.

/6/ E.Robinson, Symposium on Thermodynamics of Nuclear Materials,
     IAEA, vienna, 21-25 October 1974.

/7/ M.Van den Berg. H.E.Schmidt, High Temperatures-High Pressures,
     5 (1973), 589.

/8/ M.I.T. Staff, Electric Circuits (John Wiley, New York), 1943.

AN ANALYTICAL MODEL FOR THE MEASUREMENT OF UO_2 THERMAL CONDUCTIVITY

M.A. Aragones

Instituto Nacional de Energia Nuclear

Centro Nuclear, Salazar, Mexico

The thermal conductivity of UO_2 is one of the most important design parameters of a UO_2 nuclear fuel element. This thermal conductivity can be measured using the longitudinal heat flux method.

In this paper, a more general solution to the problem, taking into account the small radial flux always existing in the thermal conductivity measurement in the longitudinal method, is presented.

I. INTRODUCTION

The thermal conductivity of UO_2 is one of the most important design parameters of a UO_2 nuclear fuel element. It is usually measured in the laboratory using two different methods: a longitudinal or radial heat flux method.

In the longitudinal method, authors (1,2) assume that there does not exist any radial flux, which is only approximately valid.

II. THEORY

The equation of heat transport in a solid at steady state is

$$\nabla^2 T = 0 \qquad\qquad Eq.I$$

After some derivation we can get as a solution of this equation

$$T(\rho,z) = J_o(\alpha\rho) \, (\, A \exp \, (\, -\alpha z) + B \exp \, (\alpha z) \,) \quad \text{(Eq.II)}$$

or

$$T(\rho,z) = \sum J_o(\alpha_n \rho)(A_n \exp(-\alpha_n z) + B_n \exp \, (\alpha_n z) \,) \quad \text{(Eq.III)}$$

as general solution for $T(\rho,z)$.

II.1 THE BOUNDARY VALUE PROBLEM

In this paper we have kept only the first term in the above series expansion. It is expected that this would constitute a better approximation than the linear dependence of themperature with z which results in the usual longitudinal method. The temperature at three different points serve as boundary conditions to determine the three constants α, A and B. Two intermediate points where the themperatures were measured also during the experiments were used as testing points for the resulting expression. Then, equation II and the values α, A and B can be calculated as:

$$T(0,0) = A + B$$

$$T(r,0) = J_o(\alpha r)(\, A + B \,)$$

$$T(r,H) = J_o(\alpha r)(\, A \exp \, (-\alpha H) + B \exp \, (\alpha H))$$

where $T(0,0), T(r,0)$ and $T(r,H)$ are measured values. See figure 1.

II.2 THE EVALUATION OF λ_{UO_2} AND ACCURACY

The maximum accuracy of the analysis is $\pm \, 0.003$ for O/U ratio. We estimated the values of $\emptyset$ defined by the equation IV assuming the variations of temperature in the radial direction are small compared with that in the UO_2 pellets. We observed in the 304 stainless steel

$$T(0,H) \cong T(r,H)$$

so that for the 304 stainless steel

$$\emptyset \cong \lambda_{304ss} \, (\partial T / \partial z) \, \hat{z} \qquad\qquad \text{Eq.IV}$$

where:

λ_{ss} is the thermal conductivity of the 304 stainless steel, and $T(\rho,z)$ is the temperature as a function of z, and ρ. The value of the thermal conductivity of UO_2, is obtained from:

FIGURE No. 1

dimensions in cm.

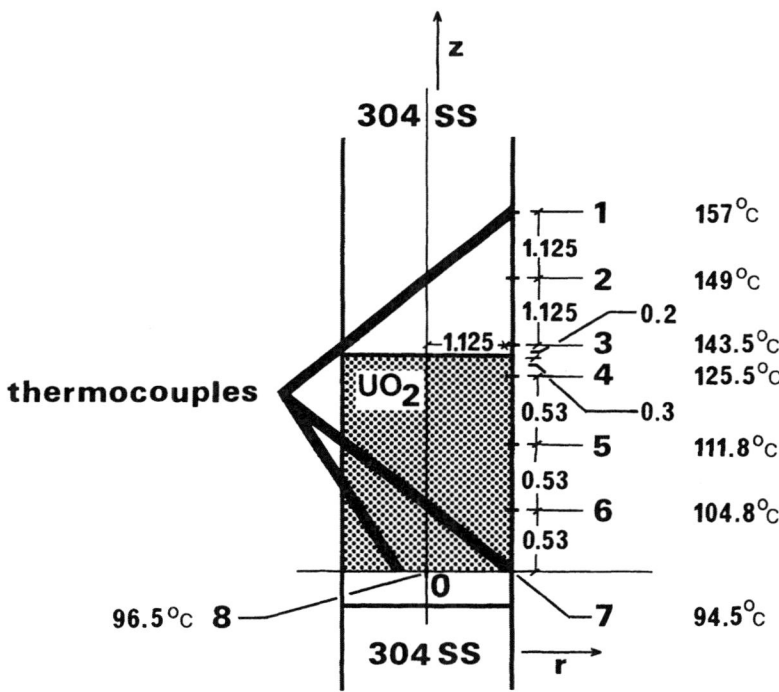

A typical thermocouples and temperatures
Distribution in UO_2 and 304 stainless steel
During the experiments.

EXPERIMENT NUMBER IV-1

$$\underline{\emptyset} = \lambda_{UO_2} \left\{ \frac{\partial T(\rho,z)}{\partial z} \ z + \frac{\partial T(\rho,z)}{\partial \rho} \ \rho \right\}$$

$$\emptyset = \lambda_{UO_2} \sqrt{\frac{\partial T}{\partial z}^2 + \frac{\partial T}{\partial \rho}^2}$$

$$\lambda_{UO_2} = \frac{|\underline{\emptyset}|}{\sqrt{\frac{\partial T}{\partial z}^2 + \frac{\partial T}{\partial \rho}^2}} \qquad \text{Eq.V}$$

III. EXPERIMENTAL

III.1 EQUIPMENT AND SPECIMEN

Figure 2 shows schematically the apparatus used. Cylindrical pellets of UO_2, of 2.54 cm diameter and very approximately the same length, with a thermocouple maximum distance of 1.6 and 2.1 cm in each case were used in two different series of experiments. Each pellet was placed between two cylindrical 304 stainless steel rods.

The upper stainless steel rod is heated by an electrical resistance, which is connected to a 115 A-C power source through an adjustable autotransformer. The maximum power into the electrical resistance is 400 W.

The lower stainless steel rod is cooled by circulating water through a cylindrical aluminium vessel surrounding the rod. The cylinder is inside a metallic sheet which has two tubes for the inlet and ountlet of argon gas.

We used iron-constantan thermocouples which are placed into orifices 3 mm and 1 mm diameter in each pellet. The same procedure is used for the 304 stainless steel rod. In both cases the and of the thermocouple rests in a bed of copper powder which surrounds the thermocouple junction in order to get good thermal contact between the pellets and the thermocouple, thereby avoiding measurement errors in the temperature.

The UO_2 pellets of each experiment were obtained from a powder which was pre-pressed at 1400 kg/cm2, granulated to 20 mesh and pressed into pellets of 2.72 cm diameter, at a pressure of 2815 Kg/cm^2 . The green pellets were sintered for five hours to 1450°C in an argon atmosphere and then centreless ground between silicon carbide wheels to 2.54 cm diameter. After each experiment the UO_2 pellet was analyzed by the combustion method in air, and its

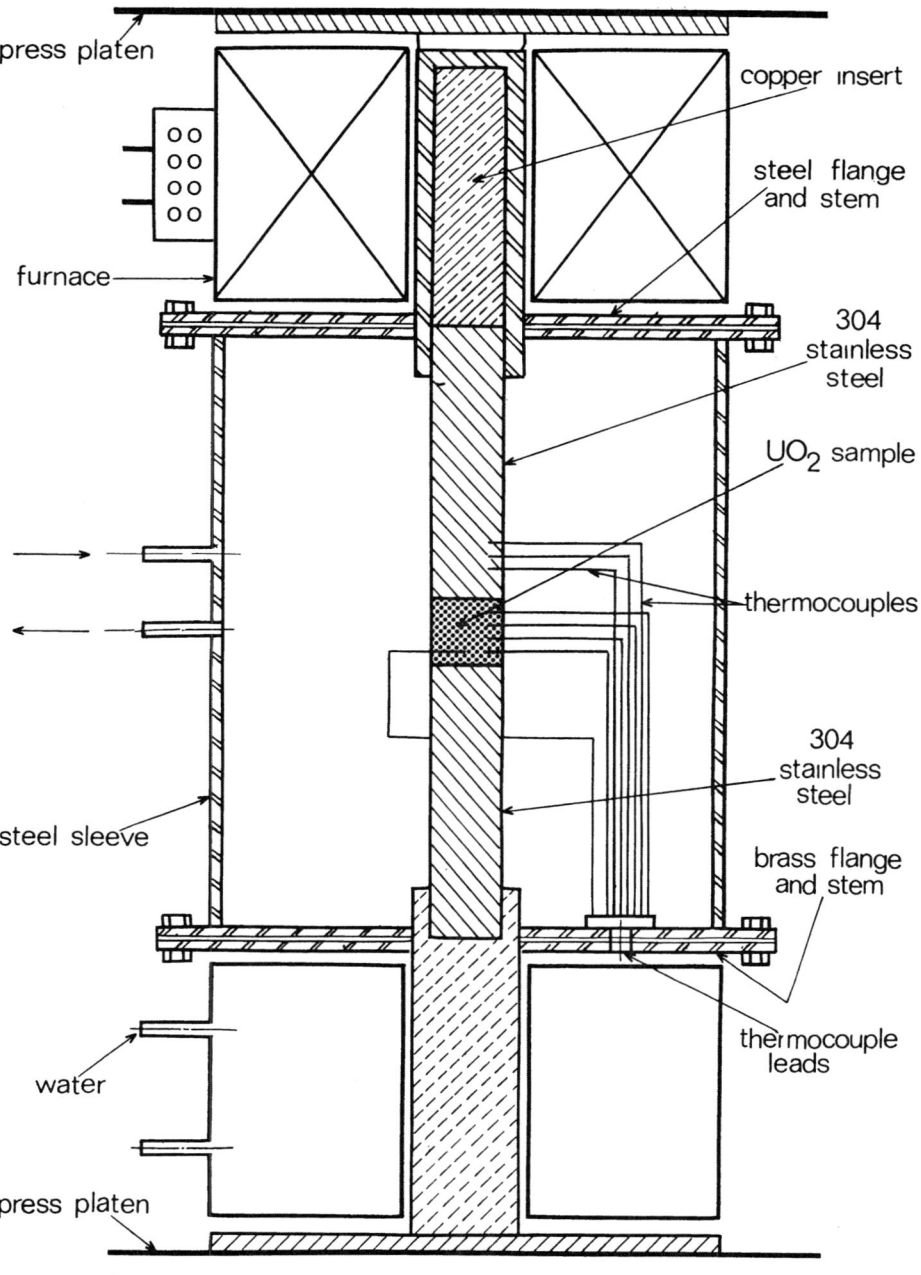

press platen

copper insert

steel flange and stem

furnace

304 stainless steel

UO$_2$ sample

thermocouples

304 stainless steel

steel sleeve

brass flange and stem

water

thermocouple leads

press platen

Cross section of thermal conductivity apparatus.

figure no 2

formula (0/U ratio) obtained with an accuracy of $\pm$ 0.003. The densities of the pellets were 10.45 and 10.48 g cm^{-3} for composition 0/U:2.045$\pm$ 0.003 and 2.066$\pm$ 0.003. The above densities represent roughly 96% of the theoretical density.

III.2 OPERATION

Each UO$_2$ pellet was fixed to the faces of the stainless steel cylinders with an aluminium silicon paint. This paint stands temperatures up to 350° C. The thermocouples were assembled in each hole of the pellets and a very small flow of argon gas passed through the apparatus in order to avoid any oxidation of the pellets. At the same time, we slowly raised the temperature. Steady state was obtained after 5 hours of operation. The maximum variation of temperature observed after the steady state is reached is 2° C/hour

Figure 1 shows a typical thermocouples and temperatures distribution in UO$_2$ and 304 stainless steel during the experiments

IV. RESULTS

Table I-1 gives the distribution of temperatures in the UO$_2$ sintered pellet and 304 stainless steel rods. We omit other results for reasons of brevity.

Table II-1 give the thermal conductivity of UO$_2$ sintered pellet for a stoichiometry and corrected for porosity(2, 3).

V. DISCUSSION AND CONCLUSIONS

As it can be seen in tables II-1, the radial flux of heat represents about 17% of the total heat flux in our experiments and we estimate a value of 4% as the maximum error for this thermal conductivity method.

We conclude that the procedure gives us the possibility of obtaining absolute accuracy with this method which has been cast into doubt by some authors (4).

Our development suggests the convenience of substituting for the over-simplified equation

$$\frac{d^2T}{dz^2} = 0 \quad \therefore \quad \frac{dT}{dz} = \text{constant} = \frac{\emptyset}{\lambda}$$

which gives a linear relation between T and z, the equation:

$$T(\rho,z) = J_o(\alpha\rho)(A \exp (-\alpha z) + B \exp (\alpha z))$$

if the pellets are not long, as in our case or

$$T(\rho,z) = \sum J_o(\alpha_n\rho)(A_n \exp(-\alpha_n z) + B_n \exp(\alpha_n z))$$

for long UO$_2$ pellets, which are difficult to produce with isotropic characteristics.

TABLE I-1

Experiment number	Distribution of temperatures in the UO$_2$ sintered pellets and 304 stainless steel. O/U ratio : 2.045 $\pm$ 0.003								
	Temperatures of the thermocouples								
	in 304 stainles			in UO$_2$ sintered pellet					
	1	2	3	4	5	6	7	8	
I-1	138.2	126.4	124.4	108.6	93.8	82.0	75.0	77.0	
I-2	143.6	132.6	130.2	113.6	98.8	85.8	78.6	80.7	
I-3	149.8	138.6	135.4	118.6	102.2	89.4	82.6	85.0	
I-4	146.4	135.4	133.3	116.2	100.6	87.8	81.8	84.0	
I-5	151.6	140.4	137.6	120.6	104.2	91.0	84.4	86.0	
I-6	151.8	140.9	138.2	120.6	104.4	91.0	85.0	87.2	
IV-1	157.0	149.0	143.5	125.5	111.8	104.8	94.5	96.5	(O/U:2.06

TABLE II-1

Experiment number	Thermal conductivity of the UO$_2$ sintered pellets O/U ratio: 2.045+ 0.003 for the experiment IV-1, O/U ratio: 2.066 $\pm$ 0.003								
	A	B	α	$\frac{\partial T}{\partial z}$ $^\circ$C/cm	$\frac{\partial T}{\partial \rho}$ $^\circ$C/cm	Q w/cm^2°C	λ m w/$^\circ$c cm	λ c w/$^\circ$c cm	T
I.1	21.9	55.1	0.280	23.8	4.84	0.925	0.038	0.042	109° C
II.1	34.3	82.4	0.300	39.7	8.62	1.395	0.034	0.038	168° C
III.1	35.4	118.9	0.284	53.9	10.31	1.703	0.031	0.035	228° C
IV.1	19.2	77.3	0.254	25.5	5.49	0.906	0.035	0.039	125° C

λ $_m$ = measured thermal conductivity

λ $_c$ = corrected thermal conductivity for porosity

REFERENCES

1. A.M.Ross.The Dependence on The Thermal Conductivity of Uranium Dioxide on Density, Microstructure, Stoichiometry and Thermal Neutron Irradiation, Atomic Energy of Canada Limited, AECL 1096 Chalk River,Ontario.Canada, September 1960

2. J.L. Daniel et al., Thermal Conductivity of UO_2, Hanford Atomic Products Operation, Richland, Washington HW 69 945, 1962.

3. M.A.Aragones et al.,The Effect of Density and Grain Size on the Thermal Conductivity of UO_2 during Irradiation.AECL 2564. Chalk River, Ont.Canada, April 1966

4. J.R.McEwan et al.,Annealing of Irradiation Induced Thermal Conductivity Changes in ThO_2-1.3% wt UO_2. The American Ceramic Society, Vol.52 No.3, March 1969 (160 - 165)

APPENDIX

The eq.I, for a cylindrical geometry and an isotropic solid can be reduced to:

$$\frac{1}{\rho}\left(\frac{\partial T}{\partial \rho} + \rho \frac{\partial^2 T}{\partial \rho^2}\right) + \frac{\partial^2 T}{\partial z^2} = 0$$

To solve this equation we assume: $T(\rho,z) = T_1(\rho)\, T_2(z)$, then

$$\frac{1}{\rho}\left(T_2(z)\dot{T_1}(\rho) + \rho T_2(z)\ddot{T_1}(\rho)\right) + T_1(\rho)T_2(z) = 0$$

$$T_2(z) = A \exp(-\alpha z) + B \exp(\alpha z) \quad \text{and}$$

$$\rho^2 \ddot{T_1}(\rho) + \rho \dot{T_1}(\rho) + \alpha^2 \rho^2 T_1(\rho) = 0$$

If we make: $\rho \alpha = x$

$$x \ddot{T_1}(x) + \dot{T_1}(x) + x T_1 = 0$$

$$T_1 = C J_o(\alpha \rho) + D Y_o(\alpha \rho), \qquad D = 0$$

$$T_1 = C J_o(\alpha \rho)$$

Then:

$$T(\rho,z) = T_2(z)\, T_1(\rho) = J_o(\alpha\rho)(A \exp(-\alpha z) + B \exp(\alpha z))$$

MEASUREMENT OF HIGH THERMAL DIFFUSIVITY VALUES

BY A NOVEL METHOD ON COPPER AND TUNGSTEN

Vladimir V. Mirkovich

Department Energy, Mines and Resources

Ottawa, Ontario, Canada

ABSTRACT

A novel apparatus is described for the measurement of thermal diffusivity. The equipment is based on the concept of an infinite cylinder and is theoretically capable of performing in a periodic temperature mode and a transitory temperature mode. Thermal diffusivities of copper and tungsten were measured to establish the measuring capacity of this apparatus in the high thermal diffusivity range and on electrically conductive specimens. The results are in good agreement with data on tungsten. However, some difficulties have been encountered in the measurements performed on copper and only incomplete thermal diffusivity data could be obtained.

INTRODUCTION

Because of the wide range of thermal diffusivities inherent in various materials no apparatus based on one method is universally suitable for measurements on all solids. Fortunately, the choice of heat flow boundary conditions in non-steady-state measurements is large, which, consequently, results in a considerable variety in experimental devices.

In broad terms, experimental methods for determining thermal diffusivity of materials can be divided into two categories: periodic temperature methods and transitory temperature methods. Depending on the shape of the specimen, each of these categories can be further subdivided into methods based on a semi-infinite rod, or a semi-infinite solid, or a thin plate, or a cylinder.

479

Among these, the flash method, a transitory temperature method, based on the concept of a thin plate, devised by Parker et al (1) gained a prominent position because of its relative experimental simplicity.

The selection of the measuring model will depend to a considerable degree on the nature of the material on which the measurement is to be performed. It should be obvious that in the case of coarse-grained materials, such as concretes or some rocks and minerals, the size of the sample used in, for example, the flash method would be too small to integrate the contribution of the individual grains toward the overall thermal diffusivity of the specimen. Also, to measure both high and low thermal diffusivities, a geometrical configuration should be chosen in which the heat pulse will undergo minimum attenuation (and thus generate signals which will not be obscured by background noise).

It was decided therefore to design and construct a thermal diffusivity apparatus which would be based on the concept of an infinite cylinder. Although cylindrical models require a relatively large quantity of material, the sample can be, in most cases, fairly readily machined or otherwise prepared. Also, a cylindrical column can be obtained by stacking discs of equal diameters. A significant feature of a cylindrical thermal diffusivity model under the condition of radial heat flow is the freedom from problems of spurious heat losses.

The novel and experimentally important feature of this apparatus is its ability to perform measurements in both transitory and periodic temperature methods. It was the purpose of this study to assess the performance of this apparatus on electrically conductive, high thermal diffusivity materials.

The mathematical boundary conditions of the theoretical thermal diffusivity models were discussed previously (2). The material is assumed to be isotropic and its thermal properties are independent of temperature over the limited temperature span in which the individual measurements are to be made.

In Model 1, which is a transitory temperature measuring method, the initial temperature of the infinite cylinder of radius a is zero, and its surface temperature is a linear function of time t. Experimentally, the temperature is measured at two points: one near the surface at radius r, and the other in the centre. By subtracting the equation representing the temperature at the centre of the cylinder from the equation representing the temperature at the point near the surface of the cylinder, an expression is derived which relates the dimensionless temperature differential V to a factor $\kappa t/a^2$. With constant-rate heating of the surface of the cylinder, the dimension-

less temperature difference V increases asymptotically from 0 to
approach 1 as the value of $\kappa t/a^2$ increases from 0 to infinity.

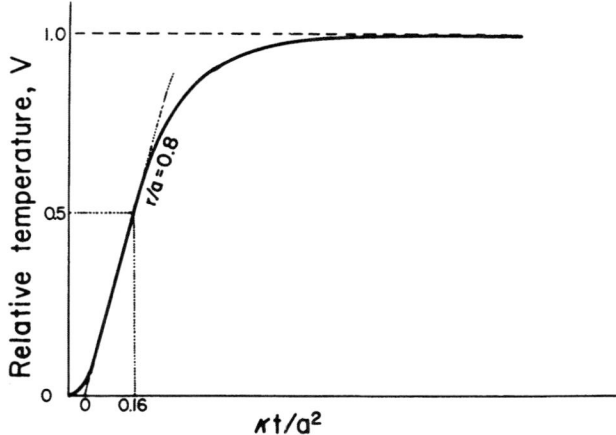

Figure 1. Plot of relative temperature, V,
versus dimensionless factor, $\kappa t/a^2$.

The position of the curve for a given V and $\kappa t/a^2$ depends on the
position of the measuring point near the surface of the cylinder,
i.e., on the dimensionless ratio r/a.

To obtain thermal diffusivity from an experimental plot, 0 and
1 are assigned to the starting and maximum temperature differentials,
respectively. Then the time t necessary to reach some convenient
value of the relative temperature differential, say V = 0.5, is
determined. As stated above, the relation between fixed values of
V and $\kappa t/a^2$ is determined by r/a. The following expression gives
$\kappa t/a$ as a function of r/a for $1 \geq r/a \geq 0.5$ for V = 0.5:

$$\kappa t/a^2 = 0.198 + 0.00747 \ (r/a) - 0.0687 \ (r/a^2)$$

As shown in the example in Figure 1, at V = 0.5, for r/a = 0.8,
$\kappa t/a^2 = 0.16$. Since t (at V = 0.5) is obtained experimentally,
the diffusivity of the material can be readily calculated.

In Model 2, thermal diffusivity of the infinite cylinder is
measured in a periodic temperature mode. The surface of the
cylinder is exposed to a sinusoidal temperature variation.
The temperature is again measured in two positions: at the centre
and near the surface of the cylinder. In Figure 2 the solid curve
is the plot of the dimensionless temperature change at the point

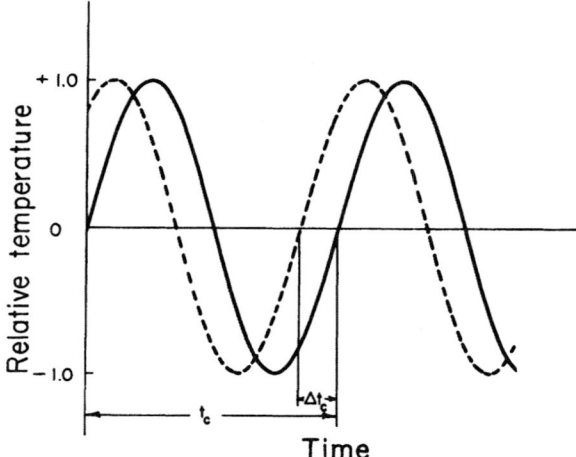

Figure 2.   Plot of the centre and surface relative
            temperatures, V, versus time.

near the surface and the dotted curve represents the dimensionless
temperature change at the centre. t_c is the time of one full cycle
and Δt_c is the time (differential) necessary for the heat wave to
travel from the surface measuring point to the centre of the
cylinder. By equating the dimensionless expressions representing
the temperatures at the point near the surface and at the centre
for some convenient value of the dimensionless temperature, say
$V = 0$, one obtains $\omega(\Delta t_c) = \theta o(\omega^1 a)$ where $\omega = 2\Pi/t_c$, $\omega^1 = (1/\kappa)^{0.5}$
$(2\Pi/t_c)^{0.5}$, and θo is a function* related to Bessel function.
Radius a in the above equation becomes the distance between the
centre of the cylinder and the measuring point near the surface,
not to the surface of the cylinder. Thermal diffusivity is obtained
from $(\omega^1 a)$, the value of which is determined from the product
$\omega(\Delta t_c)$, which is, in turn, obtained by using the measured time
differential Δt_c.

EXPERIMENTAL

A.   The Apparatus

     The apparatus is schematically shown in Figure 3. The cylin-
drical sample (B), 25 mm in diameter and 20 to 30 mm high (usually
composed of several discs, each some 5 - 10 mm thick) is held

---

*Numerical values of $\theta o(z)$ can be found, for example in "Bessel
Functions for Engineers" by N.W. McLachlan, Oxford University
Press, London, 1961, 2nd ed.

between two ceramic insulators (A) and supported on a 5 mm thick
alumina disc. This column, approximately 110 mm high is in turn
supported by three 1.25 mm diameter tungsten rods (F). Three
additional tungsten rods support an alumina ring (E) which serves
as a base for the heating element (D) and radiation shields (C).
The heating element consists of a thin fused-silica tube, 50 mm in
diameter and 100 mm high, on which two 1650-mm, 0.178-mm diameter
tungsten wires are suspended between two rows of holes at the top
and bottom. The tungsten wire heaters thus completely surround the
cylindrical sample column. One of the tungsten wire heaters provides
constant background temperature while the other supplies heat pulses.
To facilitate replacement of a burned out tungsten wire, the silica
tube can be lifted off the alumina ring. Electrical contacts are
provided in the alumina ring to supply power to the tungsten wires.
Radiation shields are made of 0.125-mm thick tantalum and surround
the heater. The tungsten rods supporting the sample column, the
heater and the radiation shields are used to minimize conductive
heat losses. They are held by a metal base (H) attached to the top
of a high vacuum unit. The measuring assembly is covered with a
glass jar (G) resting on the top of the vacuum unit.

To maintain the temperature of the specimen at a desired level,
power from a variable transformer is fed at a constant rate to the
tungsten wire wound on the outside of the silica tube. The heat
pulses are obtained by using a voltage function generator. The
output from this generator is fed to a direct current power
amplifier, whose output, in turn, is supplied to the inner tungsten
wire heater.

Temperatures in the specimens are measured with two 28 B & S
gauge chromel-alumel thermocouples. Two 0.175-mm holes were drilled
through the sample parallel to the axis: one at the centre and the
other near the perimeter of the disc. A butt-welded thermocouple
was passed through each hole so that the hot joint became wedged in
the hole.

In the case of tungsten, one leg of each thermocouple was spot-
welded in the appropriate position on the flat face of the centre
disc of the specimen of the stock making up the cylindrical specimen.
The other leg of each thermocouple was spot-welded in the corresponding position on the opposite face of same disc. The measuring
arrangement is schematically shown in Figure 4. The system is
designed to record either the absolute output of each thermocouple
or the emf difference between the two thermocouples.

To record in the absolute mode, the non-periodic portion of
the total output of each thermocouple is cancelled by means of a

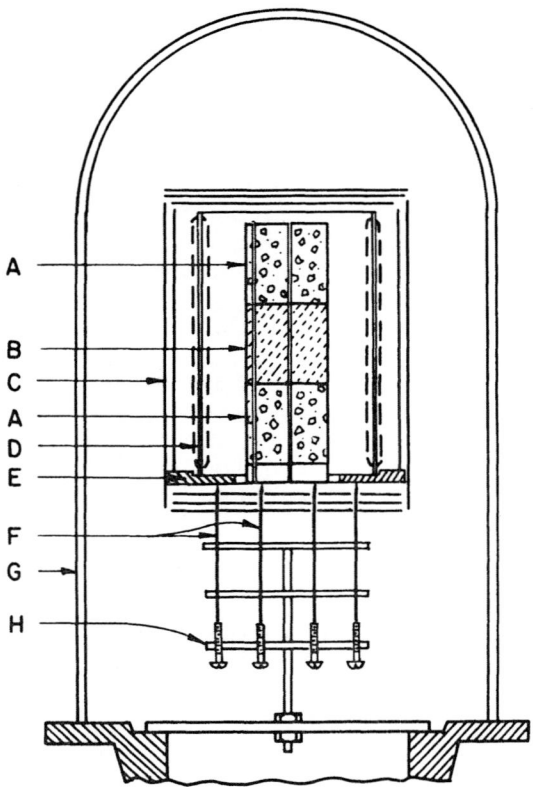

Figure 3.   Schematic of the thermal diffusivity apparatus.

voltage suppressor installed in each circuit. The periodic overflow
is fed into a low-noise amplifier, then into an isolation amplifier,
then to the emf switch, and finally to a 3-pen potentiometric type
recorder. Channels 1 and 2 are used for recording the amplified
portions of the thermocouple outputs. A direct emf signal from one
of the thermocouples is led (not shown in Figure 4) to the emf
switch and channel 3 to obtain the actual temperature of the specimen.
For measurements in the differential mode, voltage suppressors are
excluded (except for a very small voltage necessary to keep the
differential recording within the limits of the recording chart) and
by means of the emf switch the positive legs of the two circuits are
joined while the negative legs become connected to one of the
recording channels.

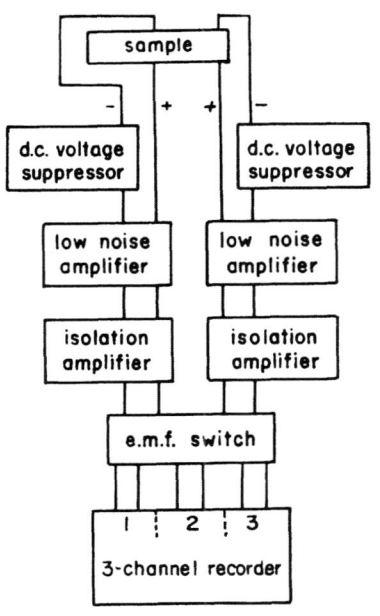

Figure 4. Block diagram of the
         measuring equipment.

Differential temperature measurements can be made with differential thermocouples if the hot joints are electrically isolated from each other. In a circuit where two thermocouples are attached to a conductor and, say, have the legs of equal polarity joined, measurements cannot be made at the terminals of opposite polarity because the differential emf is short-circuited through the conductor. As the negative side of the low noise amplifiers (which are used in the present arrangement) is grounded, the input to the amplifier is not isolated from its output, and consequently, direct differential measurement is not possible. To overcome this difficulty a one-to-one isolation amplifier is installed between the low-noise amplifier and the emf switch.

B.   Materials

The need for established standards becomes apparent to anyone who designs a new measuring apparatus. While the precision of the new equipment can readily be obtained, its accuracy can be determined only by checking it against materials with known physical properties. Unfortunately, in the field of thermal conductivity and thermal diffusivity, universally recognized standards are still not available. However, for various reasons some materials have been used by different researchers and their characteristics and values have been reported. After searching for a characterized, high thermal diffusivity, electrically conductive reference material, it was decided that both copper and tungsten should be used in this investigation.

A 99.999% pure tungsten sample having a density of 19.0 g/cm^3, manufactured by the arc-fusion method, was supplied by the Carborundum Company, Niagara Falls, N.Y. Its thermal diffusivity was measured by Naum (3) using a flash method. After machining the tungsten specimen into 6- and 12-mm thick, 25-mm diameter discs and

drilling the 0.75-mm diameter thermocouple holes, the samples were
annealed in vacuum for two hours at 1800°C.

A 99.999% certified purity, oxygen-free copper sample was
obtained from a commercial source.

RESULTS AND DISCUSSION

The results obtained from measurements on tungsten are shown in
Figures 5 and 6. The dotted curve in both diagrams represents
thermal diffusivity as measured by Naum (3). The circles in Figure 5
indicate the results obtained by the transitory method, i.e., Model 1,
while the squares denote results obtained in the periodic temperature
method, i.e., Model 2.

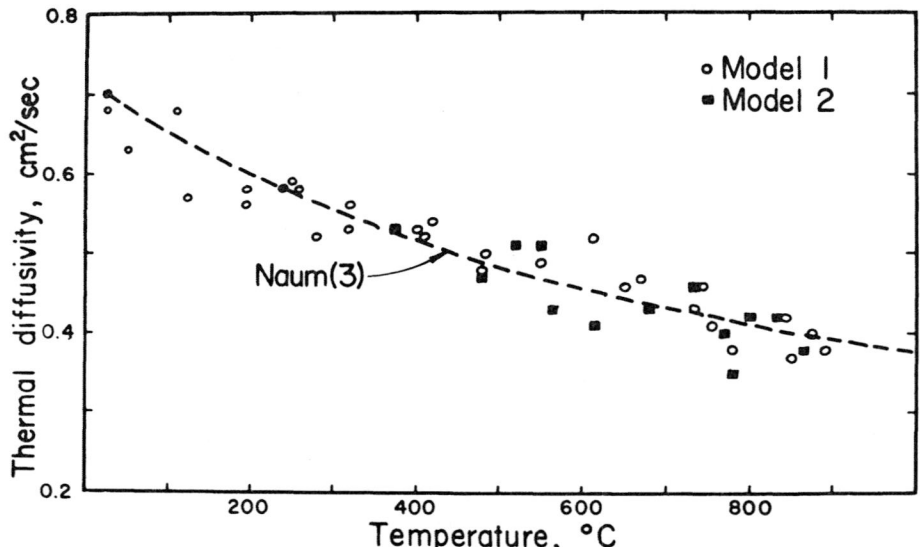

Figure 5. Thermal diffusivity of tungsten. Experimental results
obtained from Models 1 and 2.

In both cases the measurements were made up to 900°C. At this
temperature the heat losses from the measuring assembly approach
the maximum input of the heaters. The measurements in Model 1
were started at room temperature, but measurements in Model 2 could
be started only above 350°C. At and above this temperature the heat
losses from the measuring assembly were sufficiently large to permit
sinusoidal heating of the surface of the sample while maintaining a
constant average temperature.

The duration of measurements in Model 1 varied between 15 and 30 seconds, during which time the temperature of the sample rose 3° to 5°C. The actual maximum temperature difference between the points of measurement near the surface and the centre of the cylinder generally ranged between 0.2 and 0.5°C. In Model 2 the amplitude of the sinusoidal temperature wave at the outer measuring point was adjusted to 1° to 2°C and the period of one full cycle was 6.5 seconds. At longer wave lengths the time differential (Δt_c) between the heat waves at the surface measuring point and the centre was more difficult to establish, while at shorter wave lengths the wave at the centre would be attenuated too much. While the duration of a measurement in Model 2 is also relatively short, about 26 seconds to record some 4 full wave lengths, the time needed to obtain a constant average temperature level could be as much as 2 hours.

The instability of the emf signal was the major difficulty in these measurements. The probable cause for this difficulty was imperfections in the welds between the thermocouple legs and the specimen. These imperfections, not detectable at the time of welding, become apparent only during the actual measurements. Undoubtedly the scatter of data, as evident in Figure 5, must have been caused to a large degree by this instability. The standard deviation for data obtained in Model 1 is ± 3.3%. For the data obtained in Model 2 the standard deviation is ± 4.3%. The curves of best fit for the

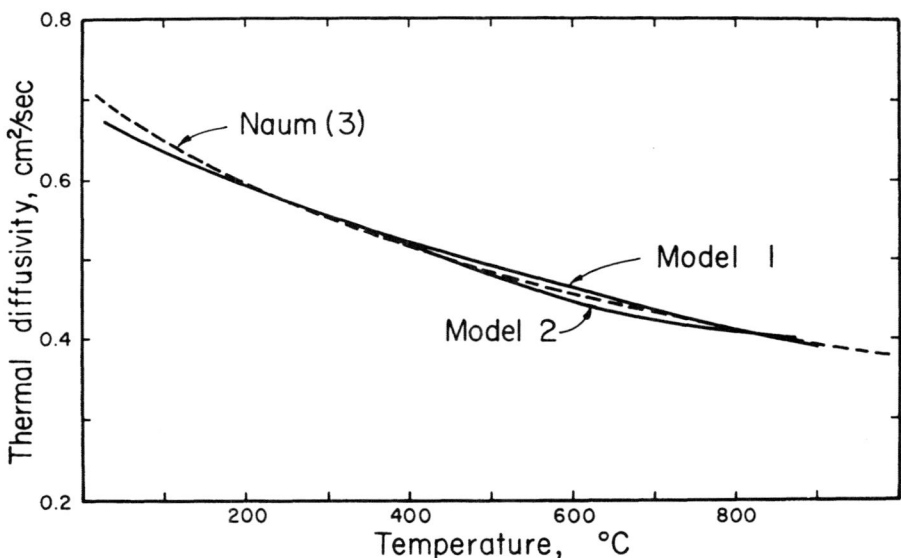

Figure 6.   Thermal diffusivity of tungsten. Least squares fit of data obtained from measurements in Models 1 and 2.

data of Model 1 and Model 2 obtained by the least squares method, are shown in Figure 6. The agreement with Naum's results is very good, with a maximum difference of less than 5% occurring at 25°C. The difference between the data of Model 1 and Model 2 cannot be considered significant.

In the course of the measurements performed on tungsten it became apparent that the measurement of the temperature-time differential (between the two measuring points) is the single most important factor for determining thermal diffusivity by this method. It is therefore essential that the response of the two thermocouples for a given heat pulse be the same. For example, if one of the thermocouples had better thermal contact with the sample or had the hot junction made so that it had a lower heat capacity than the other, the maximum relative temperature in the differential plot of Model 2 could either never be reached (it would keep on increasing) or, after reaching some maximum, it would not remain constant but would start to decline. This condition became amply evident in the measurements on copper. Since thermocouples could not be welded to copper specimens, they were attached by wedging them in the holes with slivers of copper or by pressing the thermocouple bead into a hole of smaller diameter. Unfortunately, satisfactory results could not be obtained by these procedures. It should be taken into consideration that, in the case of copper, the time to reach one-half of the maximum temperature (in Model 1), with measuring points 11.5 mm apart, is of the order of 0.3 seconds. Therefore, the smallest difference in thermocouple response, due to unequal thermal contact with the specimen, could cause the results to vary 30 to 40 percent.

Attempts were made to silver-braze the thermocouples to the specimens. Butt-welded thermocouples were pressed into thermocouple holes together with a small quantity of colloidal silver dispersed in a flux paste. The sample was then heated in vacuum to 950°C and held at this temperature for some 20 minutes. With thermocouples attached in this manner, it was possible to make a limited number of measurements in Model 1 at near-room temperatures. At temperatures above 100°C, perhaps due to the difference in the coefficients of thermal expansion of copper and silver flux, good thermal contact between the sample and the thermocouples was lost.

The data, shown in Figure 7, (Page 11) tend to support the thermal diffusivity curve for copper recommended by TPRC (4), but, naturally, they are insufficient to establish the capacity of the apparatus to perform measurements in this high thermal diffusivity range. Nevertheless, these limited data show that the nature of the problem is in the quality of the thermal contact between the thermocouple and the specimen, not in the principle on which the apparatus is based.

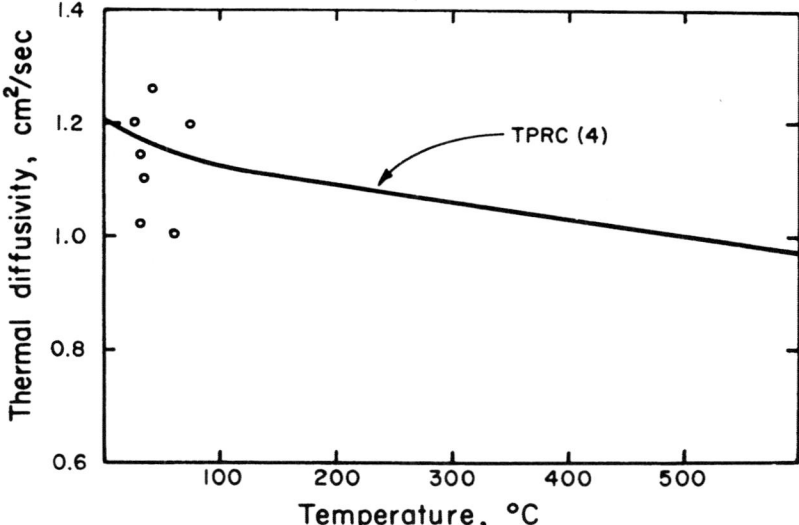

Figure 7. Thermal diffusivity of copper.

CONCLUSIONS

(1) Measurements on high thermal diffusivity, electrically conductive materials can be made with the described apparatus in either transitory temperature (Model 1) or periodic temperature (Model 2) modes, as long as good thermal contact exists between the thermocouples and the sample, and the time-temperature responses of the thermocouples are the same for a given thermal pulse.

(2) The principal factor influencing the precision of measurements is the quality of the bond or contact between the thermocouples and the specimen. The statistical precision of measurements on the tungsten sample is ± 3.3% for Model 1 and ± 4.3% for Model 2. The probable cause for this result is the difference in the mode of heat pulse generation: the precision of the linear heating rate (over the small temperature range) is greater than the precision of the generation of a sinusoidal heat wave.

(3) If the data by Naum are assumed accurate, then the accuracy of measurements of tungsten in either mode is within 2 - 3%.

REFERENCES

(1) Parker, W.J. et al.;   J. Applied Physics $\underline{32}$ (9) 1679 - 84 (1961)

(2) Mirkovich, V.V.;   Thermal Analysis, Vol. 1, Proceedings Third
        ICTA 1971, Birkhauser Verlag Basel, pages 525 - 538.

(3) Naum, R.G.;   Personal communication in a letter of Jan. 5, 1972.

(4) No, C.Y., et al.; TPRC data on elements.   Proceedings 8th
        Conference on Thermal Conductivity, Plenum Press, New York
        1969, pages 971 - 998.

ERROR ANALYSIS OF THE FLASH THERMAL DIFFUSIVITY TECHNIQUE*

Richard C. Heckman

Sandia Laboratories

Albuquerque, New Mexico 87115

An error analysis of the flash technique for thermal diffusivity measurements has been made by propagating errors in measured quantities through several data analysis techniques to determine the consequent errors in thermal diffusivity. The effects of extended pulse durations and of heat losses are considered. The results show that for most conditions actually encountered in experiments, relative measurement errors are reflected as relative errors of approximately the same size in the thermal diffusivity.

INTRODUCTION

Under ideal conditions, flash thermal diffusivity measurements[1] are made by causing a planar, instantaneous, uniform pulse of energy to be deposited on one face of a thin, flat sample and by analyzing the consequent temperature fluctuation which occurs at the opposite face (Fig. 1). Frequently, ideal conditions are not obtained and the effects of extended pulse durations or of significant heat losses must be considered.

Analysis of the temperature fluctuation to obtain the diffusivity and other parameters of the problem can be accomplished by a variety of techniques. Among the most widely employed are those which make use of the "half time", i.e., the time at which the fluctuation reaches half of its maximum value (Fig. 1, point (t_1, T_1)).

---

* This work supported by the U.S. Energy Research and Development Administration.

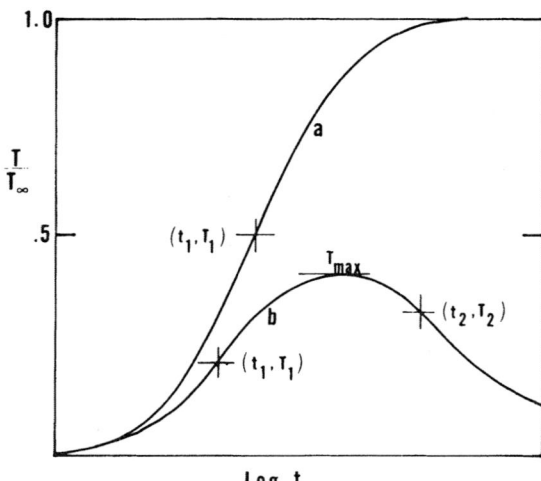

Figure 1. Rear surface excess temperature fluctuation.
a. Ideal conditions.   b. Heat losses.

Parker et al,[1] first suggested this technique for the ideal boundary
and initial conditions listed above. Subsequently, a number of
authors[2,3,4,5] provided modifications appropriate to the case of
extended heat pulses. Cowen[6] and Heckman[5] have applied the technique
to the case where heat losses are important.

     In this paper, the effects of errors on the results of flash
thermal diffusivity measurements are analyzed. This is done by propagating assumed errors of certain significant times or temperatures,
including the half time, through data reduction methods appropriate
to ideal, extended pulse or heat loss conditions.

     An alternative view of these calculations is that they consti-
tute a sensitivity analysis of the flash technique, for the particular data analysis methods considered. Consequently, they provide a
basis for comparing it on entirely theoretical grounds with other
diffusivity measurement methods.

     The word error, used in this discussion, has a particular mean-
ing which should be emphasized. It is used to mean deviations of
known magnitude and sign in the temperature-time points employed in
data analysis, and to the consequent deviations in the deduced parameters of the analysis. No consideration has been given to the
effects of random errors which would require statistical methods.

HEAT TRANSFER THEORY

Data analysis procedures require having at hand the results of heat transfer calculations for the particular set of boundary and initial conditions appropriate to the experiment. In addition, the propagation of errors requires these formulae for the calculation of certain partial derivitives. The formulae used in the present calculations are summarized below.

Ideal Case. When the ideal boundary and initial conditions mentioned earlier are met, the rear surface excess temperature, T, is given by:[1]

$$\frac{T}{T_\infty} = 1 + \sum_{n=1}^{\infty} (-1)^n \exp(-n^2 \pi^2 \tau) \tag{1}$$

where $\tau = \alpha t / \ell^2$ and T_∞ is the equilibrium excess temperature under adiabatic conditions. α is the thermal diffusivity, t is the time, and ℓ is the specimen thickness. The behavior of this function is shown graphically in Fig. 1, curve a.

Extended Pulse Effects. When a rectangular heat pulse of duration t_p is incident on the specimen, the rear surface excess temperature is given by:

$$\frac{T}{T_\infty} = \frac{\tau}{\tau_p} - \frac{1}{6\tau_p} - \frac{2}{\tau_p} \sum_{n=1}^{\infty} \frac{(-1)^n}{n^2\pi^2} \exp(-n^2\pi^2\tau) \qquad 0 \leq t \leq t_p \tag{2}$$

$$\frac{T}{T_\infty} = 1 + \frac{2}{\tau_p} \sum_{n=1}^{\infty} \frac{(-1)^n}{n^2\pi^2} \exp(-n^2\pi^2\tau) \left[\exp(n^2\pi^2\tau_p)-1\right] \quad t \geq t_p \tag{3}$$

where $\tau_p = \alpha t_p/\ell^2$. Curve a, Fig. 1, is qualitatively descriptive of eqs. (2) and (3).

Heat Loss Effects. When heat losses are significant, but otherwise ideal conditions exist, the rear surface excess temperature is given by:[5,6]

$$\frac{T}{T_\infty} = \sum_{n=1}^{\infty} Y_n(0) \ Y_n(\ell) \ \exp(-\beta_n^2\tau) \tag{4}$$

where $Y_n(x) = \dfrac{2^{\frac{1}{2}} \left[\beta_n \cos(\beta_n x/\ell)+H \sin(\beta_n x/\ell)\right]}{\left[\beta_n^2 + 2H + H^2\right]^{\frac{1}{2}}}$ and β_n (n = 1, 2,

3, ...) are the positive roots of $(\beta^2-H^2) \tan \beta = 2\beta H$. H is the Nusselt number describing the surface heat losses. The behavior of eq. (4) is shown in curve b, Fig. 1.

DATA ANALYSIS

Ideal Case. From eq. (1) for $T_1 = 0.5\ T_\infty$, numerical evaluation shows that $\tau_1 = \tau(t_1) = .1388$. The simple data reduction scheme shown in Table I then follows (ref. 1).

Extended Pulse Effects. Table II shows the details of a data reduction method suggested in ref. (5). This method is similar but not identical to those suggested in refs. (2-4). Step 5 involves the use of tables or graphs of W vs X prepared by numerical evaluation of eqs. (2) or (3).

Heat Loss Effects. The analysis of a temperature fluctuation significantly influenced by heat losses proceeds differently, primarily because there are two parameters, α and H, whose values must be found from the temperature fluctuation. Cowen's procedure[6] for this analysis is shown in slightly modified form[5] in Table III. Steps 7 and 8 are accomplished using tables or graphs constructed by numerical evaluation of eq. (4) (see ref. 5).

Table I. Ideal Case

Data Reduction	Propagation of Errors
1. Find T_∞	$\Delta T = \delta T_\infty$
2. $T_1 = 0.5\ T_\infty$	$\Delta T_1 = 0.5\ \Delta T_\infty + \delta T_1$
3. $t_1 = t(T_1)$	$\Delta t_1 = (t_T)_1\ \Delta T_1 + \delta t_1$
4. $\alpha = .1388\ \ell^2/t_1$	$\Delta \alpha = -.1388\ \ell^2\ \delta t_1/t_1^2$

Table II. Extended Pulse Effect

Data Reduction	Propagation of Errors
1. Find T_∞	$\Delta T_\infty = \delta T_\infty$
2. $T_1 = 0.5\ T_\infty$	$\Delta T_1 = 0.5\ \Delta T_\infty + \delta T_1$
3. $t_1 = t(T_1)$	$\Delta t_1 = (t_T)_1\ \Delta T_1 + \delta t_1$
4. $X = t_p/t_1$	$\Delta X = (X_t)_1\ \Delta t_1$
5. $W = W(X)$ $\quad(W = \alpha/\alpha_o)$	$\Delta W = (W_X)_1\ \Delta X$
6. $\alpha = W \cdot \alpha_o$	$\Delta \alpha = (\alpha_W)_1\ \Delta W + (\alpha_t)_1\ \Delta t_1$

Note: $\alpha_o = .1388\ \ell^2/t_1$, $(\alpha_W)_1 = \alpha_o$

Table III. Heat Loss Effects

Cowen Data Analysis	Propagation of Errors
1. Find T_{max}	$\Delta T_{max} = \delta T_{max}$
2. $T_1 = 0.5\ T_{max}$	$\Delta T_1 = 0.5 \cdot \Delta T_{max} + \delta T_1$
3. $t_1 = t(T_1)$	$\Delta t_1 = (t_T)_1\ \Delta T_1 + \delta t_1$
4. $t_2 = 5.0\ t_1$	$\Delta t_2 = 5.0\ \Delta t_1 + \delta t_2$
5. $T_2 = T(t_2)$	$\Delta T_2 = (T_t)_2\ \Delta t_2 + \delta T_2$
6. $V_2 = T_2/T_{max}$	$\Delta V_2 = (V_T)_2\ \Delta T_2 + (V_T)_{max}\ \Delta T_{max}$
7. $H = H(V_2)$	$\Delta H = (H_V)_2\ \Delta V_2$
8. $\tau_1 = \tau(H)$	$\Delta \tau_1 = (\tau_H)_1\ \Delta H$
9. $\alpha = \tau_1\ \ell^2/t_1$	$\Delta \alpha = (\alpha_\tau)_1\ \Delta \tau_1 + (\alpha_t)_1\ \Delta t_1$

ERROR ANALYSIS

The error analysis is carried out by propagating errors in the various measured quantities, i.e., δT_∞, δT_{max}, δT_1, δT_2, δt_1, and δt_2, through the data analysis schemes as shown in Tables I, II, and III. Notations such as $(t_T)_1$ represent partial derivitives, $(\delta t/\delta T)$. $(t_T)_1$ and $(T_t)_2$ are found by taking the derivitives, as appropriate, of eqs. (1), (2), (3), or (4), and numerically evaluating the resulting function. $(W_X)_1$, $(H_V)_2$ and $(\tau_H)_1$ are obtained from computer generated tables of their variables to which spline functions are fit. The derivitives are then evaluated from the spline coefficients. The remaining derivitives follow directly from the defining functions.

Those partial derivitives which require computer calculation are shown in Figs. 2 and 3. In order to make the calculations in a general form, the following substitutions have been made:
$(t_T)_1\ (T_1/t_1) = (\tau_T)_1\ (T_1/\tau_1)$ and $(T_t)_2\ (t_2/T_2) = (T_\tau)_2\ (\tau_2/T_2)$.

Ideal Case. The results of the error analysis for the ideal case are:

$$\frac{\Delta \alpha}{\alpha} = -\frac{\delta t_1}{t_1}\ ,\ \text{for } \delta T_\infty = 0,\ \delta t_1 \neq 0,\ \delta T_1 = 0.$$

$$\frac{\Delta \alpha}{\alpha} = -.769\ \frac{\delta T_\infty}{T_\infty}\ ,\ \text{for } \delta T_\infty \neq 0,\ \delta t_1 = 0,\ \delta T_1 = 0.$$

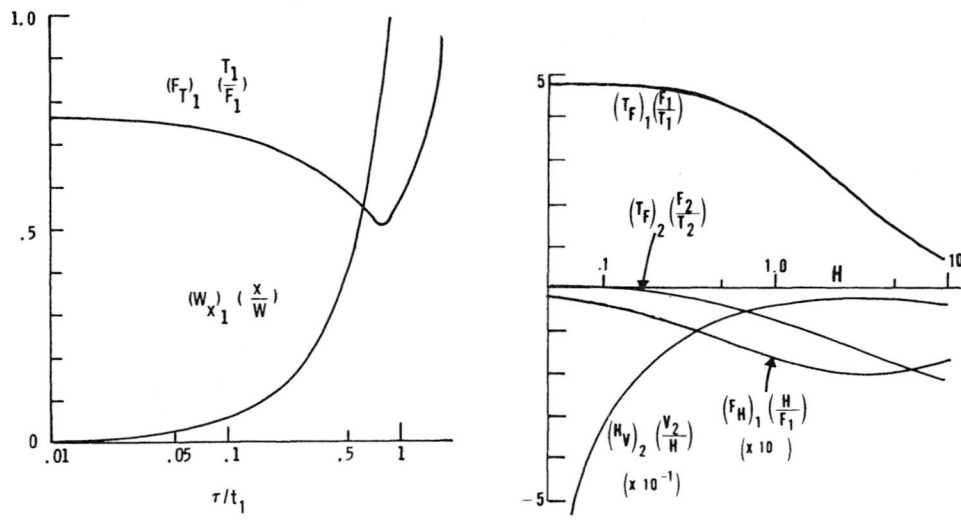

Figure 2. Normalized partial deriv-   Figure 3. Normalized partial de-
          atives vs. τ/t_1 using eqs.   rivatives vs. H using
          (1) and (3). $\ ^1$(F $\equiv$ T)   eq. (4). (F $\equiv$ T)

In both cases, relative measurement errors lead to approximately
equal magnitude relative errors in the diffusivity.

Extended Pulse Effects. The results for the case of extended
pulses are shown in Fig. 4 where the relative error in α is plotted
against τ_p/t_1. $\delta t_1/t_1 = 5\%$ or $\delta T_{max}/T_{max} = 5\%$ has been assumed.
The results show that when pulse durations are negligable compared
with diffusion times, errors do not depend on the pulse width. As
the two times become comparable, the errors increase sharply. This
conforms with the intuitive view that when the pulse duration is very
long, the experiment will be rather insensitive to the diffusivity.

Heat Loss Effects. Fig. 5 shows the results when heat losses
are significant. Errors in both α and H are shown plotted against
H. Relative errors in α display a generally weak dependence on H
and are approximately of the same magnitude as the measurement
relative errors. Relative errors in H, on the other hand, are more
strongly dependent on H and are substantially larger than those in α.

Sensitivities. It is convenient to define the sensitivity,
S_i^α, as

$$S_i^\alpha = \left| \frac{\delta P_i}{\Delta \alpha} \right|$$

where P_i is one of the measured temperatures or times, i.e., T_{max},
T_1, T_2, t_1, or t_2. According to this definition, S_i then provides
a basis on which to compare one diffusivity measurement technique

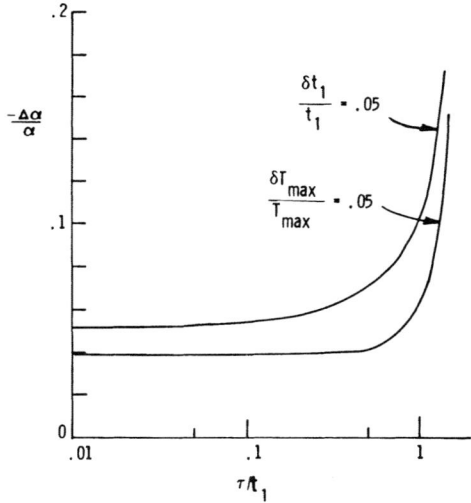

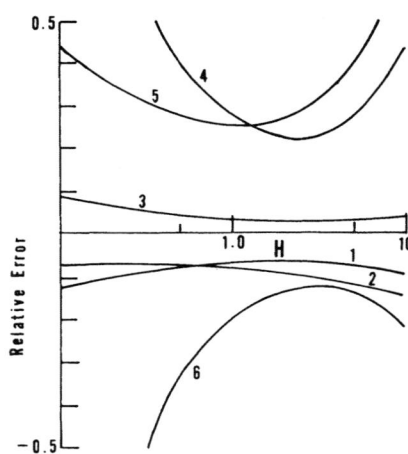

Figure 4. Relative error in α vs. τ/t_1 for +5% errors in T_{max} and t_1 (Table II).

Figure 5. Relative errors in α (curves 1, 2, 3) and H (4, 5, 6) vs. H for 5% errors in T_{max} (1, 4), t_1 (2, 5) and T_2 (3, 6) (Table III).

with another, to compare one method with itself under a different set of conditions, or to compare one data analysis method with another.

The sensitivity of the flash experiment is, for all the measured quantities and for most values of H or t_p/t_1, $S_i^\alpha \sim 1$. On the other hand, $S_i^H < .5$ for most values of H, suggesting that the flash method is a poorer technique for measurements of H than α. This is consistant with our laboratory experience where frequently in analyzing an experimental temperature-time curve, values of H outside the expected range are obtained while the same curve yields a reasonably accurate value of the diffusivity.

CONCLUDING REMARK

The foregoing results suggest that the flash method is reasonably sensitive for thermal diffusivity measurements. It should be noted, however, that the results depend critically on the data analyses procedures assumed. An error or sensitivity study based on different procedures would yield different results. This emphasizes that the method of data analysis is as important as the particular experimental technique adopted. If either is improperly selected, an insensitive overall experiment may result.

REFERENCES

1. W. J. Parker, R. J. Jenkins, C. P. Butler and G. L. Abbott,
   J. Appl. Phys. 32, 1679 (1961).

2. J. A. Cape and G. W. Lehman, J. Appl. Phys. 34, 1909 (1963).

3. R. E. Taylor and J. A. Cape, Appl. Phys. Lett. 5, 212 (1964).

4. K. B. Larson and K. Koyama, J. Appl. Phys. 38, 465 (1967).

5. R. C. Heckman, J. Appl. Phys. 44, 1455 (1973).

6. R. D. Cowen, J. Appl. Phys. 34, 926 (1963).

A NEW TECHNIQUE FOR THE MEASUREMENT OF THERMAL CONDUCTIVITY OF THIN FILMS

T.C. Boyce and Y.W. Chung

University of Hong Kong

Hong Kong

In this paper we discuss some of the inherent difficulties encountered in measurement of thermal conductivity of very thin films and present a new technique which is applicable to any electrically conducting film. The technique is to pass an electric current along the film, thus creating a temperature gradient, and deduce the thermal conductivity from the measured change in resistance. Analysis of the heat diffusion equation, under the experimental conditions imposed, shows that if the resistance change is monitored as a function of time and the half rise-time measured, then the thermal conductivity of the film can be estimated directly without knowledge of heat capacity or density; this is an immediate advantage over other transient techniques. Since the method is quite accurate and the measurement takes only a matter of seconds it has obvious advantages over steady-state techniques. Also, unlike laser pulse techniques, where there is a thickness limitation, due to heat-pulse transit times, this method is applicable to very thin films. We present here a brief outline of the theoretical analysis and experimental procedure for the new technique.

INTRODUCTION

Thermal conductivity has always been one of the most difficult transport coefficients to measure with a reasonable degree of accuracy. This is due partly to the difficulty in measuring the quantities involved and partly to the various heat losses which are usually present.

When measuring electrical conductivity, say for a metal film on glass, because the current is carried by the conduction electrons

one can safely assume that the entire electrical current is
contained within the film. While it is true that most of the heat
current in a metal film is also carried by the conduction electrons,
heat can be transmitted through the substrate and to the surrounding
medium. These effects are collectively called heat losses and must
be taken into account, either by calculation or, as in the present
case, by suitable experimental arrangement.

One problem common to most thin film techniques is that they
require accurate measurement of temperature at one or more points
on the film. In the technique to be described here the only
temperature measurement is that of the film environment, while the
thermal conductivity is determined indirectly from the change of
film resistance.

THEORETICAL ANALYSIS

Fig. 1 shows the model used for analysis in which a film of
thickness a is in perfect thermal contact with a substrate of
thickness H. With the ends of the two layer sample held at some
constant temperature a current I is passed through the film and
because of Joule heating the film resistance R changes. The problem
is then to relate the change in resistance ΔR to the thermal
conductivity K. The technique bears some similarity to those of

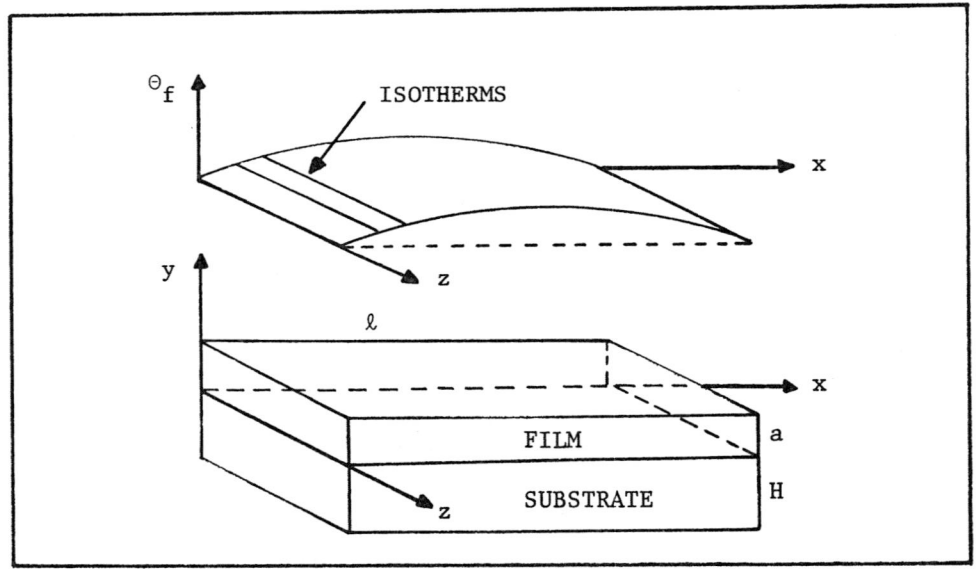

Fig. 1 The model used for theoretical analysis.

Kohlrausch and Callendar (a discussion of which can be found in reference 1) although in the present application we take account of heat conduction by the substrate and heat loss by surface radiation.

In the model the temperature is assumed to be uniform over the width of the film so that the problem reduces to heat conduction in a plane parallel to the xy plane. For the film we have the usual diffusion equation with the first two terms related to the rate of flow of heat per unit volume, the third the rate at which energy is gained per unit volume, and the fourth allows for heat generated electrically per unit volume.

$$\frac{\partial^2 \Theta_f}{\partial x^2} + \frac{K_y}{K_x} \frac{\partial^2 \Theta_f}{\partial y^2} - \frac{1}{k_f} \frac{\partial \Theta_f}{\partial t} + W(1+\alpha\Theta_f) = 0 \quad \text{for} \left\{ \begin{matrix} 0<x<\ell \\ 0<y<a \end{matrix} \right\} \quad (1)$$

Here $K_{x,y}$ is the thermal conductivity in the x and y directions, k_f the thermal diffusivity in the x direction, Θ_f the temperature rise in the film and α the temperature coefficient of resistance. The quantity $W = J^2/\sigma_x K_x$ with J the current density and σ_x the electrical conductivity in the x direction.

A similar expression, without the electrical heat generation term, can be derived for the substrate assuming an isotropic thermal conductivity K_s, a temperature rise Θ_s and thermal diffusivity k_s.

$$\frac{\partial^2 \Theta_s}{\partial x^2} + \frac{\partial^2 \Theta_s}{\partial y^2} - \frac{1}{k_s} \frac{\partial \Theta_s}{\partial t} = 0 \quad \text{for} \left\{ \begin{matrix} 0<x<\ell \\ -H<y<0 \end{matrix} \right\} \quad (2)$$

Uniform temperature initially:	$\Theta_{f,s}(x,y,t=0) = 0$	(3)
Perfect thermal contact at film-substrate interface:	$\Theta_f(y=0) = \Theta_s(y=0)$	(4)
	$K_y \frac{\partial \Theta_f}{\partial y}(y=0) = K_s \frac{\partial \Theta_s}{\partial y}(y=0)$	(5)
Both ends maintained at constant temperature:	$\Theta_{f,s}(x=0) = \Theta_{f,s}(x=\ell) = 0$	(6)
Radiative boundary condition at free film surface:	$K_y \frac{\partial \Theta_f}{\partial y}(y=a) = -E\Theta_f(y=a)$	(7)
No heat loss at free substrate surface:	$\frac{\partial \Theta_s}{\partial y}(y=-H) = 0$	(8)

Table 1. The initial and boundary conditions. The radiation loss is represented by $E = 4 S\varepsilon T^3$ where S is the Stefan constant, ε the emissivity and T the absolute temperature.

To solve these equations, subject to the boundary conditions given in Table 1, we use the method of successive transformations,[1] i.e. the time variable is removed by a Laplace transformation, followed by a Fourier transformation to deal with the space variables. For thin films the dimension a is very small and we expect little variation of temperature across the thickness of the film for any particular value of x. Hence, it makes little difference where, in the range $0<y<a$, we calculate the temperature rise in the film. For convenience we determine Θ_f at $y=0$ with the following result.

$$\Theta_f(y=0)=\frac{4WX}{\pi}\sum_{n=0}^{\infty}\frac{\sin\frac{(2n+1)\pi x}{\ell}.\left\{1-\exp.-k_st\left(\frac{(2n+1)^2\pi^2(1+X)}{\ell^2}+\frac{E}{K_sH}-\alpha WX\right)\right\}}{(2n+1)\left\{\frac{(2n+1)^2\pi^2(1+X)}{\ell^2}+\frac{E}{K_sH}-\alpha WX\right\}} \qquad (9)$$

Here, $X=K_xa/K_sH$, the film to substrate thermal conductance ratio. In obtaining this result it is assumed that the film is much thinner than the substrate ($a<<H$) and that heat diffuses more rapidly through the film than the substrate ($k_f>>k_s$);both assumptions are reasonable for conducting films on glass. To find the fractional resistance change we simply integrate the temperature rise over the length of the film.

$$\frac{\Delta R}{R}=\frac{8\alpha\sigma_xI^2R^2}{\pi^4K_x(1+1/X)}\sum_{n=0}^{\infty}(2n+1)^{-4}\left\{1-\exp.-k_st\left(\frac{(2n+1)^2\pi^2(1+X)}{\ell^2}\right)\right\} \qquad (10)$$

In this result the term ($E/K_sH-\alpha WX$) has been neglected and this can be achieved experimentally as will be discussed in a moment. Eq. 10 can be maximized to give the maximum resistance change which occurs when the steady state has been reached.

$$\left(\frac{\Delta R}{R}\right)_{max}=\frac{\alpha\ \sigma_x\ I^2R^2}{12K_x\ (1+K_sH/K_xa)} \qquad (11)$$

A useful parameter to work with experimentally is $t_{\frac{1}{2}}$, the time to reach half of the maximum resistance change. From the two previous equations and computer iteration we find

$$t_{\frac{1}{2}}=\frac{0.06876\ \ell^2}{k_s(1+K_xa/K_sH)} \qquad (12)$$

From this we can see that knowledge of certain substrate parameters (k_s, K_s and H) and measurement of film dimensions (ℓ and a) enable the film thermal conductivity K_x to be determined from a measurement of $t_{\frac{1}{2}}$.

DISCUSSION OF EXPERIMENTAL CONDITIONS AND MEASUREMENT

The assumption that the initial temperature of the film and substrate is uniform throughout can be achieved using a sample holder of the type shown in Fig. 2, in which each end of the specimen is held in good thermal contact with a large aluminium block. The aluminium blocks are electrically insulated by means of a thin mica sheet and, because of their relatively large thermal mass, the end temperatures remain constant regardless of whether or not a current is passed through the film. The middle portion of the substrate is separated from the aluminium blocks so that its lower surface is thermally insulated when in vacuum.

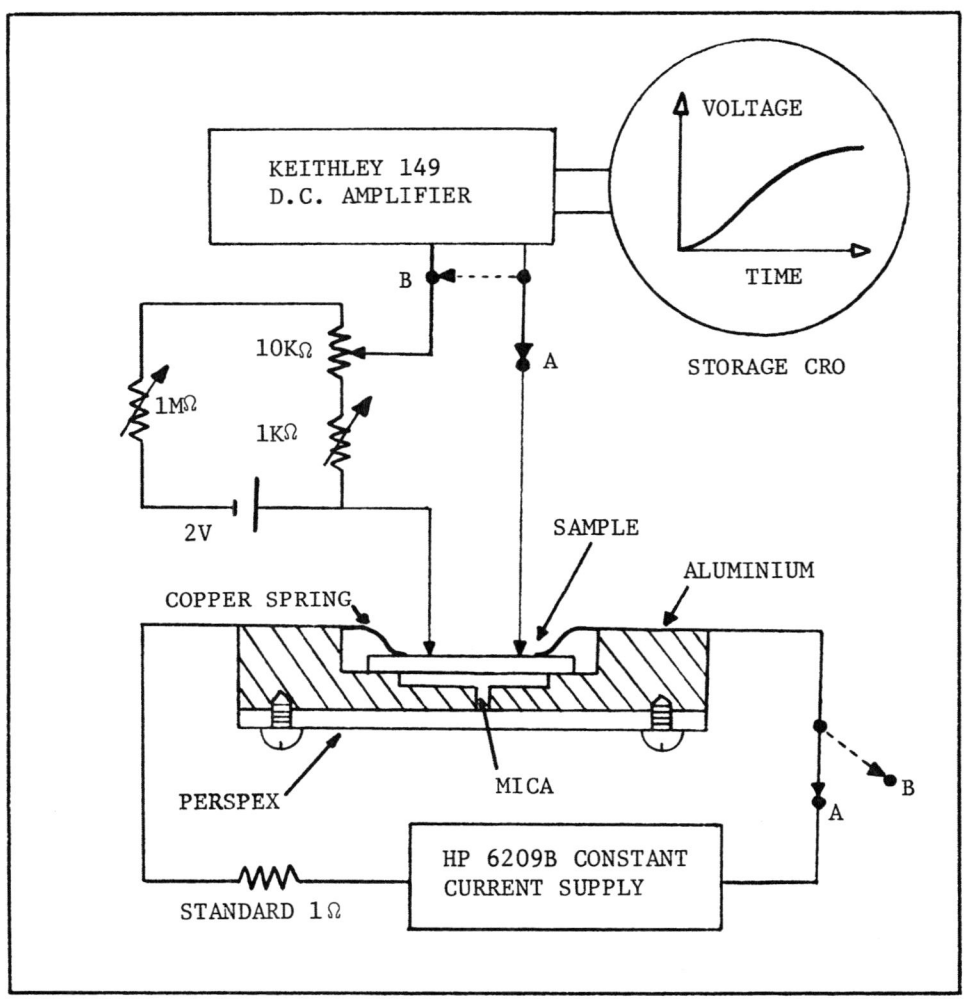

Fig. 2 Schematic of sample holder and measuring circuit.

From the previous analysis the assumption that

$$\left\{\frac{E}{K_sH} - WX\right\} << \frac{\pi^2}{\ell^2}\{1+X\}\qquad\qquad(13)$$

can be achieved experimentally in three ways;(i) the radiation term
becomes small if the experiment is performed at low temperatures
and the second term can be made small by using a small current;
unfortunately this places an upper limit on the temperature at which
the technique is applicable; (ii) by using a film of short length;
this would result in a smaller electrical resistance and since the
fractional change in resistance is proportional to R^2 the measured
signal would be considerably attenuated;(iii) by adjusting the
current such that the two terms roughly cancel. For the present
work the third arrangement was adopted since, although a number of
parameters are required in order to determine the appropriate
current, they need only be estimated approximately. Alternatively,
the appropriate current will occur when consistent results are
obtained for the thermal conductivity measurement;this approach
avoids the need for other data.

The technique was tested using polycrystalline silver films on
glass. Measurements were taken at fixed points, 77 and 295K, using
an Oxford Instrument MD4A cryostat for which the temperature
stability over one hour was better than ±0.01K. For a typical silver
film, with a resistance of 1Ω and TCR of $10^{-3}K^{-1}$, this introduces
a drift voltage of less than 0.2μV in one hour and, since voltages
measured were typically 20 to 60 μV, this error was negligible.
Because the resistance change was quite small a balance network was
used to cancel out most of the normal voltage drop along the film.
The measuring circuit is also shown in Fig. 2.

Initially, to ensure a uniform temperature throughout the
sample, helium exchange gas was allowed into the cryostat. Then,
with the switch in position A, the current was set at the estimated
value and the voltage drop along the film balanced out. Next, with
the switch in position B (current off), the cryostat was evacuated
to about 10^{-5} torr. The switch was then returned to position A and
the resistance change recorded on an oscilloscope camera. A similar
procedure was used at 77K after liquid nitrogen was introduced into
the cryostat and a stable temperature attained.

Other auxiliary measurements are as follows;film length was
about 0.02 m and was measured by travelling microscope to an
accuracy of ±10μm;film thickness was in the range 500 to 4000Å
and was measured by interference microscope to an accuracy of ±30Å;
substrate thickness was about 76μm and was measured by micrometer
to an accuracy of ±1μm;substrate thermal conductivity was taken from

data supplied by the Corning company, K_s = 0.988 J $m^{-1}K^{-1}s^{-1}$ for Corning microsheet; substrate thermal diffusivity was measured by laser pulse technique[2] to an accuracy of ±2% giving k_s = 7.3x10^{-7} m^2s^{-1} at 77K and k_s = 3.4x10^{-7} m^2s^{-1} at 295K.

       Unlike other dynamic methods the only film measurements required are the film dimensions, which are in any case required for other calculations. The combined error depends on film thickness; for films above 2000Å the accuracy is of the order of 4 to 6% while below 2000Å the accuracy becomes about 10%.

       From theoretical consideration one would expect the size-effect dependence of thermal conductivity to be similar to that of electrical conductivity however Fig. 3 shows that there is a consistent discrepancy between experimental measurements and Fuchs-

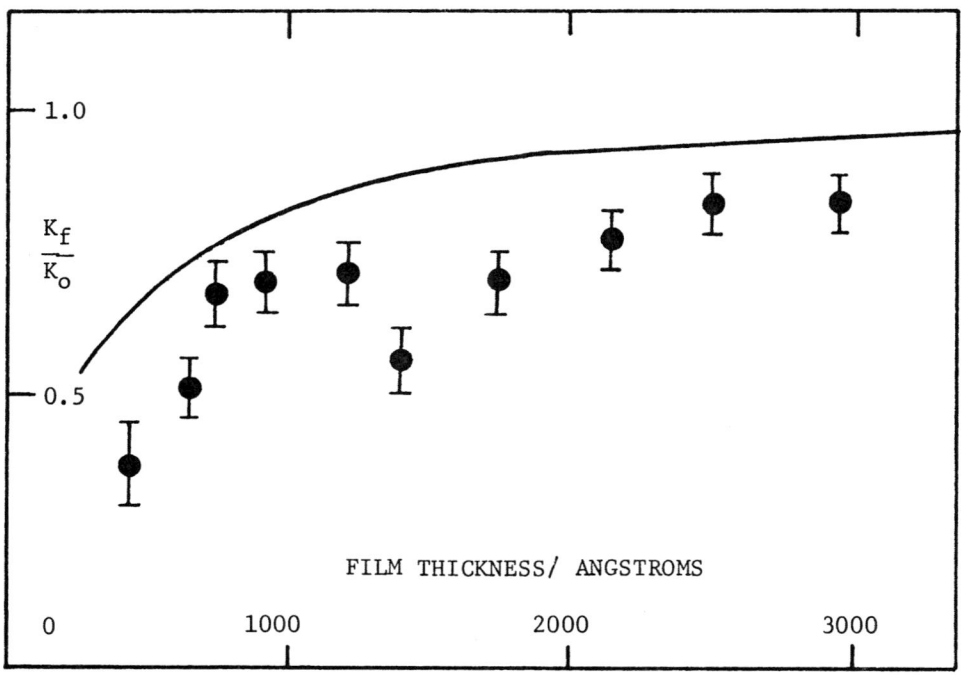

Fig. 3 Thickness dependence of reduced thermal conductivity for silver films on glass at 295K. The bulk thermal conductivity value K_O is taken from the literature[3] and the theoretical curve is drawn according to Fuchs-Sondheimer theory[4] with the surface specularity parameter p = 0.

Sondheimer theory; a similar situation has been observed for the galvanomagnetic coefficients.[5] Adjustment of the specularity parameter p is ineffective at thicknesses greater than the mean free path (530Å for silver at 295K) and it has been suggested that the effect of grain-boundary scattering[6] should be taken into account.

CONCLUSION

The technique described here has the advantage that, for fixed temperature measurements, the coefficient of thermal conductivity can be determined from a simple measurement of resistance change due to Joule heating, provided certain experimental conditions are fulfilled.

The technique is currently being used to measure the thermal conductivity of silver films on glass and it is hoped that these experiments will provide corroboration with galvanomagnetic measurements for inclusion of grain-boundary scattering in size effect studies.

[1] H.S. Carslaw and J.C. Jaeger, Conduction of Heat in Solids, Clarendon Press Oxford (1959).
[2] S. Namba, P.H. Kinoshita and T. Arai, Laser Rev. Jan. (1968).
[3] G.K. White and S.B. Woods, Phil. Trans. Roy. Soc. Lond. A251, 273 (1959).
[4] E.H. Sondheimer, Advances in Physics, 1,1 (1952).
[5] T.C. Boyce and W.H. Wong, Galvanomagnetic Size Effects in Polycrystalline Metal Films, paper presented at 1974 Thin Films Conference, Sussex U.K., to be published shortly.
[6] T.C. Boyce and W.H. Wong, Phys. Letters, 36A, 323 (1971).

AN EXPERIMENTAL STUDY OF THE

ELECTROSTATICALLY CONTROLLED HEAT SHUTTER*

W. P. Schimmel, Jr.,[1] and A. B. Donaldson[2]

Sandia Laboratories

Albuquerque, New Mexico 87115

A cooling device which uses an electrostatic field to accelerate molecules resulting in a forced convection flow field is experimentally investigated. The region adjacent to the stagnation point of a heated, horizontal right circular cylinder is visualized using a laser holographic interferometer. With no electrostatic field present, the temperature profile in this region is essentially linear, indicating a conduction dominated situation. An electrostatic probe with approximately a 6 kV potential difference with respect to the cylinder is then applied resulting in an increase of the heat flux (temperature gradient) from the heated cylinder to the air. The preliminary results reported in this paper indicate that this device might possibly be used to advantage in some situations. For example, in spacecraft, it is possible to increase the cooling capacity of a gas without resorting to solid conductors. Space situations are usually weight limited so that some weight savings might be effected. This space application of the electrostatically driven convection has been called an "Electrostatically Controlled Heat Shutter."

---

*This work was supported by the United States Energy Research and Development Administration.

[1]Member of the Technical Staff, Heat Transfer and Fluid Mechanics Division.

[2]Member of the Technical Staff, Initiating and Pyrotechnic Components Division.

INTRODUCTION

An electrostatic field has been observed to produce a cooling effect when applied to a hot, grounded object [1]**. The phenomenon is thought to be the same as that used in an ion drag pump. Ions are produced at the high voltage probe and migrate toward the grounded object because of the electrical field. The migrating ions entrain air molecules by momentum transfer and the stream impinging on the hot object produces the same effect as localized forced convection cooling. An alternate theory is that the ions themselves become electrically neutral when they strike the hot grounded object and thus leave as heated air molecules. Macroscopically, both theories produce the same cooling effect and are, therefore, considered equivalent.

This electrostatic cooling, or "ionic wind" as it has been called by some authors [2, 3] has some potentially interesting applications. For example, electrically powered components used in spacecraft typically are cooled by conduction to some external surface and thence to deep space by radiation. In the event the structural components of the system do not provide the appropriate heat sink capability, additional thermal conductors must be used. Because of the penalties associated with weight in space, the use of solid conductors or heat pipes is likely to be discouraged. It appears that it may be possible to use a gaseous conductor whose cooling capacity is increased by the ionic wind [4] as indicated in Figure 1.

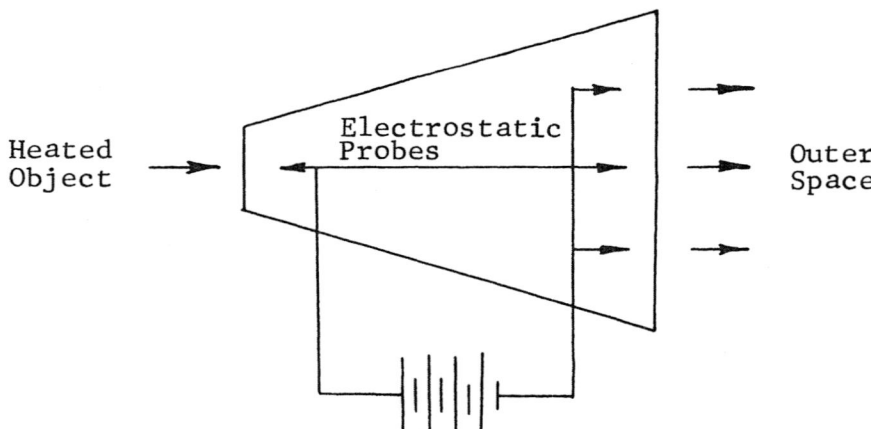

Figure 1. Electrostatic Heat Shutter

**Numbers in brackets denote references listed at the end of the paper.

The purpose of the present work is to investigate experimentally the enhancement of thermal conduction by the electrostatically driven convection. The stagnation region of a heated,
horizontal circular cylinder in air provides a conduction dominated regime. Applying an electrostatic field to this region
results in a steeper temperature gradient indicating an increased
heat flux to the air. There are other physical situations which
could be used, but the present one has the very desirable feature
of being essentially without natural convection. For a vertical
cylinder, one would have to consider the interaction of a horizontal ionic wind with a vertical boundary layer. This, of course,
is also the case for regions of the horizontal cylinder away from
the stagnation point.

Because of the intense electrical field, the use of any type
of electrical probe such as a thermocouple would produce question-
able results. In addition, the temperature field is likely to be
adversely influenced by the introduction of a measurement probe.
Optical measurements circumvent both of these restrictions with
the additional advantage that they may be considered inertialess.
This means that very rapid transients can be observed. In the
present study, however, it was found that steady-state measurements were sufficient so the transient visualization capability
was not used.

EXPERIMENTAL PROGRAM

The interferograms were produced by a two-beam laser holographic setup illuminated by a 50 mW Spectra-Physics helium-
neon laser. A schematic of the apparatus is presented in Figure 2.

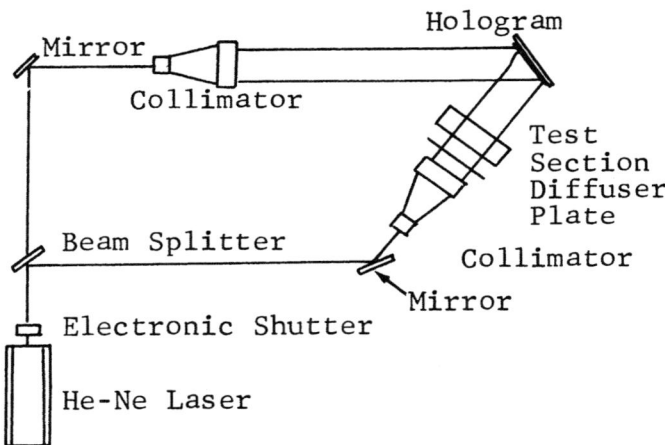

Figure 2. Laser Holographic Setup

The physical principles involved in making temperature measurements with this type instrument are discussed in references [5, 6].
 The so-called double exposure technique was used to obtain the
holographic interferograms. The hologram is exposed to both the
object and the reference beams with the test item in some initial
condition, i.e., either unheated or heated to a steady-state
condition. A second exposure is made with the test item heated
and the electrostatic field either on or off. An interference
pattern between the initial and final holograms thus results.

    The electrostatic field was generated by a regulated DC power
supply operated at about 6 kV applied through a sharp metal probe
(Figure 3). Although the cooling effectiveness or strength of the
convective flow should be a function of electrical potential
difference, only this single value (6 kV) was used. The reason is
that this corresponds to the approximate maximum which could be
applied without arcing between the probe and the cylinder. When
arcing occurred, the power supply would trip an internal circuit
breaker and terminate the experiment. Attempts to override the
breaker resulted in a poor quality interferogram because of the
optical photons in the arc. Lower values than 6 kV would naturally decrease the cooling effect.

    Finally, the heated object is a right circular cylinder 1.27
cm diameter by 7.5 cm long. Heating was provided by an electrical
resistor located at the centroid of the cylinder. Leads from the
heater were enclosed in another metal cylinder to which the high
voltage ground was attached. The experimental setup is presented
in Figure 3. In all cases reported in this paper, the total power
delivered to the resistor was 25 W. No attempt was made to use
this value since the spatial distribution of heat loss was unknown.

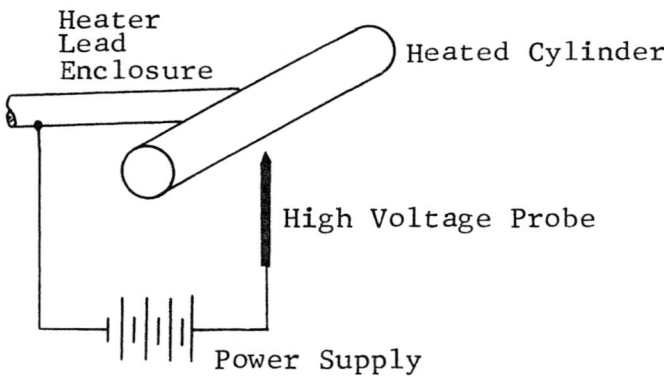

Figure 3. Experimental Setup

RESULTS

An interference photograph of the heated cylinder in air without external cooling is presented in Figure 4. The fringe pattern
is a familiar one to anyone who has ever studied interferometric
heat transfer measurements. The presence of the heater leads on
the left side of the heated cylinder perturbs the temperature
field slightly, but the right-hand side is undisturbed by them.
In Figure 5, the electrostatic field is energized at the 6 kV
level. The initial exposures were made with the cylinder at the
ambient temperature in both Figures 4 and 5. Although it is
tempting to comment on the convective flow field around the cylinder in Figure 4, only the stagnation region will be considered.
Note that the spacing between adjacent fringes decreases in
Figure 5 compared with Figure 4. Figures 6 and 7 are enlarged
views of the stagnation region for these two cases.

The procedure mentioned earlier was applied to deduce temperature profiles for the two cases. These data are presented in
Figure 8. Two interesting things can be observed from the plot.
First, the temperature profile in the region near the cylinder is
very nearly linear for both cases. This indicates that conduction
rather than convection is the dominant heat transfer mechanism.
Secondly, application of the electrostatic field results in a
steeper temperature profile which denotes a higher heat flux to
the air. In fact, the gradient is about twice as large with the
cooling as without. As mentioned earlier, this value of temperature gradient could be changed by varying the potential difference except for the breakdown characteristics of the air. In an
actual application, the dryness of the medium would probably produce a larger effect. On the day when these experiments were
performed, the relative humidity in the laboratory was about 20%
(typical for Albuquerque, NM). The dotted portion of the temperature profiles correspond to extrapolation beyond the fringe
farthest from the cylinder. Similarly, the surface temperature
of the cylinder is denoted apparent because it is difficult to
resolve fringes near the surface. Reference [7] goes into some
detail on the problems associated with interferometry near
surfaces.

CONCLUSIONS

A cooling device consisting of a sharp probe which forms the
positive pole of a high-voltage electrostatic field and a heated
grounded object have been experimentally investigated. It was
found that, below the threshold of air breakdown, this field does
indeed increase the cooling capacity of the air adjacent to the
cylinder. The experiment was designed such that conduction was

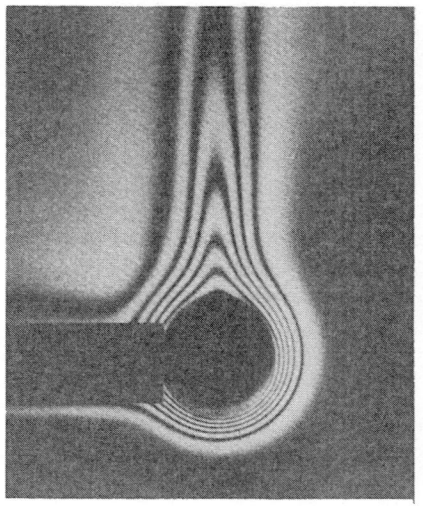

FIGURE 4
ELECTROSTATIC FIELD OFF

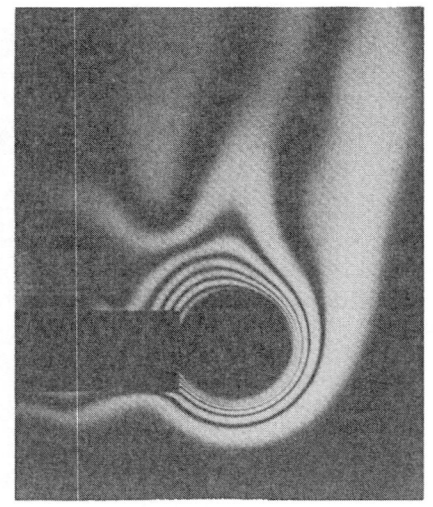

FIGURE 5
ELECTROSTATIC FIELD ON

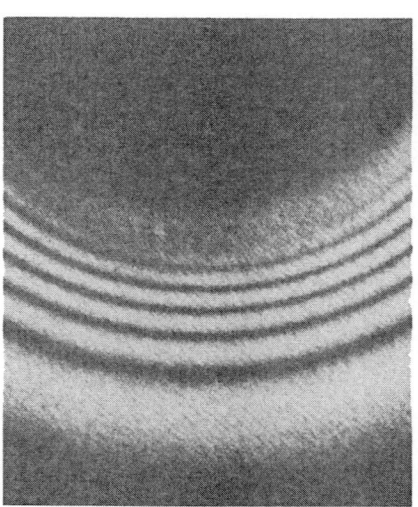

FIGURE 6
STAGNATION REGION
FIELD OFF

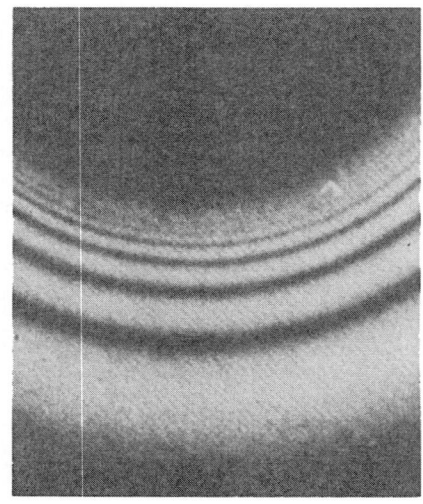

FIGURE 7
STAGNATION REGION
FIELD ON

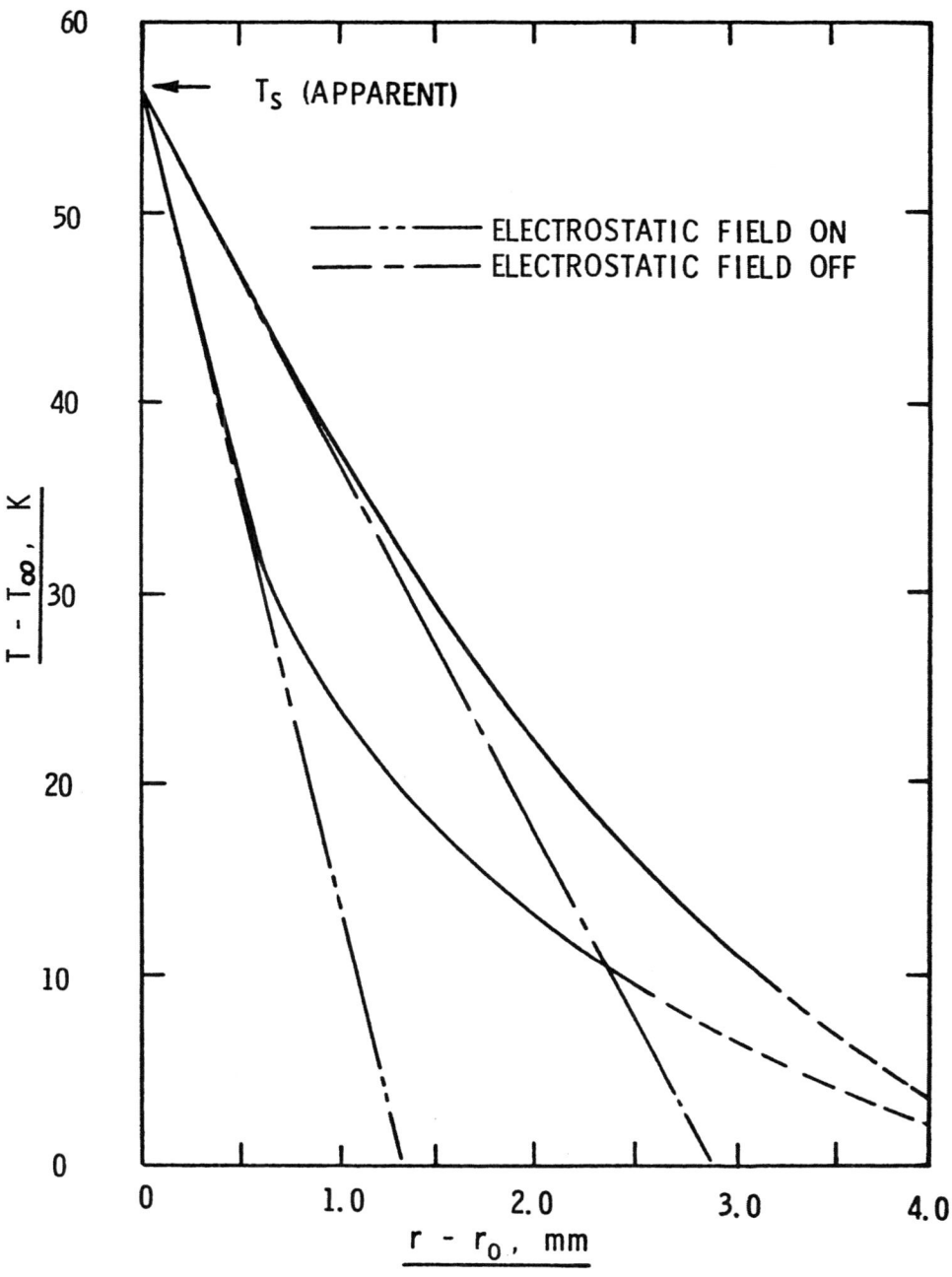

Figure 8. Experimental Temperature Profiles

the dominant energy transfer mechanism and, thus, the steepness of the temperature gradient alone could be used to evaluate the effect. Although the system provides additional cooling as expected, it is not clear that the increase warrants the use of a high voltage supply. A small fan, which admittedly has problems of its own in space applications, could provide a much greater cooling capability. Likewise, a heat pipe or some other transfer mechanism could also be more desirable than the electrostatically controlled heat shutter. In spite of arguments of this type, it does appear that the system offers a unique cooling capability for some rather unique physical situations.

REFERENCES

1. Soderholm, L., "Electrostatic Cooling," Design News, October 27, 1969.

2. Robinson, M., "Convective Heat Transfer at the Surface of a Corona Electrode, "Int. J. Heat Mass Transfer, 13, 263-274 (1970).

3. Velkoff, H. R., "An Exploratory Investigation of the Effects of Ionization on the Flow and Heat Transfer with a Dense Gas," ASD-TDR 63-842, Wright-Patterson Air Force Base, Ohio (1963).

4. "Electrostatically Controlled Heat Shutter," Mechanical Engineering, Vol. 97, No. 4, April 1975, page 47.

5. Reisbig, R. L., "A Temperature Interferometer Using Laser Holography," ISA Trans., Advances in Instrumentation, 27, 2 (1972).

6. Reisbig, R. L., Pottinger, J. M., and Quinlish, R. M., "A Procedure for Evaluating Laser Interferogram Data," ISA Trans., Proceedings of the 20th International Instrumentation Symposium, Albuquerque, New Mexico, May 1974.

7. Schimmel, W. P., Jr., Novotny, J. L. and Olsofka, F., "Interferometric Study of Radiation - Conduction Interaction," Heat Transfer, 1970, Vol. 3, Article R.2.1, Elsevier, Amsterdam, 1970.

NUMERICAL CALCULATION OF PERIODICAL TWO-DIMENSIONAL

HEAT FLOW IN COMPOSITE BUILDING WALLS

Paolo BONDI and Michele CALI'

Istituto di Fisica Tecnica

Politecnico di Torino, Italy

An analysis of heat transfer in composite building walls has shown a large variation in the thermal performance of buildings when thermal bridges, corners and structural joints are present.

The study of the general transient requires an excessive work load and since periodical temperature and heat-flux are usual for outside air conditions, the possibility of calculating stationary periodical heat-transfer as an extension of steady state conditions ensues.

The new numerical code introduces temperature and heat flow as vectors with modulus and phase, represented by complex numbers. This way the general transmission laws with the boundary conditions are solved with complex variables by a finite element method.

The interested region is divided in finite elements as usual for steady state conditions and a variational approach is used to resolve the problem.

The results are in the form of complex thermal fields (real and imaginary parts are transformed into modulus and phase lag).

A number of frequent occurrencies are analysed such as structural joints, floor-to-wall junctions etc.

Results have been suitably transformed and compared with result of simple walls in order to be used as better approximations·for the calculation of heating and cooling loads in buildings by a method long since proposed.

PHYSICAL PROBLEM

The occurrence of thermal bridges, joints etc. in the walls of buildings produces two- and three-dimensional heat-flow. The general numerical resolution of heat transfer in transient conditions for two- and three-dimensional heat-flows is rather long and complicated for large and complex systems such as buildings are. As we consider that outside conditions are usually representable by a periodical daily swing having a fundamental sinusoidal swing associated with several harmonics of suitable amplitude, a simpler method can be envisioned that resolves the thermal field in the interested region only for periodical conditions, the general form being analysed in Fourier expansion.

Such an approach has been presented for one-dimensional problems by Ferro, Sacchi and Codegone [1].

The fundamental equation of the phenomenon is the Laplace-Poisson law:

$$\alpha \nabla^2 T = \partial T / \partial \tau \tag{1}$$

with boundary conditions:

$$\lambda \partial T / \partial n + h (T - T_o) + \phi = 0$$

The solution was found separating a constant term and a time-dependent one; the last one is expressed by means of a Fourier series:

$$T = T_o + \theta \qquad \text{as} \qquad \theta = \sum_{n=1}^{\infty} \theta_n \cdot e^{j \omega_n \tau}$$

For the one-dimensional problem the solution for the n-th harmonic is expressed in the matrix form:

$$\begin{vmatrix} \theta_{en} \\ \varphi_{en} \end{vmatrix} = \begin{vmatrix} E_n & F_n \\ G_n & H_n \end{vmatrix} \cdot \begin{vmatrix} \theta_{in} \\ \varphi_{in} \end{vmatrix}$$

θ_n and φ_n are intended as amplitudes respectively of temperature and of heat-flow per unit surface of harmonics at various frequencies; "e" is for external conditions, "i" for interior of building.

PRESENT METHOD

In order to examine the occurrence of two-dimensional heat-flow, the thermal field is calculated within the interesting region of the system.

The temperature distribution is calculated with the finite element method dividing the region in sub-regions, within these temperature is supposed as a function of its values at particular points called nodes.

With some manipulation equation (1) is reduced to a system of N differential equations if N nodes are there.

$$[K] \cdot \{T\} + [C] \cdot \{\partial T / \partial \tau\} + \{F\} = 0 \qquad (2)$$

Decomposing time-dependent terms by means of Fourier analysis, each harmonic is representable by a complex number and equation (2) becomes for each harmonic:

$$([K] + \omega_n [C]) \cdot \{\theta_n\} + \{F_n\} = 0 \qquad (3)$$

This is a set of linear equations with complex numbers. In fact every variable is quantified with frequency, amplitude and phase lag; the last two parameters can be expressed with equivalent complex numbers.

The solution of (3) is a value of θ for each harmonic and each node.

A comparison of one- and two-dimensional situations can be made by calculating equivalent E_n, F_n, G_n and H_n parameters by a heat-flow balance on the whole considered modulus.

RESULTS

With the mentioned method salient features of many buildings currently used in Italy have been examined by simulating unit temperature swing on the outside.

Graphs show the lines of equal amplitude reduction and of equal phase lag in the internal points of the wall, and are relative to periods of 24 hrs.

Figure 1 shows the position of a typical module used in the calculation.

Figures 2 to 5 show the equal amplitude reduction lines for joints between floor and vertical walls. In 2 a large two-dimensional flow is evidenced, in 3 a small effect occurs.

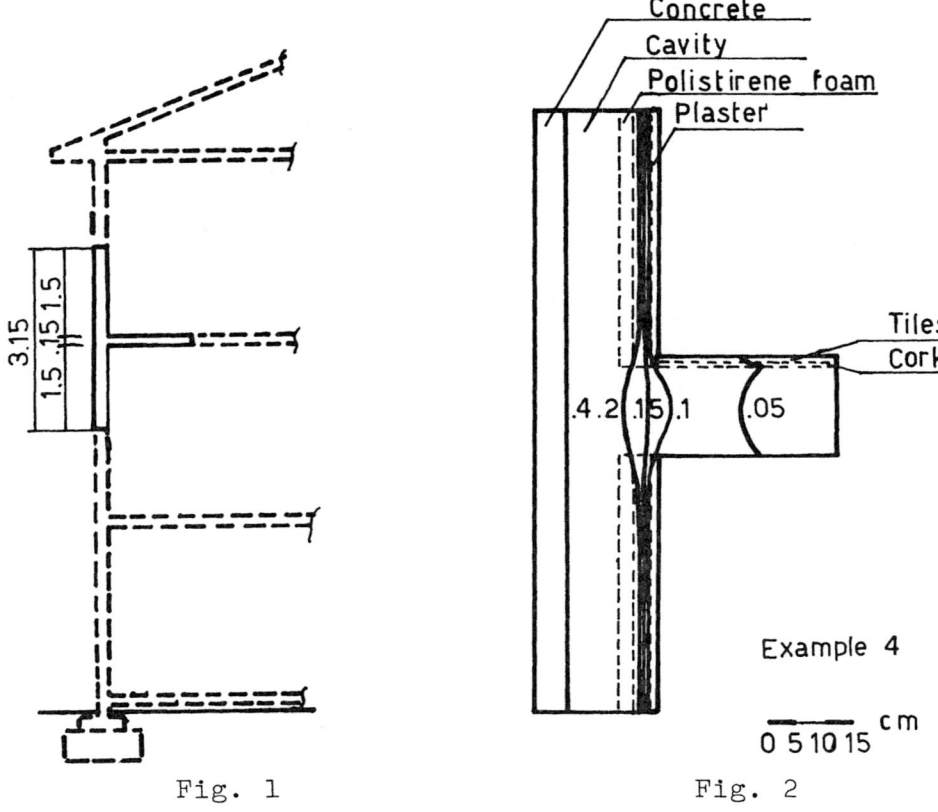

Fig. 1   Fig. 2

Figure 6 shows isothermal lines for the same situation as figure 2; a large deformation of the lines is shown; a large two-dimensional flow is present.

Many more results are ready but not presented for lack of space.

Table 1 shows the ratio of values of E, F, G, H calculated with the method accounting for the joints versus the results of the one-dimensional multilayer walls according to Ferro, Sacchi and Codegone.

E values presenting amplitude reduction of temperature swings in the singular event of zero internal heat-flow are larger when structural joints are present and F values, equivalent to an apparent thermal impedance for the considered harmonic are smaller.

In the average the joints produce a reduction of the insulating capability just like in the steady state heat transfer.

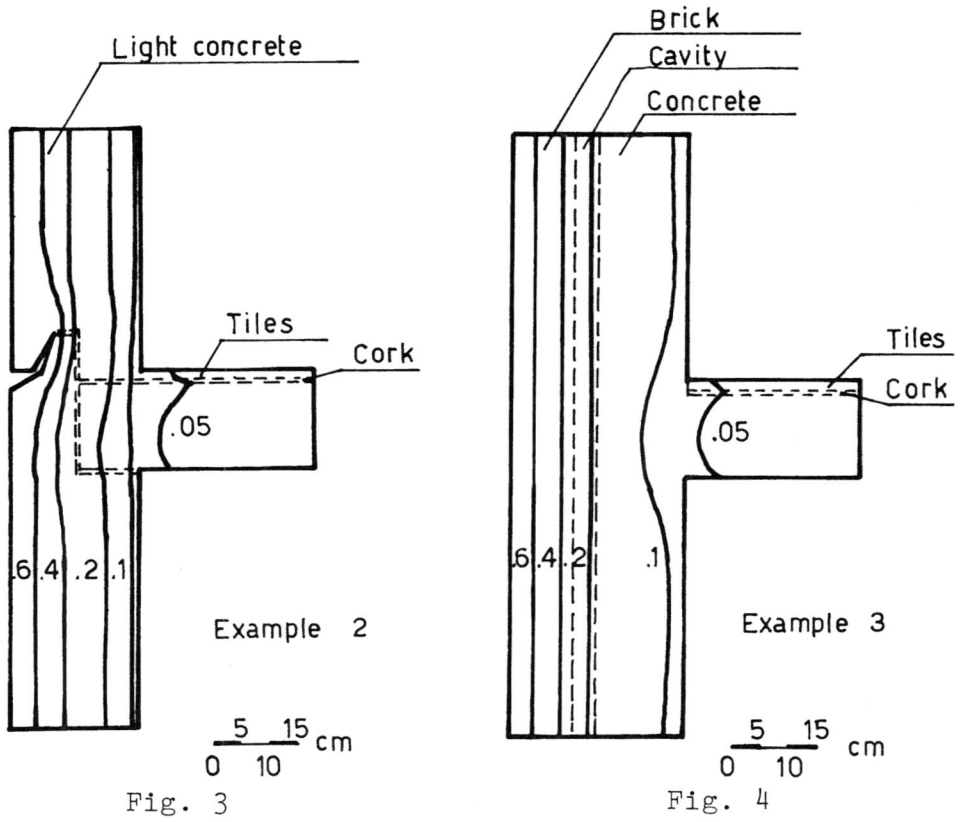

Fig. 3

Fig. 4

TABLE 1

Example	Figure n.	Ratio of			
		E	F	G	H
1	-	1.33	.45	3.11	.91
2	3	1.34	.41	2.78	.79
3	4	1.65	.45	3.34	.91
4	2	1.89	.43	3.84	.86
5	5	1.75	.45	5.84	1.48

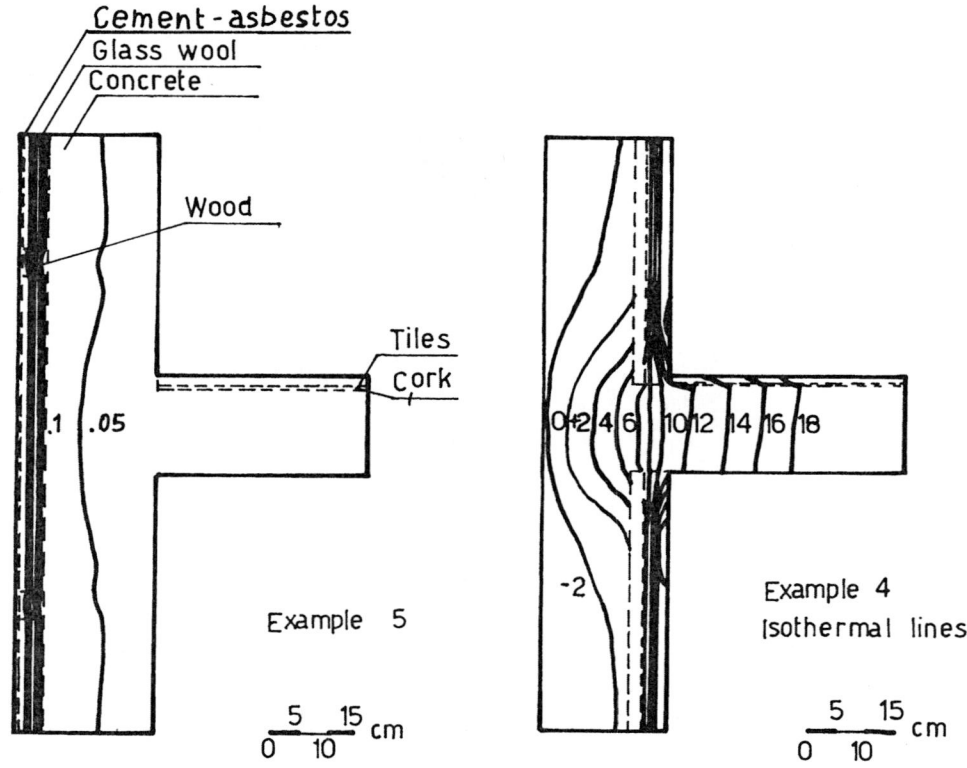

Fig. 5   Fig. 6

REFERENCES

/1/ FERRO V., SACCHI A., CODEGONE C. "Thermal attenuation
     through homogeneous and multilayer slabs in steady
     periodic conditions. Theory and experiments".
     Proceedings VIII Conference on Thermal Conductivity
     Plenum Press 1968 p. 177.

/2/ BONDI P., CALI' M., FERRO V. "Conseguenze economiche
     dei recenti orientamenti per il calcolo termico degli
     edifici".
     Proceedings XII ANDIL Conference,Firenze 1974.

A FEASIBILITY STUDY TO TEST STRUCTURE INTEGRITY BY INFRARED SCANNING TECHNIQUE*

C. K. Hsieh, M. C. K. Yang, E. A. Farber

University of Florida, Gainesville, Florida   32611

A. Jorolan

Kennedy Space Center, Florida   32815

ABSTRACT

A feasibility study is made to test structure integrity using the infrared scanning technique. The method is based on the concept that the heat flow inside a structure depends on the thermal resistance where thermal gradient exists; any flaws such as cracks, dislocations, delaminations and unbounds will increase the local thermal resistance, thereby changing the thermal patterns visualized on the surface of the structure. The instrument used for infrared scanning consists primarily of an infrared camera of AGA Thermovision System 680. To facilitate analyzing thermal patterns, the camera is operated in the isotherm mode. The transient isotherm pictures are used for analysis.

The theory of the method is discussed by using a simple model which permits a theoretical prediction of its thermal patterns. This model study is used to develop optimum test conditions promising high resolution. The technique is applied to tests of three model structures with defects simulated by slots cut at various depths and distances from the surface of the test specimen. A method of thermal pattern recognition is employed to correlate isotherm features from which a chart is developed to identify the flaws

---
*This work was sponsored by the National Aeronautics and Space Administration under contract number NAS 10-8051 with the infrared camera purchased from a National Science Foundation grant (grant number GK-38368).

as to the exact location and configuration. The study shows the
method is useful to detect large flaws present not far from the
surface with the structure made of good thermal conductors.

INTRODUCTION

Among many nondestructive methods to test structure integrity,
the infrared scanning technique is relatively new. It has received
widespread attention only recently since the real-time infrared
camera became commercially available. Application of the infrared
scanning technique has been found in the testing of bounds [1,2], and
near and subsurface defects [3-7]. With rare exception, almost all
of the past works dealt with phenomenological studies of the technique, with the presentation of thermograms as their major concern.

This paper presents a feasibility study to test structure integrity by infrared scanning technique. The study is focused on developing a new method to discern the characteristics of flaws.
Structures with simulated defects were tested in order to generate
isotherm pictures useful for analysis by means of thermal pattern
recognition. Features in the isotherm pictures were correlated with
flaws, and charts developed to identify the flaw as to its exact location and severity. It is hoped that the insight gained from this
study will be useful to stimulate further research in the area.

THEORY

The infrared scanning technique to test the structure integrity is primarily a method of transient temperature field observation. After heat injection into a structure and upon removal of
the heat source, the structure will cool due both to heat dissipation to the surroundings as well as to heat diffusion deep into the
structure. Any flaws within the structure will introduce an additional resistance to heat flow thereby causing the temperature field
to vary. This change in the temperature field can be viewed by
using an infrared camera to predict structure defects.

The governing equation describing the theory can be expressed
by the general energy equation

$$\rho c_p \frac{DT}{Dt} = \nabla \cdot (k \nabla T)$$

which is written for a heterogeneous isotropic medium without internal heat sources or sinks. ρ in the equation represents density,
c_p, specific heat at constant pressure and k, thermal conductivity.
In a real structure, the temperature distribution is three dimensional. The temperature actually monitored by an infrared scanner
is, however, two-dimensional surface temperature. For this reason,

effort has to be made to make the surface temperature variation a
weak function of the ambient condition. By careful control of the
environment free from forced convection, as well as operating the
test body at a lower temperature, both convection and radiation
heat dissipation from the surface can be minimized. This makes
the internal heat diffusion a dominant mode of heat transfer in the
structure. Any minute change in the internal thermal resistance
will cause the surface temperature pattern to vary and be useful
in identifying the flaw.

For the purpose of better understanding what might occur in a
structure during the transient heating and cooling processes and to
form a basis for developing optimum test procedures, a model structure is used for analysis. The system chosen is a 5 x 5 in. steel
plate having a magnet (1 x 1 x 1/2 in.) attached to the back center
of the plate as shown in the inset of Figure 1. Only the front surface of the plate (z=0) can receive heat injection and dissipate
heat to the surroundings, the remaining surfaces being well insulated. The thickness of the plate, being small, can be lumped in
z-direction in the analysis. In order to simulate various degrees
of sub-surface structure failures, the contact conductance between
the plate and the magnet is treated as a parameter. The range of
contact conductance chosen varies from 50 to 350 BTU hr^{-1} ft^{-2} $°F^{-1}$
equivalent to an interfacial air gap (for cracks and dislocations)
from 0.00375 to 0.00053 in. A heating period of 40 seconds is used
during which time a uniform heat flux of 442 BTU hr^{-1} ft^{-2} is
injected into the specimen which has a unit surface emissivity
value. Temperature-time response at the center of the plate is
calculated and plotted as shown in Figure 1.

As seen in the figure, the temperature at the center of the
plate upon termination of heating depends strongly on the value of
the contact conductance used in the analysis. This temperature is
higher for lower conductance curves which also show a continuous
drop in temperature with time. This can be attributed to the fact
that the magnet is still at a temperature considerably lower than
the plate upon termination of heating; since the magnet acts as a
heat sink at the center of the plate, this results in the temperature there continuing to drop over the time range of observation.
A slightly different state of affairs occurs for curves of large
conductance values. In these instances, there is little resistance
for heat flow from the plate to the magnet, thus the magnet can
reach a higher temperature. Nevertheless, the center portion of
the plate is still cooler than the edges of the plate since the
heat now flows more freely into the magnet. The subsequent temperature rise at the center of the plate is due primarily to the heat
flowing from the edges toward the center. Probably more important
is the observation of a time range where curves converge and cross
each other. This indicates a time range of poor resolution and
therefore should be avoided in the experiment where large

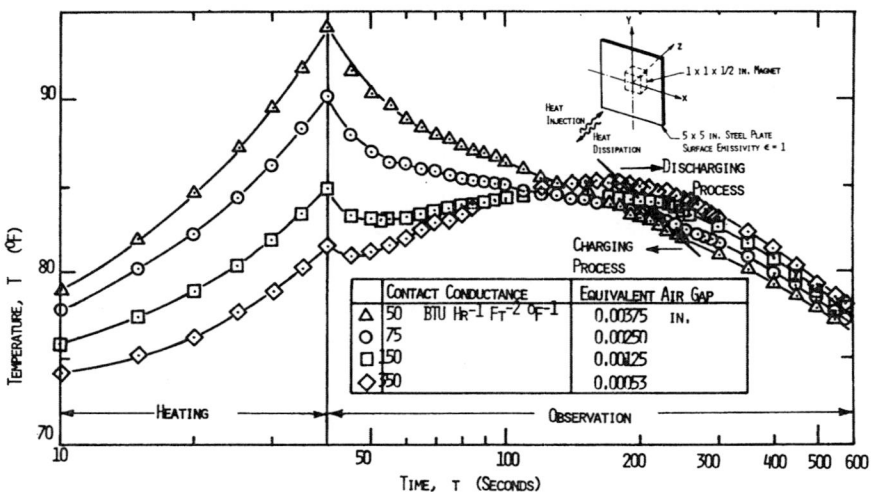

Fig. 1.   Temperature vs. Time Curves at Origin for Different
Values of Thermal Contact Conductance

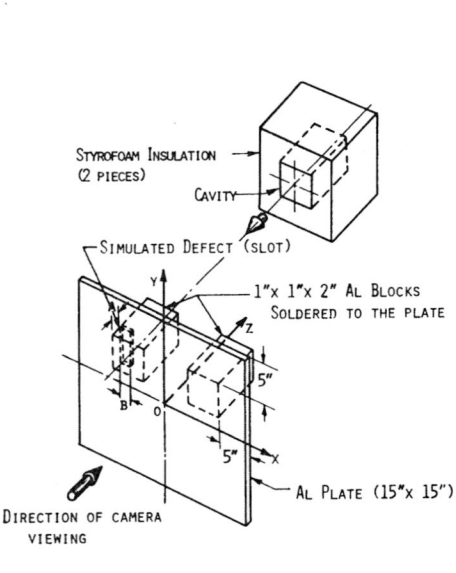

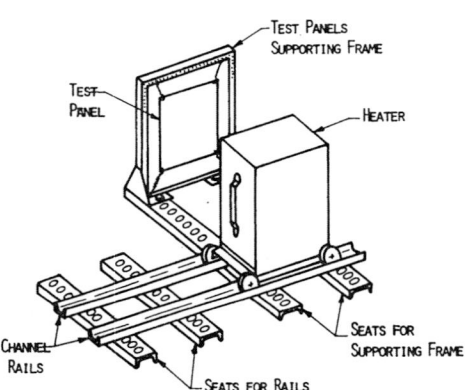

Fig. 2.   Schematic Sketch of
Experimental Set-up

A = DISTANCE OF SLOT FROM PLATE
B = DEPTH OF SLOT

Fig. 3. Sketch of a Test Panel

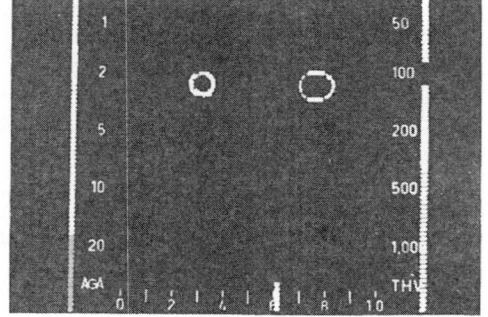

Fig. 4.   Typical Isotherm Picture

experimental error might result. An extended period of observation
will lead to another time range when the plate is cooled to a lower
temperature than the magnet, and the heat now reverses its direction
of flow. As far as the magnet is concerned, this latter event can
be termed as a discharging process for the heat sink as contrasted
to the former charging process. This is illustrated in the figure
by a slanted line which demarcates these two processes. For the
system analyzed, it is only at the early stage of the charging process the resolution is good. It is also noted that the selection of
the point of origin for observation here is purely arbitrary.
Other points could probably be used for better resolution. All the
aforementioned phenomena are only valid to the specific system
under the test conditions noted, and yet, the physics of the problem remains unchanged for such a skin-support type of structure.

EXPERIMENT

The experimental set-up is shown schematically in Figure 2.
Test panels were supported upright inside a supporting frame. A
heater consisting of nine infrared lamps (125-watt each) in a
square array was mounted on a cart which, in turn, rides on two
rails at a fixed (adjustable) distance from the test panel. The
power supplied to the lamps was closely monitored using a voltmeter
and adjusted to assure its stability. After a given heating period,
the heater was moved aside to allow for observation of the test
specimen by an infrared camera (AGA Thermovision System 680) operated in the isotherm mode.

In the experiment, the image on the electronic display was
photographed using a Minolta 135 mm camera which was set at a speed
of one fifteenth of a second. This speed was fairly close to the
picture frame frequency of the infrared camera so that each film
frame was assured to cover a complete scan of the test object with
one fifteenth overlap in the picture field. The 135 mm film was
subsequently enlarged by projecting the film on a screen for detailed picture analysis.

The test panels used in this study were 15 x 15 in. aluminum
plates with aluminum blocks (1 x 1 x 2 in.) soldered to the back
side of the plate as shown in Figure 3. Simulated defects in the
form of slots of width 1/16 in. were machined on one of the blocks
as shown. To facilitate identification of these slots, "A" was
used to denote the distance of the slot from the plate and "B", the
depth of the slot. During tests, both aluminum blocks were insulated by styrofoam to minimize heat loss. The front surface of the
plate was coated with a Nextel velvet coating of 101-C10 black paint
to improve heat absorption capability while minimizing the background reflection.

The calibration of the equipment was minimal. A black-painted aluminum plate having small holes (1/16 in. diameter) drilled in a square array was used to determine the distortion in the optics and to adjust the electronic picture display for possible picture distortion. The plate was suspended inside the supporting frame and heated to a temperature at which those small holes glow on the picture screen. The calibration of the isotherm temperature could have been made by using a black body cavity; this was not done in practice, however. Since the optimum test procedure must be established by trial and error, and the final analysis does not rely on the isotherm temperature readings in the experiment, the accuracy of the isotherm readings contributes little to the work. In practice, the isotherm temperature setting is first adjusted for better resolution and is then kept unchanged throughout the experiment.

RESULTS AND DISCUSSION

Three test panels were used in the experiment. First they were tested without defects in order to establish the reference (base) line in the correlation. An optimum test condition was found by heating the test panel for one minute and taking isotherm pictures a half minute afterwards. A typical isotherm picture is shown in Figure 4. Because of the resemblance of isotherm shapes to circles in the picture, the equivalent radii of the circles were calculated and chosen as the feature for correlation. Simulated defects were subsequently tested from specimens where A ranged from 1/2 to 1 in. and B varied from 1/8 to 7/8 in. with 1/8 in. increment. A sample correlation of two radii observed from a series of isotherm pictures taken from samples of A = 1/2 in. is shown in Figure 5. Since the magnification power of the enlarged picture on the screen is dependent on the distance of the projector from the screen, the radii are plotted in arbitrary unit. As seen in the figure, the reference line for correlation obtained from samples without defects does not fall exactly on a 45° slope. This can be ascribed to the inconsistency in workmanship in soldering the aluminum blocks to the plate. The presence of defects in the structure shifts the curves rightward. Deepening the depths of slots further increases the shift while curving the lines of correlation. This trend is representative for all A values tested and for this reason, correlation curves for A = 3/4 in. are not presented.

The shift of the curves mentioned above promises to be a distinct parameter characterizing the flaws. Shifts were thus measured from the reference line (B=0) at the value of the left isotherm radius equal to 2 (see Figure 5) and the results replotted as shown in Figure 6. Here in the figure the shift of curves shows a marked increase with deepening of the slots for a given distance A. On the

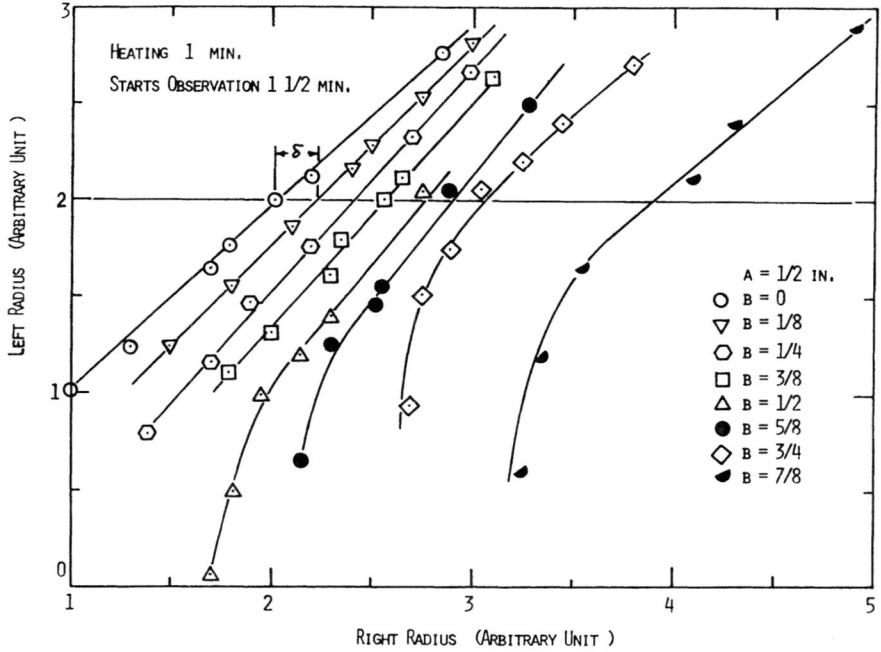

Fig. 5. Correlation of Two Radii of Isotherms for A = 1/2 inch

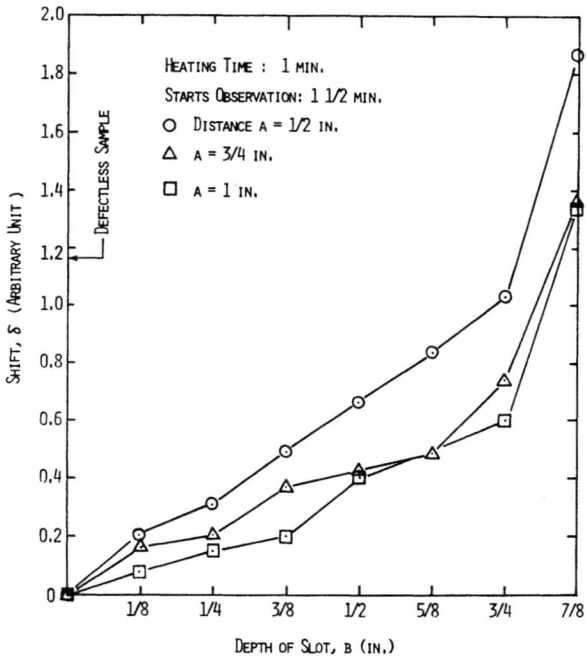

Fig. 6. Plot of Shifts vs. B for 1 min Heating Time

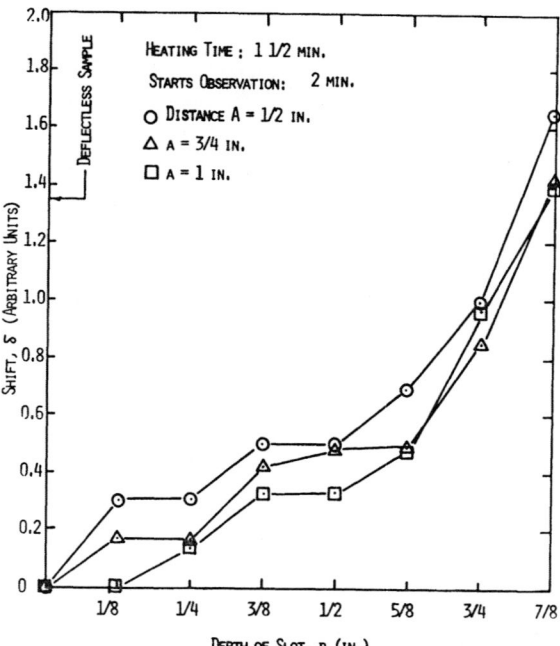

Fig. 7. Plot of Shifts vs. B for 1½ min Heating Time

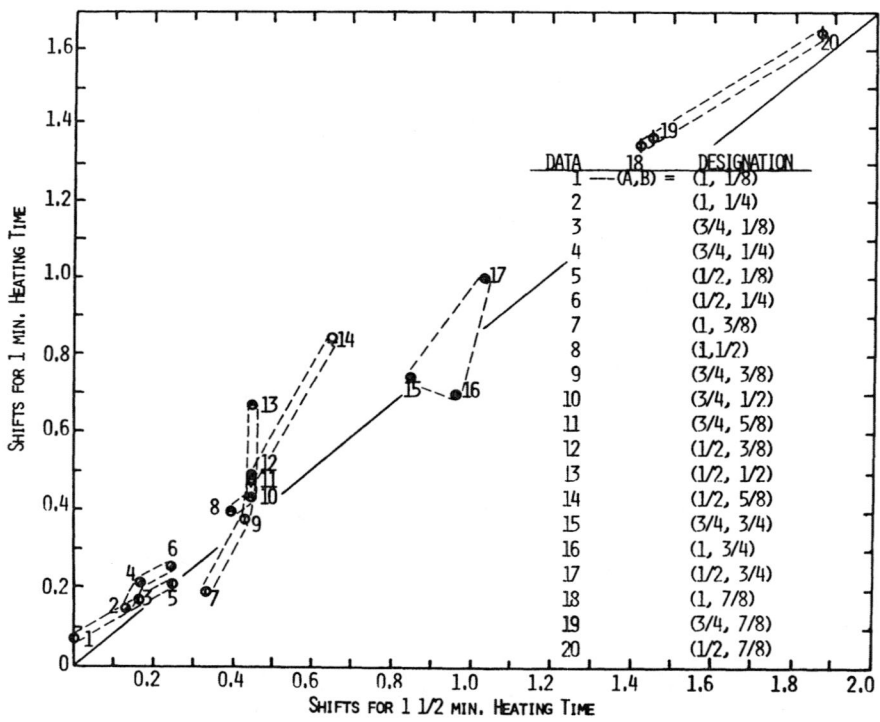

Fig. 8. Chart for Pinpointing Flaws

other hand, if the depth of the slot (B) holds constant, the shift increases with decreasing of the distance of the slot from the plate. It is not unexpected that for the specific test conditions noted, there is always a limitation in capability of the method to detect a certain class of flaws in the sense that the resolution of the method to differentiate such flaws is poor. This is exemplified by data points (A,B) = (3/4, 1/2) and (1,1/2) which fall very close to each other. However, this difficulty can be easily resolved by changing slightly the test condition. For instance, by increasing the heating time from 1 to 1 1/2 minutes and starting picture taking a half minute afterwards, the same series of experiments can be repeated to yield data as shown in Figure 7. It is seen in this figure that the resolution of the method to differentiate these two types of defects is now good under the new test condition.

The aforementioned two groups of data of two different heating times can be used to construct a chart that is useful to pinpoint structure defects. Shifts for 1 minute heating time are now plotted against shifts for 1 1/2 minutes heating time parameterized by values of A and B which characterize the defects, see Figure 8. As seen in the Figure, the uncertainty in detecting flaws of small B is getting larger as the points congregate into groups. Both of the heating times seem to be unable to pinpoint and characterize small defects. Nevertheless, some trends can be found that are useful to alleviate this difficulty. By expressing data points of same B values with the same symbol and grouping them together by dashed lines, these clusters of points fall in the vicinity of the diagonal with clusters of large B staying farther away from the origin. If each cluster of points is analyzed individually, points of smaller A always have the larger shifts for both heating times with the only exception being the cluster of B=3/4 in. where an irregularity is found. These trends are significant as they help to establish the characteristics of flaws without which repetitive experiments would have to be used to identify the flaws.

FUTURE APPLICATION

Several comments have to be made with regard to generalization of the test results to complicated structures. First, the method of isotherm pattern recognition as described in this work is useful only when two sets of isotherm pictures are available, one taken before, one after a flaw formed within a structure. Each data point in Figure 5 is taken from one isotherm picture, while the points on each curve are consecutive camera shots of the isotherm pictures. The time variable is thus totally alleviated in the correlation. In the method noted, particularly for the test panels used herein for experimentation, the isotherm for the no-fault aluminum blocks has been used as a reference with which the other isotherm is compared

and correlated. This is a valid scheme for almost all of the structures in which multiple supports are commonplace. For testing of
structure integrity of a singularly supported structure, a slightly
different scheme has to be used. This time the radius of the isotherm must be correlated with time. A problem might arise, however,
in this instance. As the liquid-nitrogen-cooled detector in the
infrared camera can hardly maintain a steady performance, a small
drift which can be observed on the picture screen as a sudden expansion or contraction of the isotherm ring is sometimes experienced
in the experiment. This drift is not detrimental to the accuracy
of the previous method, since two isotherms are compared and correlated and the drift cancels out. It will have a bearing on the
precision of the latter method when time is included in the correlation. Secondly, the supports beneath the plate have a square cross
section which results in the circular isotherms as noted earlier.
The radius was hence considered a useful feature in the process of
isotherm pattern recognition. For a real structure, the supports
can be irregular in shape. Then other parameters have to be used,
such as the curvature and derivatives of the isotherm at certain
strategic points. If a quantification of the data points is available, the capability of the flaw detection by pattern recognition
can be considerably amplified. A multiple of features can be correlated at the same time for each infrared picture. With the scope
of work greatly expanded, a computer data analysis becomes imminent.
In fact, interfacing the infrared camera with a data acquisition
system has also been made commercially available. If the data output is supplied to a digital computer for data analysis, the complete
process of testing structures can be made an automated operation
particularly attractive in field applications.

REFERENCES

1. E. W. Kutzscher and K. H. Zimmermann, Appl. Opt. $\underline{7}$(9), 1715 (1968).
2. P. E. J. Vogel, Appl. Opt. $\underline{7}$(9), 1739 (1968).
3. E. J. Kubiak, Appl. Opt. $\underline{7}$($\overline{9}$), 1743 (1968).
4. W. T. Lawrence, Mater. Evaluation, XXIX(5), 105 (1971).
5. D. R. Green, Mater. Evaluation, XXI$\overline{X(11)}$, 241 (1971).
6. R. D. Dixon, G. D. Lassahn and A. DiGiallonardo, Mater. Evaluation, $\underline{XXX}$(4), 73 (1972).
7. L. D. McCullough and D. R. Green, Mater. Evaluation, $\underline{XXX}$(4), 88 (1972).

IMPROVED HIGH TEMPERATURE CUT BAR APPARATUS

G. R. Clusener

Theta Industries, Incorporated

26 Valley Road, Port Washington, New York 11050

The temperature range has been increased from 1600°C to 2000°C; temperature sensing is done with tungsten rhenium thermocouples. Powder insulation has been replaced with Zircar insulators. The thermocouple mounting has been improved. The thermocouples are now strung around the thermodes, providing more intimate contact. They are spring loaded with teflon slides. The thermocouple reference can now be monitored with an accurate thermometer. The measuring module, which holds the sample, can be removed from the test station. Higher utilization of the test station is possible by employing several measuring modules. The accessibility of electrical and hydraulic connections is improved through a service collar. Different sized discs for the hydraulic system are interchangeable in order to accommodate the proper pressure for hard and medium hard material.

Figure 1A shows a meter bar with a known thermal conductivity, it is heated on the bottom and cooled on the top. Thermocouples sense the heat flow. Figure 1B shows the same meter bar cut open with the sample inserted. The change in the heat flow pattern is a measure of the thermal conductivity of the sample. In Figure 1C, the temperature has deformed the flat surfaces due to thermal expansion. Crushable shims overcome the interface problem as shown in Figure 1D. Figure 1E shows the complete meter bar stack with the hydraulic force mechanism.

Figure 2A shows the meter bar in more detail. The cooling head is interchangeable. The lower figure is a top view of the thermocouple attachment. The thermocouples are strung around the meter bar and the sample and are kept under tension by the holders on the left and right.

Various meter bars for different thermal conductivity and temperature ranges are shown in Figure 3.

The measuring head is shown on top of Figure 4, the three components on the bottom represent the electrical peripheral equipment namely, a millivolt meter as signal conditioner and the digital printer as recorder. A power supply serves as temperature control.

Thermally stabilized water for cooling purposes as well as for thermocouple reference is provided through a water bath. The test station, Figure 5A, has a built-in vacuum pump. The measuring head, Figure 5B, is shown here in a cross-section with the hydraulic system on the bottom. The sample (1) is surrounded by a zirconium oxide felt piece as an insulator.

Figure 6 - The heater not only must provide the high temperature, (the heater itself must exceed 2000°C in order to heat the sample to 2000°C), but the heater assembly must also support the entire meter stack assembly under pressure. The filament is surrounded by a multi-foil insulator, twenty layers of tungsten separated from each other through oxide layers. A disc made of the same material with laser drilled holes serves as a bottom plate for the chamber. The filament can be easily interchanged by removing it from the two water cooled copper bars.

Figure 7 - The actual measuring head with interchangeable measuring modules. When using several measuring modules, it is possible to load one while the other is in operation thereby reducing down-time.

Figure 8 - The high temperature module originally was insulated with powder. However, since powder such as hollow alumina spheres or carbon black is a difficult material to work with, it was replaced with zirconium oxide felt which also provided a much better insulating medium.

Figure 9 - The complete apparatus consists of the electronic recorder/controller console and test station. From top to bottom one sees the controls for pressure and water flow, controls for automatic operation, a timer for keeping the temperature at a temperature plateau, a millivolt meter with printer and power supply on the bottom.

Figure 10 - The original apparatus was built by Dr. Ralph Day in the early 60's. It is usable up to 500°C or 800°C. It is the basis for the Theta Conductronic Apparatus.

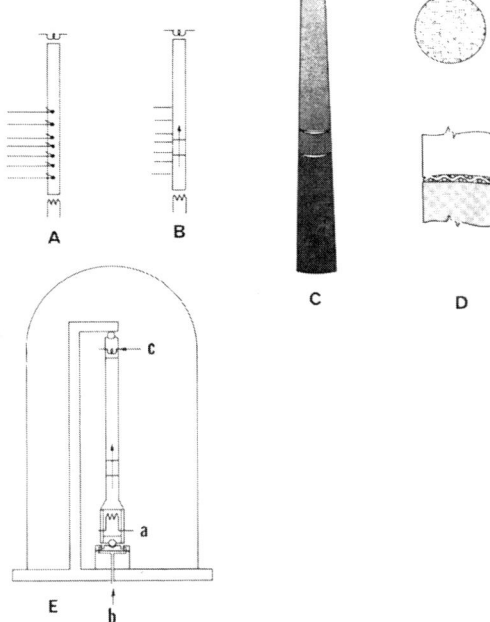

Figure 1

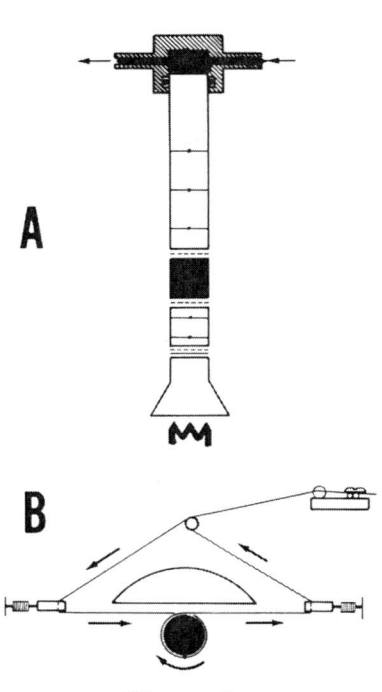

Figure 2

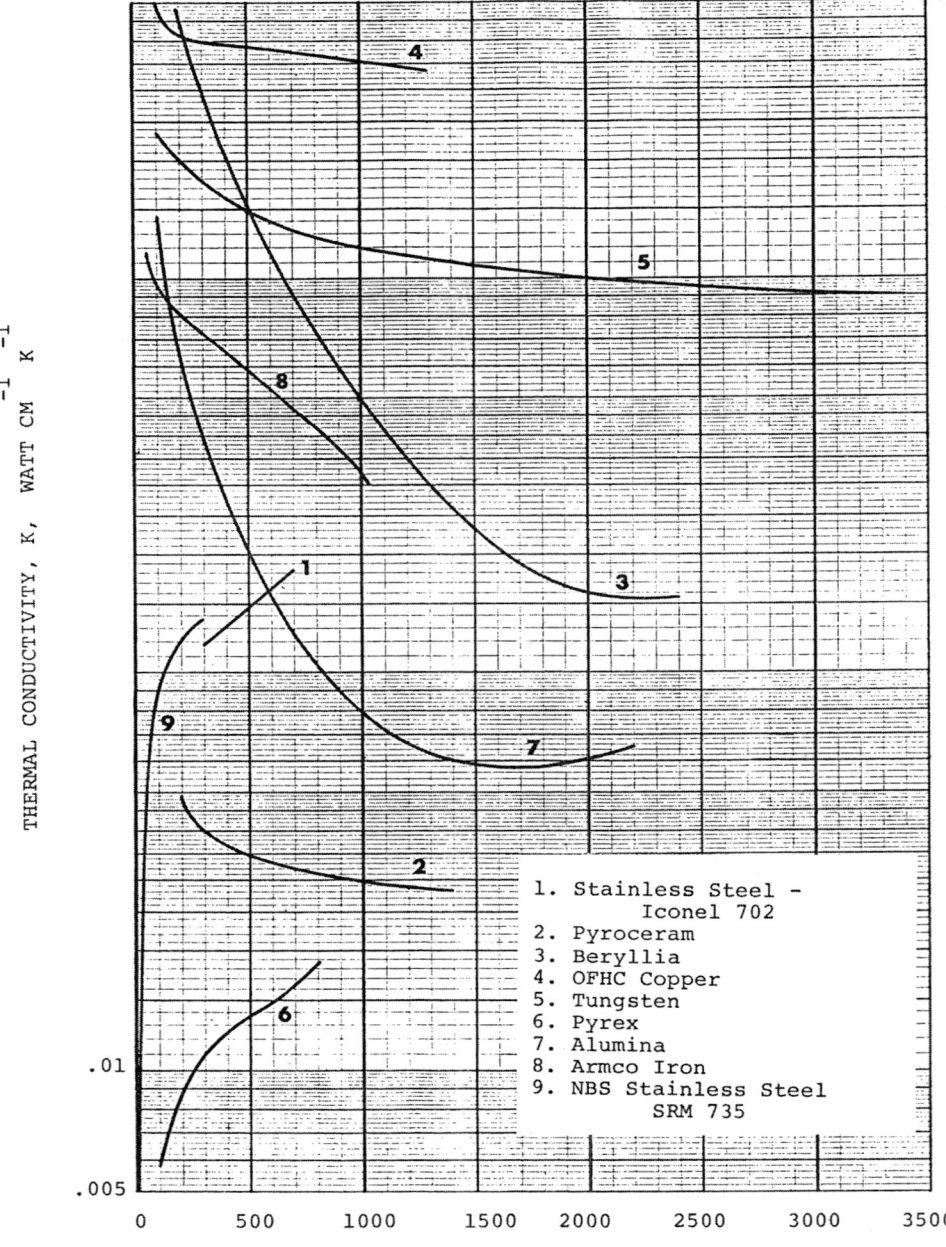

TEMPERATURE, °K

REFERENCE MATERIAL

Figure 3

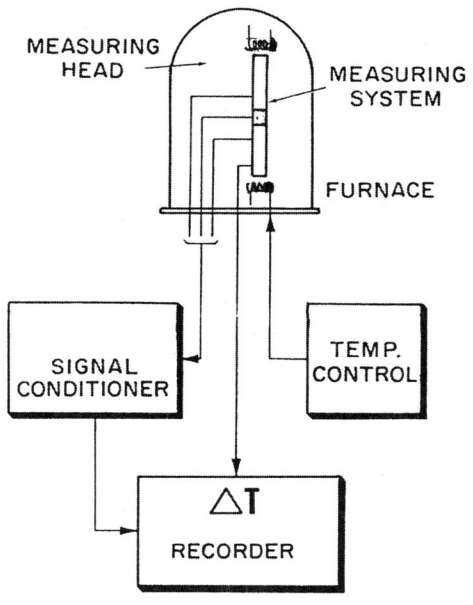

Figure 4

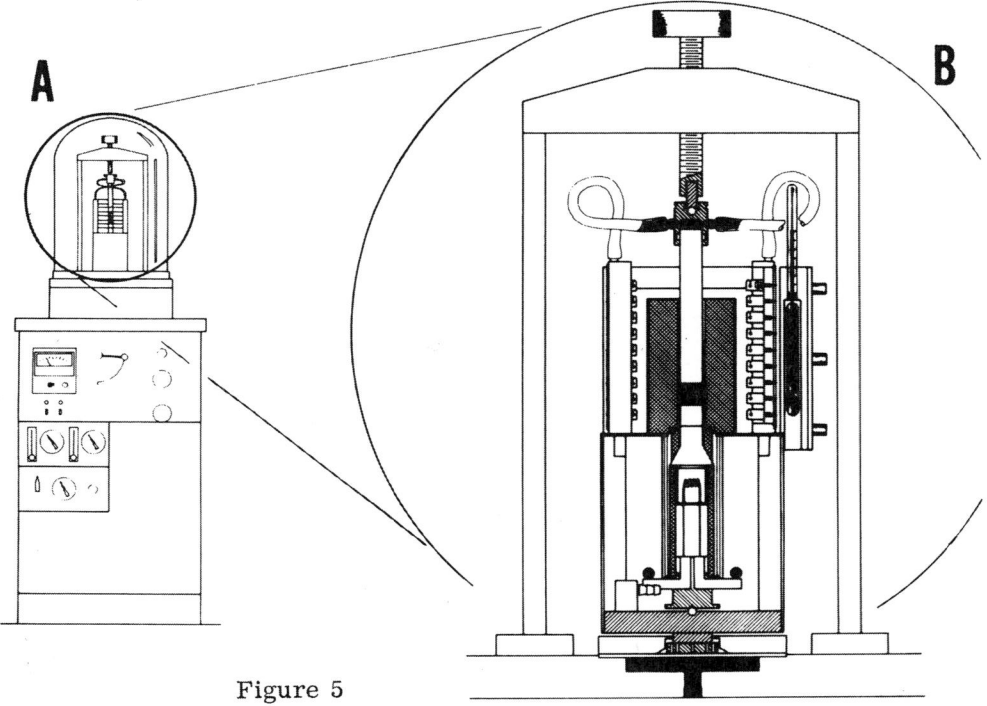

Figure 5

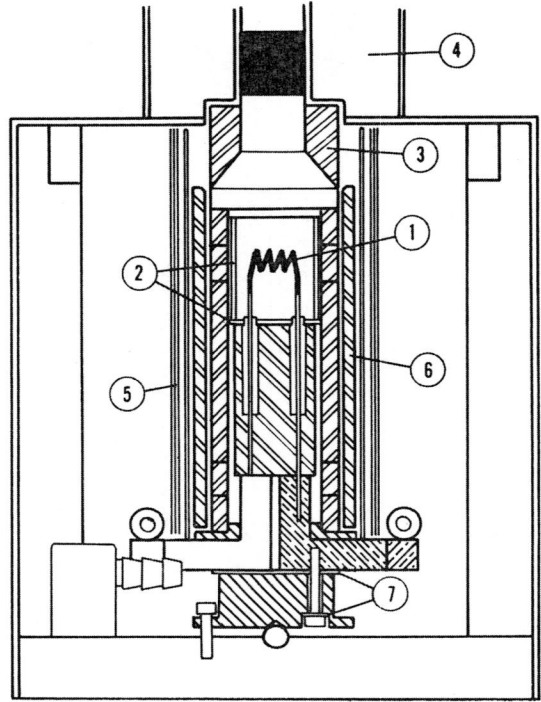

Figure 6

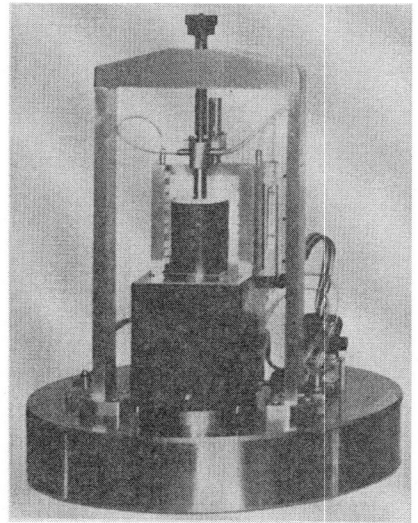

Figure 7

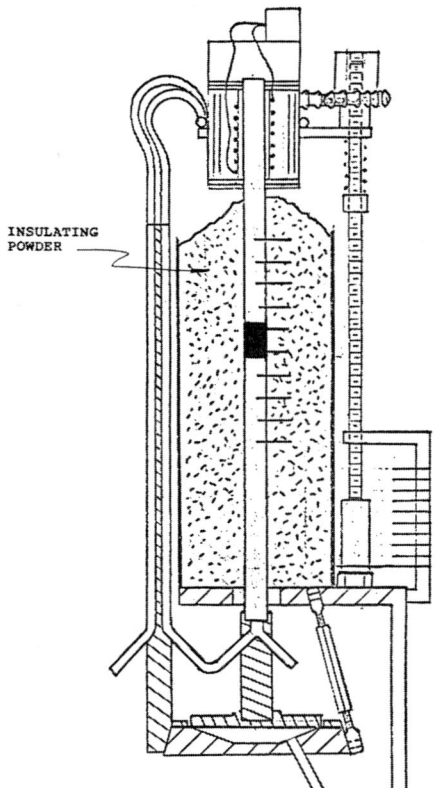

INSULATING
POWDER

Figure 8

High Temp. Module

Figure 9

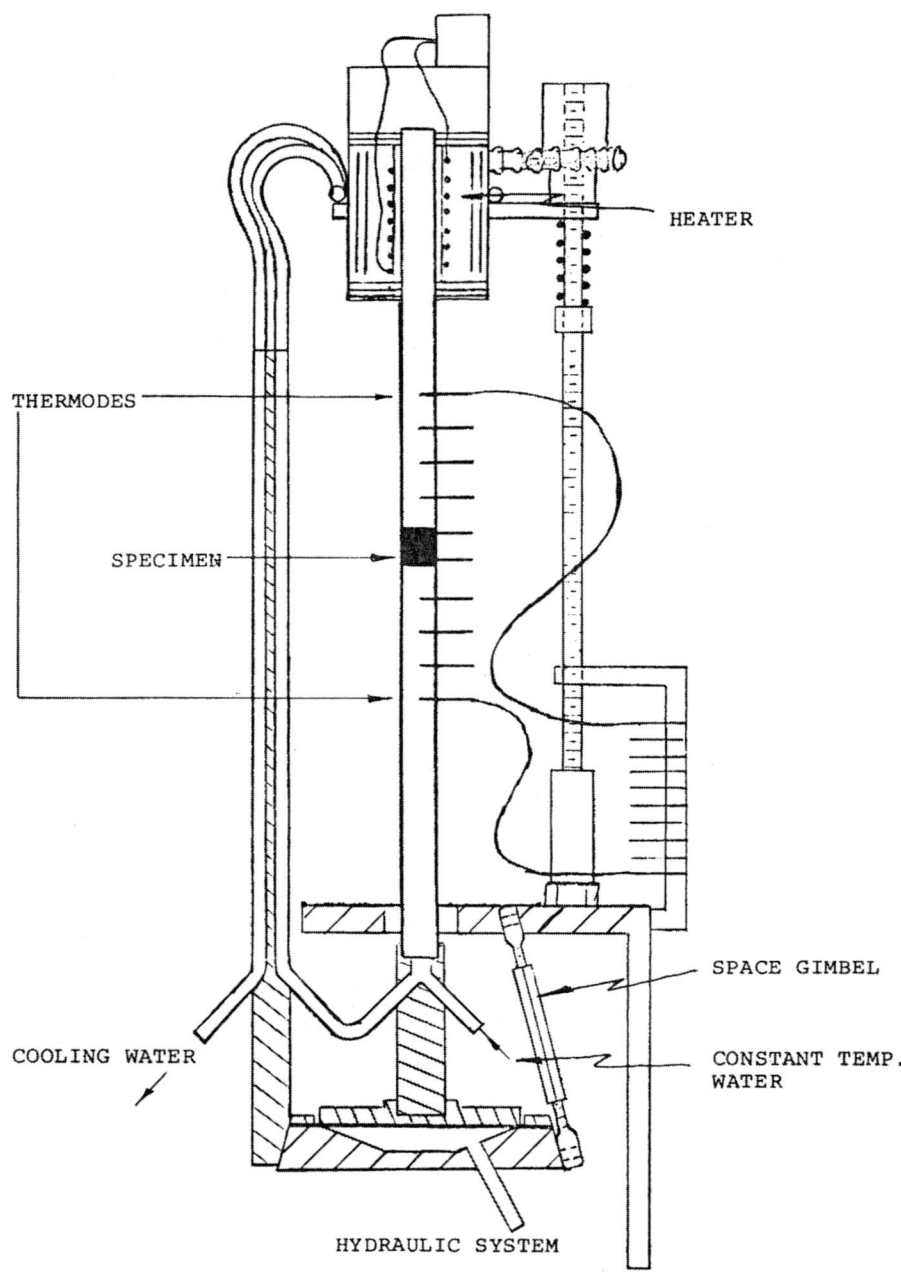

HEATER

THERMODES

SPECIMEN

SPACE GIMBEL

COOLING WATER

CONSTANT TEMP.
WATER

HYDRAULIC SYSTEM

Note: This original apparatus was built by Dr. Ralph Day in the
early 1960's.

Figure 10

COMPUTER CONTROLLED THERMOPHYSICAL TEST FACILITY - 1) HARDWARE

William C. Brown and C.J. Shirtliffe

Division of Building Research
National Research Council Canada
Ottawa, Ontario. K1A 0R6

The Division of Building Research (NRC of Canada) has developed a mini-computer centered data acquisition and control system to run a test laboratory in which the thermophysical properties of materials are measured. At present the laboratory has five guarded hot plate apparatus and two heat flow meter apparatus with provision for connecting three more apparatus to the system. The laboratory runs on a 24-hour, 7-day basis and is unattended during non-working hours. The computer system is used to control the temperatures of the apparatus, to read, process and log the test results, and to monitor the operation of the equipment for alarm conditions. The system is designed to take corrective action in the event of an alarm condition being sensed.

The computer system is part of a larger system that includes back-up analog controllers, independent over-temperature alarm sensors, and a high precision, manually operated potentiometer measuring system. A computer power failure causes the control systems to default to the analog controllers. Should the analog controllers also fail, the control functions can be taken over by an operator, but this requires direct intervention on his part. The high precision measuring system can be connected to the test stations by disconnecting the signal lines from the computer and connecting them to the potentiometer through low noise connectors. The over-temperature alarm sensors have adjustable limits that turn off the power to the test loop under alarm conditions and, at the same time, sound an alarm.

An apparatus consists of up to seven control loops and up to ten status conditions. These are all sensed and/or controlled by the computer. Analog input to the computer is through a high-speed, programmable range, analog to digital converter and through a low-speed digital voltmeter and reed relay scanner with high noise immunity. Status inputs are read through relay sensors or digital electronic sensors.

A check is maintained on the calibration of the analog input equipment by reading a high precision voltage source. The results are printed once per day. If they are outside preset limits, they are printed immediately, with an alarm message.

Analog outputs from the computer are through a digital to analog converter. These outputs are used for control purposes and to plot operator-selected variables on strip chart recorders. Status outputs are through relay output cards. Processed data are printed on a medium-speed line printer or stored on magnetic tape for further processing. A teletype unit is used to log alarm messages and to provide operator communication to the computer through specially designed software.

The computer-operated system has been found to be very reliable and accurate enough for all but the most precise standards work. With the degree of back-up incorporated, laboratory downtime has been reduced to a minimum. Productivity of the laboratory has been considerably increased and operator intervention and errors reduced. Minor changes in hardware are expected to improve the system accuracy so that all standards work can be done using the computer system.

The laboratory computer system is also being incorporated into a hierarchical computer system which consists of a powerful central computer connected by high-speed data links to several mini-computers. It is expected that this installation will increase the data handling capability of the system. It will also enable programming in Fortran IV, which in turn should allow the use of more complicated control algorithms, simplify the control of a sequence of measurements on a specimen, and allow complex data analysis to be done on-line.

COMPUTER CONTROLLED THERMOPHYSICAL TEST FACILITY - 2) SOFTWARE

William C. Brown and C.J. Shirtliffe

Division of Building Research
National Research Council Canada
Ottawa, Ontario. K1A 0R6

The Division of Building Research (NRC of Canada) operates a
laboratory for testing the thermophysical properties of materials
using a number of guarded hot plate and heat flow meter apparatus.
This laboratory operates on a 24-hour, 7-day schedule under the
supervision of a mini-computer based data acquisition and control
system. The computer system controls the apparatus, reads,
processes and logs the test data, and monitors the various test
stations for alarm conditions. Should the power to the computer
fail, the system control defaults to analog controllers and the
data can be read using a high precision, manually operated
potentiometer measuring system.

The computer software runs under the supervision of a real
time, multi-task executive. Peripheral handlers and fixed and
floating point math routines are incorporated in the system. The
various tasks in the system run the data acquisition devices,
monitor the test stations for alarm conditions (and take corrective
action when necessary), calculate and output the control signals
for the test loops, and process and log the test results. Software
routines have been designed to allow the operator to interact with
the computer in a conversational mode. Test parameters, alarm
limits, and setpoints are entered while the computer is on line and
any apparatus can be moved on or off line without disturbing the
operation of any other apparatus. An on-line debugging routine has
been written to allow debugging of new programs while the system
is running.

The system is written in modular form in assembler language.
Software development is done on a larger computer system utilizing
a cross-assembler to translate between the different languages of

the two computers. The assembled programs are transferred to the
mini-computer on paper tape and loaded into the computer using a
portable high-speed paper tape reader. The new programs are
debugged on line using the on-line debugging program and a debugged
system is saved on paper tape punched by a portable high-speed
punch. Downtime is minimized as the system is only down during the
loading of the new programs and the punching of the new system.

The conversational design of the software system allows an
operator, with a minimum understanding of computers, to control the
sample testing. The modular construction of the system permits
refinements in design to be made easily. The installation of the
computer system has decreased sample testing time and increased the
productivity of the laboratory.

IMPROVING THE ACCURACY OF THE GUARDED HOT PLATE APPARATUS

C.J. Shirtliffe

Division of Building Research
National Research Council Canada
Ottawa, Ontario. K1A 0R6

In the common forms of apparatus that have surface temperature sensors built into them, such as the guarded hot plate, the actual surface temperatures of the test specimen are not measured by the sensors. The difference in temperature between each of the sensors and the surfaces decreases as the thermal resistance of the specimen increases. As a result, an accurate measure of the thermal resistance of test specimens can only be obtained when they have a relatively high thermal resistance.

A method for eliminating the effect of the temperature differences has been developed and investigated. It involves the measurement of the thermal resistances of two sets of specimens of the same material, identical except for their thicknesses. The true thermal resistivity and thermal conductivity of the specimens can be determined from these measurements.

A simpler but less accurate procedure has been developed by making further assumptions. It is assumed that there is a thermal resistance between the sensor and the test surface and that the thermal resistance does not depend on the characteristics of the test specimen. This thermal resistance is termed the thermal resistance of the test apparatus. For a particular apparatus the value of this thermal resistance can similarly be determined from measurements on two sets of specimens of material, identical except for thickness.

Measurements on several sets of specimens in several guarded hot plate and heat flow meter apparatus confirm the validity of the assumptions.

Although other methods have been used by other investigators to eliminate from the measurements the effect of temperature difference between sensors and surfaces little justification and refinement of them was ever done. One of the most common methods was to measure the thermal conductivity of a sequence of specimens made up of differing numbers of layers of the material. The results were extrapolated to yield the thermal conductivity for a specimen with an infinite number of layers. An analysis of this method shows that the value determined is not the thermal conductivity of the material: the contact resistance between the layers in the specimens produces an error that cannot be eliminated.

This older method can be combined with the new methods to find the contact resistance between layers. Values of contact resistance have been determined from measurements on a number of materials in the apparatus mentioned.

The new method described is a simple and practical way to improve the accuracy of the results obtained from apparatus with built-in temperature sensors. The apparatus can be used to test specimens with lower thermal resistance than can normally be tested. The application of this method eliminates the error due to thermal resistance of the apparatus, thus increasing confidence in the test results.

THEORY FOR IN-SITU THERMAL DIFFUSIVITY MEASUREMENT OF THIN FILMS

BY A RADIAL HEAT FLOW METHOD*

A.B. Donaldson and W.P. Schimmel, Jr.

Sandia Laboratories

Albuquerque, New Mexico 87115

Because of physical limitations of some materials which are
used for thin films (such as thickness, stress tolerance and
strength) it is sometimes advantageous to measure their thermal
diffusivity after they have been deposited on a substrate, i.e.
in-situ. Larson and Koyama** have proposed a method for in-situ
thermal diffusivity measurement of thin films. Their method is
an extension of the standard pulse method to a two-layer
composite, i.e. one dimensional conduction following an instantaneous energy source on the surface of the film. In the data
analysis for their method, the temperature excursion at the rear
surface of the substrate is utilized. This method lacks sensitivity to the thermal properties of the film, particularly if
the substrate is thick or of low thermal diffusivity. Alternatively, we propose a method whereby a pulse of energy is
deposited uniformly over a central circular region of film surface.
In the process of thermal equilibration, locations on the surface
of the film outside the initially heated region will undergo a
temperature excursion. Through a theoretical analysis of the
problem, the characteristics of this excursion and other
(presumably known) factors can be utilized in the determination
of the thermal diffusivity of the film.

For the theoretical analysis, the assumptions are briefly:
1) axial temperature gradients in the film are neglected, 2) the
film and substrate are radially infinite and the substrate is
axially infinite, 3) the interface contact resistance, and external surface heat losses are negligible, and 4) the thermal and
physical properties of the film and substrate are independent of
temperature and position. Based on these assumptions, the

formulation of the problem results in two coupled partial
differential equations. The equation applying to the film is a
function of radial position and time, while that of the substrate
is a function of radial and axial position and time. With
appropriate boundary conditions, the Hankel transform and then
the Laplace transform are sequentially applied to the system.
The result for the substrate is an ordinary differential equation
while the film energy equation becomes an interface gradient
condition for the substrate. This system is integrated and then
the inverse Laplace transform is applied by utilizing the Residue
Theorem and the inverse Hankel transform is obtained by numerical
quadrature.

The temperature excursion in the film surface at locations
outside the initially heated region is found to be pulse shaped,
i.e. it rises from the initial value to a maximum, followed by a
decay back to the initial value. It appears that the half-time
corresponding to the rising portion of the pulse can be utilized
in data analysis. The utility of this method for accurate
thermal diffusivity measurement in thin films remains to be
demonstrated by experiment.

---

* This work was supported by the U.S. Energy Research and
  Development Administration.

** Larson, K.B. and Koyama, Karl, J. Appl. Phys. 39, No.9, 4408
  (1968).

RADIAL OUTFLOW HIGH TEMPERATURE THERMAL CONDUCTIVITY MEASUREMENTS: SOME EXPERIMENTAL DEVELOPMENTS

J. Brazel/B. S. Kennedy

Advanced Materials and Space Systems Laboratory

General Electric/RESD, Philadelphia, Penn. 19101

We have been conducting radial outflow (ROF) high temperature thermal conductivity measurements in GE/RESD's Valley Forge Space Center Laboratories since 1967, when we obtained a commercially available instrument, described in Ref. 1. At the time of acquisition, our primary need was for high temperature (1000 – 3000 K) measurements on low conductivity materials such as the chars of carbon- and silica-phenolic composites. Because of the small samples available, a small specimen size was necessary, and our original specifications called for a 2" O.D. maximum cylindrical specimen. The instrument has worked very well for low to intermediate conductivity materials at temperatures up to approximately 2000 K; however, our high temperature work has been concentrated on high conductivity materials such as the ATJ graphite series. Although the small sample size causes a loss in sensitivity at low temperatures, causing the lower temperature limit of the measurements to rise to the range 1100 – 1400 K, the more serious experimental difficulty with this class of measurements has been the electrical power measurement at high temperature, in particular the gage section voltage.

This paper describes our recent development of a markedly improved and simplified voltage measurement design for the tungsten mesh heaters used in our instrument and gives some experimental results. The description of its development may also be of use to improve other high temperature ROF instruments using refractory metal central heaters.

1. Styhr, K. and Brazel, J., "High Temperature Radial Heat Flow Measurements", Proceedings of the Seventh Thermal Conductivity Conference, NBS Special Publication 302, Washington, D.C., 1968.

SPECIFIC HEAT OF ELECTRICALLY-CONDUCTING SUBSTANCES AT HIGH TEMPERATURES MEASURED BY A PULSE HEATING TECHNIQUE

Ared Cezairliyan

National Bureau of Standards

Washington, D. C. 20234

A subsecond-duration pulse heating technique developed at NBS for the accurate measurement of specific heat of electrically-conducting substances at high temperatures (1500 K to near the melting point of the specimen) is described. The operational characteristics of the system and the sources and the magnitudes of the measurement errors are discussed. The results of the measurements performed on a number of refractory metals, refractory alloys, and graphite are presented and are compared with those reported in the literature. Unlike some of the earlier results, the present work results indicate a sharp increase in specific heat of metals and alloys near their melting points. The role of accurate specific heat data in relating thermal diffusivity to thermal conductivity is discussed.

CORRECTION FOR LASER BEAM NON-UNIFORMITY IN THE FLASH METHOD OF THERMAL DIFFUSIVITY MEASUREMENT

J. A. McKay and J. T. Schriempf

Naval Research Laboratory

Washington, D. C. 20375

Many workers have commented on the difficulty in achieving the uniform illumination required by the flash method of thermal diffusivity measurement. An analysis of the heat flow for non-uniform surface heating is presented. The errors in diffusivity values for moderately non-uniform heating are investigated. It is shown that the values obtained by the popular half-maximum method are significant, but can be sharply reduced by different methods of data analysis, even without detailed knowledge of the beam profile. Measurements on Armco iron are evaluated, with and without corrections for the known beam profile.

EXCITATION OF COHERENT AND INCOHERENT PHONONS BY MEANS OF AN INFRARED LASER

W. Grill

Institut fur Angewandte Physik II

D-69 Heidelberg, Germany

I report piezoelectric surface excitation of coherent acoustic phonons in quartz at 0.981 THz and at 2.53 THz. At 2.53 THz the effect of dispersion is visible by a time-of-flight method. At 2.53 THz and at 10.7 THz also incoherent phonons are excited by the absorption of photons and decay processes. The excited phonons are detected by small superconducting tin bolometers. The time-of-flight and the direction of propagation of the phonons had been measured.

Appendices

CONFERENCE PROGRAM

Sunday - June 1, 1975
3:00 - 6:00 P.M.
Registration at Shippee Hall
5:15 - 6:00 P.M.
Dinner: Shippee Hall Cafeteria
6:00 - 12:00 P.M.
Reception, Pequot Auditorium Shippee Hall

Monday - June 2, 1975
9:00 A.M.
Registration
Session 1a - Introductory Session
Session Chairman: Dr. R. C. Acton, Sandia Corporation

WELCOME TO THE UNIVERSITY OF CONNECTICUT
Leonid Azaroff, Professor of Physics and Director of Institute
of Materials Science at the University of Connecticut.

RADIAL OUTFLOW HIGH TEMPERATURE THERMAL CONDUCTIVITY MEASUREMENTS:
SOME EXPERIMENTAL DEVELOPMENTS J. Brazel and B. S. Kennedy,
Advanced Materials & Space Systems Laboratory GE/RESD

THERMAL CONDUCTIVITY VARIATIONS IN STEEL AND THE EROSION BY HOT
GASES A. R. Imam and R. W. Haskell, Watervliet Arsenal

KINETIC-THEORY RELATION OF THERMAL CONDUCTIVITY TO OTHER GAS
PROPERTIES E. A. Mason, Brown University

THE THERMAL CONDUCTIVITY OF DIAMONDS R. Berman and M. Martinez,
Oxford, England

THE SUPERCONDUCTING STATE AS A MAGNIFIER FOR THE STUDY OF PHONON
INTERACTIONS P. Lindenfeld and D. A. Furst, Rutgers University

TRANSPORT PROPERTIES OF SOME TRANSITION METALS AT HIGH TEMPERATURES
M. J. Laubitz and P. J. Kelley, National Research Council of
Canada

THERMAL CONDUCTIVITY AND ELECTRICAL RESISTIVITY OF COPPER AND SOME
COPPER ALLOYS IN THE MOLTEN STATE R. P. Tye, Dynatech

12:15 P.M.
Lunch - Shippee Hall Cafeteria
1:30 P.M.
Session IIa - Solids at Low Temperature
Session Chairman: Dr. M. J. Laubitz, National Research Council
of Canada

THERMAL CONDUCTIVITY OF A GLASSY METAL AT LOW TEMPERATURES
A. C. Anderson and J. R. Matey, University of Illinois

THE LOW TEMPERATURE THERMAL CONDUCTIVITY OF HIGHLY EXTENDED
POLYETHYLENE   D. Greig and S. Burgess, University of Leeds,
United Kingdom

MAGNETIC FIELD EFFECT ON THE THERMAL CONDUCTIVITY OF HIGH
PURITY COPPER AND A NiCrFe ALLOY L. L. Sparks, National Bureau
of Standards, Boulder, Colorado

MEASUREMENTS OF THE TEMPERATURE AND MAGNETIC FIELD DEPENDENCE OF
THE ELECTRICAL RESISTIVITY AND THERMAL CONDUCTIVITY OF OFHC
COPPER W. E. Nelson and A. R. Hoffman, University of Massachusetts

FREQUENCY DEPENDENCE OF THE SCATTERING PROBABILITY OF NEARLY
MONOCHROMATIC PHONONS W. E. Bron, Indiana University

THE THERMAL CONDUCTIVITY OF SOLID NEON   J. E. Clemans,
University of Illinois

THE THERMAL AND ELECTRICAL CONDUCTIVITIES OF SOME CHROMIUM ALLOYS
M. A. Mitchell and J. F. Goff, White Oak Laboratory, Silver
Springs, Maryland, M. W. Cole, Pennsylvania State University

LATTICE THERMAL CONDUCTIVITY AND COTTRELL ATMOSPHERE IN COPPER-
GERMANIUM ALLOYS T. K. Chu and N. S. Mohan, University of
Connecticut

1:30 P.M.
Session IIb — Liquids and High Temperatures
Session Chairman: Dr. Paul Wagner, Los Alamos

THERMAL CONDUCTIVITY OF SILICONE OILS OF THE POLYMETHYLPHENYL
SILOXANE TYPE D. T. Jamieson and J. B. Irving

THE THERMAL CONDUCTIVITY OF SIMPLE DENSE FLUID MIXTURES
W. McElhannon and E. McLaughlin, Louisiana State University

AN EXPERIMENTAL STUDY OF THE ELECTROSTATICALLY CONTROLLED HEAT
SHUTTER W. P. Schimmel and A. B. Donaldson, Sandia Laboratories,
New Mexico

THERMAL TRANSPORT IN REFRACTORY CARBIDES R. E. Taylor, Purdue
University and E. K. Storms, Los Alamos

AN OSCILLATOR TECHNIQUE TO DETERMINE THE THERMAL CONTACT CONDUC-
TANCE BETWEEN FUEL AND CLADDING IN AN OPERATING NUCLEAR REACTOR
FUEL ROD H. E. Schmidt, M. Van den Berg and M. Wells, EURATOM

Institute, Karlsruhe, Germany

5:30 - 7:00 P.M.
Dinner, Shippee Hall Cafeteria
8:00 P.M.
Informal Evening Session
Session Chairman: Dr. D. H. Damon, University of Connecticut

Tuesday - June 3, 1975
8:45 A. M.
Session IIIa - Gases
Session Chairman: Dr. E. A. Mason, Brown University

THE THERMAL CONDUCTIVITY OF GASES AND THEIR MIXTURES
J. Kestin, Brown University

QUANTUM-MECHANICAL CORRECTIONS TO TRANSPORT CROSS-SECTIONS FOR
HARD SPHERES J. G. Solomon and S. Larsen, Temple University

ON ESTIMATING THERMAL CONDUCTIVITY COEFFICIENTS IN THE CRITICAL
REGION OF GASES H. J. M. Hanley, National Bureau of Standards,
Boulder, J. V. Sengers, University of Maryland, and J. F. Ely,
Rice University

A THEORY FOR THE COMPOSITION DEPENDENCE OF THE THERMAL CONDUC-
TIVITY OF DENSE BINARY MIXTURES OF MONATOMIC GASES R. DiPippo,
Southeastern Massachusetts University, J. R. Dorfman, University
of Maryland, J. Kestin, H. E. Khalifa, and E. A. Mason, Brown
University

MOLECULAR ORIENTATION EFFECTS ON THERMAL CONDUCTIVITY AND THERMAL
DIFFUSION L. Biolsi, University of Missouri and E. A. Mason,
Brown University

VISCOSITY AND HEAT CONDUCTIVITY OF FLUORINE GAS: COMPUTED VALUES
FROM 100 - 2500K AND .25 - 2.0 ATMOSPHERES K. I. Oerstavic and
T. S. Storvick, University of Missouri

DETERMINATION OF THERMAL CONDUCTIVITY OF GASES FROM HEAT TRANSFER
MEASUREMENTS IN THE TEMPERATURE JUMP REGIME S. C. Saxena and
B. J. Jody, University of Illinois

THE THERMAL CONDUCTIVITIES OF GASEOUS MIXTURE OF HELIUM WITH THE
HYDROGEN ISOTOPES AT 77.6 AND 283.2K A. A. Clifford, L. Colling,
E. Dickinson, and P. Gray, University of Leeds, United Kingdom

THE THERMAL CONDUCTIVITY OF THE REFRIGERANT 13($CClF_3$) J. E. S.
Venart, N. Mani and R. V. Paul, University of New Brunswick, Canada

8:45 A.M.
Session IIIb-Solids at Intermediate Temperatures
Session Chairman: Dr. R. E. Taylor, Purdue University

STANDARD REFERENCE MATERIALS FOR THERMAL CONDUCTIVITY AND
ELECTRICAL RESISTIVITY J. G. Hust, National Bureau of Standards,
Boulder

THE THERMAL AND ELECTRICAL CONDUCTIVITY OF ALUMINUM J. G. Cook,
National Research Council of Canada, J. P. Moore, Holifield
National Laboratory, Oak Ridge, Tennessee, T. Matsumura and M. P.
Van Der Meer, National Research Council of Canada

EVIDENCE FOR SUB-THERMAL PHONON DISTRIBUTION TO THE THERMAL CONDUC-
TIVITY OF A RHOMBOHEDRAL CRYSTAL J. P. Issi, J. P. Michenaud and
J. Heremans, Universite Catholique de Louvain, Belgique

LATTICE THERMAL CONDUCTIVITY OF PbTe: INFLUENCE OF A DISPERSION
ANOMALY J. F. Goff and B. Houston, White Oak Laboratory, P. G.
Klemens, University of Connecticut

INFLUENCE OF PRESSURE ON THE THERMAL CONDUCTIVITY OF GASES R. S.
Frost, R. Y. S. Chen and R. E. Barker, University of Virginia

ANISOTROPIC LATTICE THERMAL CONDUCTIVITY OF α -QUARTZ AS A
FUNCTION OF PRESSURE AND TEMPERATURE D. M. Darbha and H. H.
Schloessin, University of Western Ontario, Canada

ANOMALOUS THERMAL BEHAVIOR OF AN AMORPHOUS ORGANIC SEDIMENT
R. McGaw, U. S. Army Cold Regions Research and Engineering
Laboratory

PRESSURE DEPENDENCE OF THE THERMAL CONDUCTIVITY OF COMPLEX
DIELECTRIC SOLIDS Micheline C. Roufosse, University of Connecticut

EXPERIMENTAL DETERMINATION OF THE THERMAL PROPERTIES OF FUSED
QUARTZ POLY-AND MONOCRYSTALLINE Al_2O_3 S. R. Atalla, A. A.
El-Sharkawy and R. P. Yourchack, Moscow State University USSR

12:15 P.M.
Lunch - Shippee Hall Cafeteria
1:30 P.M.
Session IVa - Metals and Alloys at Low Temperatures
Session Chairman: Dr. A. C. Anderson, University of Illinois

THE THERMAL CONDUCTIVITY OF HIGH PURITY VANADIUM
W. D. Jung and G. C. Danielson, Ames Laboratory ERDA

THE IMPORTANCE OF THERMAL CONDUCTIVITY IN INTERPRETING THERMOPOWER
DATA: ALLOYS WITH ATOMIC ORDER-DISORDER TRANSITIONS   C. L. Foiles,
Michigan State University

LATTICE THERMAL CONDUCTIVITY AND LORENZ FUNCTION OF COPPER NICKEL
AND SILVER PALLADIUM ALLOY SYSTEMS M. W. Ackerman, K. Y. Wu and
C. Y. Ho, CINDAS, Purdue University

LATTICE THERMAL CONDUCTIVITY OF COPPER AND ALUMINUM ALLOYS
N. S. Mohan, A. C. Bouley and D. H. Damon, University of
Connecticut

THERMAL AND ELECTRICAL TRANSPORT PROPERTIES OF HIGH PURITY SINGLE
CRYSTALS OF BERYLLIUM W. E. Nelson and A. R. Hoffman, University
of Massachusetts

ON THE POSSIBILITY OF DETECTING PHONON DRAG DUE TO AN ELECTRONIC
HEAT CURRENT R. Fletcher, Queen's University, Canada

THE TRANSPORT PROPERTIES OF POTASSIUM AND ELECTRON-ELECTRON
SCATTERING M. J. Laubitz and J. G. Cook, National Research
Council of Canada

ANISOTROPIC TRANSPORT DISTRIBUTION FUNCTIONS FOR POTASSIUM C. R.
Leavens and M. J. Laubitz, National Research Council of Canada

1:30 P.M.
Session IVb - Experimental Techniques and Applications
Session Chairman: Dr. R. E. Willett, Anaconda Brass Company

THE NEED FOR REFERENCE MATERIALS OF LOW THERMAL CONDUCTIVITY
R. P. Tye, Dynatech

A NEW TECHNIQUE FOR THE MEASUREMENT OF THERMAL CONDUCTIVITY IN
THIN FILMS T. C. Boyce and Y. W. Chung, University of Hong Kong

THEORY FOR IN-SITU THERMAL DIFFUSIVITY MEASUREMENTS OF THIN FILMS
BY RADIAL HEAT FLOW METHOD A. B. Donaldson and W. P. Schimmel,
Sandia Laboratories

NICROSIL II AND NISIL THERMOCOUPLE ALLOYS: PHYSICAL PROPERTIES
AND BEHAVIOR DURING THERMAL CYCLING TO 1200K J. P. Moore, R. S.
Graves, M. B. Hershovitz, K. R. Carr and R. A. Vandermeer, Oak
Ridge National Laboratory

IMPROVING THE ACCURACY OF THE GUARDED HOT PLATE APPARATUS C. J.
Shirtliffe, National Research Council of Canada

ANALYSIS OF EDGE-HEAT-LOSS OF A GUARDED-HOT-PLATE APPARATUS
M. C. I. Siu, National Bureau of Standards, Washington, D. C.

IMPROVED HIGH TEMPERATURE CUT BAR APPARATUS   G. R. Clusener,
Theta Industries, Inc.

MEASUREMENT OF HIGH THERMAL DIFFUSIVITY VALUES BY A NOVEL METHOD
ON COPPER AND TUNGSTEN V. V. Mirkovich, Department of Energy,
Mines and Resources, Canada

AN ANALYTICAL MODEL FOR THE MEASUREMENT OF UO_2 THERMAL CONDUC-
TIVITY M. Aragones, Instituto Nacional De Energia Nuclear,
Mexico

6:15 P.M.
Cocktail Hour and Banquet at the Faculty Alumni Center,
University of Connecticut

Wednesday - June 4, 1975
8:45 A.M.
Session Va - Experimental Techniques and High Temperature
Measurements
Session Chairman: Dr. J. U. Trefney, Wesleyan University

DETERMINATION OF THERMOPHYSICAL PROPERTIES OF LAYER COMPOSITES
BY FLASH METHOD H. J. Lee and R. E. Taylor, Purdue University

INVESTIGATION OF THE IMPROVEMENT OF THERMAL CONTACT CONDUCTANCES
BETWEEN A BASEPLATE AND A HONEYCOMB PLATFORM USING VARIOUS INTER-
STITIAL MATERIALS A.Wild, HSD, England and J. P. Bouchez, ESTEC,
The Netherlands

EXACT SOLUTIONS FOR THERMAL CONDUCTANCES OF PLANAR AND CIRCULAR
CONTACTS T. N. Veziroglu, M. A. Huerta and S. Kakac, University
of Miami

CORRECTION FOR LASER BEAM NON-UNIFORMITY IN THE FLASH METHOD OF
THERMAL DIFFUSIVITY MEASUREMENT J. A. McKay and J. T. Schriempf,
Naval Research Laboratory

THE EFFECT OF DELTA-PHASE STABILIZERS ON THE THERMAL DIFFUSIVITY
OF PLUTONIUM H.D.Lewis, J. F. Kerrick and K. W. R. Johnson, Los
Alamos Scientific Laboratory

ERROR ANALYSIS OF THE FLASH THERMAL DIFFUSIVITY TECHNIQUE R. C.
Heckman, Sandia Laboratories

NUMERICAL CALCULATION OF PERIODIC TWO-DIMENSIONAL HEAT FLOW IN
COMPOSITE BUILDING WALLS M. Cali and P. Bondi, Instituto di

Fisica Tecnica, Italy

SPECIFIC HEAT OF ELECTRICALLY CONDUCTING SUBSTANCES AT HIGH
TEMPERATURES MEASURED BY A PULSE HEATING TECHNIQUE   A. Cezairliyan,
National Bureau of Standards

8:45 A.M.
Session Vb - Solids
Session Chairman: Dr. G. A. Slack, General Electric Company

FOUR-PHONON PROCESSES AND THE THERMAL EXPANSION EFFECTS AS
CONTRIBUTORS TO THE THERMAL RESISTIVITY OF CRYSTALS AT HIGH
TEMPERATURES   D. J. Ecsedy, University of Connecticut

EXCITATION OF COHERENT AND INCOHERENT PHONONS BY MEANS OF AN
INFRARED LASER   W. Grill, Institut fur Angewandte Physik II,
Heidelberg, Germany

THERMAL CONDUCTIVITY OF HIGHLY ORIENTED PYROLYTIC BORON NITRIDE
E. K. Sichel and R. E. Miller, RCA Laboratories

WIEDEMANN-FRANZ RATIO OF MAGNETIC MATERIALS   K. V. Rao, S. Arajs
and D. Abukay, Clarkson College

THERMAL CONDUCTIVITY OF COPPER-NICKEL AND SILVER-PALLADIUM ALLOY
SYSTEMS   C. Y. Ho, M. W. Ackerman, and K. Y. Wu, CINDAS, Purdue
University

MOLECULAR DYNAMICAL CALCULATIONS OF THE THERMAL DIFFUSIVITY OF
A PERFECT LATTICE   R. A MacDonald and D. H. Tsai, National
Bureau of Standards, Washington, D. C.

THE THERMAL CONDUCTIVITY OF A 1 MeV ELECTRON-IRRADIATED Al_2O_3
F. P. Lipschultz, D. Strom, G. Wilham and P. Saunders,
University of Connecticut

MEASUREMENT OF THE ELECTRICAL AND THERMAL CONDUCTIVITY
COEFFICIENTS OF As_2Se_3 - As_2Te_3 GLASSY ALLOYS   M. F. Kotkata,
S. R. Atalla and M. K. El-Mously, Ein Shams University, Cairo,
Egypt

12:15 P.M.
Lunch - Shippee Hall Cafeteria
2:00 P.M.
Special Joint Session with the Thermal Expansion Conference
Chairman: Dr. R. O. Simmons, University of Illinois

NATURE OF SCIENTIFIC AND TECHNICAL INFORMATION
Y. S. Touloukian, Purdue University

CONDUCTION PROPERTIES AND THERMAL EXPANSION P. G. Klemens,
University of Connecticut

COMPUTER CONTROLLED THERMOPHYSICAL TEST FACILITY- 1) HARDWARE
C. J. Shirtliffe and W. C. Brown, National Research Council
of Canada

COMPUTER CONTROLLED THERMOPHYSICAL TEST FACILITY- 2) SOFTWARE
C. J. Shirtliffe and W. C. Brown, National Research Council
of Canada

MEASUREMENTS AT HIGH TEMPERATURES BY TRANSIENT TECHNIQUES
(or some related topic) A. Cezairliyan, National Bureau of
Standards, Washington, D. C.

A DISCUSSION ON RECENT TECHNICAL ADVANCES
Discussion Leader, R. E. Taylor, Purdue University

LIST OF AUTHORS

SUBJECT INDEX